CUSTOM MEMORY MANAGEMENT METHODOLOGY
Exploration of Memory Organisation for Embedded Multimedia System Design

CUSTOM MEMORY MANAGEMENT METHODOLOGY
Exploration of Memory Organisation for Embedded Multimedia System Design

FRANCKY CATTHOOR
IMEC, Leuven, Belgium
Also professor at Katholieke Universiteit Leuven

SVEN WUYTACK
IMEC, Leuven, Belgium

EDDY DE GREEF
IMEC, Leuven, Belgium

FLORIN BALASA
Currently with Rockwell Intnl. Corp., Newport Beach, CA

LODE NACHTERGAELE
IMEC, Leuven, Belgium

ARNOUT VANDECAPPELLE
IMEC, Leuven, Belgium

KLUWER ACADEMIC PUBLISHERS
BOSTON / DORDRECHT / LONDON

A C.I.P. Catalogue record for this book is available from the Library of Congress.

ISBN 978-1-4419-5061-1

Published by Kluwer Academic Publishers,
P.O. Box 17, 3300 AA Dordrecht, The Netherlands.

Sold and distributed in North, Central and South America
by Kluwer Academic Publishers,
101 Philip Drive, Norwell, MA 02061, U.S.A.

In all other countries, sold and distributed
by Kluwer Academic Publishers,
P.O. Box 322, 3300 AH Dordrecht, The Netherlands.

Printed on acid-free paper

Preface

The main intention of this book is to give an impression of the state-of-the-art in system-level memory management (data transfer and storage) related issues for complex data-dominated real-time signal and data processing applications. The material is based on research at IMEC in this area in the period 1989-1997. In order to deal with the stringent timing requirements and the data dominated characteristics of this domain, we have adopted a target architecture style and a systematic methodology to make the exploration and optimization of such systems feasible. Our approach is also very heavily application-driven which is illustrated by several realistic demonstrators, partly used as red-thread examples in the book. Moreover, the book addresses only the steps above the traditional high-level synthesis (scheduling and allocation) or compilation (traditional or ILP oriented) tasks. The latter are mainly focussed on scalar or scalar stream operations and data where the internal structure of the complex data types is not exploited, in contrast to the approaches discussed here. The proposed methodologies are largely independent of the level of programmability in the data-path and controller so they are valuable for the realisation of both hardware and software systems.

Our target domain consists of signal and data processing systems which deal with large amounts of data. This happens both in real-time multi-dimensional signal processing (RMSP) applications like video and image processing, which handle indexed array signals (usually in the context of loops), and in sophisticated communication network protocols, which handle large sets of records organized in tables and pointers. Both classes of applications contain many important applications like video coding, medical image archival, advanced audio and speech coding, multi-media terminals, artificial vision, ATM networks, and LAN/WAN technology. For these applications, we believe (and we will demonstrate by real-life experiments) that the organisation of the global communication and data storage, together with the related algorithmic transformations, form the dominating factors (both for area and power) in the system-level design decisions. Therefore, we have concentrated mainly on the effect of system-level decisions on the access to large (background) memories, which require separate access cycles, and on the transfer of data over long "distances" (i.e. which have to pass between source and destination over long-term main memory storage).

The cost functions which we have incorporated for the storage and communication resources are both area and power oriented. Due to the real-time nature of the targeted applications, the throughput is normally a constraint.

The material in this book is partly based on work in the context of several research projects, including especially the Basic Research Projects No.3280 and 6632 NANA (Novel Parallel Algorithms and New Real-time VLSI Architectural Methodologies), No.3281 ASCIS (Architecture Synthesis for Complex Integrated Systems) and the more industrially oriented main-stream ESPRIT project No.2260 SPRITE (Synthesis for signal processing systems). All these projects have been sponsored by the ESPRIT program of Directorate XIII of the European Commission. In addition, several research projects have been sponsored by the Flemish IWT, such as HASTEC (HArdware System design TEchnology for concurrent control-oriented TElecom Components), IT-TELEVISIE, and the MEDEA SMT project (System level Methods and Tools), including partial support of the industrial partners Alcatel Telecom, Philips, and Frontier Design. Finally, also the ESA sponsored SCADES-3 project (WSM/PS/066) has partly contributed to the demonstrator applications.

A major goal of the system synthesis work within these projects has been to contribute systematic design methodologies and appropriate tool support techniques which address the design trajectory from *real system behavior* down to the detailed architecture level of the system. In order to provide complete support for this design trajectory, many problems must be tackled. In order to be effective, we believe that the design methodology and the supporting techniques have to be (partly) domain-specific, i.e. targeted. This book illustrates this claim for a particular target application domain which is of great importance to the current industrial activities in the telecommunications and multi-media sectors: data-dominated real-time signal and data processing systems. For this domain, the book describes an appropriate methodology partly supported by efficient and realistic design techniques embedded in prototype CAD tools. We do not claim to cover the complete system synthesis path, but we do believe we have contributed to the solution of the most crucial problems in this domain, namely the ones related to data transfer and storage exploration (DTSE).

We therefore expect this book to be of interest in academia, both for the overall description of the methodology and for the detailed descriptions of the system-level methodologies and synthesis techniques and algorithms. We also provide a view on the many important but less widely known issues which must be addressed to arrive at industrially relevant results.

All projects which have driven this research, have also been application-driven from the start, and the book is intended to reflect this fact. The real-life applications are described, and the impact of their characteristics on the methodologies is assessed. We therefore believe that the book will be of interest as well to senior design engineers and CAD managers in industry, who wish either to anticipate the evolution of commercially available design tools over the next few years, or to make use of the concepts in their own research and development.

It has been a pleasure for us to work in this research domain and to co-operate with our project partners and our colleagues in the architecture- and system-level synthesis community. In addition to learning many new things about system synthesis and related issues, we have also developed close connections with excellent people. Moreover, the pan-European aspect of the projects has allowed us to come in closer contact with research groups with a different background and "research culture", which has led to very enriching cross-fertilization.

We would like to use this opportunity to thank the many people who have helped us in realizing these results, both in IMEC and at other places. In particular, we wish to mention: Ivo Bolsens, Jan Bormans, Koen Danckaert, Gjalt de Jong, Hugo De Man, Jean-Philippe Diguet, Michel Eyckmans, Frank Franssen, Mark Genoe, Martin Janssen, Stefan Janssens, Chidamber Kulkarni, Chen-Yi Lee, Paul Lippens, Miguel Miranda, Hans Samsom, Peter Slock, Jef van Meerbergen, Michael van Swaaij, Ingrid Verbauwhede.

We finally hope that the reader will find the book useful and enjoyable, and that the results presented will contribute to the continued progress of the field of system-level synthesis for both hardware and software systems.

Francky Catthoor, Sven Wuytack, Eddy De Greef, Florin Balasa,
Lode Nachtergaele, Arnout Vandecappelle
June 1998.

Glossary

1-D	one-dimensional
ACU	Address Calculation Unit
ADT	Abstract Data Type
ADOPT	ADdress OPTimisation
AGU	counter-based Address Generation Unit
ASIC	Application Specific Integrated Circuit
ASIP	Application Specific Instruction set Processor
ASU	Application Specific Unit
ATOMIUM	A Tool-box for Optimizing Memory and I/O communication Using geometric Modelling.
CAD	Computer aided design
CDFG	Control/Data-Flow Graph
CDO	Conflict Directed Ordering
CG	Conflict Graph
DCT	Discrete Cosine Transform
DFG	Data-Flow Graph
DRAM	Dynamic Random-Access Memory
DSP	Digital Signal Processing
DTSE	Data Transfer and Storage Exploration
ECG	Extended Conflict Graph
HIMALAIA	HIgh-level Memory ALlocation And I/O Assignment
HLDM	High Level Data path Mapping
HLMM	High Level Memory Management
IFDS	Improved Force-Directed Scheduling
ILP	Integer Linear Program
I/O	Input/Output
M-D	Multi-Dimensional
ME	Motion Estimation
MHLA	Memory Hierarchy Layer Assignment
MILP	Mixed Integer Linear Program
MIMD	Multiple Instruction, Multiple Data
PAM	Pointer-Addressed Memory
PDG	Polyhedral Dependence Graph
RAM	Random-Access Memory
RMSP	Real-time Multi-dimensional Signal Processing
ROM	Read-Only Memory
RSP	Real-time Signal Processing
SBO	Storage Bandwidth Optimisation
SCBD	Storage Cycle Budget Distribution
SDRAM	Synchronous Dynamic Random-Access Memory

SIMD	Single Instruction, Multiple Data
SPP	Segment Protocol Processor
SRAM	Static Random-Access Memory

Contents

1 OPTIMISATION OF GLOBAL DATA TRANSFER AND STORAGE ORGANISATION FOR DECREASED POWER AND AREA IN CUSTOM DATA-DOMINATED REAL-TIME SYSTEMS

Embedded systems have always been very cost-sensitive. Up till recently, the main focus has been on area-efficient designs meeting constraints due to performance and design time. During the last couple of years, power dissipation has become an additional major design measure for many systems next to these traditional measures. More and more applications become portable because this is felt as an added value for the product (e.g., wireless phones, notebook computers, multimedia terminals). The less power they consume, the longer their batteries last and the lighter their batteries can be made. Higher power consumption also means more costly packaging and cooling equipment, and lower reliability. The latter has become a major problem for many high-performance (non-portable) applications. As a consequence, power efficient design has become a crucial issue for a broad class of applications.

Many of these embedded applications turn out to be data-dominated, both in the multi-media domain and in the telecommunication domain. Experiments have shown that for these applications, a very large part of the power consumption is due to data storage and data transfer. Also the area cost is for a large part dominated by the memories. Hence, we believe that the memory architecture should be optimized as a first step in the design methodology for this type of applications. This has been formalized in our Data Transfer and Storage Exploration (DTSE) methodology.

This book provides a summary of the research activities in the context of the system-level DTSE at IMEC. In this introductory chapter, only a high-level view is provided which will be refined in the subsequent chapters.

1.1 CONTEXT AND PROBLEM FORMULATION

For most real-time signal processing applications many ways exist to realize them in terms of a specific algorithm. As reported by system designers, in practice this choice is mainly based on "cost" measures such as the number of components, performance, pin count, power consumption, and the area of the custom components. Currently, due to design time restrictions, the system designer has to select — on an ad-hoc basis — a single promising path in the huge decision tree from abstract specification to more refined specification (figure 1.1). To alleviate this situation, there is a need for fast and early feedback at the algorithm level *without* going all the way to assembly code or hardware layout. Only when the design space has been sufficiently explored at a high level and when a limited number of

1

promising candidates have been identified, a more thorough and accurate evaluation is required for the final choice (figure 1.1).

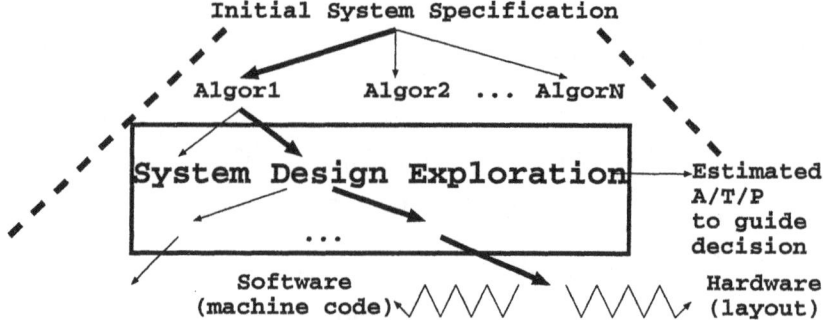

Figure 1.1.   System exploration environment: envisioned situation

Experience shows that the initial specification heavily influences the outcome of the underlying architectural estimation or mapping tools (e.g. for data-path allocation, memory allocation, address generation). Therefore, transforming this specification is one of the most prominent tasks during the early system-level exploration of cost measures. This task is both very difficult to explore globally using ad hoc strategies, and very tedious and error-prone if done fully manually.

To remove this time bottle-neck, we are developing systematic methodologies, partly supported with automatable steering methods, for the set of system-level transformations that have the most crucial effect on the system exploration decisions. Such transformations change the control/data-flow graph or other specification models on which the subsequent estimation/synthesis tasks work. For the most important categories of transformations in our target domain, an appropriate model has been identified which allows to (as much as possible globally) explore the effect of these transformations on the cost measures relevant to the system designers. Based on this model, efficient steering techniques are established. These guide the transformations such that the system designer becomes less dependent on the original specification description while still arriving at a reasonable cost estimate for a particular figure of merit, without having to explore all the possible transformations manually. This approach will make the system designer much less dependent on the way the original specification is described.

Most research and development efforts in digital electronics have focused on increasing the speed and integration of digital systems on a chip while keeping the silicon area as small as possible. This has resulted in a powerful, but power hungry, design technology, which enabled a whole series of new applications such as real-time 3D rendering, multi-media terminals, video compression, speech recognition, and so on. While focusing on speed and area, power consumption has long been ignored.

This situation is, however, rapidly changing [335]. The main reason is the increasing demand for portable systems in the areas of communication (e.g., cellular phones and pagers), computation (e.g., notebook computers and personal digital assistants), and consumer electronics (e.g., multi-media terminals and digital video cameras). These systems require sophisticated and power-hungry algorithms for high-bandwidth wireless communication, video compression and decompression, handwriting recognition, speech processing, and so on. As relatively little improvement in battery technology is anticipated (expected battery life time will increase with no more than 30 to 40% over the next 5 years), portable devices will suffer from either very short battery life or unreasonably heavy battery packs unless a low power design approach is adopted. Hence, for portable applications, *average* power consumption has become a critical design concern which is starting to replace speed and area as the most important implementation constraints.

Portability is, however, not the only driving force behind the move towards low-power design. Heat dissipation is another one. The cost associated with packaging and cooling high-performance devices such as modern microprocessors is becoming prohibitive. Since core power consumption must be dissipated through the packaging, increasingly expensive packaging and cooling techniques are required as the *root mean squared (RMS)* chip power consumption increases. Therefore, there is a clear cost advantage in reducing the power consumption of high-performance systems. Next to the direct implementation cost, there is the issue of reliability: it is a well known fact that every 10°C increase in operating temperature roughly doubles an integrated circuit's failure rate. Both *RMS* and *peak* power consumption are important here. Clearly, also in the design of high-performance systems, power starts to play a crucial role.

All of this has resulted in much interest in low-power design over the last five years, while area and performance of course remain crucial also. By now, it is commonly agreed that low power design requires optimizations at all levels of the design hierarchy, i.e., technology, device, circuit, logic, architecture, algorithm, and system level [335, 66]. Even though the largest gains in power consumption can be realized at the highest abstraction levels, very little research has addressed the system and algorithm levels except for some very recent contributions. Most low power research still focuses on the lower abstraction levels. Therefore, we have decided to target our research on the *system architecture* and *algorithm levels*, being the *most promising* and *least explored* areas.

We cannot achieve this ambitious goal for general applications and target architectural styles. So we only target data-dominated applications, and we have selected a number of reasonable assumptions discussed further on. We will show that with a largely common methodology, both power and area cost can be significantly reduced, while still meeting given real-time performance constraints in the embedded system.

In this book we look at two important application domains fitting in the category of cost-sensitive data dominated applications: the Real-time Multi-dimensional Signal Processing (RMSP) domain and the network component domain (e.g., ATM applications). Although they are both data-dominated, there are important differences between them. These differences lead to partly different optimization techniques. Therefore, IMEC is developing two system-level exploration environments: ATOMIUM for RMSP applications and MATISSE for network component applications. In the book, we will focus on the RMSP domain but some links will be provided to the directly related steps in the other domain too.

1.2   REAL-TIME MULTI-DIMENSIONAL SIGNAL PROCESSING

A very important application domain is that of the RMSP applications. This domain includes applications like video and image processing, medical imaging, artificial vision, real-time 3D rendering, advanced audio and speech coding, multimedia terminals, . . .
This subsection presents a typical RMSP application, lists the main characteristics of RMSP applications, and introduces the ATOMIUM exploration environment developed for optimizing them.

1.2.1   Typical example: H.263 video decoder

An H.263 video conferencing decoder is a typical example of an RMSP application. It is meant for coding video for transmission over narrow telecommunication channels at less than 64 Kbit/s [343]. The coding/decoding is implemented with a block based algorithm that exploits spatial and temporal redundancy. The spatial redundancy is exploited through conversion of the picture data to the frequency domain by means of a Discrete Cosine Transform (DCT). These frequency components are then quantized which allows for compression of the data because fewer bits are required to code the quantized frequency components. To get the best visual quality for a given compression ratio, this quantization is made non-uniform: larger quantization steps (and hence fewer bits) are used for those frequency components for which the human eye is less sensitive. The temporal redundancy in a video sequence is exploited by encoding only the difference between consecutive frames. Due to the similarity of consecutive frames, the difference between them can usually be encoded much more effectively than the frames themselves. To increase the effectiveness of the compression, motion estimation is used to accurately predict a given frame from previously encoded frames. This results in very small differences to be coded and hence a very high compression.

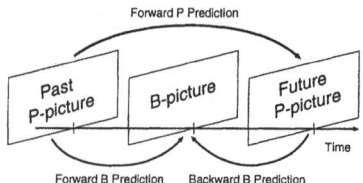

Figure 1.2.   Forward P, forward B, and backward B predictions.

An H.263 decoder can be in one of three modes: I, P, or PB mode. In I (intra) mode, only spatial redundancy in the picture is used to encode it. In P (predicted) mode, a future picture is predicted from a past picture by means of motion estimation.

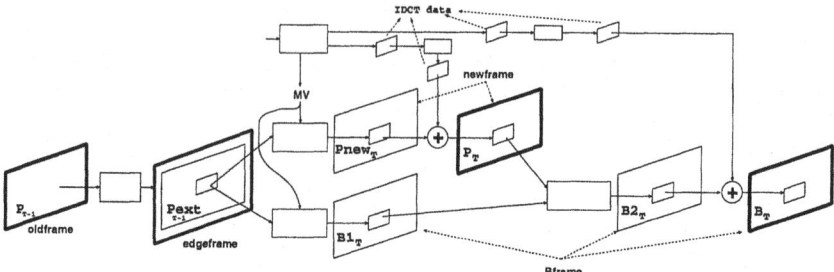

Figure 1.3.   Data-flow for decoding PB-frames in H.263.

In PB (predicted and bidirectional) mode, two pictures are encoded together: one of them is a P-picture predicted from a past picture, the other one is a B-picture predicted from the interpolation between the P-pictures immediately before and after it (see figure 1.2). The PB mode is obviously the most complicated mode in H.263. The data-flow for decoding a PB-frame with emphasis on the main multi-dimensional signals is shown in figure 1.3.

1.2.2   Characteristics

The main characteristics of RMSP applications are listed below:

1. *Real-time nature* They include tight timing and ordering constraints on inputs and outputs, and in many cases also on important internal nodes. Several sampling rates can be present.

   H.263 example: The global frame rate is usually 25 or 30 frames/s depending on the context. Also other timing constraints are present.

2. *Control-flow*

   - *nested loops*

     RMSP applications typically contain many irregularly nested loops for processing the multi-dimensional data.

     H.263 example: The H.263 code contains lots of nested loops for processing the 2D frames.

   - *mostly manifest conditions and indices*

     Most of the conditions and indexing are manifest. They are function of the loop iterators and hence known or derivable at compile time. In many cases, these are also affine expressions (i.e. linear plus constant) which heavily simplifies the modeling and the input specification analysis (see chapters 5 and 6).

However, more and more RMSP applications contain also non-linear data-dependent conditions and indexing (e.g., MPEG 2 and 4). The data-dependent conditions are usually globally scoped (e.g., support for different modes). Fortunately, extensions are possible for the most commonly occurring piece-wise affine and data-dependent conditions, as introduced in chapters 5 and 6.

H.263 example: The H.263 code contains manifest conditions for treating pixels near the border differently. It also contains a few data-dependent conditions for the mode in which the decoder is working (I, P, or PB mode).

3. *Data types*

- *strong data typing*

  The initial specification is usually strongly typed, including bit-true semantics.

  H.263 example: The images and other signals exhibit specific fixed-point formats and bit-true shift and round operations are frequently specified.

- *multi-dimensional arrays*

  RMSP applications by definition work on multi-dimensional arrays of data. However, also large one-dimensional arrays are often encountered. So they are data-dominated.

  H.263 example: Most data in the H.263 are 2D arrays representing image data: frames, macro blocks, ...

- *(mostly) statically allocated data*

  Most data is statically allocated. Its exact storage location can be optimized at compile time. Even if the exact number of data is (partly) variable, the worst-case can usually be statically determined and based on that, compile-time optimisations can be entered in the code.

  H.263 example: In H.263 it is known at compile time how much and what data has to be stored for processing one frame in a given mode. The memory for storing the data can therefore be allocated statically.

  MPEG4 example: For very data-dependent RMSP code as in some parts of e.g. the new MPEG4 coding standard [374, 373], a last recourse is to foresee several compile-time optimisation alternatives in the code which are then conditionally executed dependent on a run-time check. This has been verified to work very well for the video motion estimation of MPEG4 [49].

- *temporary data*

  Most of the data is temporary, only alive during one or at most a few iterations of the algorithm. When the data is not alive anymore, it can be overwritten by other data.

  H.263 example: Most data values in H.263 are only alive for at most a few iterations of the algorithm.

4. *Many pages of complex code* In order to deal with this complexity, heavy pruning of the initial specification is typically required (see chapter 7).

  H.263 example: The video motion compensation alone, comprises already about 30 pages of complex high-level C code.

1.2.3 The ATOMIUM *exploration environment*

The ATOMIUM system-level data transfer and storage exploration environment is being developed at IMEC for the type of applications described in this section. ATOMIUM bridges the gap from a system specification to an optimized embedded single-chip HW/SW implementation. It supports the memory architecture exploration step before going to traditional tools for HW synthesis and/or SW compilation. The methodology is introduced in more detail in chapter 3.

1.3 NETWORK COMPONENT APPLICATIONS

Another important class of applications considered in this book is that of the network component applications operating at layers 3 to 6 in the OSI network layer model (e.g., ATM network applications and LAN/WAN technology). This section presents a typical network component application, lists the main characteristics of network applications, and introduces the MATISSE exploration environment for optimizing them.

1.3.1 Typical example: Segment Protocol Processor (SPP)

Alcatel's Segment Protocol Processor (SPP) [389] is a typical example of a data-dominated network component application. It has been studied very intensively by us in the context of the IWT HASTEC project with Alcatel Telecom.

The SPP is a crucial part of a connectionless server on top of a connection oriented Asynchronous Transfer Mode (ATM) network [114]. It supports data communication between geographically distributed computers or local area networks (LANs). The communication service offered is connectionless, meaning that a sender does not have to set up a connection prior to sending data, but can start transmitting data immediately. The connectionless server is responsible for storing incoming data, finding out where to route it to, and forwarding it.

The connectionless server is an implementation of the Switched Multi-megabit Data Service (SMDS) [39]. The SMDS protocol features packets of variable length, which are transported by one or more fixed size ATM cells. In addition to the original packet data, the ATM cells also carry various fields for control and management. Two fields incorporated in the ATM cells are the LID and MID fields. The LID (Local ID) identifies, for a given server, a connection from a user to that server. The MID (Multiplexing ID) identifies an application program which makes use of that connection, allowing ATM cells of packets from different applications to be interleaved.

The function of the SPP is to store and forward user and internal cells, to perform a number of checks on them, to issue requests (e.g., routing requests) to other components in the connectionless server, to process routing replies, and to do some background processing for garbage collection.

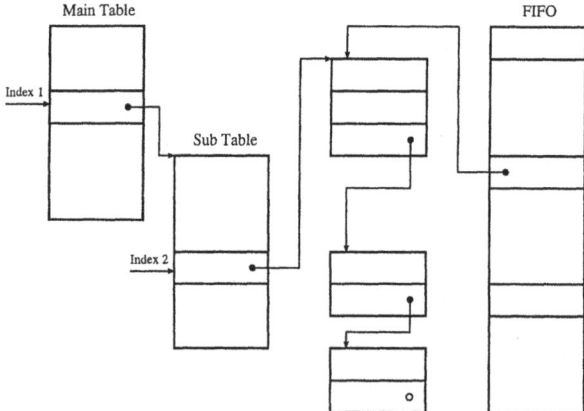

Figure 1.4.   Table structure of the Segment Protocol Processor.

At the heart of the application are a number of stored data structures shown in figure 1.4. A FIFO is used to buffer the incoming cells. The dynamically allocated packet records contain information about the packets being transported by the connectionless server, such as the number of cells received so far and the time the first cell was received. They also contain a pointer to a linked list of routing records. The packet record associated with an incoming user cell is accessed through two levels of tables with the LID and MID fields of the cell as keys.

1.3.2 Characteristics

The main characteristics of data-dominated network component applications are:

1. *Control-flow*

   ■ *(almost) no loops*
   Apart from the implicit time loop, almost no loops are present. Sometimes very small loops are present that can easily be unrolled.

   SPP example: Next to the time loop, no loops are present in the code.

   ■ *data-dependent conditions and indices*
   The applications contain lots of data-dependent conditions. Most indexing is also data-dependent (often via pointers).

   SPP example: Conditions: Most conditions depend on the contents of incoming cells; *Indexing:* All tables are accessed with data-dependent indexes. The dynamically allocated records are accessed with run-time dependent pointers (see figure 1.4).

2. *Data types*

   ■ *tables, lists, and records*
   The data structures are relatively complicated: they consist of tables, lists, and records, linked together via pointers.

   SPP example: Figure 1.4 shows the tables, lists, and records present in the SPP.

   ■ *dynamically allocated data*
   Most of the data is dynamically allocated and deallocated by a dynamic memory manager (new and delete operations). Therefore the exact location of the data in memory is non-manifest: it is only known at run-time. Still, much can be done about the internal organization of each record and about how the records are managed.

   SPP example: Figure 1.4 shows that both static and dynamically allocated data structures are present in the SPP.

   ■ *semi-permanent data*
   Most of the data is semi-permanent: it stays alive between different iterations of the implicit time-loop. During each iteration of the algorithm (a very small) part of the data is altered by the algorithm, the rest of the data remains the same.

   SPP example: Most of the data-structures such as the LID-table, MID-tables, and IPI-records contain data that stays alive during many iterations of the algorithm. During every iteration only a very small part of the data contained in these data structures is updated.

1.3.3 The MATISSE *exploration environment*

The MATISSE system-level exploration environment is being developed at IMEC for the type of applications described in this section. MATISSE bridges the gap from a system specification, using a concurrent object-oriented language, to an optimized embedded single-chip HW/SW implementation. It supports step-wise exploration and refinement, memory architecture exploration, and gradual incorporation of timing constraints before going to traditional tools for HW synthesis, SW compilation, and HW/SW interprocessor communication synthesis. In this book only the data storage and transfer related steps from the MATISSE design flow are considered. A description of the other steps can be found in [97] and will be briefly reviewed in chapter 14.

1.4 WHY CONCENTRATE ON STORAGE AND TRANSFERS?

In this section we show that for data-dominated applications the system power consumption is dominated by the data storage and transfer. Also the area is heavily influenced by the size of the memory organisation.

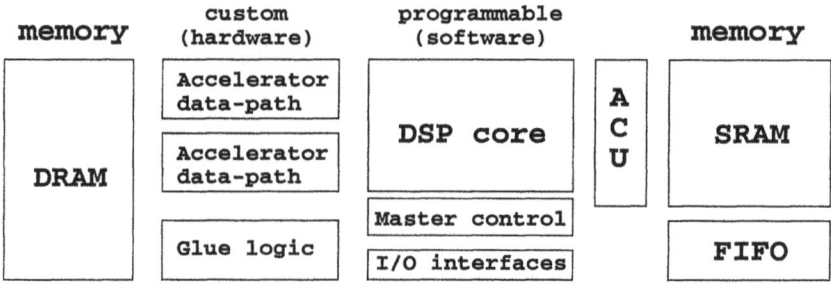

Figure 1.5.   Typical heterogeneous embedded system architecture.

1.4.1 Analysis of system power consumption distribution

From the previous section, we know that data-dominated applications need lots of memory to store the data being processed. They also transfer huge amounts of data back and forth between the memories and the data paths. This naturally leads to the top-level view of a typical heterogeneous system architecture as illustrated in figure 1.5. It includes custom hardware (application-specific accelerator data-paths and logic), programmable hardware (DSP core and controller), and a distributed memory organisation which is usually expensive in terms of power and area cost. For this context, figure 1.6 shows that both in hardware (HW) and software (SW) realizations, data transfer and memory access operations consume much more power than a data-path operation. For instance, fetching an operand from an off-chip memory for an addition consumes 33 times more power than the addition itself in case of a HW processor. If also the other operand and the result have to be transferred to the off-chip memory, this means that 100 times more power is spent in transferring the data than in doing the actual computation! Luckily, not all data has to be stored in *off-chip* memories. Still, even a transfer to/from an *on-chip* memory consumes about 4 to 10 times more power than an addition. Obviously, the data transfer and storage dominates the power consumption for data-dominated applications in HW realizations.

We have demonstrated this for RMSP applications in [435] and this has later been confirmed by other studies elsewhere [264, 291]. For network component applications, an experiment with the Segment Protocol Processor has demonstrated that the maximal power consumption for the 9 off-chip memories was 6 W, while the ASIC containing all data paths, controllers, interfaces, and local memories consumed about 2 W. In these figures the power dissipated in off-chip interconnect is not yet included. Clearly, also for this type of applications the system power is dominated by the storage and transfer of data.

Also for software realizations the power is usually dominated by the data storage and transfer (see figure 1.6) [393, 164], although the difference in power consumption between transfers and data path operations is less pronounced because of the larger overhead of SW processors compared to HW processors.

1.4.2 Target architecture model

Our target architecture model is illustrated in figure 1.7. It focuses on the background memory organisation and address hardware embedded in a global heterogeneous embedded system architecture. Only the ATOMIUM labelled shaded box is targeted by the DTSE approach in this book. The ADOPT labelled box is the focus of our address generation related research presented elsewhere[1]. Note that in general, large frame memories (residing potentially off-chip) will be single-port because our methodology will try to reduce the number of ports on the large (most costly) memory components.

---

[1] ASU= Application-Specific data-path Unit; ACU= data-path based Address Calculation Unit; AGU= Address Generation Unit based on counters and logic.

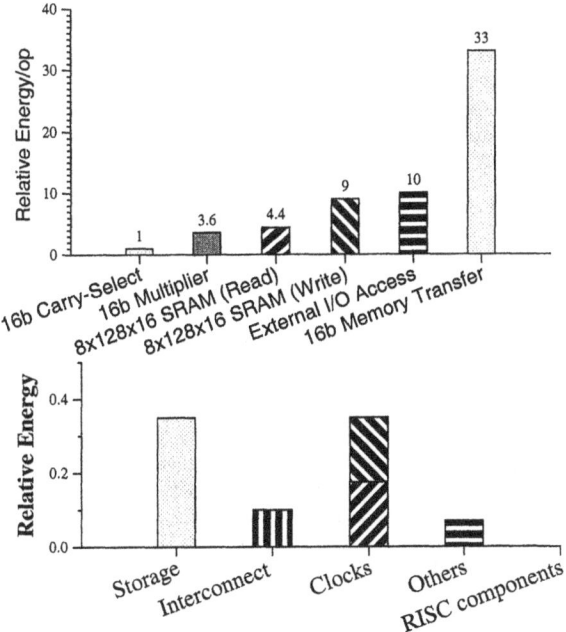

Figure 1.6.   Demonstration of dominance of transfer and storage over data-path operations both in hardware [264] and in software [164, 393].

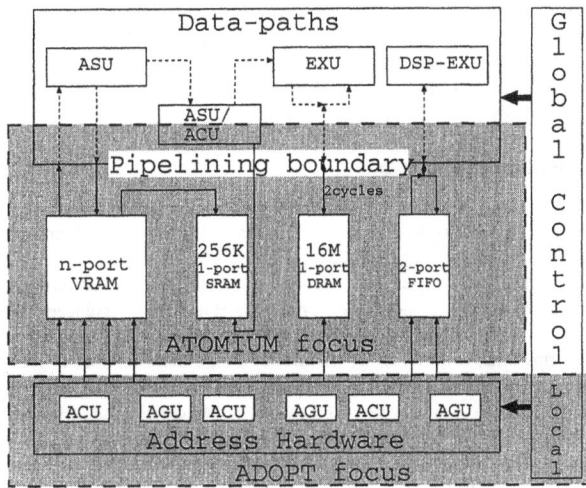

Figure 1.7.   Illustrative instance of target architecture model for ATOMIUM and ADOPT.

The multi-port memories will only be used for the smaller background memories which are closer to the data-path processing and which reside on-chip.

From the previous, we can conclude that it is important to focus on data transfer and memory power consumption when optimizing the overall system power consumption. Hence, the first stage in our design flow is to optimize the data storage and transfer of an application, *before* doing the data path, address generator, and controller synthesis [402, 54]. This heavily simplifies the overall design task. The "phase" coupling between these 2 stages is relatively small because of the dominance of the data storage and transfer related costs in our target domain, which allows to ignore the other issues in the first stage.

Another important issue is the presence of the system pipeline, introduced by our focus on data transfers. Because data-path issues are postponed, extensive use of (hardware and software) pipelining is required in our target architecture model (see figure 1.7) to end up with a feasible design. The full cycle budget (with a small "interface tolerance") is allocated to the memory and I/O access. When the memory and communication organisation is fixed, this will provide sufficient data at the appropriate "cycles" to the ports of the data-path black box. It is possible, however, that the time when data arrives at the data-path ports does not coincide with the moment the operation using it is scheduled. This introduces the need of a system pipeline at the data-path ports.

Within the bandwidth and ordering constraints derived for the memory architecture, the data-path task can then be performed. If this task starts with the appropriate data-path and parallelism oriented transformations (see [54] for motivation, model and examples), the search space available to the data-path synthesis is still sufficient to arrive at both feasible and efficient (near-optimal) architectures, even with stringent timing requirements. This reduces the phase coupling factor further.

1.5   DATA TRANSFER AND STORAGE EXPLORATION

The goal of the Data Transfer and Storage Exploration (DTSE) stage is to determine an optimal execution order for the data transfers together with an optimal memory architecture for storing the data of the given application. The cost functions which currently incorporate for the storage and communication resources are both power and area oriented [53, 375]. Due to the real-time nature of the targeted applications, the throughput is normally a constraint.

Even though both RMSP and network applications are data-dominated, there are some important differences between them. This is also reflected in our DTSE optimization methodologies for both application domains: some steps are very well suited for one application domain and not as useful for the other, other steps are useful for both application domains but require different optimization techniques. Figure 1.8 shows all the steps in our DTSE methodology that are useful for data-dominated applications. Steps that are discussed in detail in this book are annotated with the chapter that is devoted to them. The following subsections briefly describe each of the steps.

1.5.1   Global data-flow optimization

One main goal of the global data-flow optimization step is to reduce the number of bottle-necks in the algorithm that prevent optimizing code restructuring transformations from being applied. Another objective is to remove access redundancy in the data-flow.

We have classified the set of system-level data-flow transformations that have the most crucial effect on the system exploration decisions [61]. They consist mainly of advanced signal substitution avoiding unnecessary copies of data, modifying computation order in associative chains enabling certain loop transformations, shifting of "delay lines" through the algorithm to reduce the storage requirements, and recomputation issues to reduce the number of transfers and storage size. Typically also data-flow bottle-neck removal by means of e.g. look-ahead transformations is considered, as an enabler for the subsequent steps. In addition, a formalized steering methodology has been developed.

This has a positive effect on the power and the area consumption by reducing the amount of data transfers and/or the amount of data to be stored and by acting as an enabling transformation for the next step in the methodology, i.e., control-flow optimization.

The data-flow optimization step is most useful for RMSP applications as it mainly acts as an enabling step for the control-flow optimization step which is especially useful for RMSP applications.

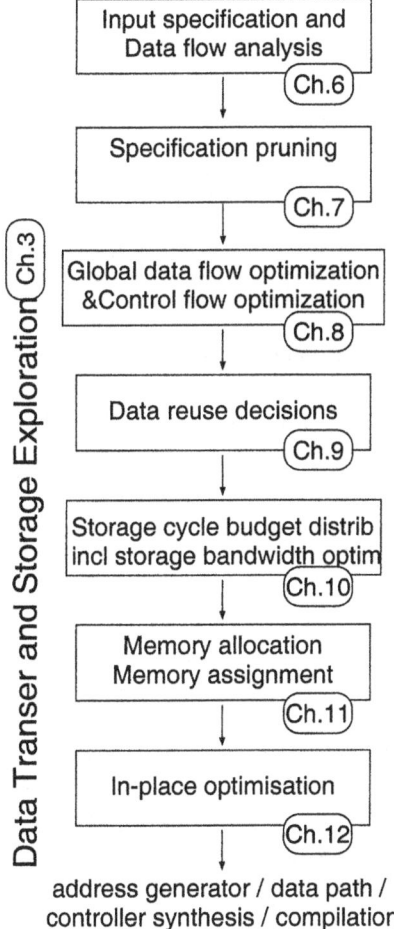

Figure 1.8.   Storage management for data-dominated applications.

1.5.2   *Global loop and control-flow optimization*

The goal of the global loop and control-flow optimization step is to reduce the global life-times of the signals and to increase the locality and regularity of the data accesses.

This has a positive effect on the power and the area consumption because it removes system-level copy overhead in buffers and it is a crucial enabler for allowing the data to be stored in smaller memories (less power consumption in memories) closer to the data paths (less power dissipation in interconnect).

This step is most useful for RMSP applications, because they process their data in a regular and manifest way, whereas network component applications process their data mostly in an irregular and data dependent way. Hence, optimizing the order in which the data has to be processed is both easier and more effective for RMSP applications.

1.5.3 Data reuse decisions

The goal of the data reuse decisions step is to better exploit a hierarchical memory organization to benefit from the available temporal locality in the data accesses such that frequently accessed data can be read from smaller and thus less power consuming memories.

This obviously has a positive effect on the power consumption because the most frequently accessed data is then read from less power consuming memories. Also the smaller memories can then be closer to the data paths thereby reducing the dissipation in the interconnect, especially if off-chip memory accesses are replaced by on-chip memory accesses.

This step is only useful for applications that process their data in a regular and predictable way. Hence it is especially useful for RMSP applications, and less useful for network component applications, except when sufficiently accurate profiling information indicates that some data access locality is present. It is explained in detail in chapter 9.

1.5.4 Storage cycle budget distribution

The goal of the Storage Cycle Budget Distribution (SCBD) step is to ensure that the (usually stringent) real-time constraints are met with a minimal cost penalty. The major substep involves Storage-Bandwidth Optimization (SBO) to determine which data should be made simultaneously accessible in the memory architecture such that the real-time constraints can be met with minimal memory bandwidth related costs.

The storage-bandwidth optimization step determines for a large part how the cycle budget is distributed over the different loop nests. The more cycles are assigned to a given loop nest, the less storage bandwidth is needed to access the data during the execution of that loop nest. Giving more cycles to one loop nest, however, means that less cycles are available for the others, resulting in a higher storage bandwidth requirement for those. Therefore, optimizing the storage bandwidth requires a careful distribution of the available cycle budget over the different loop nests. The execution frequency of each loop nest is a very important parameter during the distribution process.

The positive effect of storage-bandwidth optimization (which reduces the number of parallel memory ports required) on power is perhaps not clear at first sight. Indeed, allocating more and hence smaller memories usually results in less power consumption. However, in practice, the number of memories is limited because of routing overhead, extra design effort (e.g., floor planning), and more expensive testing when many memories are present in the design. Storage-bandwidth optimization takes the effect on power dissipation into account when determining which data should be made simultaneously accessible through different memories: the data which dominate the power are split up preferably to the less important data. It also minimizes the number of multi-port memories required for heavily accessed data.

The storage-bandwidth optimization step is important for both RMSP and network component applications, but the two domains require partly different optimization techniques. This material is explained in detail in chapter 10.

1.5.5 Memory allocation and assignment

The goal of the memory allocation and assignment step is to determine an optimal memory architecture for the background data. This has a direct effect on area (it determines the sizes of the memories) and an indirect effect on power consumption (heavily accessed data are preferably assigned to smaller memories, consuming less power). This step is important for both RMSP and network component applications. Moreover, they can easily share the same optimization techniques.

1.5.6 In-place optimization

The goal of the in-place optimization step is to find the optimal placement of the data inside the memories such that the required memory size is minimal. Again several substeps are important to achieve this but the main ones are related to storage order optimization which involves both intra- and inter-signal related substeps.

This has an effect on the power consumption [402, 109], as smaller memories usually consume less power than bigger ones. Moreover for on-chip memories: the smaller the memories, the less area, the shorter the on-chip interconnect, and hence the smaller the power dissipation in the on-chip interconnect.

This step is most useful for RMSP applications that deal with a lot of temporary data. It is not useful for network component applications when they only contain (semi-) permanent data.

1.5.7   Signal partitioning issues

In many cases, both for RMSP and network components, signals have to be partitioned compared to their initial specification to allow more freedom in the search space of the DTSE stage. The orignal signal definitions can indeed contain too large clusters of data. In the sequel, we will use the term basic group for such partitions. A detailed discussion will be provided in section 10.3.1. In the RMSP domain, the analysis to identify potentially interesting basic groups, is based on a so-called basic set analysis [32] (see section 6.2). For network components, another analysis scheme has to be provided [376] (section 10.8).

It should be stressed however that partitioning the original signals should only be performed if there is a need (to create a feasible solution space) or a motivated use (to enlarge the search space) for it. If there is no clear advantage in the basic group partitioning, the complexity for the respective DTSE steps and the final addressing and controller realisation are needlessly increased.

1.5.8   Extensions

Most of this activity has been aimed at application-specific architecture styles, but recently also predefined processors (e.g. DSP cores) are envisioned [55, 102, 104, 103, 110, 171, 280, 292]. Moreover, our methods and prototype tools are also useful in the context of global communication of complex data types between multiple (heterogeneous) processors [170], like the current generation of (parallel) multi-media processors [105, 89], but this is a topic of current and future research (see chapter 14). Also in a software/hardware codesign context a variant of our approach provides much better results than conventional design practice [90].

1.6   RELATED WORK

Up to now, little design automation development is available to help designers with the data storage and transfer exploration problem. A detailed summary of the literature will be provided in chapter 2 but a summary is presented here already.

Commercial EDA tools, such as SPW-HDS (Alta-Cadence Design), Monet-System Design Station (Mentor Graphics), and the COSSAP environment (CADIS-Synopsys), support system-level specification and simulation but are not geared towards design exploration and optimization of memory or communication-oriented designs. Indeed, all of these tools start from a procedural interpretation of the loops where the memory organisation is largely fixed. Moreover, the actual memory organisation has to be indicated by means of user directives or by a partial netlist. In the CASE area, represented by e.g. Statemate (I-Logix), Matrix-X (ISI) and Workbench (SES), these same issues are not addressed either. In the parallel compiler community, much research has been performed on loop transformations for parallelism improvement (see e.g. [13, 36, 330, 434]). In the scope of our multi-media target application domain, the effect on memory size and bandwidth has however been largely neglected or solved with a too simple model in terms of power or area consequences, even in recent work.

Within the system-level/high-level synthesis research community, the first results on memory management support in a hardware context have been obtained at Philips (Phideo environment [245]) and IMEC (prototype ATOMIUM environment [420, 155, 58]). Phideo is mainly oriented to stream-based video applications and focuses on memory allocation and address generation. Recently, a few other initiatives focusing on point tools have been started up too [339, 215].

Most research on power oriented methodologies has focussed on data-path or control logic, clocking and I/O. We have shown [435], however, that for data-dominated applications much (more) power can be gained by reducing the number of accesses to large frame memories and buffers. Also other groups have made similar observations [264] for video applications. Prior to our work, however, no systematic approach has been published to target this important field. Indeed, most effort up to now has been spent either on data-path oriented work (e.g., [66]), on control-dominated logic, or on programmable processors (see [375] for a good overview).

1.7 CONTRIBUTIONS

This section briefly describes the main contributions of this book.

- *Memory management related literature survey:* In chapter 2 a quite extensive survey is provided of the main related work in the domain of this book. It is organized per topic, addressed in a hierarchical way.

- *DTSE methodology:* A systematic methodology for data transfer and storage exploration is proposed in chapter 3. It allows to cover a much larger search space than what is typically feasible with current ad hoc exploration approaches (with the same amount of design effort). It can be applied manually but for real-life applications it still requires too much effort for a thorough exploration. Therefore, we believe that tool support is extremely valuable for the most crucial and time-consuming (sub)steps.

- *Memory oriented model extraction:* The techniques proposed in this book, can only be applied using geometric data models which differ quite heavily from traditional scalar data models. Also the parallelizing compiler models cover only part of the required functionality, so we have proposed significant extensions, as described in chapter 5.

- *Memory oriented data-flow analysis:* Novel array data-flow analysis techniques in order to effectively deal with the characteristics of RMSP applications are described in chapter 6.

- *Input specification pruning:* In order to deal with large applications, it is necessary to select from the original specification only the most relevant data and constructs for the actual exploration. Systematic pruning techniques to achieve this are described in chapter 7.

- *Global data-flow transformations:* We have classified the set of system-level data-flow transformations that have the most crucial effect on the system exploration decisions, and we have developed a formalized methodology for applying them. Two main categories exist. The first one directly optimizes the important DTSE costs factors. The second category serves as enabling transformation for the subsequent steps because it removes the data-flow bottle-necks wherever required. This step is described in chapter 8.

- *Global loop and control-flow transformations:* These aim at improving the data access locality for multi-dimensional (M-D) signals and at removing the system-level buffers introduced due to mismatches in production and consumption ordering between different subsystems. A systematic methodology to steer the application of such transformations is described in chapter 8.

- *Data reuse decisions:* A crucial step in our DTSE methodology is to make efficient use of data reuse opportunities present in the algorithm by exploiting temporal locality in the data accesses. In this step, we have to decide on the exploitation of the memory hierarchy including bypasses wherever they are useful. Important considerations here are the distribution of the data (copies) over the hierarchy levels as these determine the access frequency and the size of each of the resulting memories. Our work has heavily contributed to a systematic methodology for data reuse exploration and decision early in the design flow, as described in chapter 9.

- *Storage cycle budget distribution:* Determining optimal accessibility constraints for the data such that the timing constraints can be met with an as cheap as possible memory architecture, both in terms of power and area is another crucial step in our DTSE methodology targeting real-time applications. The main contributions of this work are the correct formulation of this task and practical methodologies for solving this task for network component applications and (briefly also for) RMSP applications. The formalized exploration technique has been implemented in a prototype support tool. The results are described in chapter 10.

■ *Memory allocation and assignment:* In this step, the necessary number of memory units and their type are determined, matching the required parallelism in the memory access from the previous step. In addition, the multi-dimensional signals are assigned to these background memory units and ports. The main contributions of this work are again the correct formulation of this task and practical methodologies for solving it for both network component applications and RMSP applications (see chapter 11).

■ *In-place optimization:* In RMSP applications, most signals have a limited life-time which can be exploited to reduce the (background) memory cost heavily. Up till now, no data models or exploration techniques were available to solve this while taking into account the internal organisation of the complex data types (multi-dimensional arrays). We have shown that substantial gains can be obtained while exploiting this, and we have also proposed effective data models and practical solution methodologies for solving this task (see chapter 12). The formalized exploration technique has been implemented in a prototype support tool.

1.8   OVERVIEW OF THE BOOK

The rest of the book is organized as follows:

Chapter 2 gives an overview of the related work in the domain covered by this book.

Chapter 3 introduces all the steps in the DTSE methodology and illustrates these on a small but representative real-life example: a 1-D wavelet kernel used in the context of a video coding application.

Chapter 4 gives an overview of the cost models (power, size, area, and band-width) used in this book.

Chapter 5 proposes the geometrical models underlying the entire methodology.

Chapter 6 describes the input specification analysis including novel array data-flow analysis techniques.

Chapter 7 describes our input specification pruning strategies which allow to substantially reduce the complexity of applying our methodology and techniques. In addition, several other preprocessing substeps will be discussed also.

Chapter 8 presents a systematic methodology for global data-flow and control-flow transformations.

Chapter 9 presents a systematic methodology for data reuse exploration and decisions.

Chapter 10 explains what storage cycle budget distribution is, with emphasis on the storage-bandwidth optimization step, and illustrates its effect and importance on examples. In addition, our storage-bandwidth optimization technique is proposed.

Chapter 11 presents a systematic methodology for memory allocation and signal-to-memory assignment. Also techniques to automate the latter substep are discussed and illustrated.

Chapter 12 explains what in-place optimization is, with emphasis on the storage order optimization substeps, and illustrates its effect and importance on examples. In addition, our storage order optimization techniques are proposed.

Chapter 13 substantiates the claims made in this book on several real-life application demonstrators for custom memory architecture realisations.

Chapter 14 concludes the book and points out some interesting areas for future research.

2 RELATED WORK ON DATA TRANSFER AND STORAGE MANAGEMENT

In this chapter, an extensive summary is provided of the main related work in the domain of this book. It is organized in a hierarchical way where the most important topics receive a separate discussion, with pointers to the available literature. Wherever needed, a further subdivision in subsections or paragraphs is made.

2.1 SCALAR ORIENTED HARDWARE MEMORY MANAGEMENT APPROACHES

Almost all published techniques for dealing with the allocation of storage units are scalar-oriented and employ a *scheduling-directed* view (see e.g. [398, 224, 26, 168, 7, 350, 381, 362]) where the control steps of production/consumption for each individual signal are determined beforehand. This applies also for memory/register estimation techniques (see other section).

This strategy is mainly due to the fact that applications targeted in conventional high-level synthesis contain a relatively small number of signals (at most of the order of magnitude 10^3). The control/data-flow graphs addressed are mainly composed of potentially conditional updates of scalar signals.

Therefore, as the major goal is typically the minimization of the number of registers for storing scalars, the scheduling-driven strategy is well-fitted to solve a *register allocation* problem, rather than a *background memory allocation* problem. In that case, (binary) ILP formulations [7, 26, 362] and (iterative) line packing [224, 168], graph colouring [381], or clique partitioning [398] techniques have provided satisfactory results for register allocation, signal-to-register assignment, and signal-to-port assignment, under the usually implicit assumptions mentioned above.

In the compiler literature, the register allocation problem has also lead to many publications. One of the first techniques is reported in [62], where colouring of the interference (or conflict) graph is applied. Many other variants of this colouring have been proposed. Usually also spilling from foreground to background memory is incorporated. The fact that scheduling and register allocation are heavily interdependent has lead to a number of compiler approaches where register allocation is done in 2 phases: once before and once after scheduling [396], or where scheduling is first done for minimizing the registers and then again after register allocation for the other cost functions [166]. Sometimes, scheduling and register allocation are fully combined [47]. Practically all these techniques are restricted however to the scope of basic blocks. Recently, more general approaches (like [323]) have been proposed which offer a more general framework and heuristics to deal also with across basic-block analysis and inter-procedural considerations.

Several more recent techniques in the architecture synthesis community look at the allocation of larger memories but the approaches still consider the data to be stored as if they were "scalar units": partly overlapping life-times or exploiting the shape of the index space addressed by M-D signals is not done yet. Examples of this are the memory selection technique in GAUT [366] and the synthesis of intermediate memories between systolic processor arrays [362].

2.2   ARRAY-ORIENTED HARDWARE MEMORY ORGANISATION APPROACHES

Both the scalar-oriented strategy and the related techniques present serious shortcomings for most real-time multi-dimensional signal processing applications for several reasons. First, scheduling *must* precede (background) memory management in the conventional high-level synthesis systems. However, by switching the order of the memory accesses, the dominant background memory cost can be reduced further [402, 404] and the freedom for data-path allocation and scheduling remains almost unaffected [54]. Furthermore, within the scalar view, many examples are untractable because of the huge size of the ILP formulations.

Exceptions to this focus on scalar applications have been initiated by early work at IMEC and Philips. PHIDEO [245, 417] at Philips is oriented to periodic stream based processing of video signals. It uses a systematic methodology with the following tasks: data-path mapping, stream scheduling [412], memory allocation, and internal memory organisation combined with address generation.

At IMEC, extensive research has been performed in this array-oriented custom memory organisation management direction. Originally, this was focussed mainly on in-place optimization for which the problem was formulated [402]. The first attempt to tackle this problem is described in [288]. This work has been further formalized and has resulted in the definition of the concept of an address reference window for page mode memory organisation in Cathedral-2 [413]. Later on, the work has been oriented to automated memory allocation and assignment *before* the scheduling or the procedural ordering are fully fixed [28, 32, 31, 376]. This has lead to a significantly extended search space for memory organisation solutions, which is effectively explored in the HIMALAIA tool. More recent work has lead to a solid theoretical foundation for the in-place optimization of data in memories and to the development of promising heuristics to solve this very complex problem in a reasonable CPU time in a 2-stage approach for M-D signals [105, 107, 109]: intra-signal windowing [108], followed by inter-signal placement [106]. These techniques have been implemented in an efficient prototype tool for evaluating in-place storage of M-D signals. Finally, the important storage cycle budget distribution step has been added, where we determine the bandwidth requirements and the balancing of the available cycle budget over the different memory accesses in the algorithmic specification [438, 441]. A complete methodology for custom background memory management (or data transfer and storage exploration as we call it) has been proposed in the DTSE methodology, with a small variation for memory size/area [290, 59] and power [53, 58, 59] costs. Also extensions to the network component (e.g. ATM) application domain [376] and to mapping on embedded processor cores [292] have been developed.

Recently also several other approaches for specific tasks in memory management oriented to non-scalar signals have been published. The MeSA [339] approach has been introduced at U.C.Irvine. Here, the emphasis is on memory allocation where the cost is based both on a layout model and on the expected performance, but where the possibility of M-D signals to share common storage locations when their life times are disjoint is neglected. Strategies to mapping arrays in an algorithm on horizontally and vertically partitioned dedicated memory organisations have been proposed too, at CMU [361]. Also some manual methodologies oriented to image processing applications have been investigated at the Univ. of Milano [20].

In the context of systolic arrays, the reduction of M-D signals in applicative languages to less memory consuming procedural code has been tackled, especially at IRISA [428, 430]. The principle of memory reuse through projection of multi-dimensional arrays is described, together with an analysis model and a set of necessary and sufficient constraints that have to be satisfied. But no strategy for obtaining a good projection is presented. Moreover, memory reuse between different arrays is not considered, in contrast to our approach. Nevertheless, this projection approach is very valuable and it can can be combined with our more general storage order optimizing (or "in-place optimization") approach [109]. Also in the parallel compiler context, similar work has been performed (see software oriented memory management section) with the same focus and restrictions.

Finally, also in the context of DSP data-flow applications, work has been performed on optimizing the buffer requirements between interacting data-flow nodes for a given schedule. This has been in particular so for the static and cyclo-dynamic data-flow models [2, 3] but it is based on data-flow production/consumption sequence analysis which poses restrictions on the modeling.

In addition, array-oriented memory estimation [27, 405] has been a topic of recent research in the system-level synthesis community (see other section).

2.3   SOFTWARE-ORIENTED MEMORY MANAGEMENT APPROACHES

Recently, the interaction between RISC architecture features and the (instruction-level) parallelism and compilation issues have been emphasized heavily, with a range of optimisations from pure static compile-time transformations to run-time transformations executed on the processor hardware [4]. Also the memory interaction has been identified as a crucial bottle-neck. Still, the amount of effort at the compiler level to address this bottle-neck show a focus on a number of areas while leaving big holes in other (equally important) research issues.

2.3.1   Dynamic memory management issues

In most work on parallel MIMD processors, the emphasis in terms of storage has been on hardware mechanisms based on cache coherence protocols. Work in this direction has focussed especially on safe but more efficient protocols and also on analysis of cache behaviour. Examples include the work at U.S.C. [121], Stanford [159], IBM [246], Lund [378] and Sun [153]. The relation with the compiler optimisation is sometimes incorporated, as in the work at Illinois [72], but even then it remains run-time oriented. In this context, also work has been going on optimal policies for page fetching and replacement, i.e. dynamic memory management policies [429]. Up to now this has been focussed mostly on policies for dynamically allocated data sets as in large data-bases, e.g. at the Univ. of Davis [71]. Interesting requirements for the multi-media applications are reviewed in [379]. A formal search space representation for dynamically allocated data and systematic methodologies to explore this for power and access count related cost functions have been proposed by us in [97, 436]. Both abstract data type refinement [436, 439] and virtual memory manager allocation [96] have been addressed.

2.3.2   Memory organisation issues

Several papers have analysed memory organisation issues in processors, like the number of memory ports in multiple-instruction machines [277], or the processor and memory utilisation [131, 395]. Also the deficiencies of RISC processor memory organisations have been analysed (see e.g. [23, 273]). This is however only seldom resulting in a formalizable method to guide the memory organisation issues. Moreover, the few approaches are usually addressing the "foreground" memory organisation issues, i.e. how scalar data is organized in the local register files. An example of this is a theoretical strategy to optimally assign scalars to register-file slots [22]. Some approaches are quite ad hoc, e.g. dedicated memory storage patterns for numerical applications on supercomputers have been studied in [252].

Some approaches address the data organisation in processors for programs with loop nests. Examples include a quantitative approach based on life-time window calculations to determine register allocation and cache usage at IRISA [45], a study at McMaster Univ. for the optimisation of multiple streams in heavily pipelined data-path processors [260], and work at Rice on vector register allocation [9]. Also the more recent work on minimizing access time or data pattern access in SIMD processors [11] is relevant in this context. It is however limited to the SIMD context and only takes into account the minimisation of the cycle cost. Moreover, it does not support data-dependent access. The recent work at Ecole des Mines de Paris [18] on data mapping is closer already to what is desired for multi-media applications. It tries to derive a good memory allocation in addition to the partitioning of operations over a distributed memory MIMD processor. It does not support optimisations inside the processor memory however, so the possibility to map signals in-place on top of one another in the memory space is not explored yet and every processor has a single memory.

Also in an ASIP software synthesis context, the approaches are practically always "scalar", or "scalar stream" oriented. Here the usually heavily distributed memory organisation is fixed and the target is to assign the variables to registers or memory banks [242] and to perform data routing to optimize code quality (e.g. [21, 230, 384]).

In a software/hardware co-design context, several approaches try to incorporate the memory modules. Up till now however, they are restricting themselves mostly to modeling the memory transfers and analyzing them by software profiling (e.g. [182]). Preliminary work on memory allocation in a GNU compiler script for minimizing memory interface traffic during hardware/software partitioning is described in [203].

In a systolic (parallel) processor context, several processor allocation schemes have been proposed. Recently, they also incorporate I/O and memory related costs (albeit crude), e.g. with a solution approach based on a combination of the LSGP and LPGS approaches [124].

2.3.3   Data locality and cache organisation related issues

Data locality optimizing algorithm transformations have been studied relatively well already in the past, e.g. at Illinois/INRIA [149] and Stanford [432]. The focus lies on the detection of spatial and temporal data reuse exploitation in a given procedural code. Tiling (blocking) combined with some local (applied within a single loop nest) uniform loop transformations is then used to improve the locality. This work forms a good basis for work in this research domain. Recently, this has been extended and implemented in the SUIF compiler project at Stanford [19, 181] to deal with multiple levels of memory hierarchy but it still is based on conventional loop nest models. Also the work at IRISA, Rennes [397] which distributes data over cache blocks is relevant here. It uses a temporal correlation based heuristic for this but relies mostly on the hardware cache management still and is oriented mostly to scalars (or indivisible array units).

At U.C.Irvine, also recent work has focused on multi-media application related memory organization issues in an embedded processor context [310]: distributing arrays over multiple memories with clique partitioning and bin packing [307], optimizing the main memory data layout for improved caching based on padding and clustering signals in the main memory to partly reduce conflict misses [306, 308], selecting an optimal data cache size based on conflict miss estimates [311] and distributing data between the cache and a scratch-pad memory based on life-time analysis to determine access conflicts in a loop context [309]. All of this is starting from a given algorithm code and the array signals are treated as indivisible units (i.e. similar as scalars but with a weight corresponding to their size).

Bypassing the cache for specific fetches is supported by several multi-media oriented processors. Exploiting this feature for instruction caches can be steered in the compiler based on simple usage counts of basic blocks [205]. Related write buffer oriented compiler issues have been addressed in [255]. Even the disk access from processors is a topic of recent studies. Similar transformations as used for cache optimisation are applied here [209].

In addition, there is a very large literature on code transformations which can potentially optimize this cache behaviour for a given cache organisation which is usually restricted to 1 level of hardware caching (see other section).

2.3.4   Data storage organisation and memory reuse

In the context of systolic arrays and later also for parallel compilation, the reduction of M-D signals in applicative languages to less memory consuming procedural code has been tackled, especially at IRISA [428, 430, 334]. This is a restricted form of in-place optimization. Also in the work of PRISM on storage management for parallel compilation [234], the emphasis is again on the removal of overhead for static control (applicative) programs. Similarly, in the context of parallel architectures and compilation, also some related work on so-called update-in-place analysis has been reported for applicative languages (see e.g. [140] and its refs). That research only focuses on the removal of redundant copy nodes introduced by the applicative language rewriting.

2.3.5 Related work at IMEC

At IMEC, we have worked on several issues of programmable processor oriented memory organisation optimization. A complete methodology for data transfer and storage exploration (DTSE) as a precompiler stage to traditional (parallel) compilers is under development in the ACROPOLIS research project [57]. This effort is based on our background in custom memory management (or DTSE) in the context of our ATOMIUM research project [53, 59]. First, we have shown that the system-level storage organization for the multi-dimensional (M-D) signals in multi-media applications should be addressed as the first step in the overall methodology, even before the processor parallelization and load balancing [89, 91], or the software/hardware partitioning [90]. We have also shown that hybrid task-/data-parallel solutions are very effective for obtaining power-efficient data transfer and storage related organisations [89, 91]. The transformation steps in the DTSE methodology developed for programmable multi-media processors in the ACROPOLIS project, are for the moment shared with the ATOMIUM project results for custom architecture targets. The latter have been addressed in another section. Here, only the new cache and main memory organisation related issues will be summarized. A solid theoretical foundation for the in-place optimization of data in memories has been laid, including the development of promising heuristics to solve this very complex problem in a reasonable CPU time in a 2-stage approach for M-D signals [105, 107, 109]: intra-signal windowing [108], followed by inter-signal placement [106]. These techniques are used as the basis for advanced cache allocation schemes with more optimal reuse of the cache memory locations. We have obtained promising results for both software controlled caches [223] and more traditional hardware controlled ones [222].

2.4 MEMORY SIZE ESTIMATION APPROACHES

The problem of computing storage requirements in high-level synthesis has been tackled in the past mainly as the optimization of register allocation/assignment in RT programs [148].

The problem can be formulated as a *vertex colouring* problem [8], p.545, and solved heuristically for software compilers. The approach employed in the FACET design system [398] is based on *clique partitioning*, basically equivalent to the vertex colouring method. For nonrepetitive unconditional schedules, the register allocation/assignment is equivalent to interval-graph colouring [165, 224] — which can be optimally carried out, in polynomial time, with the "line-packing" or "left-edge" algorithms — used for channel routing problems in layout design [183]. An extension of the approach for given repetitive and conditional schedules has been proposed in [167], although the method — based on life-time analysis — does not always yield optimal solutions. Stok solves optimally the register allocation/assignment problem, allowing slight modifications of the given schedule [381]. An optimal solution for repetitive schedules, employing circular graphs, has been more recently proposed in [313]. Lower bounds are derived in [113].

When the operation scheduling is not yet fully fixed, a lower-bound on the number of registers required to implement control/data-flow graphs can be computed with a force-directed approach [320]. An ILP formulation for scheduling problems, minimizing life-times of scalars – implicitly reducing the number of registers, has been proposed in [191].

Background memory size estimation for M-D signals in RMSP algorithms has only recently gained attention in high-level synthesis. Note that the previous scalar-oriented techniques at RT level could compute the *exact* values of storage requirements. Because background memory size determination for M-D signals has to handle problems significantly larger (in terms of number of scalars), and because the specifications are often nonprocedural, the techniques described in the sequel are usually computing *approximative* values of storage requirements.

The first estimation methods have tried to apply the techniques from register allocation/assignment, by flattening the M-D signals at the scalar level. For instance, the *symbolic evaluation* consists in enumerating all indexed signals for all index combinations [289], and mapping them in-place [404].

The tool $s2p$ from the CATHEDRAL 2/3 system computes windows having a size of the form 2^k for every indexed signal (in order to ensure a low-cost addressing hardware), trying to minimize the memory requirements for each individual M-D signal [413].

These methods are valid only for procedural specifications. Moreover, they all suffer from complexity problems when loop boundaries are large.

More recently, novel results – less dependent on the size of iterator ranges – have been obtained for the case when the algorithm specification is nonprocedural. Modifications of the loop hierarchy and the sequence of execution as specified in the source code are used to optimize the storage requirements [422]. Retaining only the data-flow constraints, a new control-flow is provided by placing polytopes of signals in a common space (with an ILP technique), and searching for an ordering vector in that space.

For solving the memory size estimation problem in the nonprocedural case, data-flow analysis has been consistently employed as the main exploration strategy in [27, 32]. In a more recent work, data dependence information provided by the Omega test [329] has also been applied in memory size estimation [405]. Although the method is very appealing in terms of speed, the assumptions regarding loop hierarchy and control-flow – e.g., a nest of loops is executed in the sequence as specified in the source code, and a sequence of loops is executed in the order specified in the source code – only partly remove the procedural constraints.

2.5 DATA-FLOW ANALYSIS FOR ARRAY SIGNALS

Most data-flow analysis approaches are oriented to the affine case, based on polyhedral theory [34] with convex polyhedral sets. These approaches have been extended to more general affine indices leading to linear bounded lattice domains [390]. The most powerful approaches at this moment appear to be the environments of Univ. of Maryland (the Omega system [329, 330, 332]), the Univ. of Versailles (PIP [134]), and Stanford (SUIF [258, 259]). Each of these have strong and weak points in terms of flexibility and efficiency. The different data dependence abstractions required to solve particular tasks have been classified too [17]. Experiments to determine which classes typically appear are e.g. presented in [322].

In many applications however, also more general index expressions have to be handled. Extensions towards more general data-flow modelling have been proposed for conditional data-dependent indexing [143], for WHILE loops [76, 172], for piece-wise affine indices [29], and for data-dependent indices by lazy flow dependence [256] or fuzzy data-flow analysis [77] or for mixed procedural/functional models [78]. The extension towards multi-media type requirements and data storage order modeling has been performed also [109], including efficient counting of points in polytopes [33] to steer the cost functions. General symbolic non-linear expressions have been analyzed also [44]. Also interprocedural data-flow analysis has been investigated within specific assumptions [83, 84]. This has even been applied to analyze virtual functions in an object-oriented context [24]. In a systolic array context, also reducible integral recurrences [341], splitting up affine recurrences into uniform recurrences [173] and the transformation of affine into quasi-uniform recurrences [443] have been studied.

Finally, a large class of applications like numerical computing and 3D modelling problems cannot be accurately analyzed at compile-time. For this purpose, run-time environments are required. A good overview is provided in [80].

2.6 PARALLELISM AND DTSE OPTIMIZING LOOP AND DATA-FLOW TRANSFORMATIONS

It has been recognized quite early in compiler theory (for an overview see [35]) and high-level synthesis [402] that in front of the memory organisation related tasks, it is necessary to perform transformations which optimize mainly the loop control-flow. Otherwise, the memory organisation will be heavily suboptimal.

2.6.1 Interactive loop transformations

Up to now most work has focused on the loop control-flow. Work to support this crucial transformation task has been especially targeted to interactive systems of which only few support sufficiently general loop transformations.

In the **parallel compiler domain**, interactive environments like Tiny [433, 434], Omega at U.Maryland [210], SUIF at Stanford [13, 19, 181], the Paradigm compiler at Univ. of Illinois [37] (and earlier work [325]) and the ParaScope Editor [261] at Univ. of Rice have been presented. Also non-singular transformations have been investigated in this context [210, 239]. These provide very powerful environments for interactive loop transformations. In practice however, textual editing is restricted and only possible with general-purpose text editors.

In the **regular array synthesis domain**, an environment on top of the functional language $Alpha$ at IRISA [115] has been presented.

In the **high-level synthesis and formal languages community**, several models and environments have been proposed to allow incremental transformation design (see e.g. SAW at CMU [392] and Trades at T.U.Twente [266]). These are however not oriented to loop or data-flow transformations on loop nests with indexed signals, and they assume that loops are unrolled or remain unmodified. An exception to this is SynGuide [354], a flexible user-friendly prototype tool which allows to interactively perform loop transformations. It been developed at IMEC to illustrate the use of such transformations on the applicative language Silage and its commercial version DFL (of EDC/Mentor).

2.6.2 Automated loop transformations

In addition, research has been performed on (partly) automating the steering of these loop transformations. Many transformations and methods to steer them have been proposed which **increase the parallelism**, in several contexts. This has happened in the array synthesis community (e.g. at Saarbrucken [390] (mainly intended for interconnect reduction in practice), at Versailles [135, 137] and E.N.S.Lyon [94] and at the Univ. of SW Louisiana [368]) in the parallelizing compiler community (e.g. at Cornell [239], at Illinois [304], at Stanford [431, 13], at Santa Clara [371], and more individual efforts like [400] and [70]) and even in the high-level synthesis community (at Univ. of Minnesota [312] and Univ. of Notre-Dame [315]). Some of this work has focussed even on run-time loop parallelisation (e.g. [237]) using the Inspector-Executor approach [353]. None of these approaches work globally across the entire system, which is required to obtain the largest impact for multi-media algorithms.

Efficient parallelism is however partly coupled to **locality of data access** and this has been incorporated in a number of approaches. Therefore, within the parallel compiler community work has been performed on improving data locality. Most effort has been oriented to dealing with *pseudo-scalars* or signal streams to be stored in local caches and register-files. Examples of this are the work at INRIA [125] in register-file use, and at Illinois for vector registers [9].

Some approaches are dealing also with *array signals* in loop nests. Examples are the work on data and control-flow transformations for distributed shared-memory machines at the Univ. of Rochester [74], or heuristics to improve the cache hit ratio and execution time at the Univ. of Amherst [262]. At Cornell, access normalisation strategies for NUMA machines have been proposed [238]. Partitioning or blocking strategies for loops to optimize the use of caches have been studied in several flavours and contexts, in particular at HP [130] and at Toronto [220, 221, 254, 253]. Recently, also multi-level caches have been investigated [204]. The explicit analysis of caches is usually done by simulation but an analytical approaches to model the sources of cache misses have been recently presented also [160, 207]. This allows to more efficiently steer the transformations. Finally, also in the context of DSP data-flow applications, work has been performed on "loop" transformations to optimize the buffer requirements between interacting data-flow nodes. This has been in particular so for the SDF model [42] but it is based on data-flow production/consumption sequence analysis and sequence merging which poses restrictions on the modeling. The main focus of these memory related transformations has been on performance improvement in individual loop nests though and not on power or on global algorithms.

Some work is however also directly oriented to storage or transfer optimisation *between the processor(s) and their memories* to reduce the memory related cost (mainly in terms of area and power). At IMEC, successful attempts in this direction have been made starting from applicative descriptions of image and video processing applications. This has lead to a proposal for an automated loop transformation technique which is based on a cost function modeling a combination of area and power [420, 144]. A model and formal methodology for globally ordering and overlapping the loop nests to optimize the required memory bandwidth has been added to this [438, 441]. For power, we have developed a systematic methodology oriented to removing the global buffers which are typically present between subsystems and on creating more data locality [103, 110]. An extended model and problem formulation to introduce locality allowing more aggressive transformations to achieve in-place storage of array signals, have been introduced at IMEC too [105]. Finally, we have developed novel data reuse (copy) transformations based on a data reuse decision step where we have to decide on the exploitation of the memory hierarchy including bypasses wherever they are useful [437, 440]. We have proposed a formalized methodology to steer this, which is driven by estimates on bandwidth and high-level in-place cost [118, 440]. The recent work at Ecole des Mines de Paris [18] on data mapping is also relevant in the context of multi-media applications. It tries to derive a good memory allocation in addition to the transformation (affine scheduling and tiling) of loop nests on a distributed memory MIMD processor. It does not support optimisations inside the processor memory however, so the possibility to map signals in-place on top of one another in the memory is ignored.

In the *high-level synthesis community*, also some other recent work has started in this direction. An example is the research at U.C.Irvine on local loop transformations to reduce the memory access in procedural descriptions [214, 215]. In addition, also at the Univ. of Notre Dame, work has addressed multi-dimensional loop scheduling for buffer reduction [317, 318]. Within the Phideo context at Philips and T.U.Eindhoven, loop transformations on periodic streams have been applied to reduce an abstract storage and transfer cost [408, 410, 411].

2.6.3 Data-flow transformations

Also data-flow transformations can effect the storage and transfer cost to memories quite heavily. This has been recognized in compiler work [8, 265] but automated steering methods have not been addressed there. The specific subject of handling associate scans or chains has been addressed in somewhat more depth in the context of systolic array synthesis [347] and recently also in parallel array mapping [38], but mainly with increased parallelism in mind. Look-ahead transformations have been studied in the context of breaking recursive loops in control [312], algebraic processing [251] and (scalar) signal processing [312], but not in the context of memory related issues. At IMEC, we have classified the set of system-level data-flow transformations that have the most crucial effect on the system exploration decisions and we have developed a formalized methodology for applying them in real-time image and video processing applications [56, 61]. The effect of signal recomputation on memory transfer and data-path related power has been studied too but only in the context of flow-graphs with fixed power costs associated to the edges [152].

2.6.4 Parallel processor models

To evaluate all these transformations effectively and accurately, a good abstraction of the detailed architecture features behind the very diverse world of parallel processor architectures is required. Several attempts have been made in the past to come up with such a model. Examples of this are the Bulk Synchronous Parallel (BSP) model of [401] and the Block Distributed Memory (BDM) model of [199]. Both of them have only few parameters and are relatively simple to use but practical architectures in the (embedded) multi-media domain do not meet the requirements which are imposed by these models. Also models to represent the memory hierarchy impact have been proposed [198].

2.7 MIMD PROCESSOR MAPPING AND PARALLEL COMPILATION APPROACHES

In most work on compilation for parallel MIMD processors, the emphasis has been on parallelisation and load balancing, in domains not related to multi-media processing. The storage related issues are assumed to be mostly solved by hardware mechanisms based on cache coherence protocols (see other section). Several classes of parallel machines can be identified, depending on several classification schemes [184, 193].

Several manual methodologies have been proposed to map specific classes of multi-media applications to multi-processors in terms of communication patterns [87, 263, 380]. These are however not automatable in a somewhat more general context. Whether this can/should be automated and how far is still under much discussion in the potential user community [82].

In terms of design automation or parallel compilation, especially parallelism detection [44] and maximalisation [304] (see loop transformation section) have been tackled. For this purpose, extensive data-flow analysis is required which is feasible at compile-time or at run-time (see other section). Once introduced, this parallelism is exploited during some type of scheduling stage which is NP-complete in general [40]. A distinction is made between distributed memory machines where usually a data parallel approach is adopted and shared-memory machines where usually more coarse-grain tasks are executed in parallel. In principle though, both these levels can be combined. In addition, there is a lower, third level where parallelism can be exploited too, namely at the instruction level. Also here several approaches will be reviewed below.

Recently, also the mapping on (heterogeneous) clusters of workstations has attracted much attention. Here the focus is still mainly on the general programming environment [52, 98, 141] and ways to ease the communication between the nodes in such a network, like the PVM and MPI libraries. The importance of multi-media applications for the future of the parallel processing research community has been recognized also lately [147, 218], because the traditional numerical processing oriented application domains have not had the originally predicted commercial break-through (yet).

2.7.1 Task-level parallelism

In terms of task parallelism, several scheduling approaches have been proposed some of which are shared-memory oriented [10, 243], some are not assuming a specific memory model [127, 154, 232, 217], and some incorporate communication cost for distributed memory machines [189, 305]. In addition, a number of papers address task-level partitioning and load balancing issues [6, 68, 146, 188]. Interesting studies in image processing of the potential effect of this data communication/storage versus load balancing trade-off at the task level are available in [89, 174, 383].

2.7.2 Data-level parallelism

Data parallelism in a distributed memory context requires an effective data parallelism extraction (see other section) followed by a data partitioning approach which minimizes the data communication between processors as much as possible. This is achieved especially in relatively recent methods. Both interactive approaches oriented to difficult or data-dependent cases [67, 15] and more automated techniques focussed on regular (affine) cases [190] within predefined processor arrays have been proposed. The automated techniques are oriented to shared memory MIMD processors [5], systolic arrays [138] or to uniform array processors [369, 69, 370, 119]. The minimal communication cost for a given level of parallelism forms an interesting variant on this and is discussed in [117]. Also clustering of processes to improve the access locality contributes to the solution of this problem [249]. All the automatable approaches make use of regular "tiling" approaches on the data which are usually distributed over a mesh of processors. Also in the context of scheduling this tiling issue has been addressed, already since Lamport's hyperplane method [229]. Since then several extensions and improvements have been proposed including better approaches for uniform loop nests [93], affine loop nests [135, 136], non-singular cases [442], trapezoid methods [399], interleaving for 2D loop nests [316] and a resource-constrained formulation [120]. Also in the array synthesis context, such tiling approaches have been investigated for a long time already (see e.g. [50]). Recently, they also incorporate I/O and memory related costs (albeit crude) and are based on a combination of the LSGP and LPGS approaches [124]. Most methods do this locally within a loop nest. Only some recent work also performs the required analysis across loop boundaries [257, 175, 192].

None of these approaches consider however the communication between the processors and their memories or the size of the required storage units. In some cases however, there is also a data locality improvement step executed during compilation (see other section). An interesting study in image processing of the potential effect of this data communication/storage versus load balancing trade-off at the data level is available in [89].

The final production of the communication code necessary to transfer the data in the parallel processor [342] is addressed rarely.

2.7.3 Instruction-level parallelism

A huge amount of literature has been published on scheduling and allocation techniques for processors at the instruction level. Here, only a few techniques oriented to signal processing algorithms are mentioned. One of the first approaches is the work of Zeman et al. [447] which was very constrained. Later on extensions and improvements have been proposed for scheduling based on cyclo-static models [364], list scheduling including loop folding [165], pipeline interleaving [231], pipelined multi-processors [211], and the combination of scheduling and transformations [424, 423]. An evaluation of several scheduling approaches has been described in [186].

In addition, instruction-level parallelism (ILP) has become a major issue in the RISC compiler community. Examples of this can be found in the ILP workshop embedded in Europar'96 [150]. Notable issues at the compilation side are advanced software pipelining [12], global code motion on the inter-BB (basic block) level [248], and the fact that advanced code manipulation is required to obtain the required speed-up potential [326].

Very recently, the interaction between RISC architecture features and the (instruction-level) parallelism and compilation issues has been emphasized heavily, with a range of optimisations from pure static compile-time transformations to run-time transformations executed on the processor hardware [4].

2.8 HARDWARE AND SOFTWARE ARCHITECTURES FOR MULTI-MEDIA PROCESSING

Many architectures have been proposed in the direction of audio, image and video processing, and graphics in the last decades (see e.g. [300]). In recent years, the emphasis within the image and video processing part of multi-media processing has been on novel, mostly parallel architectures. Actually, a major trend in the market for parallel computing is towards such multi-media applications [147].

In general, the trend towards a more distributed memory organisation to accommodate the massive amount of data transfers needed, is clearly present both in hardware and software solutions [64, 359]. More and more of this will be on-chip memory [295, 359, 444]. Also the emphasis on low-power realisations is growing [64, 295, 359]. Several studies have shown that both area [403] and power [53, 240, 264, 435] in such custom architectures are dominated by the storage and transfer of complex array signal handling. This is also true for the newer generation of programmable processors [164, 194]. An overview of power oriented work at the architecture and system-level design automation is provided in [375].

Within the customized hardware architectures all this is especially witnessed by the evolution in the domain of MPEG like subsystems. A good survey of this is provided in [324]. Nice examples of MPEG2 coders are: the low area decoder of Hitachi [302], the low power encoder of NEC [275], and the partly programmable encoders of Philips [407] and Matsuhita [274].
H.263 related custom codecs are reported e.g. in [46, 267].
Wavelet and DCT co-processor units for programmable processors have been proposed also [301]. Initial investigations for modules in MPEG4 have been performed also [386]. The importance of a quite global memory/buffer reordering in order to achieve cost effective realisations for embedded applications is confirmed in the MPEG1 decoder architecture of [247].

For programmable processors, a good overview of recent work is available in [180] and [346]. Examples include the TI C80 [30], the Philips Trimedia, the Chromatic Mpact, the MicroUnity Mediaprocessor and other recent announcements [180]. A very low voltage architecture has been proposed by TI for wireless communication applications [233]. For MPEG like applications, several very powerful multi-media processors have been proposed [197, 446], several of them based on SIMD architectures [445].

Also on such processors, the power appears to be dominated by the data transfers of complex data types [393, 394]. In burst-mode type of applications, power can be gained by variable supply voltage operation [51, 176] or with multiple supply voltages [65].

The instruction fetch related power can be heavily reduced by appropriate caching [25], especially if the instruction cache and decoder are distributed and placed very locally to the parallel FUs in the processor [60].

Architecture and circuit solutions have also been proposed to reduce the energy dissipation in the data cache hierarchy [212].

For the data transfer over large busses, also several solutions have been proposed at the architecture levels, mainly focussed on reduced voltage swings and low-power data encodings (see [377] for an overview).

2.9 LOW POWER STORAGE MODULES AND CIRCUITS

Many RAM circuits and architectures have been published where apart from speed and reliability/noise issues, also power considerations play an important role. Already a decade ago, the growing performance gap between the processor and the off-the-shelf (D)RAM evolution became a major issue (see overview of evolution in [79] and [328]). Also the increasing power consumption (up to several Watt for large DRAM and SRAM chips) became a major concern. Since then, a significant effort has been invested in better solutions at the circuit and process technology levels.

Issues directly related to the reduction of the energy-delay product in DRAMs have focussed especially on the following [340, 60, 195, 196]: the more aggressive partitioning in hierarchical memory planes (with usually more than 32 divisions for RAM sizes above 16Mb), wide memory words to reduce the access speed [14], multi-divided arrays (both for word-line and data-line) with up to 1024 divisions in a single matrix [385], low on-chip V_{dd} (up to 0.5 V [444]) with $V_{dd}/2$ precharge [444], special voltages on the word-line to reduce the leakage current, special NAND decoders [283], reduced bit/word-line swing (up to only 10% of the V_{dd}) [195], differential bus drivers and charge recycling in the I/O buffer [283].

Because of all these principles to distribute the power consumption from a few "hot spots" to all parts of the architecture, the end result is indeed a very optimized design for power where every piece consumes about an equal amount (see e.g. [365]). It is expected however that not much more can be gained because the "bag of tricks" is now containing only the more complex solutions with a smaller return-on-investment. Note however that the combination of all these approaches indicates a very advanced circuit technology which clearly outperforms the current state-of-the-art in data-path and logic circuits for low power design. Hence, It can be expected that the relative power in the non-storage parts can be more drastically reduced still towards the future (on condition that similar investments are done). Combined with the advance in process technology, all this had lead to a remarkable reduction of the DRAM related power: from several Watt for the 16-32 Mb generation to about 100 mW for 100 MHz operation in a 256 Mb DRAM (see e.g. [287]).

The decrease of the power consumption in fast random-access memories is not as advanced yet as in DRAMs but also that one is saturating, because many circuit and technology level tricks have already been applied also in SRAMs. As a result, fast SRAMs keep on consuming on the order of Watts for high-speed operation around 500 MHz [48, 294]. A general view on modeling and optimization is available in [129].

From the process technology point of view this is not so surprising, especially for submicron technologies. The relative power cost of interconnections is increasing rapidly compared to the transistor (active circuit) related components. Clearly, local data-path and controllers themselves contribute little to this overall interconnect compared to the major data/instruction busses and the internal connections in the large memories. Hence, if all other parameters remain constant, the energy consumption (and also the delay or area) in the storage and transfer organisation will become even more dominant in the future, especially for deep submicron technologies. The remaining basic limitation lies in transporting the data and the control (like addresses and internal signals) over large on-chip distances, and in storing them.

Attempts have been made also to address the off-chip access bottle-neck by technological innovations. Several recent activities (see e.g. the Mitsubishi announcements [351] and the IRAM initiative of U.C.Berkeley [319]) motivate that the option of embedding logic on a DRAM process leads to a reduced power cost and an increased bandwidth between the central DRAM and the rest of the system. This is indeed true for applications where the increased processing cost is allowed. However, it is a one-time drop after which the widening energy-delay gap between the storage and the logic will keep on progressing, due to the unavoidable evolution of the interconnect contributions relative to the arithmetic and control logic. So on the longer term, the bottle-neck should be broken also by other means. In another section, it is shown that this is feasible with quite spectacular effects at the level of the system design methodology.

In addition to the data RAM units themselves, also other components related to storage can consume much power on programmable processors. This is in particularly so for program memories, Table Look-aside Buffers (TLBs) [206] and prefetch memories. Approaches to reduce the power related to program memories include: compression of the object code [446, 394, 233], specific techniques oriented to ROM circuits [99], instruction scheduling [394, 233].

Power models to be used at the high-level have been proposed by several groups, e.g. the PowerPlay model of U.C.Berkeley [241] and many other foundry related models which are usually confidential (at IMEC we have several of those from companies like Motorola, TI, VTI and Alcatel).

For caches, specific power models have been developed which try to incorporate the behaviour of the cache [160, 207] without requiring a too time-consuming analysis. This is essential for steering of transformation and optimization techniques to reduce the cache related power.

3 DATA TRANSFER AND STORAGE EXPLORATION METHODOLOGY

In this chapter, our DTSE methodology is introduced and illustrated on a simple but representative real-life application. The starting point of the DTSE methodology [58, 59] is a system specification with accesses on multi-dimensional (M-D) signals which can be statically ordered (single thread of control). The output is a net-list of memories and address generators (see figure 1.7), combined with a transformed specification where the background memory accesses are heavily reorganized. This new code is the input for the architecture (high-level) synthesis when custom realizations are envisioned, or for the software compilation stage in the case of predefined processors.

3.1 MOTIVATION OF STEPS FROM POWER VIEWPOINT

This section motivates the steps of our DTSE methodology, first mainly from a viewpoint of low power consumption, but afterwards also the area effects will be discussed. In order to achieve a significant drop of the actual energy consumed for the execution of a given application, we have to examine the power formula for memories: $F_{real} \cdot V_{dd} \cdot V_{swing} \cdot C_{load}$. Note that the real access rate F_{real} should be provided and not the maximum frequency F_{cl} at which the memory can be accessed (see chapter 4).

We can reduce both F_{real} and the effective C_{load} in the way we use the storage and interconnect devices. The goal should be to arrive at a memory and transfer organisation with the following characteristics.

1. Reduced redundancy in data transfers.

2. Introduction of more temporal locality in the accesses, so that more data can be retained in registers local to the data-paths.

3. Use of a hierarchical memory organisation where the smaller memories (with reduced C_{load}) are accessed the most and the larger ones are accessed the least.

4. Avoidance of the use of multi-port memories if 1-port alternatives with a reasonable cost can be used instead, because more area but also more interconnect (and hence energy-delay) is required for these multi-port components. Also the energy per memory access is significantly increased.

These principles can be enabled by applying an appropriate system design methodology. General principles are proposed in papers like [336] but for our particular situation we require a more cus-

tomized approach. A systematic methodology for supporting our objective has been developed at IMEC for this purpose [53, 58, 437].
The main steps involve (see also figure 3.1):

1. Reducing the required data bit-widths for storage by applying optimizing algorithmic transformations. These can change the actual algorithm choice, e.g. in digital filters or in the choice DFT versus FFT. Much can be done also in the step from floating-point data types to fixed-point types (see e.g. [58]).

2. Reducing transfer redundancy by data-flow transformations which do not modify the global input-output functionality [61].

3. Increasing the access locality and regularity by global loop transformations.

4. Exploiting memory hierarchy better by an optimized decision on reuse of multi-dimensional array data.

5. Balancing distribution of the globally available cycle budget over the loop nests to reduce the minimally required band-width. This ensures that fewer multi-port memories are required and also fewer memories overall.

6. Allocating a more distributed memory organisation than conventionally used and assigning the array signals cost-effectively to the different memories at each hierarchical level.

7. Exploiting the restricted life-times of parts of the array signals to overlap them in the address space (in-place optimisation). This enables a better exploitation of the caches and reduces the overall size of the required memories (and hence C_{load}).

3.2 APPLICATIVE VERSUS PROCEDURAL STARTING POINT

The applicative nature of the internal data model is essential for us to create sufficient freedom to perform effective memory management. If this model would be directly derived from a specification in a procedural language where the procedural constructs (like data variables and pointer variables) are fully exploited, it would in general not be possible to extract the data-flow. Extensive data-flow analysis is however essential to allow a full modification of the execution and storage order. Moreover, C specifications have a compiler semantic which implies in the worst-case a fixed execution order[1] and row-major storage order. This semantic is also (mis)used by the designers writing the C specification. If we would accept these (sometimes even implicit) semantics, then only the memory allocation/assignment and the address generation stage [270] would remain fully useful. The static in-place optimisation step [105, 108] (see figure 3.4) performed at compile-time would be reduced to operating on signals which have different names because the actual data dependencies within an indexed array can in general not be sufficiently analyzed. Also dynamic in-place management (as performed by some C compilers at run-time) can still be done but with much less optimisation scope.

To avoid this situation and to provide more cost-effective realisations, more formal analysis and a wide search space are required. Therefore, the proposal is to start from an applicative single-assignment data model (as a subset of a procedural language like C or C++) where the user constraints are explicitly expressed only at the place where they are needed (and not embedded in the unnecessary constraints of in a procedural specification). This means that the designer imposes ordering and timing constraints on I/O and internal signals. In the internal data models (like control/data-flow graphs) and polyhedral dependence graphs [421, 109]) these should then be accurately represented. This allows formal analysis and verification and it is also the basis for interactive or partly automated exploration/optimisation. Some of the smaller examples in this book will be presented in the purely applicative language Silage [187] to stress this issue.

---

[1] In the compiler the sequencing can be modified in principle if the difference cannot be observed after running the program, but in the presence of procedural constructs this is in general impossible to analyze and guarantee in practice.

On the other hand, there is no harm in representing the system modules which are "irrelevant" for DTSE in another form (procedural in C or even VDHL if co-simulation is feasible).

On the longer term, more general procedural language constructs will be allowable when more advanced array data-flow analysis techniques become mature (like the PIP activities at Univ. de Versailles [134], the Omega system at Univ. of Maryland [330], or the SUIF compiler of Stanford [13]). In order to deal with realistic applications, these will then have to handle pointer accesses, and variables which are used in a complex data-dependent loop context.

3.3 OVERVIEW OF FULL METHODOLOGY

Our research results on systematic DTSE methodologies, formalized data models and techniques, and on prototype support tools[2] which have been obtained within the context of the ATOMIUM research project [58, 59] at IMEC in the period 1989 – 1997 are briefly discussed now (see also figure 3.1). The underlying DTSE methodology can be used for simulation and for processor-level hardware/software synthesis. This methodology is partly supported with prototype tools in our ATOMIUM system exploration environment. The last task on address optimisation/generation is embedded in a separate high-level address optimisation and generation methodology, partly supported in our prototype tool-box ADOPT. More details are available in the subsequent chapters and in the cited references.

3.3.1 Memory oriented data-flow analysis and model extraction

A novel data/control-flow model [109, 419, 421, 422] has been developed, aimed at memory oriented algorithmic reindexing transformations, including efficient counting of points in polytopes [33] to steer the cost functions. Originally it was developed to support irregular nested loops with manifest, affine iterator bounds and index expressions. Extensions are however possible towards WHILE loops and to data-dependent and regular piece-wise linear (modulo) indices [142, 143, 29, 422]. A synthesis backbone with generic kernels and shared software routines is under implementation.

3.3.2 Global data-flow transformations

We have classified the set of system-level data-flow transformations that have the most crucial effect on the system exploration decisions [56, 61], and we have developed a formalized methodology for applying them. Two main categories exist. The first one directly optimizes the important DTSE costs factors and consist mainly of advanced signal substitution (which especially includes moving conditional scopes), modifying computation order in associative chains, shifting of "delay lines" through the algorithm, and recomputation issues. The second category serves as enabling transformation for the subsequent steps because it removes the data-flow bottle-necks wherever required. An important example of this are advanced look-ahead transformations. No design tool support has been addressed as yet.

3.3.3 Global loop and control-flow transformations

Global-scope loop and other control-flow transformations aim at improving the data access locality for M-D signals and at removing the system-level buffers introduced due to mismatches in production and consumption ordering. The effect of these system-level transformations is shown with two simple examples below. Both of them have been written in single-assignment form, as motivated earlier.

---

[2]All of these prototype tools operate on models which allow run-time complexities which are dependent in a limited way on system parameters like the size of the loop iterators, as opposed to the scalar-based methods published in conventional high-level synthesis literature. Usually, there is a dependence on the number and structure of the loop nests of course.

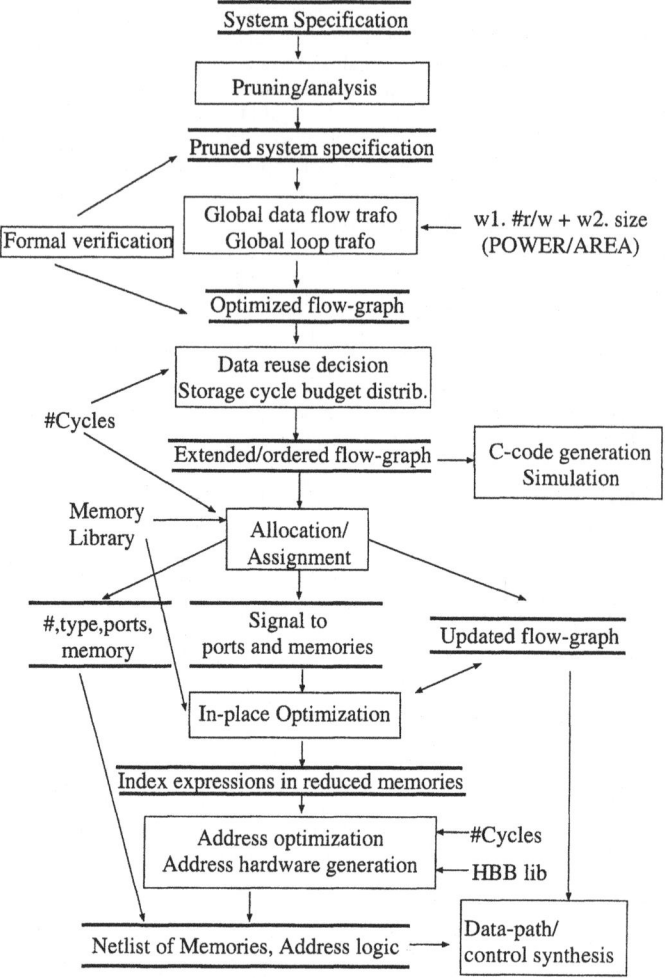

Figure 3.1.   DTSE methodology for data transfer and storage exploration.

Ex.1: Effect on storage cycle budget distribution and memory allocation (assume 2N cycles):

```
FOR i:= 1 TO N DO              FOR i:= 1 TO N DO
    B[i]:=f(A[i]);                BEGIN
FOR i:= 1 TO N DO                 B[i]:=f(A[i]);
    C[i]:=g(B[i]);                C[i]:=g(B[i]);
                                 END;
```

⇒ 2 background memory ports   ⇒ 1 background port + 1 foreground register

In this first example, the intermediate storage of the B[] signals in the original code is avoided by merging the two loop bodies in a single loop nest at the right. As a result, the background access bandwidth is reduced and only 1 background memory port is required, next to the register to store the intermediate scalar results.

Ex.2: effect on potential for in-place optimisation (assume 1 memory and first element of each
row A[i][] initialized to 0):

FOR j:= 1 TO M DO	FOR i:= 1 TO N DO
FOR i:= 1 TO N DO	BEGIN
BEGIN	FOR j:= 1 TO M DO
A[i][j]:=g(A[i][j-1]);	A[i][j]:=g(A[i][j-1]) ;
END;	OUT[i]:= A[i][M];
FOR i:= 1 TO N DO	END;
OUT[i]:= A[i][M];	

⇒ N locations (background)   ⇒ 1 location (foreground)

In example 2, a function of the A[] values is iteratively computed in the first loop nest, which is expressed by the j loop iterating over the columns in any row of A[][]. However, only the end results (N values in total) have to be stored and retrieved from background memory for use in the second loop nest. This intermediate storage can be avoided by the somewhat more complex loop reorganisation at the right. Now only a single foreground register is sufficient, leading to a large reduction in storage size.

In order to provide design tool support for such manipulations, an interactive loop transformation engine (SYNGUIDE) has been developed that allows both interactive and automated (script based) steering of language-coupled source code transformations [354]. It includes a syntax-based check which captures most simple specification errors, and a user-friendly graphical interface. The transformations are applied by identifying a piece of code and by entering the appropriate parameters for a selected transformation. The main emphasis lies on loop manipulations including both affine (loop interchange, reversal, and skewing) and non-affine (e.g. loop splitting and merging) cases.

In addition, research has been performed on loop transformation steering strategies. For power, we have developed a systematic methodology oriented to remove the global buffers which are typically present between subsystems and to create more data locality [103, 110]. This can be applied manually. Also an automatable CAD technique has been developed, partly demonstrated with a prototype tool called MASAI, aiming at total background memory cost reduction with emphasis on transfers and size. An abstract measure for the number of transfers is used as an estimate of the power cost and a measure for the number of locations as estimate for the final area cost [144]. This prototype tool is based on an earlier version [420, 421]. The current status of this automation is however still immature and real-life applications cannot yet be handled. Future research is needed to solve this.

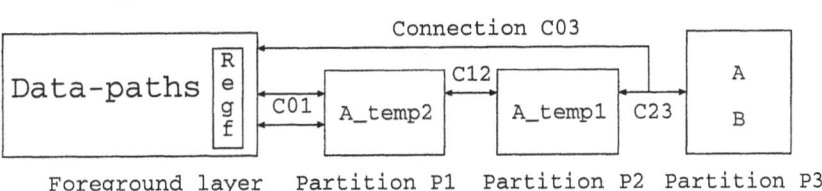

Figure 3.2.   Memory hierarchy illustration.

3.3.4 Data reuse decision in a hierarchical memory context

In this step, we have to decide on the exploitation of the memory hierarchy including bypasses wherever they are useful [437, 440]. Important considerations here are the distribution of the data (copies) over the hierarchy levels as these determine the access frequency and the size of each of the resulting memories. Obviously, the most frequently accessed memories should be the smallest ones. This can be fully optimized only if a memory hierarchy is introduced. We have proposed a formalized methodology to steer this, which is driven by estimates on bandwidth and high-level in-place cost [118, 440]. Based on this, the background transfers are partitioned over several hierarchical memory levels to reduce the power and/or area cost. An illustration of this is shown in figure 3.2. The foreground memory next to the data-paths consists of registers and register files; the intermediate buffers (partition $P1$ and $P2$) typically consist of fast embedded synchronous SRAMs with many ports to the lower level where potential "signal copies" from the higher levels are stored; the top layer (partition $P3$ in this example) consists of a slower mass storage unit with low access bandwidth. The figure shows two signals, A and B, stored in background memory. Signal A is accessed via two intermediate signal copies (A_temp1 and A_temp2) so its organisation involves 3 levels, while signal B is accessed directly from its storage location, i.e. located at hierarchy level 1. Remark that A and B are stored in the same memory, but A is at hierarchy level 3 and B at hierarchy level 1. This illustrates that a certain memory can store signals at a different level in the hierarchy. Therefore we group the memories in partitions ($P1$, $P2$, and $P3$) instead of "levels". The data reuse and memory hierarchy decision step selects how many partitions are needed, in which partition each of the signals (most of which are only intermediate data in RMSP applications) and its temporary copies will be stored, and the connections between the different memory partitions. An important task at this step is to perform transformations which introduce extra transfers between the different memory partitions and which are mainly reducing the power cost. In particular, these involve adding temporary values – to be assigned to a "lower level" – wherever a signal in a "higher level" is read more than once. This involves clearly a trade-off between the power lost by adding these transfers, and the power gained by having less frequent access to the larger memories in the higher level.

At this stage of the overall methodology, the transformed behavioral description can already be used for more efficient simulation or it can be further optimized in the next steps.

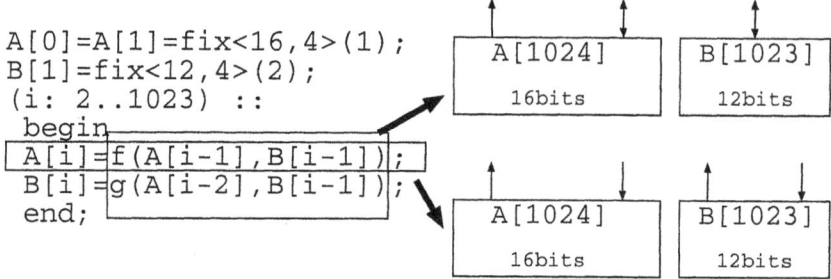

Figure 3.3.   Illustration of two alternatives for the SCBD and memory allocation steps for a simple single-assignment specification in Silage.

3.3.5   Memory organisation issues

This task is illustrated in figure 3.3, where it is assumed that two cycles are available to execute the inner loop body (and that everything has to be stored in background memory). The result is based on an storage bandwidth optimisation decision. Either the two statements are each executed in a separate cycle (bottom) or the right-hand sides are evaluated in cycle 1, followed by the left-hand side stores in cycle 2 (top). Clearly the top solution is more efficient as it requires a single-port memory for B instead of a dual-port one.

In general, given the cycle budget, the goal is to allocate memory units and ports (including their types) from a memory library and to assign the data to the best suited memory units. This happens in several steps. In a first step, we determine the bandwidth requirements and the balancing of the available cycle budget over the different memory accesses in the algorithmic specification (**storage cycle budget distribution** or SCBD) [438, 441]. A prototype tool has been built to support the main substep, namely storage-bandwidth optimisation (or SBO), for flat graphs [438, 441]. Extensions for hierarchical graphs and nested loops have been proposed also [441]. Then, the necessary number of memory units and their type are determined matching this balanced flow-graph and the required parallelism in the memory access (**memory allocation**) [32]. In a last step, the multi-dimensional signals are assigned to these background memory units and ports (**memory assignment**) [32, 376]. Again, this results in an updated flow-graph specification. The cost function for each of these steps is based on a combination of area and power related criteria. A prototype tool, called HIMALAIA, supporting the memory allocation and assignment substeps, has been developed to illustrate the feasibility of our approach [27, 28, 32, 31]. It also includes a high-level but accurate estimation of the memory cost, under the assumption of an applicative (non-procedural) input specification [27], which can be used in the early steps of the DTSE methodology. An extended version of the allocation and assignment prototype tool has been developed recently which now incorporates conflict information from the storage cycle budget distribution step and which includes several powerful memory model extensions [376].

3.3.6 In-place optimization

Initial methods have been investigated for deciding on in-place storage of multi-dimensional signals [403, 404, 413]. Recently, further experiments have been performed to allow even more reduced storage requirements and to identify extensions for predefined memory organisations as in programmable DSP's [102, 104, 103, 110]. This has lead to a solid theoretical foundation for the in-place optimisation task and the development of promising heuristics to solve this very complex problem in a reasonable CPU time in a 2-stage approach for M-D signals [105, 107, 109]: intra-signal windowing [108], followed by inter-signal placement [106].

For the intra-signal windowing, illustrated in figure 3.5, the polyhedral modeling and analysis is illustrated at the top of the figure. That analysis is needed to provide the intersection information for the intra-signal windowing. Two different consumptions of signal a[] are accessing different ranges, analyzed by computing the intersection between the definition domain in loop nest 1 and the operand (consumption) domains in loop nests 2 respectively 3. In the actual in-place windowing example at the bottom of figure 3.5, it is shown that this intersection on itself (without further optimisation) can lead to pessimistic ranges (namely 20 scalars) if applied to signals defined iteratively. In that case, a "snake-like" in-place window of 1 row (5 scalars) is sufficient as buffer for the signals which are still "alive" at the different steps in the iteration. Each iteration, the last element of the window becomes available to be overwritten by a newly produced element. This is found by our intra-signal windowing substep of the in-place optimisation.

For the inter-signal placement, illustrated in figure 3.4, in the worst-case, all signals require separate storage locations (left-hand side solution). However, if the life-times of signals $B[]$ and $C[]$ are not overlapping with the life-time of signal $A[]$ (right-hand side, bottom), the space reserved in the memory for these groups can be shared (right-hand side, top).

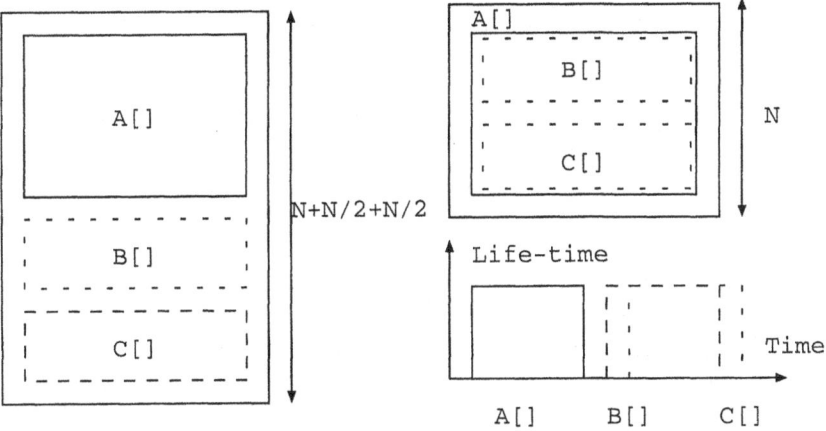

Figure 3.4.   Illustration of inter-signal in-place optimisation task.

Both of these techniques have been implemented in a prototype tool for evaluating in-place storage of multi-dimensional signals. Very promising results for realistic applications have been obtained with a prototype tool supporting the proposed technique [108, 106, 107, 109].

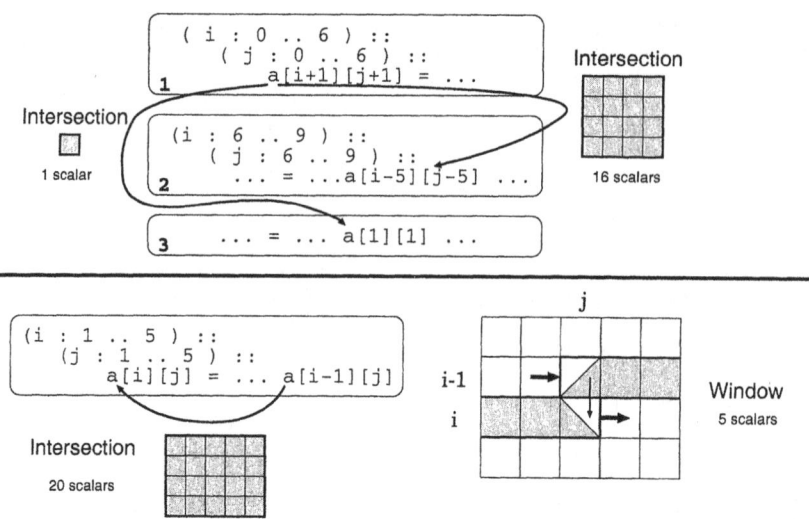

Figure 3.5.   Illustration of simplified intra-signal in-place optimisation.

3.3.7   End result of applying the DTSE methodology.

The result at this stage is the output of the data transfer and storage exploration. It is important to realize that it does not yet lead to a final scheduling, or even the complete allocation and assignment of the background memories. What is produced are:

1. a transformed control/data-flow graph (and corresponding code) to be used as input for the subsequent custom architecture synthesis or processor compilation stage

2. constraints on the access schedule: only a partial ordering is fixed, not only due to the absence of constraints for the arithmetic operations but even for the read/write accesses to the background memories. So many feasible schedules still remain on the absolute time axis. Hence, after DTSE, new code should be produced. The basic code generation is especially based on a substep called "scanning polyhedra" in literature. Several algorithms have been described for this purpose (see e.g. [16, 94] and their references). It is important also to detect which loops should be constrained (fixed) and which can be left free for DTSE purposes. The remaining freedom is needed to extract sufficient parallelism during address optimisation and generation, and during the final handling of the arithmetic and control operations. This freedom should be detected by a polyhedral analysis routine. The generated code should then also reflect the essential partial ordering constraints.

3. constraints on the background memory allocation: minimal number and size and type of background memories which can later on be increased still due to scalar spilling and other foreground signal storage (see e.g. [230]).

4. constraints on the M-D signal to background memory assignment: the full assignment is not yet fixed because also other signals can be assigned to the allocated memories later on, e.g. during data routing and spilling or other low-level memory management decisions.

The final decisions on all of these issues will have to be fixed afterwards, during the actual architecture synthesis stage including custom data-path synthesis, low-level memory management and control synthesis steps, or during their equivalents in a conventional compiler approach. Therefore, as motivated several times already, the DTSE stage should *not* be seen as a parallel step with these architecture synthesis or compilation steps: it should be done beforehand. Hence, DTSE can be better seen as part of the system exploration (or system synthesis) stage.

If the lower stages (either manual or automated) do not allow to take these issues as constraints to extend upon, the decisions which have been taken during DTSE can also be seen as final. In that case, the complete background memory organisation and the ordering on the reads and writes to the M-D array signals become fixed.

3.3.8 Extensions to other contexts

In addition to RMSP applications, we have also investigated extensions of the methodology towards network applications as occurring, e.g. in Layer 3-4 of ATM [97, 111, 376]. The methodology for the steps related to the bandwidth requirements and balancing the available cycle budget over the different memory accesses in the algorithmic specification (flow-graph balancing for non-hierarchical graphs within a single loop body) during background memory allocation, has been worked out in detail [438, 441]. Also the combination with memory allocation and assignment has been explored with very promising results [376].

Finally, we have also demonstrated that an extension of the methodology can be effectively employed for power-efficient system-level reuse of hardware and software IP (Intellectual Property) modules for data-dominated applications (see [418] for information).

In this book, we will focus mainly on the RMSP domain and custom memory organisations but the extensions will be mentioned wherever it is useful.

3.4 RELATED STEPS

A number of tasks are related to the DTSE methodology:

1. *System-level validation by formal verification of the global data-flow and loop transformations:* a methodology to tackle this crucial validation task has been developed for a very general context including procedural (e.g. WHILE) loops, general affine indices (including the so-called "lattices" of the form $[a.i+b]$), and data-dependent signal indices. Such a formal verification stage avoids very CPU-time and design time costly resimulation. A prototype verification tool is operational, dealing with affine loop transformations like loop interchange, reversal and skewing on manifest loop nests with affine signal indices [355, 357, 356]. It demonstrates the feasibility of our approach. Recently, we have added heuristics to deal with very large applications [86] and we have also developed a methodology to deal with data-flow transformations and more general cases of data dependencies. The necessary (prototype) tool support for these extensions is also being added.

2. *High-level address optimisation and address unit synthesis:* These tasks are very crucial for data-dominated applications. They are error-prone and tedious if done manually for RMSP applications. Therefore, a methodology has been developed which is partly supported in IMEC's ADOPT environment [270, 272], which is embedded in a separate prototype tool-box. The subtasks addressed in this toolbox are complementary to the traditional high-level synthesis techniques as employed e.g. in Synopsys BC, so ADOPT can be seen as a high-level preprocessing stage.

   The methodology is illustrated in figure 3.6 [272]. After algebraic optimisation of the address expressions, they are efficiently shared on hardware units within the available cycle budget. Two target architecture styles are supported: logic/counter based or custom data-path (ASU) based. In the former case, address sequences are generated and realized as a counter modified by a two- or multi-level logic filter. In the ASU case, the address expressions are realized with custom arithmetic building blocks selected from a library. A prototype tool GATE has been developed for optimally transforming the (address) arithmetic operations in flow-graph expressions [201]. A model and automated steering techniques have been developed which allow to minimize the weighted operation cost, allowing invariance to structural changes in the specification. Other prototype tools allow merging [201, 202] and/or sharing [155] M-D index expressions such that the address cost is reduced [269]. This includes a measure for selecting the target architecture, i.e. logic/counter based [415, 271] or custom data-path (ASU) based [155, 157, 156, 158, 298, 299].

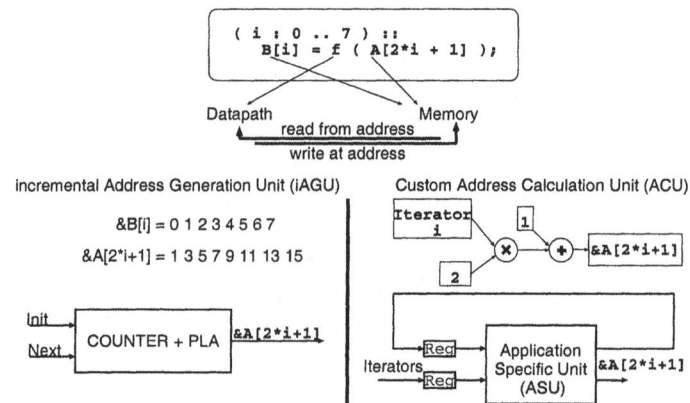

Figure 3.6.   Illustration of address optimisation and hardware mapping task.

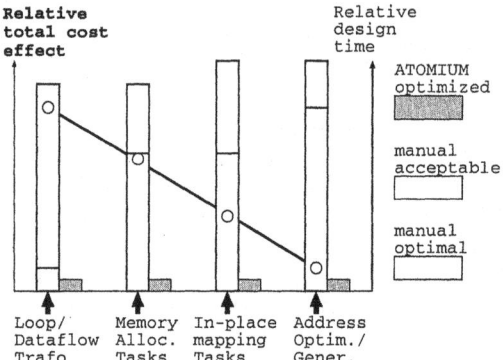

Figure 3.7. Motivation of need for design support of each of the 4 main mapping and optimization steps in the DTSE methodology. Only a qualitative analysis is provided here so the above data only provides a trend.

3.5 MOTIVATION OF NEED FOR DESIGN TOOL SUPPORT

The need for design support of each of these different mapping and optimization tasks is motivated in figure 3.7.

The bars indicate the relative design times for different scenarios. They are not based on absolute measured figures but give an indication about our own experience in design effort. The white (largest) bars represent the effort needed to come to an "optimal" result manually. For several of the tasks it is however possible to arrive with much less effort to a correct "acceptable" solution, within the application constraints (like timing). This is not true for the address optimisation though where the main effort is in book-keeping and making certain that the time budget is met. Also the memory allocation (see subsection 3.3.5) and in-place optimisation stages (see subsection 3.3.6) involve much book-keeping and error-prone work for the medium-size and smaller array signals present in most applications. What we hope to achieve is that these design times can be significantly reduced, while keeping the design quality, when the design tasks are supported with tools. This can happen either by fully interactive tools for error-prone tasks or partly automated tools when much book-keeping is involved.

The circles connected by a line represent the relative total cost effect of each of the tasks. It can be seen that the tasks which take less time to produce an acceptable result also have the largest cost

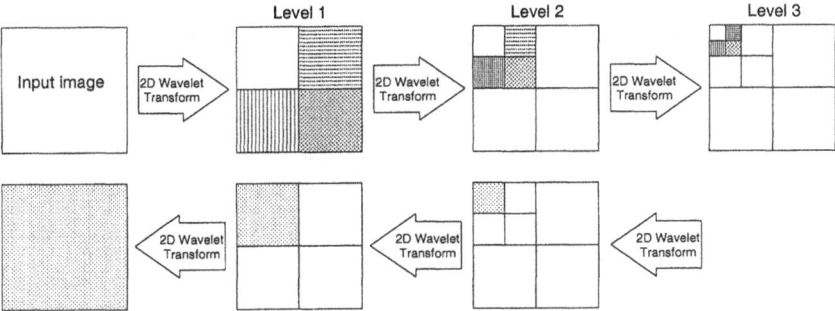

Figure 3.8.   The two-dimensional wavelet transform is applied upon the whole image (depicted at the upper left) in order to decorrelate the input date. After each two-dimensional (2D) wavelet transform, three new subband are created (the shaded parts). The final result of a three-level wavelet transform is the subband structure depicted at the upper right. The subbands contain combined frequency and spacial information.

effect and vice versa. For this reason, the overall conclusion of this qualitative analysis is that there is sufficient motivation to develop some level of design support for all of the 4 main stages.

It should be stressed however, that only for the most crucial and error-prone/tedious subtasks, mature CAD tool support should be developed. Hence, a number of subtasks are left for the algorithm/system designer, with a manually applied systematic methodology. Some of the subtasks also require only interactive tool support, and not fully automated tools. The latter are typically only needed to provide fast, less accurate feedback for decisions taken at a higher level subtask in the methodology.

3.6   ILLUSTRATION ON 1-D WAVELET TRANSFORM EXAMPLE

Everyday, millions of people decompress images while browsing on the internet. Most natural images are compressed with the JPEG standard [321]. This successful standard employs a block based compression scheme. For each block, only the frequency information that is important to obtain a good visual representation is retained.

Modern natural image compression schemes promise even better coding efficiency (especially at the lower bitrates) by considering the whole image at once instead of block by block. A new digital signal processing tool, called the wavelet transform, is applied at the image to decorrelate it. The result is a subband structure depicted on the upper right of figure 3.8.

From the transformed image, the three smallest subbands are combined with the smallest rectangle information in the upper left of the transformed image (i.e. the so called DC band). Together they produce the dotted area after an inverse wavelet transform. Repeating this process results in the recovered original image.

The wavelet transforms decorrelate the original image. Next the transformed image is coded. Therefore, so-called zero trees are extracted [372]. Zero trees consist out of one DC pixel, four subband pixels form the smallest subband, sixteen pixels from the next subband, and so on. Examples of consecutive two zero trees in the diagonal direction and one in the horizontal direction are depicted in figure 3.9.

In figure 3.8, the subbands that are produced last in the transform process are used first to form the zero trees. Hence, most implementations store the complete wavelet transformed image *before* extracting zero trees for encoding. Hence, an extra memory that is able to hold a complete image is required in the encoder. Applying our DTSE methodology allows for a significant reduction of the storage requirements of the intermediate signals. This is an important breakthrough because the implementation complexity of wavelet based image coding systems is heavily reduced as well and reaches a level where it can compete with the traditional block based approaches.

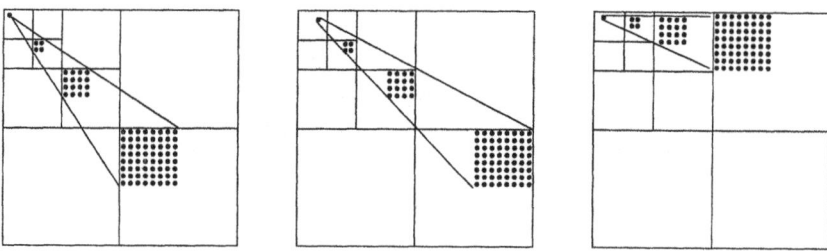

Figure 3.9.   The two zero trees on the left are two consecutive diagonal zero trees. The zero tree on the right is the first zero tree in the horizontal direction. Zero trees exist in the vertical direction as well but they are not depicted in this figure.

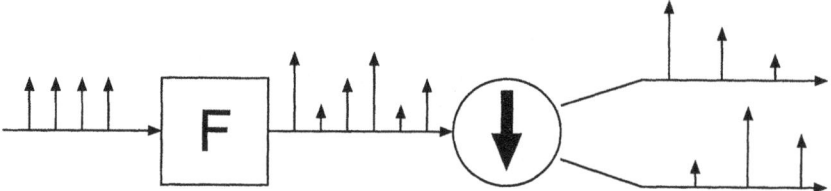

Figure 3.10.   Filtering and subsampling are basic signal processing operators in a wavelet transform

To give an intuitive feeling about the steps in the DTSE methodology described above, it is now illustrated on a wavelet - zero tree system. To not overload the example, a one dimensional (1D) version [276] is chosen. The complexity of this example is somewhere between trivial and complex. Hence, not too much effort is required from the reader to grasp what is going on but the result should reveal (almost) the full power of this methodology.

To start, the data dependencies of a simplified 1D wavelet transform are extracted (Section 3.6.1).

3.6.1   Memory oriented data-flow analysis and model extraction

3.6.1.1   Introduction.   It this subsection, the basic data dependencies for a 1D wavelet transform are derived. First, a C++ program fragment that performs a filter operation is presented. From this C++ program, a pruned notation that keeps the essential information for deriving data dependencies is discussed. A similar notation for down- and up-sampling is introduced as well. Based on these notations, the data dependencies for one branch of a three-level wavelet decomposition will be derived.

A wavelet transform employs two basic signal processing operations:

- Finite Impulse Response (FIR) filtering and

- sampling.

These operations are depicted in figure 3.10. The filtering is depicted with the rectangular box and a letter to identify the filter coefficient set. The "down-sampling by two" operator is represented by the circle with a downward arrow in it.

Basically, a wavelet decomposition can be seen as a repeated application of a sampling operation followed by a filtering. A three-level wavelet decomposition tree is depicted in figure 3.11.

3.6.1.2   Data dependencies of a filter operation.   A filter operation, on the N signals in array I, with a mask of length $2 \times M + 1$ is calculated in the following C++ program:

```
1    const unsigned int N = 128;
2    const int M = 2;
```

```
3     const unsigned int nb_coeffs = 2*M + 1;
4     const int coeff[nb_coeffs] = {1, 2, 4, 2, 1};
5     int I[N+nb_coeffs], A[N];
6
7     // Symmetrical extention of array I
8     for (unsigned int i = 0; i < M; i++) {
9         I[M-i] = I[M+i]; I[M+(N-1)+i] = I[M+(N-1)-i];
10    }
11    // Convolute I with vector coeff and store result in A
12    for (i=0; i < N; i++) {
13        for (int k=-M; k <= M; k++)
14            sum += coeff[M+k] * I[M+(i+k)];
15        A[i] = sum;
16    }
```

Because only the data dependencies between the array signals are needed for DTSE, the following notation captures the essential information and hides the details:

for $i = 0 \ldots N - 1$ **do**
   $A[i] = I[i \pm M]$;
end for

Read the statement in the loop body as follows: to produce $A[i]$, the data $I[i - M], I[i - M + 1], \ldots, I[i + M - 1], I[i + M]$ are needed. It expresses that $A[i]$ depends on $I[i \pm M]$.

3.6.1.3 Data dependencies of a sampling operation.
Subsampling signal I with a factor of 2 is expressed as:

for $i = 0 \ldots N/2 - 1$ **do**
   $I0[i] = I[2i]$;
   $I1[i] = I[2i + 1]$;
end for

Array $I0[]$ contains phase 0 of signal I and array $I1[]$ contains phase 1.
If something depends on both multi-rate phases of I, we express it using the $\ldots$ operator:

for $i = 0 \ldots N/2 - 1$ **do**
   LHS $= I[2i + 0 \ldots 1]$;
end for

Subsampling with a factor M is expressed as:

for $i = 0 \ldots N/M - 1$ **do**

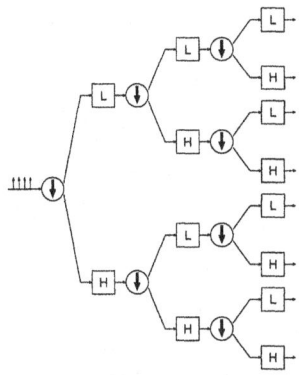

Figure 3.11.   Three levels of a 1D wavelet decomposition

LHS $= I[M \times i + 0 \dots M - 1]$;
end for

3.6.1.4 Data dependencies between a wavelet transform and a zero tree. One level of the wavelet transform, as depicted in figure 3.10, is written as:

{Convolute and subsample}
for $i = 0 \dots N/2 - 1$ **do**
$A[i] = I[2i \pm M]$;
end for

One path of a wavelet tree of three levels is then constructed as follows:

{Subsample and filter}
for $i = 0 \dots N/2 - 1$ **do**
$A[i] = I[2i \pm M]$;
end for
for $i = 0 \dots N/4 - 1$ **do**
$B[i] = A[2i \pm M]$;
end for
for $i = 0 \dots N/8 - 1$ **do**
$C[i] = B[2i \pm M]$;
end for

The path is indicated by the thick line in figure 3.12.

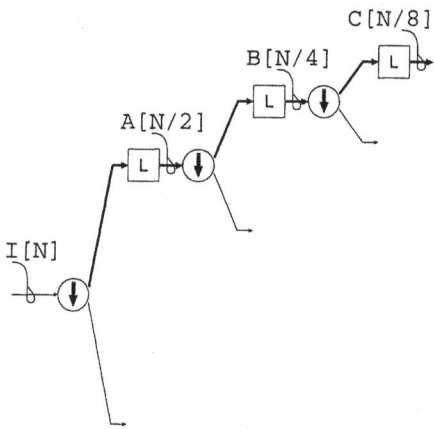

Figure 3.12.   One path in a three-level 1D wavelet decomposition

From the wavelet transform of the input signal, zero trees are extracted for compression. An explanation about the wavelet zero tree algorithm is found in [372] and its references. Here only the data dependencies are considered.

Each zero tree is constructed from 1 element of signal C, two elements of signal B and four elements of signal A. This is expressed as:

$$ZT = C[i], B[2i + 0 \dots 1], A[4i + 0 \dots 3];$$

Hence, the DTSE optimisations of a 1D wavelet transform and zero tree coder must operate on the following program:

{Subsample and filter}
for $i = 0 \dots N/2 - 1$ **do**
$A[i] = I[2i \pm M]$;

end for
for $i = 0 \ldots N/4 - 1$ **do**
    $B[i] = A[2i \pm M]$;
end for
for $i = 0 \ldots N/8 - 1$ **do**
    $C[i] = B[2i \pm M]$;
end for
{Zero tree compression}
for $i = 0 \ldots N/8 - 1$ **do**
    ZT = C[i], B[2i+0...1], A[4i+0...3];
end for

The above program captures all the relevant information needed to study how transformations can be applied to reduce the power consumption of the algorithm.
The corresponding C++ program of the above derived model is:

```
1    #include <iostream.h>
2    #include <stdlib.h>
3
4    // Number of samples
5    const unsigned int N = 1<<7;
6    // Length of one side of the filter
7    const int M = 4;
8    // Number of filter coeffs
9    const unsigned int nb_coeffs = 2*M + 1;
10   // Length of a zero tree
11   const unsigned int ZT_LENGTH = 1 + 2 + 4;
12   const int coeff[nb_coeffs] = {1, 8, 28, 56, 70, 56, 28, 8, 1};
13
14   void zero_tree(int ZT[ZT_LENGTH]) {
15       for (unsigned int i=0; i <= ZT_LENGTH-1; i++)
16           cout << "ZT[" << i << "] = " << ZT[i] << endl;
17       cout << endl;
18   }
19
20   int main() {
21   int I[N+nb_coeffs], A[(N/2)+nb_coeffs],
22       B[(N/4)+nb_coeffs], C[N/8], ZT[ZT_LENGTH];
23   int i, di, k;
24
25   srand(10);
26   for (i=0; i <= N-1; i++)
27       I[M+i] = rand();
28
29   for (i = 1; i <= M; i++) {
30       I[M-i] = I[M+i]; I[M+(N-1)+i] = I[M+(N-1)-i];
31   }
32   for (i=0; i <= N/2-1; i++) {
33       int sum = 0;
34       for (k=-M; k <= M; k++)
35           sum += coeff[M+k]*I[M+(2*i+k)];
36       A[M+i] = sum>>9;
37   }
38   for (i = 1; i <= M; i++) {
39       A[M-i] = A[M+i]; A[M+((N/2)-1)+i] = A[M+((N/2)-1)-i];
40   }
41   for (i=0; i <= N/4-1; i++) {
42       int sum = 0;
43       for (k=-M; k <= M; k++)
```

```
44          sum += coeff[M+k]*A[M+(2*i+k)];
45      B[M+i] = sum>>9;
46  }
47  for (i = 1; i <= M; i++) {
48      B[M-i] = B[M+i]; B[M+((N/4)-1)+i] = B[M+((N/4)-1)-i];
49  }
50  for (i=0; i <= N/8-1; i++) {
51      int sum = 0;
52      for (k=-M; k <= M; k++)
53          sum += coeff[M+k]*B[M+(2*i+k)];
54      C[i] = sum>>9;
55  }
56  for (i=0; i <= N/8-1; i++) {
57      ZT[0] = C[i];
58      for (di=0; di <= 1; di++)
59          ZT[1+di] = B[M+2*i+di];
60      for (di=0; di <= 3; di++)
61          ZT[1+2+di] = A[M+4*i+di];
62      zero_tree(ZT);
63  }
64  }
```

The output of the above C++ program will be used to check if the transformed programs are still correct.

3.6.2   Analysis of the initial memory requirements

Starting from the original specification of the data dependencies, the memory needed to store intermediate results while performing a 1D wavelet transforms of three levels is:

$$\frac{N}{2} + \frac{N}{4} + \frac{N}{8} + 2 \times (2 \times M + 1) = N(1 - \frac{1}{8}) + 4 \times M + 2 \approx N \qquad (3.1)$$

where N is the number of input samples and $(2 \times M + 1)$ is the number of taps of the filter. Indeed, array signal A is $\frac{N}{2}$ words, array signal B is $\frac{N}{4}$ and array signal C is $\frac{N}{8}$. Array signals A and B are symmetrically extended hence $2 \times (2 \times M + 1) = 4 \times M + 4$ extra words are needed to store the border. Summing all these gives the total number of words to be stored (see eq. 3.1). The number of words is slightly less then N when $N \gg M$.

Table 3.1 tabulates the number of words needed to store the intermediate data, of the wavelet-zero tree algorithm, in function of the parameter N when parameter M is set to four.

Table 3.1.   Initial memory requirement

N	16	32	64	128	256	512	1024	2048	4096
Memory	32	46	74	130	242	466	914	1810	3602

3.6.3   Example of a global data-flow transformation

The C++ implementation of the FIR filter, symmetrically extends the input signal to avoid illegal index values at the borders of the signal. This introduces extra transfers and requires extra storage. A relatively well-known data-flow transformation is to introduce a function that operates on the index values itself and provides the symmetrical extension functionality without the extra transfers and storage. This is an example of "advanced signal substitution and propagation". This index function is called wrap() in the following code excerpt:

```
1    const unsigned int N = 128;
```

```
2    const int M = 2;
3    const unsigned int nb_coeffs = 2*M + 1;
4    const int coeff[nb_coeffs] = {1, 2, 4, 2, 1};
5
6    int I[N], A[N];
7
8    unsigned int wrap(int i) {
9        int new_i = (i >=0 ) ? i : -i;
10        return (new_i < N) ? new_i : 2*N-new_i;
11    }
12
13    for (unsigned int i=0; i < N; i++) {
14        for (int k=-M; k <= M; k++)
15            sum += coeff[M+k] * I[wrap(i+k)];
16        A[i] = sum;
17    }
```

3.6.4 Global loop Transformations

The size of the memory can be reduced by altering the control-flow by means of loop transformations. The main goal is to increase the locality of reference. This could be achieved by merging the loops that produce the intermediate signals A,B and C with the loop that consumes them to generate zero-trees.

3.6.4.1 Step one : merge last the two loop nests.

> **for** $i = 0 \ldots N/2 - 1$ **do**
> $A[i] = I[2i \pm M]$;
> **end for**
> **for** $i = 0 \ldots N/4 - 1$ **do**
> $B[i] = A[2i \pm M]$;
> **end for**
> **for** $i = 0 \ldots N/8 - 1$ **do**
> $C[i] = B[2i \pm M]$;
>   ZT = C[i], B[2i+0...1], A[4i+0...3];
> **end for**

Merging the loops that produces array signal A and B is not straightforward. The reason is that the loops have a different number of iterations. This can be solved by a tiling transformation.

3.6.4.2 Step two : tiling the first two loops. In order to merge the first two loops with the last one, we have to tile the first loop with a factor 4 and the second loop with a factor 2:

> **for** $i = 0 \ldots N/(2*4) - 1$ **do**
> $A[4i + 0 \ldots 3] = I[2(4i + 0 \ldots 3) \pm M]$;
> **end for**
> **for** $i = 0 \ldots N/(4*2) - 1$ **do**
> $B[2i + 0 \ldots 1] = A[2(2i + 0 \ldots 1) \pm M]$;
> **end for**
> **for** $i = 0 \ldots N/8 - 1$ **do**
> $C[i] = B[2i \pm M]$;
>   ZT = C[i], B[2i+0 ... 1], A[4i+0 ... 3];
> **end for**

Now the three loops have the same number of iterations and merging is straightforward (at first sight!).

3.6.4.3 Step three : merge loop one, two and three.

> **for** $i = 0 \ldots N/(2*4) - 1$ **do**
> $A[4i + 0 \ldots 3] = I[2(4i + 0 \ldots 3) \pm M]$;

$$B[2i + 0 \ldots 1] = A[2(2i + 0 \ldots 1) \pm M];$$
$$C[i] = B[2i \pm M];$$
$$ZT = C[i], B[2i + 0 \ldots 1], A[4i + 0 \ldots 3];$$
end for

Unfortunately, the last step violates some data dependencies. This can be understood by considering the data needed in iteration $i = 0$. Filling in zero for i in the last statement shows that the signals $C[0]$, $B[0 \ldots 1]$ and $A[0 \ldots 3]$ contribute to the first zero tree symbol. This is exactly what is produced by the first three statements in the loop body. But to be able to calculate $C[0]$, the values of $B[(2 \times 0) - M]$ to $B[(2 \times 0) + M]$ need to be known. Because signal B is symmetrically extended, this means that signals $B[0 \ldots M]$ need to be available. But the second statement only calculates $B[2 \times 0 + 0 \ldots 1]$. This means that when $M > 1$, data dependencies are violated.

The border effect ripples through. This is so because to calculate the signals $B[0 \ldots M]$, we need $A[-M \ldots 2M + M]$ and the first statement only produces $A[0 \ldots 3]$.

Thus, before starting the merged loop, some values of array signal A, B and C must be filled in first. How this is done is explained in the next section.

3.6.4.4 Step four : extract iterations that produce the first zero tree. To calculate $C[0]$, we need $B[0..M]$. One instance of the second statement of the loop, generates two new values. This means that $\lceil M/2 \rceil$ iterations of second statement must be completed before the loop can be started. Therefore we extract $\lceil M/2 \rceil$ instances of the second statement.

To explain what we mean by extracting, consider a loop that executes N times the statements $S1$, $S2$ and $S3$:

for $i = 0 \ldots N - 1$ **do**
  $S1(i)$;
  $S2(i)$;
  $S3(i)$;
end for

Extracting α iterations of statement $S2$ from the above loop results in two loops:

for $i = 0 \ldots \alpha - 1$ **do**
  S2(i);
end for
for $i = 0 \ldots N - 1$ **do**
  S1(i);
  if $i \leq (N - 1 - \alpha)$ **then**
    S2($\alpha + i$);
  end if
  S3(i);
end for

The first loop is executed α times. The second loop is still executed N times. However the condition that now guards the statement S1, allows only $N - \alpha$ executions of statement S1.

We now apply this extraction transformation three times in order to isolate the loops that are needed to generate the first zero tree. We start by extracting $\alpha_0 = 1$ iteration instances of the last two statements in the loop:

for $i = 0 \ldots N/8 - 1$ **do**
  $A[4i + 0 \ldots 3] = I[2(4i + 0 \ldots 3) \pm M]$;
  $B[2i + 0 \ldots 1] = A[2(2i + 0 \ldots 1) \pm M]$;
  $C[i] = B[2i \pm M]$;
  ZT = C[i], B[2i+0 … 1], A[4i+0 … 3];
end for

The result is:

$\alpha_0 = 1$;
for $i = 0 \ldots \alpha_0 - 1$ **do**

$C[i] = B[2i \pm M]$;
$ZT = C[i], B[2i + 0 \ldots 1], A[4i + 0 \ldots 3]$;
end for
for $i = 0 \ldots N/8 - 1$ **do**
  $A[4i + 0 \ldots 3] = I[2(4i + 0 \ldots 3) \pm M]$;
  $B[2i + 0 \ldots 1] = A[2(2i + 0 \ldots 1) \pm M]$;
  if $i \leq (N/8 - 1) - \alpha_0$ **then**
    $C[\alpha_0 + i] = B[2(\alpha_0 + i) \pm M]$;
    $ZT = C[(\alpha_0 + i)], B[2(\alpha_0 + i) + 0 \ldots 1], A[4(\alpha_0 + i) + 0 \ldots 3]$;
  end if
end for

We now extract $\alpha_1 = \lceil M/2 \rceil + 1$ iterations of the statement that produces B:

$\alpha_0 = 1$;
$\alpha_1 = \lceil M/2 \rceil + 1$;
for $i = 0 \ldots \alpha_1 - 1$ **do**
  $B[2i + 0 \ldots 1] = A[2(2i + 0 \ldots 1) \pm M]$;
end for
for $i = 0 \ldots \alpha_0 - 1$ **do**
  $C[i] = B[2i \pm M]$;
  $ZT = C[i], B[2i + 0 \ldots 1], A[4i + 0 \ldots 3]$;
end for
for $i = 0 \ldots N/8 - 1$ **do**
  $A[4i + 0 \ldots 3] = I[2(4i + 0 \ldots 3) \pm M]$;
  if $i \leq (N/8 - 1) - \alpha_1$ **then**
    $B[2(i + \alpha_1) + 0 \ldots 1] = A[2(2(i + \alpha_1) + 0 \ldots 1) \pm M]$;
  end if
  if $i \leq (N/8 - 1) - \alpha_0$ **then**
    $C[\alpha_0 + i] = B[2(\alpha_0 + i) \pm M]$;
    $ZT = C[(\alpha_0 + i)], B[2(\alpha_0 + i) + 0 \ldots 1], A[4(\alpha_0 + i) + 0 \ldots 3]$;
  end if
end for

To be able to calculate this extracted loop, we already need some signals of A. The first signal, needed when $i = 0$, is $A[2(2 \times 0 + 0) - M]$. The last signal, needed when $i = \alpha_1 - 1$, is $A[2(2(\alpha_1 - 1) + 1) + M]$. Because four values of A are calculated at once, we need to extract $\alpha_2 = \lceil (2(2(\alpha_1 - 1) + 1) + M)/4 \rceil + 1$ instances of the definition of A:

$\alpha_0 = 1$;
$\alpha_1 = \lceil M/2 \rceil + 1$;
$\alpha_2 = \lceil (2(2((\alpha_1 - 1) + 1) + M)/4 \rceil + 1$;
for $i = 0 \ldots \alpha_2 - 1$ **do**
  $A[4i + 0 \ldots 3] = I[2(4i + 0 \ldots 3) \pm M]$;
end for
for $i = 0 \ldots \alpha_1 - 1$ **do**
  $B[2i + 0 \ldots 1] = A[2(2i + 0 \ldots 1) \pm M]$;
end for
for $i = 0 \ldots \alpha_0 - 1$ **do**
  $C[i] = B[2i \pm M]$;
  $ZT = C[i], B[2i + 0 \ldots 1], A[4i + 0 \ldots 3]$;
end for
for $i = 0 \ldots N/8 - 1$ **do**
  if $i \leq (N/8 - 1) - \alpha_2$ **then**
    $A[4(\alpha_2 + i) + 0 \ldots 3] = I[2(4(\alpha_2 + i) + 0 \ldots 3) \pm M]$;
  end if
  if $i \leq (N/8 - 1) - \alpha_1$ **then**

Table 3.2.   Analytical derivation of number of read and writes

Signal	No. of writes	No. of reads
I	N	$4 \times (2 \times M + 1) \times (N/8)$
A	$4 \times (N/8)$	$(2 \times (2 \times M + 1)) + 4) \times (N/8)$
B	$2 \times (N/8)$	$((2 \times M + 1) + 2) \times (N/8)$
C	$N/8$	$(N/8)$
ZT	$(1 + 2 + 4) \times (N/8)$	$(1 + 2 + 4) \times (N/8)$

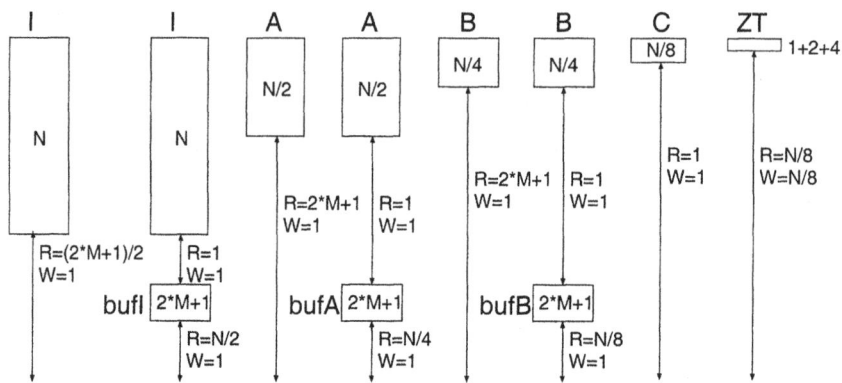

Figure 3.13.   Data reuse tree of wavelet - zero tree example

$$B[2(\alpha_1 + i) + 0 \ldots 1] = A[2(2((\alpha_1 + i) + 0 \ldots 1) \pm M];$$
end if
if $i \le (N/8 - 1) - \alpha_0$ **then**
$$C[\alpha_0 + i] = B[2(\alpha_0 + i) \pm M];$$
$$ZT = C[(\alpha_0 + i)], B[2(\alpha_0 + i) + 0 \ldots 1], A[4(\alpha_0 + i) + 0 \ldots 3];$$
end if
end for

3.6.5 Data reuse decision

Allocation of small buffers to store intermediate results over successive filter steps reduces the transfers to the main input frame signal. This is systematically explored in our data reuse decision step. When applied to the wavelet kernel, we obtain the following results.

The main signals to store are the input signal I, the intermediate wavelet coefficients A, B, C and the zero tree signal ZT. In table 3.2 the number of reads and writes to these signals are calculated.

It can be observed that without intermediate buffering, many duplicate accesses are present on the filter coefficients I, A, and C. In the column of the reads in table 3.2 a common factor $2 \times M + 1$ pops up. This is due to the filtering operation with a filter that has $2 \times M + 1$ taps. By allocating an extra memory hierarchy level to store the values to be filtered, the number of accesses to the memory that contains all the signals can be reduced drastically as shown in table 3.3. Therefore, it makes sense to introduce a data reuse tree (see chapter 9) with intermediate buffer signals *bufI*, *bufA* and *bufB*.

The introduction of these extra filter buffers increases both the total number of transfers and the total size. However, since the extra memory of $3 \times (2 \times M + 1)$ words will be much smaller than the original memories (because $M \ll N$) and because the majority of accesses are now to this smaller memory, it is expected that the total power budget will decrease.

Table 3.3.   Analytical derivation of number of read and writes, after data reuse

Signal	No. of writes	No. of reads
I	N	$4 \times (N/8)$
A	$4 \times (N/8)$	$(2+4) \times (N/8)$
B	$2 \times (N/8)$	$(1+2) \times (N/8)$
C	$N/8$	$(1) \times (N/8)$
ZT	$(1+2+4) \times (N/8)$	$(1+2+4) \times (N/8)$
$bufI$	$8 \times (N/8)$	$8 \times (2 \times M + 1) \times (N/8)$
$bufA$	$4 \times (N/8)$	$4 \times (2 \times M + 1) \times (N/8)$
$bufB$	$2 \times (N/8)$	$2 \times (2 \times M + 1) \times (N/8)$

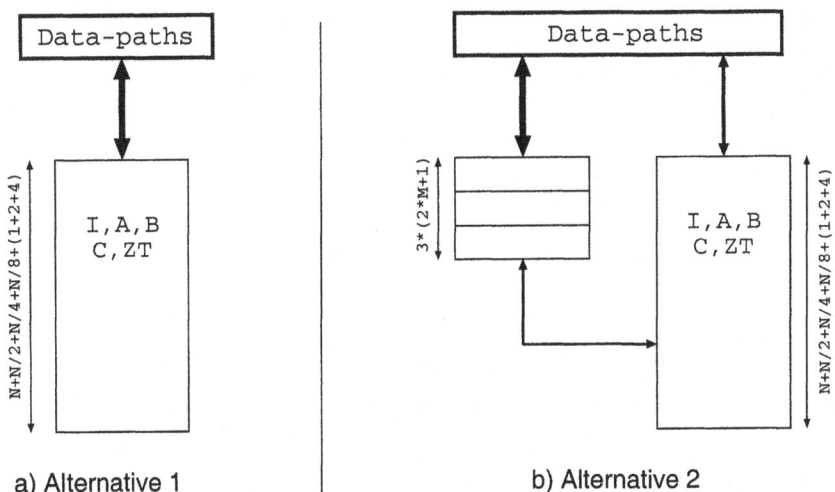

a) Alternative 1   b) Alternative 2

Figure 3.14.   Allocation and assignment alternatives

3.6.6   SCBD and memory allocation/assignment

Based on the information about number of accesses, size of the signals, and cycle budget in table 3.2, storage cycle budget distribution followed by memory allocation and assignment can be decided.

Two alternatives will be studied. In case the signals I, A, B, C, and ZT are stored in 1 memory with 1 read-write port, the number of words of the memory is $N + N/2 + N/4 + N/8 + (1 + 2 + 4)$ and the number of bits in a word is the maximum number of bits of all stored signals. A lower bound on the number of cycles in the case of 1 read-write port is the number of transfers. This solution is depicted at the left as alternative 1 of figure 3.14.

The introduction of the solution with memory hierarchy, i.e. with the extra filter buffers $bufI$, $bufA$ and $bufB$ increases the total number of transfers. The case depicted on the right of figure 3.14 as alternative 2 shows that an extra memory of $3 \times (2 \times M + 1)$ words is allocated and all extra buffers are assigned to it. However, as already mentioned, the total power is expected to be significantly lower.

Many other alternative data reuse decisions, allocations, and assignments are possible. A full exploration is outside the scope of this overview example. A similar optimisation can be applied for the filter operation on every level of the full two-dimensional wavelet transform but it is not worked out here either.

3.6.7 In-place optimisation

The first loop stores all elements of signal A needed to calculate the first zero tree. Once the first tree is formed, 4 elements of A are not needed any longer. These four elements can be filled up again with the four next ones. Hence we need only place to store A's produced by the loop:

$$\alpha_0 = 1;$$
$$\alpha_1 = \lceil M/2 \rceil + 1;$$
$$\alpha_2 = \lceil (2(2((\alpha_1 - 1) + 1) + M)/4 \rceil + 1;$$
for $i = 0 \ldots \alpha_2 - 1$ **do**
$\quad A[4i + 0 \ldots 3] = I[2(4i + 0 \ldots 3) \pm M];$
end for

The window of array A is thus obtained by substituting the upper bound of the loop ($\alpha_2 - 1$) into the index expression $4i + 3$:

$$4(\alpha_2 - 1) + 3 = 4\alpha_2 - 1$$

To verify that the window is large enough, the largest distance between production and consumption of array A in the main loop:

for $i = 0 \ldots N/8 - 1 - \alpha_2$ **do**
$\quad A[4(\alpha_2 + i) + 0 \ldots 3] = I[2(4(\alpha_2 + i) + 0 \ldots 3) \pm M];$
$\quad B[2(\alpha_1 + i) + 0 \ldots 1] = A[2(2(\alpha_1 + i) + 0 \ldots 1) \pm M];$
$\quad C[\alpha_0 + i] = B[2(\alpha_0 + i) \pm M];$
$\quad ZT = C[\alpha_0 + i], B[2(\alpha_0 + i) + 0 \ldots 1], A[4(\alpha_0 + i) + 0 \ldots 3];$
end for

is computed. The distance between production index expression $4(\alpha_2 + i) + 3$ and consumption index expression $2(2(\alpha_1 + i) + 0 \ldots 1) - M$ is:

$$\begin{aligned}
(4(\alpha_2 + i) + 3) - (2(2(\alpha_1 + i) + 0) - M) + 1 &= \\
4\alpha_2 + 4i + 3 - 4\alpha_1 - 4i + M + 1 &= \\
4(\alpha_2 - \alpha_1) + M + 4
\end{aligned}$$

Whether this distance is smaller or bigger than the window $4\alpha_2 - 1$ depends on the particular value of parameter M. If we subtract the distance between production and consumption from the window size of A, we get:

$$\begin{aligned}
(4\alpha_2 - 1) - (4\alpha_2 - 4\alpha_1 + M + 4) &= \\
-4 + 4\alpha_1 - M - 4 &= \\
4(\lceil M/2 \rceil + 1) - M - 8
\end{aligned}$$

If $4(\lceil M/2 \rceil + 1) - M - 8$ is positive it means that the window is large enough to store the intermediate values between the production and consumption of A in the merged loop. Hence, the size of the window for signal A is determined by the extracted loop. If $4(\lceil M/2 \rceil + 1) - M - 8$ is negative, it means that the window of signal A is determined by the buffer needed in the merged loop.

The sign of the expression $4(\lceil M/2 \rceil + 1) - M - 8$ can be calculated if it is assumed that parameter M is even. Then the expression simplifies to:

$$2M + 4 - M - 8 \quad = \quad M - 4$$

This is positive if $M \geq 4$. In case $M = 4$, the distance in the merged loop equals the distance in the extracted loop for the production and consumption of array signal A.

Hence the window of signal A, called WA, is (provided that M is even):

$$\begin{aligned}
4(\alpha_2 - \alpha_1) + M + 4 \quad &\text{if } M < 4 \\
4\alpha_2 - 1 \quad &\text{if } M \geq 4
\end{aligned}$$

In general (M odd or even) the window is:

$$WA = \max(4(\alpha_2 - \alpha_1) + M + 4, 4\alpha_2 - 1) + \{0, 1\}$$

Depending on the schedule, the window is one bigger then the maximum distance. In this particular case, the schedule requires to take the inplace buffer one word bigger then maximum distance between the read and write operations.

For the array B, the story is a bit different. Again we must at least provide memory locations to store the results of the extracted loop:

$\alpha_0 = 1;$
$\alpha_1 = \lceil M/2 \rceil + 1;$
$\alpha_2 = \lceil (2(2((\alpha_1 - 1) + 1) + M)/4 \rceil + 1;$
for $i = 0 \ldots \alpha_1 - 1$ **do**
$\quad B[2i + 0 \ldots 1] = A[2(2i + 0 \ldots 1) \pm M];$
end for

By filling in the upper-bound of the loop ($\alpha_1 - 1$) into the index expression of the production of B it is calculated that $2(\alpha_1 - 1) + 1$ locations are needed.

Unfortunately, this is not enough for our main merged loop:

for $i = 0 \ldots N/8 - 1 - \alpha_2$ **do**
$\quad A[4(\alpha_2 + i) + 0 \ldots 3] = I[2(4(\alpha_2 + i) + 0 \ldots 3) \pm M];$
$\quad B[2(\alpha_1 + i) + 0 \ldots 1] = A[2(2(\alpha_1 + i) + 0 \ldots 1) \pm M];$
$\quad C[\alpha_0 + i] = B[2(\alpha_0 + i) \pm M];$
$\quad ZT = C[\alpha_0 + i], B[2(\alpha_0 + i) + 0 \ldots 1], A[4(\alpha_0 + i) + 0 \ldots 3];$
end for

In this loop, $B[2(\alpha_0 + i) \pm M]$ and $B[2(\alpha_0 + i) + 0 \ldots 1]$ are the signals that are consumed. The loop produces $B[2(\alpha_1 + i) + 0 \ldots 1]$.

It is clear that when signals $B[2(\alpha_0 + i) \pm M]$ are alive, signals $B[2(\alpha_0 + i) + 0 \ldots 1]$ are also alive (provided that $M \geq 1$).

Hence, we only have to calculate the distance between the production index function $2(\alpha_1 + i) + 0 \ldots 1$ and the consumption index function $2(\alpha_0 + i) \pm M$:

$$
\begin{aligned}
WB &= (2 \times (\alpha_1 + i) + 1) - (2 \times (\alpha_0 + i) - M) + 1 + 0, 1 \\
&= 2\alpha_1 + 2i + 1 - 2\alpha_0 - 2i + M + 1 + \{0, 1\} \\
&= 2\alpha_1 - 2\alpha_0 + M + 2 + \{0, 1\} \\
&= 2 \times (\alpha_1 - \alpha_0) + M + 2 + \{0, 1\}
\end{aligned}
$$

Filling in the definition of $\alpha_0 = 1$ and $\alpha_1 = \lceil M/2 \rceil + 1$ gives:

$$
\begin{aligned}
2(\lceil M/2 \rceil + 1) - 2 \times 1 + M + 2 &= \\
2(\lceil M/2 \rceil + 1) + M &= \\
2\lceil M/2 \rceil + M + 2 &
\end{aligned}
$$

Assuming that M is even, this simplifies to:

$$M + 2 + M = 2M + 2 \tag{3.2}$$

Applying the in-place optimisation with window WA on signal A and window WB on signal B and converting back to C++ gives:

```
1    #include <iostream.h>
2    #include <stdlib.h>
3
4    // Number of samples
```

```
5    const unsigned int N = 1<<7;
6    // Length of one side of the filter
7    const int M = 4;
8    // Number of filter coeffs
9    const unsigned int nb_coeffs = 2*M + 1;
10   // Length of a zero tree
11   const unsigned int ZT_LENGTH = 1 + 2 + 4;
12   const int coeff[nb_coeffs] = {1, 8, 28, 56, 70, 56, 28, 8, 1};
13
14   unsigned int wrap(int i, const int max) {
15       int new_i = (i >=0 ) ? i : -i;
16       return (new_i < max) ? new_i : 2*max-new_i-2;
17   }
18
19   void zero_tree(int ZT[ZT_LENGTH]) {
20       for (unsigned int i=0; i <= ZT_LENGTH-1; i++)
21           cout << "ZT[" << i << "] = " << ZT[i] << endl;
22       cout << endl;
23   }
24
25   int main() {
26   const unsigned int a0 = 1;
27   const unsigned int a1 = (M/2)+1;
28   const unsigned int a2 = (2*(2*(a1-1)+1)+M)/4+1;
29   const unsigned int WA = 4*a2-1+1;
30   const unsigned int WB = 2*(a1-a0)+M+2;
31
32   int I[N], A[WA], B[WB], C;
33   int i, di, k;
34
35   srand(10);
36   for (i=0; i <= N-1; i++)
37     I[i] = rand();
38
39   for (i=0; i <= a2-1; i++) {
40      for (di=0; di <= 3; di++) {
41         int sum = 0;
42         for (k=-M; k <= M; k++)
43            sum += coeff[M+k]*I[wrap(2*(4*i+di)+k,N)];
44         A[(4*i+di)%WA] = sum>>9;
45      }
46   }
47   for (i=0; i <= a1-1; i++) {
48      for (di=0; di <= 1; di++) {
49         int sum = 0;
50         for (k=-M; k <= M; k++)
51            sum += coeff[M+k]*A[(wrap(2*(2*i+di)+k,N/2))%WA];
52         B[(2*i+di)%WB] = sum>>9;
53      }
54   }
55   i=0; int sum = 0;
56   for (k=-M; k <= M; k++)
57     sum += coeff[M+k]*B[(wrap(2*i+k,N/4))%WB];
58   C = sum>>9;
59
60   const unsigned int ZT_LENGTH = 1 + 2 + 4;
61   int ZT[ZT_LENGTH];
62
```

```
63  ZT[0] = C;
64  for (di=0; di <= 1; di++)
65      ZT[1+di] = B[(di)%WB];
66  for (di=0; di <= 3; di++)
67      ZT[1+2+di] = A[(di)%WA];
68  zero_tree(ZT);
69
70  for (i=0; i <= N/8-1; i++) {
71      if (i <= (N/8-1-a2)) {
72          for (di=0; di <= 3; di++) {
73              int sum = 0;
74              for (k=-M; k <= M; k++)
75                  sum += coeff[M+k]*I[wrap(2*(4*(a2+i)+di)+k,N)];
76              A[(4*(a2+i)+di)%WA] = sum>>9;
77          }
78      }
79      if (i <= (N/8-1-a1)) {
80          for (di=0; di <= 1; di++) {
81              int sum = 0;
82              for (k=-M; k <= M; k++)
83                  sum += coeff[M+k]*A[(wrap(2*(2*(a1+i)+di)+k,N/2))%WA];
84              B[(2*(a1+i)+di)%WB] = sum>>9;
85          }
86      }
87      if (i <= (N/8-1-a0)) {
88          int sum = 0;
89          for (k=-M; k <= M; k++)
90              sum += coeff[M+k]*B[(wrap(2*(a0+i)+k,N/4))%WB];
91          C = sum>>9;
92          ZT[0] = C;
93          for (di=0; di <= 1; di++)
94              ZT[1+di] = B[(2*(a0+i)+di)%WB];
95          for (di=0; di <= 3; di++)
96              ZT[1+2+di] = A[(4*(a0+i)+di)%WA];
97          zero_tree(ZT);
98      }
99  }
100 }
```

The output of the above C++ program was verified by simulation against the output of the original C++ program listed in Section 3.6.1.4 for various parameter values for N and M. So the global input-output behavior is not changed by applying the DTSE methodology. It also nicely demonstrates that no information is destroyed by using the notation proposed in this chapter since a fully operational C++ program can be derived again at the end.

3.6.8 Summary of results

The result of applying the DTSE transformations is that the memory requirements for storing the temporarily signals is reduced significantly. Straightforward implementation of the initial description would require $N/2 + N/4 + N/8 + 2 \times (2 \times M + 1)$ memory locations. After DTSE, the required number of memory locations is reduced to $WA + WB + 1$, where:

$$
\begin{aligned}
\alpha_0 &= 1 \\
\alpha_1 &= \lceil M/2 \rceil + 1 \\
WB &= 2(\alpha_1 - \alpha_0) + M + 2 \\
\alpha_2 &= \lceil (2(2(\alpha_1 - 1) + 1) + M)/4 \rceil + 1 \\
WA &= \max(4\alpha_2 - 1, 4(\alpha_2 - \alpha_1) + M + 4) + 1
\end{aligned}
$$

Table 3.4.   The column labeled with O is the original number of temporary signals needed to be stored for calculating the 1D wavelet transform of a signal of length N (first column) (assuming a filter length of 4). In the columns labeled with the length of the filter M (from 1 to 14), the optimised number of signals to be stored ($= WA + WB + 1$) is tabulated.

N	O	1	2	3	4	5	6	7	8	9	10	11	12	13	14
16	32	15	23	25	29	31	39	41	45	47	55	57	61	63	71
32	46	15	23	25	29	31	39	41	45	47	55	57	61	63	71
64	74	15	23	25	29	31	39	41	45	47	55	57	61	63	71
128	130	15	23	25	29	31	39	41	45	47	55	57	61	63	71
256	242	15	23	25	29	31	39	41	45	47	55	57	61	63	71
512	466	15	23	25	29	31	39	41	45	47	55	57	61	63	71
1024	914	15	23	25	29	31	39	41	45	47	55	57	61	63	71
2048	1810	15	23	25	29	31	39	41	45	47	55	57	61	63	71

and filtering is done with a filter of $2 \times M + 1$ coefficients. The initial memory requirement is compared with the optimised result in table 3.4 for several filter lengths. Table 3.4 shows that for $N = 16$, a small number of input signals, the optimisation does not reduce the memory requirements unless $M \leq 5$. For $N = 32$, the memory is reduced if $M \leq 8$. Starting from a more realistic $N = 64$, the optimisation reduces the memory requirements for all cases listed in the table.

Similar optimisation results can be achieved for the full 2D wavelet - zero tree algorithm, where they are much more tedious and error-prone to execute though. The key issue with the optimisations presented in this chapter is that they are obtained by applying a systematic methodology which allows to effectively explore a huge search space even manually. The general techniques for this are the topic of the rest of this book. The formalisation is also a requirement to be able to support the most critical optimisations with CAD tools in the end.

3.7   CONCLUSIONS

In this chapter, first an overview has been provided of the DTSE methodology. Next the steps in this methodology have been illustrated on a simple but representative example, namely a 1-D wavelet transform. This example also shows that our system-level DTSE methodology has the potential to significantly reduce the data transfer and storage requirements of real-life applications.

However, the extra requirement to reduce the power of a design introduces extra design steps that increase the design time. Hence, CAD tools that support application of the optimisation techniques for low power are necessary to reduce design time. Therefore, we have developed formalized data models and techniques suited to support the most crucial steps in the methodology. These can be applied manually (to reduce the design effort) or be the basis for actual CAD tools. The main results of this research will be the topic of the subsequent chapters.

4 COST MODELS AND ESTIMATION

This chapter provides an overview of the cost models in terms of size, area, band-width, and power, as used in this book. In practice, timing issues are addressed as constraints and the cost function is mainly composed of area and power related terms. For memories, it is usually feasible to obtain closed-form formulas for the area and power in terms of the main parameters: number of bits, number of words, postdecoding (folding) factor, number of ports (R, W and R/W). Of course this model will heavily depend on the type of memory: SRAM, DRAM, SDRAM, Pointer-Addressed Memory, Video-RAM and so on. The parameters are not accurately available in each step of the DTSE methodology but they can always be estimated, albeit only crudely in the early steps.

Memory area estimates depend on the size (see section 4.1) and the number of ports (see section 4.2). The combined formula is discussed in section 4.3. Note that these estimates do not require the full knowledge of the relative ordering of the accesses as opposed to the detailed size/area computation determined in the final in-place optimisation step (see chapter 12). Hence, they are also useful as cost feedback mechanism for any of the higher-level steps in the methodology. In addition, the on-chip interconnect contribution will be ignored because in the current n-layer metal technologies, the area overhead of these global busses has become much less pronounced. Off-chip interconnect can of course not be ignored but this can only be incorporated in the board and package cost.

For the power, also several cost models are derived for the different components in data transfer and storage, namely memory units and global interconnection (see section 4.4 and following). Also here, it is motivated that the contribution of the on-chip interconnection in the current technologies can still be largely ignored as a second order effect, at least for high-level estimates to be used in relative comparisons. Also here, off-chip interconnect should not be ignored.

4.1 MEMORY SIZE MODELS

The evaluation of the memory area required by an RMSP algorithm implies three distinct aspects (see chapter 4 of [31] and also [27]):

1. assessment of the number of storage locations, which determines the total number of memory cells;

2. assessment of the number and type (*read, write,* or *read/write*) of memory ports, which heavily influences the area cost of a single cell (see section 4.2).

3. evaluation of the actual silicon area occupied by a background memory having given characteristics, like type (e.g., SRAM, DRAM), word-length, storage locations, port configuration (see section 4.3).

The evaluation of a distributed memory architecture, with modules of possible different characteristics, then proceeds as follows: the total number of locations is provided by the first assessment, the total bandwidth by the second one, and the total area is computed as the sum of individual module areas, estimated as in item 3.

4.1.1 Estimation problem formulation

The main goals of the memory estimation are (1) to trade off different descriptions of a RMSP algorithm in a very early design stage, and (2) to predict the memory characteristics of designs without synthesizing them. For the sake of computation efficiency, and taking also into account these main objectives, it is assumed for the time being that all M-D signals share the same global memory, having the word-length equal to the maximum signal bit size. If a more accurate distributed memory evaluation has to be available, the more complex and expensive problem of background memory allocation has to be solved (see chapter 11).

Finding the minimum number of memory locations/registers when the operation ordering is that one provided by the order of the signal productions, and the nesting of the loops in the algorithm code (or when the scheduling of the operations was previously accomplished), has already been solved. As discussed in chapter 2, knowing the detailed sequence of operations as in scalar memory management allows to use an extended "left-edge" type algorithm to find out both the minimum storage requirement and the assignment of n individual signals to memory locations.

Unfortunately, the same problem becomes significantly harder when the operation ordering is still not fixed. In that case, the problem belongs to the NP class [151], even in the absence of conditionals. Data-flow analysis constitutes an effective strategy for estimating the number of memory locations when the operation scheduling is still not decided. As RMSP algorithms contain usually a huge amount of scalars, a polyhedral data-flow analysis operating with *groups of signals* (called *basic sets*), rather than a flattened one operating with *individual signals*, is developed in chapter 6.

This section will mainly deal with storage size estimation based on this polyhedral model of data-flow analysis. Subsection 4.1.2 will thoroughly discuss the basic concept of memory bounds in applicative programs. It will introduce the in-place optimization of signals when the detailed operation ordering is still unknown, as opposeed to the detailed in-place optimisation step in our overall methodology. Relying on these concepts, the assessment of the minimal memory size of a RMSP algorithm is carried out by means of a heuristic traversal of the polyhedral data-flow graph, created as described in section 6.3 .

The estimation of the port requirements (number and type) is achieved by a partial ordering of the $read/write$ operations within a given cycle budget (see section 4.2). The evaluation of the bandwidth is done taking into account also the partial ordering constraints derived from the DFG structure (section 6.3) and traversal (subsection 4.1.5), It does not incorporate the accurate conflict graph models introduced in chapter 10 however because these require a too detailed analysis at the early steps in the methodology.

Section 4.3 will briefly present the basic features of the layout model [286] employed by our memory estimation and allocation tool when evaluating the silicon area occupied by a background memory.

4.1.2 Estimation of storage requirements

As shown in chapter 2, the current memory estimation approaches are either only suited for scalar data types, or can only handle procedurally interpreted specifications. The extended background memory estimation approach presented in this section extends on that.

After partitioning the signals from the given RMSP algorithm (subsection 6.2.1), and after accumulating all the possible dependency information at the level of basic sets (section 6.3), the subsequent step is to obtain an accurate evaluation of the minimal memory size (locations) compatible with the resulting data-flow graph.

Even the simpler problem of finding the minimum number of memory locations necessary to compute a directed acyclic graph has been proven to be an NP-complete problem [367]. Structurally, the DFG's determined as in section 6.3 can be more complex: e.g., they may contain cycles, as 2 *groups* of signals may contain 2 subsets with opposite dependencies to the other group[1]. In [31] (chapter 4), it is proven that also finding the minimum number of locations necessary to compute a data-flow graph — as explained in section 6.3 — is NP-complete. Consequently, the assessment of the minimal memory size is achieved basically by means of a heuristic traversal of the data-flow graph. It must be emphasized that the goal of this approach is to introduce only a *partial* operation ordering — necessary to reduce the storage requirements — while a proper scheduling implies a *total* ordering, which is unnecessary for our problem. This DFG traversal provides a data-flow which is equivalent to a certain reorganization of the code (see the example in subsection 4.1.6). The procedural execution of this functionally equivalent code entails a low (eventually minimum) number of storage locations for computing the respective data-flow graph.

Subsection 4.1.3 provides an intuitive understanding of the concept of memory upper-bound when computing a group of scalars in an applicative language. Subsection 4.1.4 introduces the methodology of in-place optimization, incorporated in the traversal approach. Subsection 4.1.5 presents the basic features of the DFG traversal. The main ideas in this section will be illustrated in subsection 4.1.6.

4.1.3 Memory bounds in applicative languages

The problem is to find out the minimum number of memory locations necessary to carry out the computation.

Example Assume we start from the illustrative piece of Silage[187] code in figure 4.1, where signal A is not consumed elsewhere, but all scalars $B[i][j]$ are necessary in the sequel of the code.

```
#define W fix<16,4>

(k: 0 .. 9)::
      begin
                A[2*k] = W(0);     A[2*k+1] = W(1);
      end;

(i: 0 .. 9)::
      (j: 0 .. 9)::  B[i][j] = A[i+j] + A[2*i+1];

      . . . . . . . . . . . . . . . . . .
```

Figure 4.1.   Silage code illustrating memory lower- and upper- bounds

It must be emphasized from the very beginning that the formulation of the problem is *ambiguous*, as Silage is an applicative and thus non-procedural language.

Assume for the moment a procedural interpretation of the code in figure 4.1. Then, it can be easily verified that no more than 101 locations are really necessary. Indeed, at each execution of the i-loop, $B[i][0]$ can be stored in the location occupied previously by $A[i]$, as this scalar is no longer used. Consequently, the execution of each of the first 9 i-loops will increase the necessary memory with 9 locations. The last i-loop does not require any extra storage location, as every $B[9][j]$ can overwrite $A[9 + j]$.

Two essential implicit assumptions result whenever the code is interpreted procedurally: (1) the instructions are executed *sequentially*, and (2) the order of execution is directly derived from the iterations in the code. As Silage is an applicative language, the assumptions mentioned above are no longer valid.

---

[1] In such a case, node clustering is required.

First, parallel computations may occur whenever the possibility exists [36]. For instance, in the example in figure 4.1, all scalars $B[i][j]$ can be computed in parallel, and the minimum storage requirement becomes 100 locations, as signal B can overwrite signal A. However, if perfect parallelism cannot be carried out – in the sense that some scalars $B[i][j]$ may be computed and have to be stored before the reading of the A operands is completely finished, then a more conservative answer – i.e., 120 locations – should be given. But they can be also partitioned in groups of e.g. 5, and each group computed in parallel. In this latter case, the minimum number of locations depends both on the partitioning and on the group ordering.

Second, even in the case of sequential execution, there are 100! possible orders of computation[2]. A challenging reformulation of the problem is: "Assuming the computations are carried out sequentially, what is the *absolute* minimum storage requirement, for all the possible computation orders?" Or, in other words, "what is the *lower-bound* of the minimum storage?" Another way of reformulating the problem is the following: "Assuming the computations are carried out sequentially in the final (as yet unknown) code, what is the minimum number of storage locations, *independent* of any computation order?" Or, in other words, "what is the *upper-bound* of the minimum storage?"

It is clear now that the first formulation of the problem can imply different interpretations, and therefore, different answers. The solutions for the sequential case will be discussed in the sequel.

4.1.3.1   Lower-bound of the minimum storage.   Notice that the computation of signal B requires at least 100 locations, as there are 100 scalars $B[i][j]$ to be stored. As shown before, the procedural interpretation of the code implies a minimum of 101 locations. Is there any computation order for the example in figure 4.1 leading to 100 locations? The answer is affirmative: there are several ways of computing sequentially the scalars $B[i][j]$ employing no more than 100 locations. One possible ordering is displayed in figure 4.2.

```
           . . . . . . . . . . . . . . . . .

(i: 0 .. 9)::
   (j: 0 .. 9)::
      B[i][j] = if ( i!=j+1 & j!=i+1 )  -> A[i+j] + A[2*i+1];

(i: 0 .. 9)::
   (j: 0 .. 9)::
      B[i][j] =       if ( i==j+1 )          -> A[i+j] + A[2*i+1];

(i: 0 .. 9)::
   (j: 0 .. 9)::
      B[i][j] =       if ( j==i+1 )          -> A[i+j] + A[2*i+1];

           . . . . . . . . . . . . . . . . .
```

Figure 4.2.   Equivalent procedural Silage code requiring an absolute minimum storage

When the number of scalars to be produced is larger than the number of *dying*[3] scalars (as in the example in figure 4.1), the minimum storage situation is met when each production consumes for the last time *no more than one scalar*. Only in this situation the memory occupied by the dying operand domains can be completely overwritten by the new definition domain (see polyhedral model in chapter 5).

In figure 4.2, for instance, each of the 9 signal productions in the third loop nest consume for the last time one scalar: $A[1]$, respectively $A[3]$, ... , $A[17]$. The last signal production in the second loop nest consumes for the last time $A[19]$ (when $B[9][8]$ is produced). The scalars $A[\ ]$ having even indices are consumed for the last time in the first loop nest. As each signal production contains at most one operand having an even index, it is not possible that two scalars $A[\ ]$ are simultaneously consumed for the last time.

---

[2]This is a very large number: it has 24 zero's.
[3]i.e., consumed for the last time.

When the number of scalars to be produced is smaller than or equal to the number of dying scalars, the definition domain can be stored in-place (can overwrite the memory occupied by dying operand domains) if the computation order is such that each signal production consumes for the last time *at least one scalar*.

Unfortunately, there is no known algorithm of reasonable complexity able to detect the situations described above, and to generate the corresponding procedural code. Furthermore, it is not known how to determine the reachable lower-bound of the minimum storage. As Silage is a single-assignment language, each program can be flattened in a directed acyclic graph (the nodes being the scalars, and the arcs data dependencies). As mentioned already in the introduction of subsection 4.1.2, Sethi[367] proved that the problem of finding the minimum number of storage locations for the computation of a directed acyclic graph is NP-complete.

4.1.3.2   Upper-bound of the minimum storage.   As already mentioned, the computation of signal B requires no more than 120 locations (for storing simultaneously both signals A and B). However, this upper-bound is too conservative. Indeed, as there are 20 scalars of $A[\ ]$ consumed for the last time, and each signal production has two operands, there are at least 10 signal productions (and at most 20) consuming scalars for the last time. The signals $B[i][j]$ produced during these signal productions can be stored in-place, overwriting the dying scalars $A[\ ]$. In conclusion, no more than 90 scalars $B[i][j]$ need to be stored in locations different from those occupied by $A[\ ]$, and therefore, no sequential ordering needs more than 20+90=110 memory locations.

Is there any computation order for the example in figure 4.1 leading to 110 locations? The answer is affirmative: there are in fact several ways of computing sequentially the scalars $B[i][j]$ which have to employ 110 locations. One possible ordering is displayed in figure 4.3.

```
        . . . . . . . . . . . . . . . . . .

(i:  0 .. 9)::
   (j:  0 .. 9)::
      B[i][j] =   if ( i!=j )   -> A[i+j] + A[2*i+1];

(i:  0 .. 9)::
   (j:  0 .. 9)::
      B[i][j] =   if ( i==j )   -> A[i+j] + A[2*i+1];

        . . . . . . . . . . . . . . . . .
```

Figure 4.3.   Equivalent procedural Silage code requiring a maximal minimum storage

As it can be easily noticed, the 10 signal productions in the second loop nest consume for the last time two scalars A each: $(A[0], A[1])$, respectively $(A[2], A[3])$, and so on. Consequently, the scalars $B[i][i]$ can overwrite the scalars $A[\ ]$. The 90 scalars produced in the first loop nest, $B[i][j]$ $(i \neq j)$, have to be stored in different memory locations. Therefore, this computation order requires 110 locations.

However, in general, it is not possible to decide whether the upper-bound of the minimum storage is the best one: it is possible that the reachable upper-bound is even lower. But it is very hard to determine automatically the best upper-bound, and to generate the procedural code.

In order to preserve as much freedom of decision as possible for the subsequent steps, the earliest DTSE steps should employ storage upper-bounds. Gradually, more ordering decisions will be imposed on the code, lowering the upper-bound on the storage requirement but removing some freedom. Only after the SCBD step most of the ordering freedom for the main background memory accesses is removed, but even then some freedom remains for the subsequent architecture synthesis and compilation stages. Deriving good upper-bounds is crucial in order to prevent an important overestimation of the memory size.

4.1.4 High-level in-place cost estimation

In the example from figure 4.1, it was assumed that signal A is consumed for the last time while computing signal B. In large programs, this situation has to be detected. Moreover, it usually happens that only *parts* of indexed signals are consumed for the last time.

The polyhedral data-flow analysis presented in chapter 6, allows to determine which parts of an M-D signal are consumed for the last time when a signal definition domain is produced. More specifically, the partitioning of indexed signals (subsection 6.2.1), and the detection of dependencies between partitions (*basic sets*) provide the solution. This information, together with the partition sizes (subsection 6.2.3.2) and the number of dependencies between them (subsection 6.3.1), allows to detect the possibility of storing signals in-place independent of the computation order, and at the same time, to derive tight upper-bounds for the storage currently occupied when the groups of signals are produced.

The problem of in-place optimization refers to the possibility that signals in a RMSP algorithm share the same memory locations during algorithm execution. This problem may be approached from two viewpoints:

(1) *high-level* in-place estimation, referring to the memory sharing due to the data-flow, hence independent of any detailed operation ordering;

(2) *low-level* in-place optimization, which refers to the same issue but dependent on the full operation ordering [404, 105]. Moreover, this form of in-place must be carried out separately for each memory module. The low-level in-place thus depends on the distributed memory architecture and on the signal-to-memory assignment. Therefore, it is the last step of the DTSE methodology.

The latter problem can sometimes be tackled with a symbolic or scalar-oriented technique (see subsection 2.4), but it usually requires more advanced techniques for array signals (see chapter 12). The high-level version is even more difficult as it requires the computation of tight upper-bounds on the (usually huge) set of valid operation orderings.

The data-flow graphs, constructed as shown in section 6.3, contain sufficient embedded information in order to derive tight memory upper-bounds, *independent* of the production order, which will be decided later.

Several cases of upper-bound determination will be considered in the sequel. These cases represent essential situations encountered when computing a data-flow graph. The more complex ones can be reduced to, or can be solved by simple extensions of the cases below.

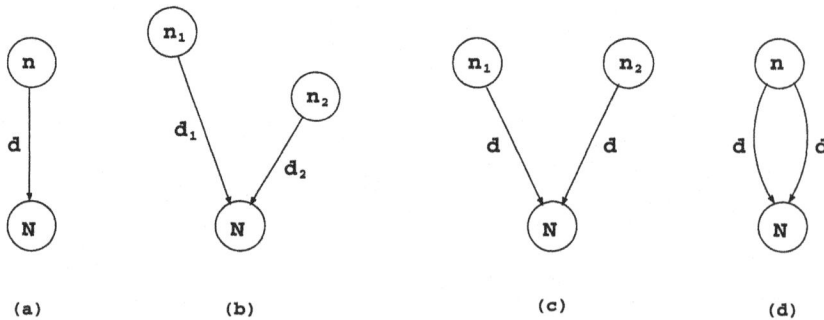

Figure 4.4.   Different types of production of basic sets in a data-flow graph

(a) Suppose a DFG contains two basic sets (see section 6.2) of n, respectively N, scalars each, and there are d scalar-level dependencies between these basic sets (figure 4.4a). Suppose also that no other basic set in the DFG depends on basic set n – already produced. Analyzing the variation of the memory size when basic set N is produced[4], a conservative observation is that the occupied memory could have a relative growth of *MemoryIncrease* $= N$ locations (see figure 4.5a). After the production of the N scalars (no matter in which order, as the code is applicative), the n scalars consumed for the last time are no longer necessary, and the corresponding storage can be freed: *MemoryDecrease* $= n$.

Taking a closer look, two situations may occur:
(1) if $d \geq n$, only $d - n$ scalars from the basic set N need new storage locations. Also, possibly, the $N - d$ scalars that do not depend on the basic set n) (see figure 4.5b) need to be stored.
(2) if $d < n$, it follows that $n - d$ scalars in the basic set n are no longer necessary and they can be eliminated immediately — *before* the production of the N scalars. After this memory decrease, a situation as described previously in (1) is obtained (as the number of dependencies is now equal to the number of scalars in n). At this moment, a relative memory increase of $N - d$ is expected (see figure 4.5c).

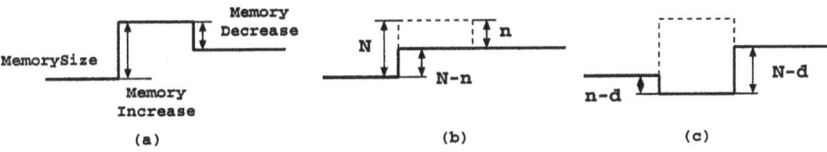

Figure 4.5.   Variation of the memory size when a basic set is produced

In both cases, the relative increase of the occupied storage is:
MemoryIncrease $= \max\{N - n, 0\}$, instead of N, as we would be tempted to answer at a quick glance.

Example In the DFG from figure 4.7, the computation of the basic set A0 $(A[j][0], j = \overline{0,5})$, which depends on the basic set $in0$ (the input scalar in), implies a relative increase of the occupied memory of $\max\{6 - 1, 0\}=5$ locations. Indeed, the last scalar $A[j][0]$, whichever it is, can overwrite the input scalar in (see line (1) in the Silage code from figure 4.6).

(b) Suppose the basic set N depends on two basic sets containing n_1, respectively n_2 scalars, the number of dependencies being d_1 and, respectively, d_2. There are two possibilities:
(1) the definition domain covering the basic set N is produced while consuming a single operand domain. In this case, the scalars in the basic sets n_1 and n_2 are consumed *sequentially* (see figure 4.4b), and the relative increase of the memory is

$$MemoryIncrease = (d_1 - n_1) + (d_2 - n_2) + (N - d_1 - d_2) = N - n_1 - n_2$$

(2) the computation of the definition domain relies on two operand domains. In this case, the scalars in the basic sets n_1 and n_2 are consumed *simultaneously* (see figure 4.4c); therefore $d_1 = d_2 \overset{def}{=} d$, and the growth is upper-bounded by:

$$MemoryIncrease = \min\{d - n_1, d - n_2\} + (N - d) = N - \max\{n_1, n_2\}$$

Example $(i : 1..n) ::$
    $(j : 1..m) :: \quad B[i][j] = A[0] + A[j] ;$
The DFG is of the form displayed in figure 4.4c, where $n_1 = 1$, $n_2 = m$ (corresponding to $A[0]$, respectively $A[j]$), $N = nm$ (corresponding to $B[i][j]$), and $d_1 = d_2 = nm$. It can be easily noticed that, independent of any computation order of $B[i][j]$, *MemoryIncrease* $= (n - 1)m$. But also $N - \max\{n_1, n_2\} = nm - \max\{1, m\} = (n - 1)m$.

---

[4]Producing/consuming a basic set means here producing/consuming *the signals* within the basic set.

```
#define n 6
#define W fix<16,4>

func main (in : W)  B : W[]=
begin
        (j : 0 .. n-1) ::
        begin
(1)       A[j][0] = in;
          (i : 0 .. n-1)::
(2)           A[j][i+1] = A[j][i] + W(1);
        end;
        (i : 0 .. n-1) ::
        begin
(3)       alpha[i] = A[i][n+i];
          (j : 0 .. n-1) ::
(4)           A[j][n+i+1] = if (j < i) -> A[j][n+i]
(5)                            || A[j][n+i] + alpha[i]
                             fi;
        end;
(6)     (i : 0 .. n-1) :: B[i] = A[i][2*n];
end;
```

Figure 4.6.   Illustrative Silage code

Remark It must be emphasized that $N - \max\{n_1, n_2\}$ is an upper bound which might not be necessarily reached:

Example $(i : 1 .. n) ::$
$$(j : n+1 .. n+m) :: \quad B[i][j] = A[i] + A[j] ;$$

requires a relative memory increase of $(n-1)(m-1) = nm - (n+m-1)$, independent of any computation order. But $n + m - 1 > \max\{n, m\}$ if $n, m > 1$.

The situation described here can be extended without any difficulty to the case when the basic set N depends on several basic sets n_i, consumed for the last time.

A special case for the situation presented in figure 4.4c is displayed in figure 4.4d, where the two basic sets n_1 and n_2 coincide. The upper-bound of the relative memory growth is

$$MemoryIncrease = \left\lfloor \frac{2d - n}{2} \right\rfloor + (N - d) = N - \left\lceil \frac{n}{2} \right\rceil$$

Example $(i : 0 .. 10) :: \quad B[i] = A[i] + A[10 - i] ;$

The DFG is of the form displayed in figure 4.4d, where $N = n = d = 11$. As it can be easily noticed, the upper bound of the memory increase is

$$MemoryIncrease = N - \left\lceil \frac{n}{2} \right\rceil = 11 - \left\lceil \frac{11}{2} \right\rceil = 5$$

and it is effectively reached when, for instance, the code is executed procedurally.

(c) More difficult situations appear when nodes in the DFG have self-dependency arcs. Such cases are due to recurrence relations between the elements (scalars) of the same indexed signal, as shown in figure 4.8a, where the basic set A3 represents the index space A3=$\{ x = i \mid 8 \geq i \geq 3 \}$. Analyzing the memory variation while computing the DFG in figure 4.8a, it is clear that the basic sets A0, A1, A2

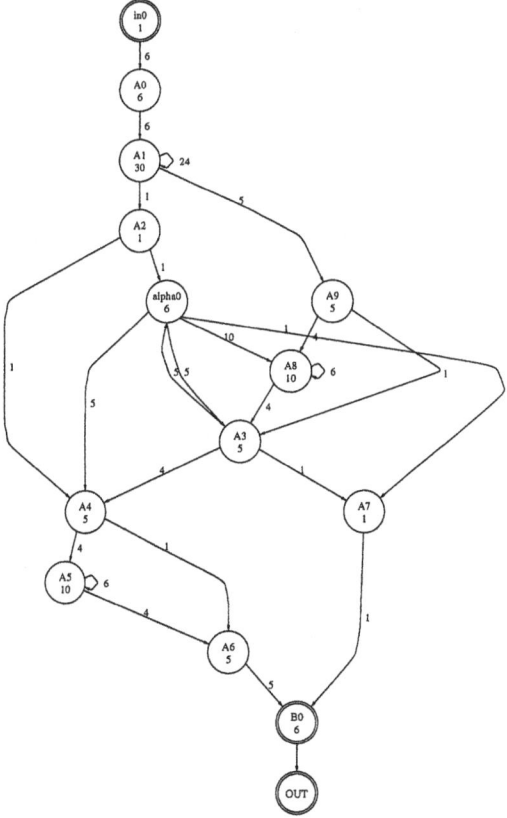

Figure 4.7.   The data-flow graph for the illustrative example in figure 4.6

have to be computed first (in any order), and the required storage so far is 3 locations. The production of A3 is carried out with one operand domain. There are 3 dependencies emerging from A3 towards other basic sets, which means that at most 3 scalars from A3 are needed to proceed the computation of the graph, the others being necessary for the self-computation of A3. This implies that, in the worst case, 2 extra locations are needed by the production of A3: indeed, the first scalar of A3 can overwrite one of the previous basic sets, say A0; the next two scalars of A3 could require new locations; but afterwards, the next scalars can overwrite either A1, A2, or the older elements of A3. In conclusion, the upper-bound of the relative storage increase is 2 locations (for the whole DFG being 5 locations).

It must be emphasized that this is the best estimation we can get from the DFG in figure 4.8a. At this granularity level (see [31], section 2.4), the insufficient information concerning the computation of A3 — the internal dependencies of A3, as well as the dependencies between its components and the other basic sets — do not allow a more precise estimation of the memory requirement. Increasing the granularity level, the DFG expanded one loop level (figure 4.8b) offers sufficient information to derive one of the 6 best computation orders:

$$A[0] \; = \; in \; ; \quad (i : \; 3 \mathbin{..} 9 \mathbin{..} 3) :: \; A[i] \; = \; A[i-3] \; ; \quad out[0] \; = \; A[9] \; ;$$

and so on, which requires only 2 locations — the absolute minimum value, for any computation order. It has to be remarked that, for the simpler recurrence relation

```
func  main (in: fix<16,4>)  out:  fix<16,4>[] =
begin
    A[0] = in ;      A[1] = in ;      A[2] = in ;

    (i: 3 .. 11)::   A[i] = A[i-3] ;

    out[0] = A[9] ;  out[1] = A[10] ;  out[2] = A[11] ;
end;
```

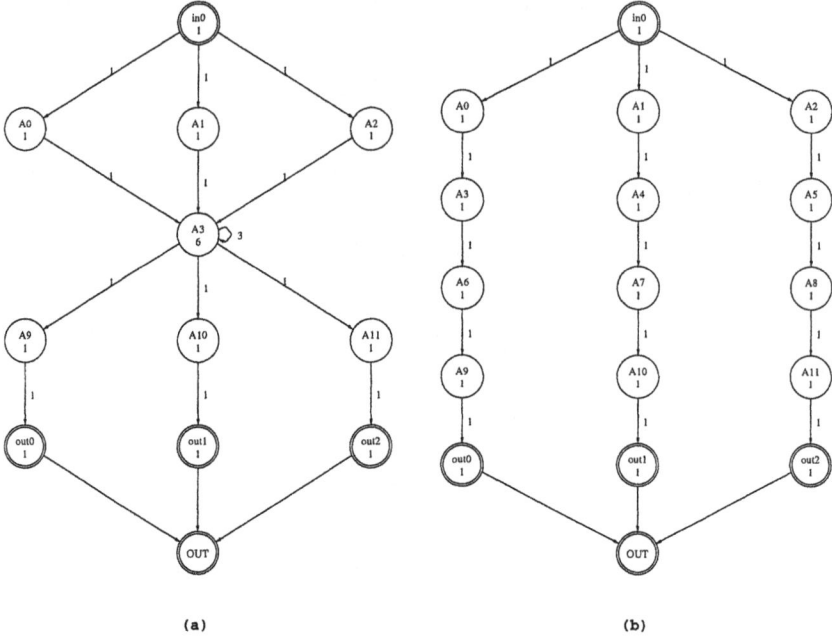

(a)   (b)

Figure 4.8.   (a) Data-flow graph with self-dependency arc; (b) expanded one loop level

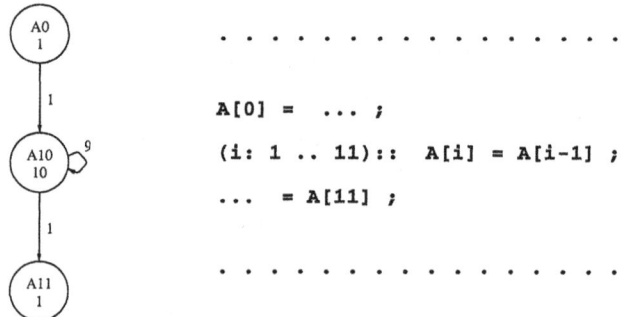

```
A[0] =  ... ;

(i: 1 .. 11)::  A[i] = A[i-1] ;

... = A[11] ;
```

Figure 4.9.   Simpler data-flow graph with self-dependency arc

$A[i] = A[i-1]$, the minimal storage requirement of one location is detected even on the granularity level 0 (see figure 4.9).

Multiple recurrence relations between elements of the same indexed signal can create multiple self-dependency arcs (see figure 4.10a). With a similar reasoning as above, the upper-bound of the relative memory increase is 1 location for the production of the basic set A36 (the scalars $A[3]$, $A[4]$, $A[5]$, $A[6]$), and respectively 1 location for the production of A78 (the scalars $A[7]$, $A[8]$). For the whole DFG in figure 4.10a, the upper-bound of the storage requirement is 5 locations. Going one granularity level deeper (see figure 4.10b), it can be noticed that 3 locations are enough for any valid computation ordering.

All the cases discussed allow to compute the *local*[5] memory upper-bounds in data-flow graphs by applying simple formulas. The more complex cases can be solved similarly: for instance, the computation of a domain of N scalars with k nonoverlapping operands (rather than 2 in figure 4.4c) requires at most $N - \max\{n_1, \ldots, n_k\}$ locations, where n_i denotes the number of scalars of the i-th operand.

```
func  main (in: fix<16,4>)  out:  fix<16,4>[] =
begin
        A[0] = in ;        A[1] = in ;        A[2] = in ;

        (i: 3 .. 9)::    A[i] = A[i-1] + A[i-3] ;

        out[0] = A[9] ;
end;
```

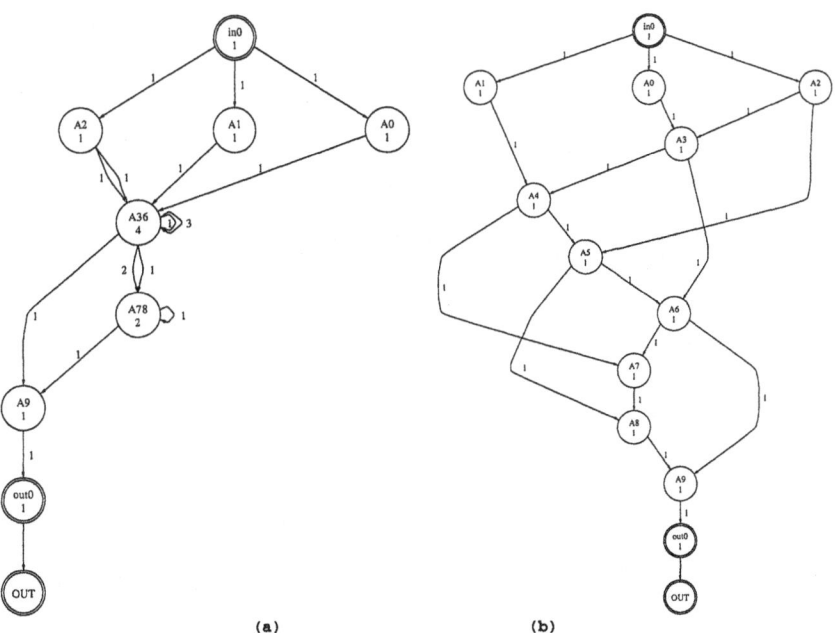

(a)   (b)

Figure 4.10.   (a) Data-flow graph with self-dependency arcs; (b) expanded one loop level

---

[5]For the computation of one basic set of signals.

4.1.5 Data-flow graph traversal

The DFG traversal is a fast heuristic which attempts to find the best possible ordering in which the basic sets of signals should be produced such that the memory size is kept as low as possible. Obviously, a basic set can be produced only after the production of all basic sets having dependencies to it. A more time-consuming but also more accurate modeling and decision process for this problem is performed during the SCBD step.

Remark It is assumed in the sequel that basic sets of signals are produced *sequentially*. However, it is possible to extend this model in order to allow also *parallel* production of basic sets. The difficulty does not consist in the determination of memory upper-bounds for parallel computation. The main problem is to determine *beforehand* which basic sets should be computed in parallel – as such a decision entails a potential increase of the data-path (unknown at this design stage). A global solution is:

1. to extract the (largely hidden) parallelism from the initial source code in the form of parallel **do_all** loops;

2. to determine the lowest degree of parallelism for the **do_all** loops needed to meet the throughput and hardware requirements. In this way, the basic sets of signals to be computed in parallel can be determined at this stage.

While producing a basic set, an increase of the memory size occurs (see figure 4.11a). This increase is followed by an eventual decrease, representing the total size of basic sets consumed for the last time (having no other dependency relations towards basic sets still not produced). The production of a basic set may not affect the current value of the maximum memory size – registered after starting the traversal (figure 4.11a), or may lead to exceeding this value (*MaxMemorySize*) which has to be therefore updated (figure 4.11b). The amount of the memory is yielded by the worst-case high-level in-place estimation (see subsection 4.1.4).

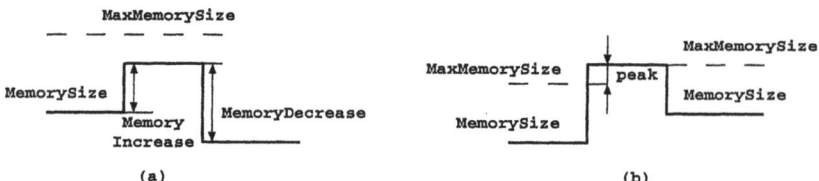

Figure 4.11.   Variation of the memory size during the DFG traversal

The traversal takes into account that if delayed signals are present from previous sample periods, then these are already stored in the memory. The aggregates of constants, usually stored in ROM's, are permanently residing in the memory.

Taking into account the remarks and assumptions mentioned above, the data-flow graph traversal works according to the following general scheme:
select all basic sets having nonzero delay value, and declare them *produced*;
select all basic sets corresponding to blocks of constants, and declare them *produced*;
initialize the values of *MemorySize* and *MaxMemorySize*;
       to the total size of the basic sets *produced* so far

do {
    for *all basic sets still not produced, but having all the ancestors produced*
        compute the cost function:
           $\lambda_1 \cdot peak + \lambda_2 \cdot (MemoryIncrease - MemoryDecrease)$
           where $peak = max\{0, MemorySize + MemoryIncrease$
           $-MaxMemorySize\}$
    select (any) one basic set having the minimum cost, and declare it *produced*

update the current values of *MemorySize* and *MaxMemorySize*

$$MemorySize \quad + = \ MemoryIncrease - MemoryDecrease$$
$$MaxMemorySize \ + = \ peak$$

} **until** *all basic sets are produced*

This is a relatively greedy approach, but it is based on a thorough global analysis for deriving the tightest storage upper-bounds (see subsection 4.1.4) which decide on the order of the basic set production.

Remark Branch-and-bound or look-ahead techniques can also be applied in order to find a better (or even optimal) computation ordering at basic set level. However, the size of the data-flow graphs and the large number of possibilities for realistic applications can imply a serious computation overhead, without significant storage improvements.

4.1.6 Example of storage assessment

The memory size evaluation methodology is exemplified for the Silage code in figure 4.12.

```
     #define W fix<16,4>       /*  m , n , M , N  are predefined constants  */

     func main (A: W[M+m+1][N+n+1]) opt: W =

     begin
(1)      optDelta[0] = W(0);

         (i: m .. M)::
         (j: n .. N)::
           begin
(2)            Delta[i][j][0] = W(0);

             (k: i-m .. i+m)::
             (l: j-n .. j+n)::
(3)              Delta[i][j][(k-i+m)*(2*n+1)+1-j+n+1] = A[i][j] - A[k][l]@1
                                    + Delta[i][j][(k-i+m)*(2*n+1)+1-j+n];

(4)          optDelta[(i-m)*(N-n+1)+j-n+1] = min(Delta[i][j][(2*m+1)*(2*n+1)],
                                   optDelta[(i-m)*(N-n+1)+j-n]);
           end;
(5)
         opt = optDelta[(M-m+1)*(N-n+1)];
     end;
```

Figure 4.12.   Silage code extracted from a motion detection algorithm

The number of memory locations required for the algorithm execution is approximated with the number of locations necessary to compute its data-flow graph. For instance, assume we have to compute the number of locations required for the computation of the DFG in figure 6.9.

According to the traversal scheme (subsection 4.1.5), the basic sets are produced/loaded in the order: A0@1, A1@1, A0, Delta0, Delta2, Delta1, optDelta0, optDelta2, optDelta1, opt0, A1 . This traversal yields (taking into account the in-place optimization) $MemorySize = (M + m + 1)(N + n + 1) + 2(M - m + 1)(N - n + 1) = 1203$ for the chosen values of the parameters ($M = N = 20$, $m = n = 4$). Only two partial constraints are essential to guide the subsequent scheduling stage: the basic set *optDelta0* must be produced after *Delta1*, and the input basic set *A1* must be loaded after the production of *opt0* . The equivalent source code compatible with the proposed ordering is shown in figure 4.13.

A partial expansion of the loops, resulting in a lower granularity of signal structure, leads to a data-flow analysis with smaller groups of signals ([31], section 2.4). This provides lower values for the memory size at the expense of the gradual expansion of the loop organization. At the same

```
        #define W fix<16,4>

        func main (A: W[M+m+1][N+n+1]) opt: W =
        begin
            (i: m .. M)::
                (j: n .. N)::
                    begin
(2)                     Delta[i][j][0] = W(0);

                        (k: i-m .. i+m)::
(3)                         (l: j-n .. j+n)::
                                Delta[i][j][(k-i+m)*(2*n+1)+1-j+n+1] = A[i][j] - A[k][l]@1
                                      + Delta[i][j][(k-i+m)*(2*n+1)+1-j+n];
                    end;
(1)
            optDelta[0] = W(0);
            (i: m .. M)::
(4)             (j: n .. N)::
                    optDelta[(i-m)*(N-n+1)+j-n+1] = min(Delta[i][j][(2*m+1)*(2*n+1)],
                                          optDelta[(i-m)*(N-n+1)+j-n]);
(5)
            opt = optDelta[(M-m+1)*(N-n+1)];
        end;
```

Figure 4.13.   The Silage code compatible with the DFG traversal

time, more constraints are conveyed to the scheduler. The memory size evaluation for the data-flow graph corresponding to granularity level 1 yields a storage requirement of 660 locations. The variance of the memory during the graph traversals is displayed in figure 4.14 for the DFG's of granularity levels 0 and 1. Similar evaluations for granularity levels 2, 3, and 4 (the scalar level) yields the same memory size, i.e., 627 locations. The analysis of the code in figure 4.12 shows that $(M + m + 1)(N + n + 1) + 2 = 627$ locations represents indeed the absolute lower-bound on storage.

This example also shows that it is not necessary to descend to the scalar level in order to obtain a significant reduction for the memory size. Furthermore, even more crude evaluations for lower granularity levels (as that one corresponding to level 1) may be considered good enough for this stage of the synthesis. As the amount of constraints is relatively reduced, there is more freedom for scheduling, low-level in-place optimization, and data-path allocation. Further memory adjustments can be done afterwards, taking into account the detailed ordering and scheduling decisions.

4.2   HIGH-LEVEL ESTIMATION OF PORT REQUIREMENTS

Once a partial computation order at basic set level is introduced by means of the data-flow graph traversal (subsection 4.1.5), the next step is the bandwidth estimation, subject to a given budget of cycles necessary to accomplish *read/write* operations from/into the background memory. Note that the data-path is still unknown, and therefore, the global throughput information cannot be directly employed.

If all basic sets are assumed to be stored in a single multi-port memory (which is reasonable during very early stages of the methodology), the problem can be obviously reduced to the determination of the maximum simultaneous *read* and *write* operations, taking into account: (1) the computation order of the basic sets of signals, and (2) the maximum number of cycles intended for memory operations.

A simple heuristic scheduling-type algorithm is employed. The number of simultaneous *read/write* operations is gradually increased — starting from $\lceil \#RW/cycle\_budget \rceil$ (where $\#RW$ is the total number of memory accesses) — until the detection of a computation solution, complying with the two constraints mentioned above.

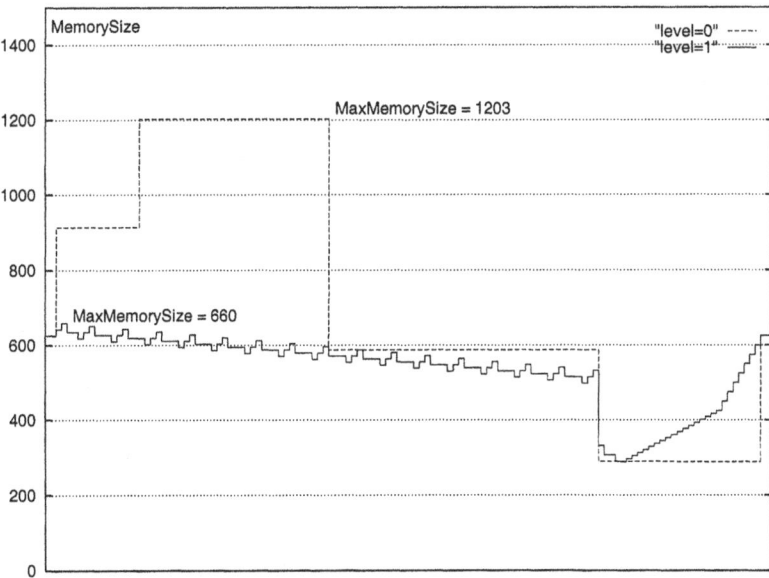

Figure 4.14. Trace of memory size during the DFG's traversals for the Silage code in figure 4.12 (granularity levels 0 and 1)

In order to ensure a reasonable stopping criterion, the bandwidth is potentially increased up to a threshold value, derived from the maximum desired computation parallelism. For instance, if a sequential computation is desired (see the remark in subsection 4.1.5), the threshold value is equal to one plus the maximum number of operands in the RMSP algorithm necessary to produce a definition domain. This choice is justified by the fact that this number of ports is sufficient to allow in one cycle the reading of all operands of any signal instance, followed by its production and writing into the memory. If no global computation solution can be found under these circumstances, then the cycle budget has to be increased, or more parallelism has to be introduced.

4.3 ESTIMATION OF THE LAYOUT AREA OF RAMS

For the assessment of the actual silicon area occupied by the global background memory, the model presented in [286] is employed. Unlike simpler models proposed in the past (see the references in [286]), which have an acceptable accuracy only for relatively large memories, the Mulder model incorporates also the *overhead area*, like drivers, sense amplifiers, address decoder, and control logic. If the parameters are carefully tuned in advance, this area model has proven to be suitable for comparing the size of (small- to medium-size) on-chip memories of different organizations, yielding no more than 10% error when verified against real memories [286].

According to this model, the actual silicon area occupied by a RAM is given by the formula:

$$A = TechnoFactor \cdot bits \cdot (1 + \alpha Ports) \cdot (N + \beta) \cdot [1 + 0.25(Ports + Ports_{rw} - 2)] \quad (4.1)$$

where:

> *TechnoFactor* is a technology scaling factor, equal to $(min\_geometry\,[\mu]/\,2)^2$;
> *bits* - the width of a memory location in bits;
> N - the number of storage locations;

Ports - the total number of ports (*read*, *write*, and *read/write*);

Ports$_{rw}$ - the number of *read/write* ports;

α and β – constants, which recommended values are 0.1, respectively 6 ÷ 12, in benchmark tests [286].

The area is expressed in an empirical unit, equal to the area of a register cell of $37 \times 55\mu m$, in a CMOS technology with 2 μm minimum geometry. Scaling factors of 0.6 for SRAMs, respectively 0.3 for DRAMs, are also recommended [286].

Evaluating the number of storage locations by the data-flow graph traversal, as shown in subsection 4.1.5, and the number and type of ports, as shown in section 4.2, an estimation of the layout area occupied by the background memory can be computed with the formula (4.1). During the high-level estimation, all M-D signals are assumed to share the same global memory, having the word-length equal to the maximum signal bit size. The distributed memory allocation approach, based on the similar mathematical concepts and data-flow view, will disregard this assumption (chapter 11).

To evaluate this approach, the area values obtained from the layout model [286] embedded in the memory estimation/allocation tool have been compared against the values obtained building the memory floorplan with an existing IMEC module generator of 1993. The tests have been carried out assuming CMOS technologies with 1.2 μm, respectively 1.0 μm, minimum geometry. Figure 4.15 displays the variation of the memory area – for RAMs constructed with the module generator – as a function of bit capacity[6]. The results obtained by our method, indicated with arrows on the figure, fit quite accurately on those curves (less than 10%), confirming the validity of the layout model for RAMs within a range of 4 and 64 kbits[7].

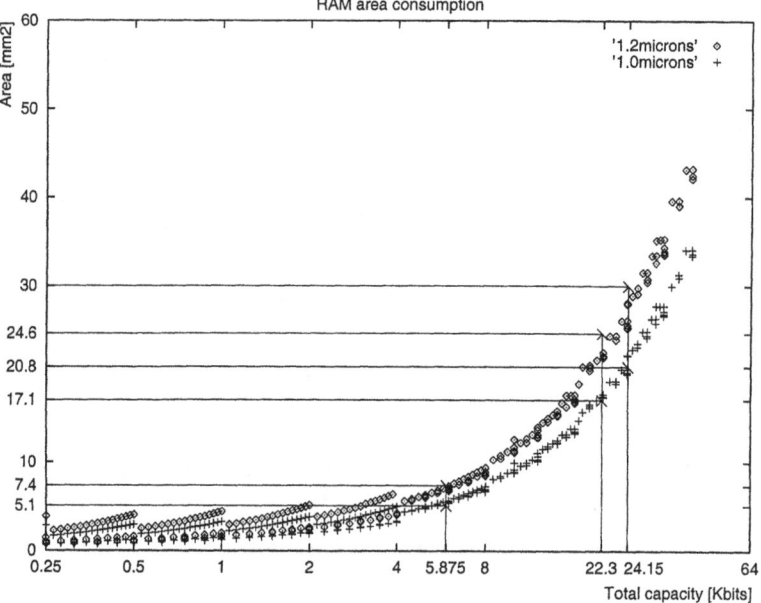

Figure 4.15.   RAM area vs. bit capacity for different CMOS technologies

---

[6]The discontinuities are due to the fact that the curves were drawn for different values of memory word-length.

[7]The errors have higher values for bit capacities outside this range. The memory area seems to have an increase rate (function of the bit capacity) higher than that given by the layout model. Consequently, the memories larger than 64 kbits, provided by the module generator, have areas of higher values than those predicted by the Mulder's model.

It has to be mentioned that, because of the limitations of the module generator, the comparative tests have been carried out only for RAMs with one R/W port. Also the RAM library employed by PHIDEO is limited to single-port memories, and (1 *read* + 1 *write*) memory ports (see the remark in [245], p. 733]). More recently, a new layout memory model has been proposed in [338]. Unfortunately, the model was not verified by comparison with real layouts[8].

4.4 POWER RELATED MODELS

Optimizing data storage and transfer related power consumption requires adequate power models for the three components of a memory architecture: on-chip memories, off-chip memories, and interconnect. In the subsequent subsections, we present the power models used in this book.

Two sources of power dissipation exist in CMOS circuits [63, 426]:

1. Static dissipation, mainly due to leakage current, *and*

2. Dynamic dissipation, due to:

    (a) short circuit current occurring when both n-channel and p-channel devices are on during a switching event, *and*

    (b) charging and discharging of load (and parasitic) capacitors.

Usually, the dynamic dissipation due to the charging and discharging of load capacitors dominates the total power consumption of CMOS circuits. Therefore, the other two components are often ignored when estimating power dissipation. Anyhow, they can be greatly reduced by proper circuit design: short circuit power can be minimized by managing transition rates, and leakage power is dictated by the technology employed (provided that circuit design is done properly) and is very small compared to the dynamic components.[9] Hence, in this book, only the dynamic power dissipation due to charging and discharging load capacitors is taken into account.

The dynamic power consumption of a node in a CMOS circuit with capacitance C is then given by:

$$P = \frac{1}{2} \cdot C \cdot V_{dd} \cdot V_{swing} \cdot f_{real} \tag{4.2}$$

where V_{dd} is the supply voltage, V_{swing} is the difference in voltage between a high and low level signal on the node, and f_{real} is the *real* switching frequency of the node.

This formula can be refined for the following three categories of circuits used in this book:

1. *Logic* For logic circuits, V_{swing} usually equals the supply voltage V_{dd}, and the real access frequency is given by:

$$f_{real} = a \cdot f_{clk} \tag{4.3}$$

where a is the activity factor of the node, i.e., the probability that the node will toggle, and f_{clk} is the frequency at which the circuit is clocked. This results in the following power estimation formula for a node in a logic circuit:

$$P = \frac{1}{2} \cdot a \cdot C \cdot V_{dd}^2 \cdot f_{clk} \tag{4.4}$$

2. *Buses*

For long buses, V_{swing} is often smaller than V_{dd} to save power. Since there is not necessarily a transfer every clock cycle, the activity factor can conceptually be split into two factors: a_{access}, which is the fraction of the clock cycles on which a transfer is being initiated, and a_{toggle}, which is

---

[8]For instance, the more than quadratic dependency on the number of ports is doubtful for state-of-the-art RAM generators.
[9]This is true except for very low voltage-technologies.

the probability that a bit line toggles between two consecutive transfers. The product $a_{access} \cdot f_{clk}$ then corresponds to the access frequency (f_{access}) of the resource. This leads to a more natural splitting of f_{real} for this context:

$$
\begin{aligned}
f_{real} &= a_{toggle} \cdot a_{access} \cdot f_{clk} \\
&= a_{toggle} \cdot f_{access}
\end{aligned}
\tag{4.5}
$$

Eq. 4.2 then becomes:

$$
P = \frac{1}{2} \cdot a_{toggle} \cdot C \cdot V_{dd} \cdot V_{swing} \cdot f_{access}
\tag{4.6}
$$

3. *Memories*

For memories, V_{swing} of the bitlines is often much smaller than V_{dd}. The power dissipation of memories is almost independent from the actual data values being read or written. This is due to differential logic and/or pre-charge at $V_{dd}/2$ to maximize the access speed. Consequently, the dynamic power estimation formula for memories does not contain an a_{toggle} factor and hence $f_{real} = f_{access}$. Usually, the V_{swing}/V_{dd} ratio is incorporated in the effective capacitance value of the memory. The power estimation formula for memories then becomes:

$$
P = \frac{1}{2} \cdot C_{eff} \cdot V_{dd}^2 \cdot f_{access}
\tag{4.7}
$$

Note that the real access rate f_{access} is used and not the maximum frequency f_{max} at which the RAM can be accessed. The maximal rate is only needed to determine whether enough bandwidth is available for the peak access rates. If the background memory is not accessed, it is assumed to be in power-down mode.[10]

Throughout this book, a power supply voltage $V_{dd} = 5$ V is assumed. If a lower supply voltage can be allowed by the process technology, the appropriate scaling has to be taken into account. It will however be (realistically) assumed that V_{dd} is fixed in advance as low as possible within the process constraints and noise tolerance for the memory organisation, and that it cannot be lowered any further by architectural considerations.

The rest of this chapter is organized as follows. Section 4.4.1 presents the related work. Sections 4.4.2 and 4.4.3 present the power models used for on- and off-chip memories respectively. Section 4.4.5 presents the power models used for the interconnect. Section 4.4.6 discusses the consequences for high-level power estimation models.

4.4.1   Related work on power models

In his Ph.D. dissertation [128], Robert Evans presents five approaches for modeling the energy consumption of static RAMs. A summary of this work can be found in [129]. Except for the first one, which is only suitable for relative comparisons, the models require lots of detailed information about the memory circuit, layout, and process technology. These models are therefore not suited for our high-level power estimations.

Dake Liu and Christer Svensson present in [246] a power model for on-chip memories. The power formula is the sum of five terms: bit line pre-charging, row decoding, row driving, column selection, and sense-amplification plus load driving. The bit line pre-charging is said to be the dominant term. It is proportional to the number of memory cells. Hence, in practice, it is very similar to Paul Landman's model which encompasses a black-box capacitance model for on-chip memories, and which is discussed in the next subsection. Since we have measured values for the parameters in Landman's model and not for Liu's model, we have chosen to use the Landman model for our experiments.

---

[10]This statement is true for any modern low-power RAM [195].

4.4.2 On-chip Memories

On-chip memories are usually generated by module generators from user supplied parameters such as the required bit width and word depth. As a consequence, for on-chip memories it is relatively easy to find models expressing power consumption in function of memory size, at least for recent libraries.

Traditionally only Static RAM (SRAM) was used for on-chip memory. State-of-the-art process technology, however, allows to combine also Dynamic RAM (DRAM) with logic on a single chip (e.g., the Mitsubishi [351] and the IRAM initiative of U.C.Berkeley [319]). Since this option is still very new, no public quantitative information is available about it. Therefore, only *on-chip* SRAM will be considered in this section. The following three subsections present the power models for on-chip SRAMs available to us.

4.4.2.1 Landman's Model. Landman has presented a number of black-box capacitance models for architectural power analysis [230]. Given its capacitance, the power consumption of a memory is calculated with Eq. 4.7.

The capacitance models for SRAMs (one for read and one for write operations) take two input parameters: the number of words in the memory, W, and the bit width, N. The capacitance models for a memory containing $W \times N$ storage cells are:

$$C_{read} = C_{read\,0} + C_{read\,1} \cdot W + C_{read\,2} \cdot N + C_{read\,3} \cdot W \cdot N \qquad (4.8)$$

$$C_{write} = C_{write\,0} + C_{write\,1} \cdot W + C_{write\,2} \cdot N + C_{write\,3} \cdot W \cdot N \qquad (4.9)$$

The coefficients for Berkeley's 1.2 μm SRAM are given in Table 4.1.

Table 4.1.   Capacitance values obtained for Berkeley's 1.2 μm SRAM.

	C_{x0} [fF]	C_{x1} [fF]	C_{x2} [fF]	C_{x3} [fF]
C_{read}	9707	108	1126	6
C_{write}	7994	117	759	9

The capacitance models are in fact effective capacitances, which means that the V_{swing}/V_{dd} ratio is incorporated. Therefore, the power formula combining both read and write accesses is given by:

$$P = \frac{1}{2} \cdot V_{dd}^2 \cdot (C_{read} \cdot f_{readaccess} + C_{write} \cdot f_{writeaccess}) \qquad (4.10)$$

The read model experiences larger errors than the write model, but still remains within 7.9% of switch level simulations, for memories ranging in size from 16 × 4 to 128 × 32. The read and write power in function of memory size are shown in figure 4.16 and figure 4.17 respectively. The memory power consumption increases linearly with increasing memory size for this Berkeley library.

4.4.2.2 Texas Instruments' Model. We have also used power models of Texas Instruments for their single and dual port on-chip memories. These models also take the word depth and the bit width of the memory as input. They estimate the power consumption per MHz for the given memory size. These values have to be multiplied with the real access frequency of the memory to get the power consumption in Watts. Unfortunately, we are not allowed to publish any absolute figures obtained with these proprietary models. We are allowed to publish relative figures, though. Figures 4.18 and 4.19 show the relative power dissipation in function of memory size for single and dual port memories respectively. Both figures are drawn on the same scale. Hence, these figures show that a dual port memory consumes roughly twice the amount of power of a single port memory. Compared to Landman's model, the power consumption of TI's memories are less dependent on memory size than Berkeley's SRAMs.

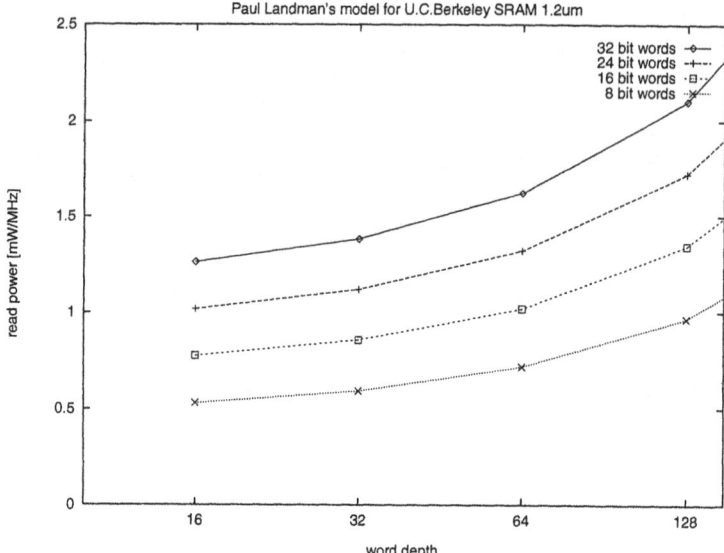

Figure 4.16.   Read power in function of memory size with Landman's model.

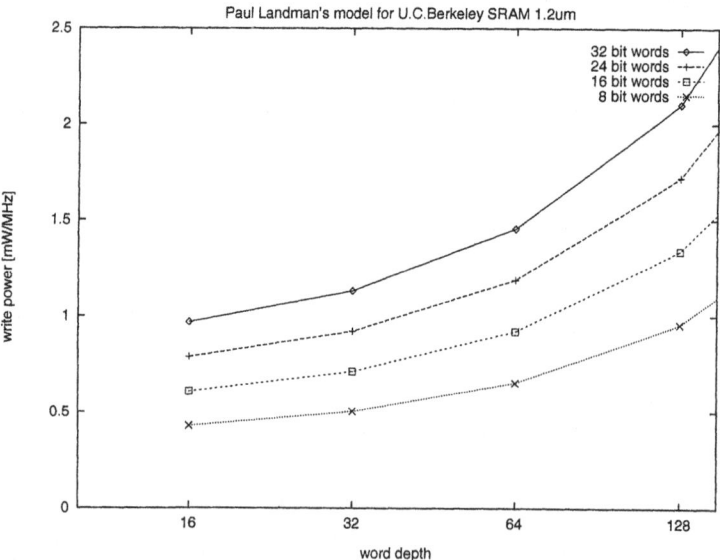

Figure 4.17.   Write power in function of memory size with Landman's model.

4.4.2.3 VLSI Technology Inc.'s Model.   VLSI Technology Inc. (VTI) presents tables of the power consumption of SRAMs generated with their RAM module generator in the data books for their 0.6 μm cell library operating at 5 V. The tables list the power consumption per MHz in function of the word depth and bit width of the memory. These figures have to be multiplied with the real access frequency of the memory to get the power consumption in Watts. Figure 4.20 shows the power in function of

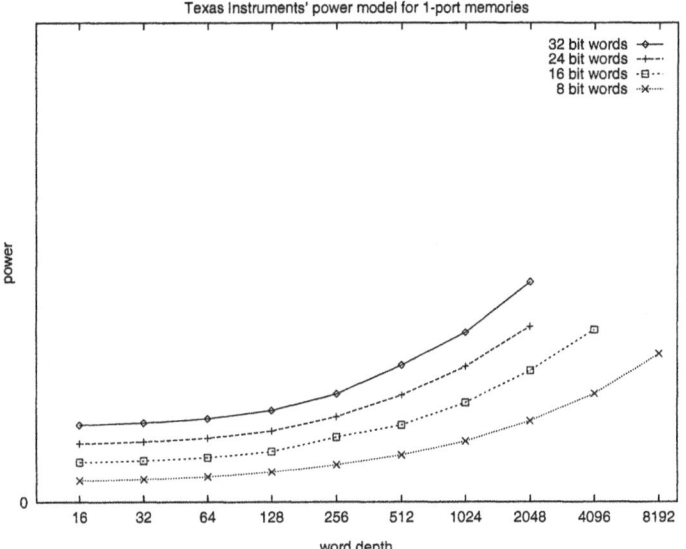

Figure 4.18.   Power vs. memory size with TI's model for single port memories.

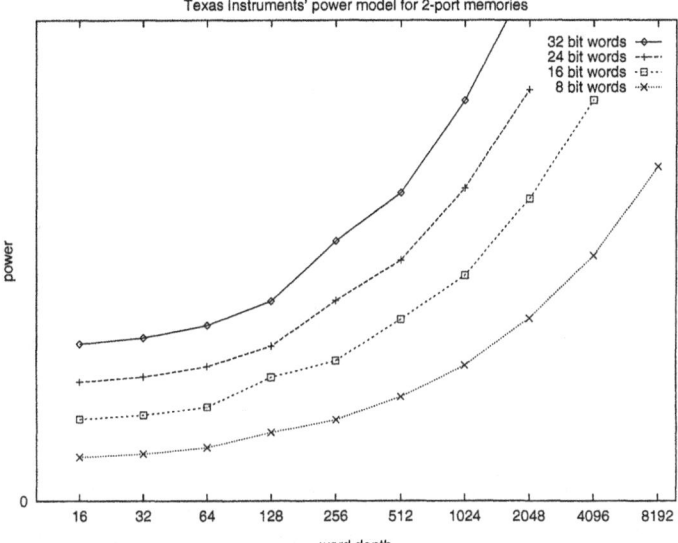

Figure 4.19.   Power vs. memory size with TI's model for two-port memories.

memory size for VTI's on-chip memories. The dependency on the word depth appears to be much less than the other two models presented in this chapter: only a near logarithmic curve is obtained which means that the memory matrix is almost not contributing and the main power contribution stems for the periphery.

In addition to the power consumption of the memory core, the model also provides data on the input capacitances of the memories which can be used in calculating the interconnect power.

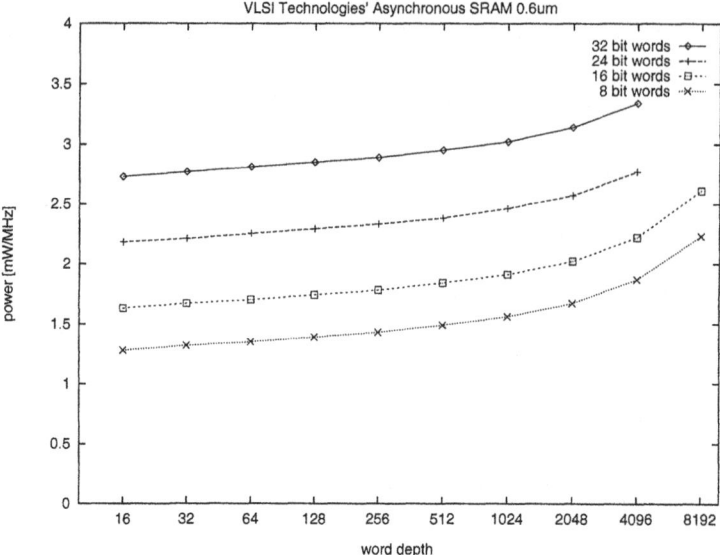

Figure 4.20.   Power in function of memory size with VTI's model.

4.4.3 Off-chip Memories

In practice, off-chip memories are off-the-shelf components that are available in a relatively small number of sizes: only a few different bit widths are commonly available and the word depth is usually a power of two. Modeling the power consumption for these components in function of their size is very difficult for the following reasons:

■ Memories are designed by different companies. This makes comparisons between designs made by different companies more difficult.

■ Memory designs are done on different moments in time: every time a new technology becomes available, the memory vendors design a new memory with a larger storage capacity than their previous designs. However, they usually do not redesign their previous smaller memories. Therefore, as a rule of thumb, the larger the the memory, the later it has been designed.

■ Memories of different size usually use a different processing technology: the larger the memory, the more advanced the processing technology.

■ Memories of different size use a different design: the larger the capacity, the more advanced the design. Also the design criteria change over time. In the past, power was not a big issue. Therefore, not much effort was spent in minimizing power consumption. Nowadays, however, memories are designed in a very power conscious way.

Because larger memories are often processed in a better technology, use more advanced designs, and are designed in a more power conscious way, they often consume the same or even less power than smaller ones. However, if the smaller memories would be redesigned in the same way, they would obviously consume less power than the bigger ones.

As a conclusion, we can assume that the power of off-chip memories is relatively independent of their actual size (except for very large size factors).

4.4.4 SRAM memories

For the large off-the-shelf units on a separate chip, we have assumed SRAMs because at the time this research was done, SRAMs consumed the least amount of power [195] and quantitative data about their power consumption was more readily available. As long as the specialized access modes of modern DRAMs (e.g., page mode and burst mode) are not used, there is not much difference between the power consumption models of SRAMs and DRAMs: only the resulting power figures are somewhat different.

For the SRAMs, we have used the model of a Fujitsu low-power memory [365]. It leads to 0.26 W for a 1 Mbit SRAM operating at 100 MHz at 5 V. Because this low-power RAM is however internally partitioned, the power will not really be significantly reduced by considering a smaller memory, as long as the memory remains larger than the partitions. The power budget [365] clearly shows that about 50% of the power in this low power version is consumed anyhow in the peripheral circuitry, which is much less dependent on the size. We use a power budget of 0.26 W for 100 MHz operation in all off-chip SRAMs. For lower access frequencies, this value will be scaled linearly.

4.4.5 Interconnect

The power dissipated in the interconnect is the third major component of the memory-architecture power consumption. With interconnect we mean here the data buses connecting the data paths and the memories, the address buses connecting the address generators to the memories, and the control buses connecting the controllers to the memories. In case of a hierarchical memory architecture, it also includes the data buses connecting memories on different layers in the architecture.

The on-chip interconnect power is difficult to estimate before the final placement and routing of the ASIC for the following reasons:

- The interconnect capacitance depends heavily on the wire length, which is unknown before placement and routing and extremely difficult to estimate because of the huge variance in practical designs.

- The interconnect capacitance depends on the metal layer on which the bus is being routed. Moreover, a bus does not have to be routed on a single metal layer, which complicates matters even more.

Also the off-chip interconnect is not easy to estimate early in the design:

- The placement and routing of the external components (ASIC + off-chip memories) on the printed circuit board or multi chip module should be known. Also here buses can be routed on different layers, which complicates estimating the bus wire capacitances.

- Fortunately, the distances depend more on packaging sizes for the ASIC and its directly related memories such that an estimate should be easier to obtain provided that the technology parameters are made available by the vendors.

For these reasons, only a very rough power model can be used before placement and routing, as is the case for the high-level optimizations considered in this book.

4.4.5.1 Total interconnect power. The power dissipated by the interconnect, $P_{Interconnect}$ [mW], is simply the sum of the power dissipated by each bus in the memory architecture:

$$P_{Interconnect} = \sum_{b \in B} P_{Bus}(b) \qquad (4.11)$$

where:

- B is the set of all buses b of the memory architecture, *and*

- $P_{Bus}(b)$ [mW] is the power dissipated by bus b.

4.4.5.2 Bus power. The power dissipated by a specific bus b, $P_{Bus}(b)$ [mW], is approximated by:

$$P_{Bus}(b) = W_{Bus}(b) \cdot a_{toggle}(b) \cdot f_{Access}(b) \cdot (C_{Driver}(b) + C_{Load}(b)) \cdot V_{dd}^2 \cdot 0.001 \qquad (4.12)$$

where:

- $W_{Bus}(b)$ [-] is the width of bus b,

- $C_{Driver}(b)$ [pF] is the capacitance of the buffer that drives bus b,

- $C_{Load}(b)$ [pF] is the total load capacitance of bus b, *and*

- the 0.001 factor is included to obtain the power in mW.

This is a simplified model. For instance, the voltage swing on the buses is usually no longer the full V_{dd} range, as is assumed in this formula. This is especially so for off-chip buses and for the largest on-chip buses.

4.4.5.3 Load capacitance. The load capacitance of a single wire of bus b can be estimated with the following formula:

$$
\begin{aligned}
C_{Load}(b) \;=\;\; & L_{On-chip}(b) \cdot C_{1mmon-chip} + C_{Pin} \\
+\;\; & L_{Off-chip}(b) \cdot C_{1mmoff-chip} + \sum_{m \in \mathcal{M}(b)} C_{In}(m)
\end{aligned}
\qquad (4.13)
$$

where:

- $L_{On-chip}(b)$ [mm] is the on-chip length of bus b,

- $C_{1mmon-chip}$ [pF/mm] is the capacitance per mm on-chip interconnect,

- $L_{Off-chip}(b)$ [mm] is the off-chip length of bus b,

- $C_{1mmoff-chip}$ [pF/mm] is the capacitance per mm off-chip interconnect,

- C_{Pin} [pF] is the capacitance of an I/O pin of the ASIC,

- $\mathcal{M}(b)$ is the set of memories connected to bus b, *and*

- $C_{In}(m)$ [pF] is the input-pin capacitance of memory m connected to b.

The I/O pin capacity, C_{Pin}, only has to be included for buses that go off-chip.

4.4.5.4 Wire length estimation. The wire length estimation is probably one of the most inaccurate parts in this model if the placement and routing is unknown. Consider the following cases:

1. **Placement and routing known.** In this case the exact wire lengths are known and can be used directly.

2. **Placement known, routing unknown.** In this case we can use the semi-perimeter method [352], which is an efficient and widely used method to estimate the wire length of a net during placement. The method consists of finding the smallest bounding rectangle that encloses all the pins of the net to be connected. The estimated wire-length of the interconnects is half the perimeter of this bounding rectangle. Assuming no winding of paths in actual routing, for two and three pin nets this is an exact approximation.[11] This method provides a good estimate for the most efficient wiring scheme, which is the Steiner tree [352]. For heavily congested chips this method always underestimates the wiring length.

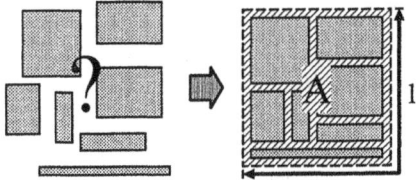

Figure 4.21.   Estimation of wire-length when placement and routing is unknown.

3. **Placement and routing unknown.** As the placement of these terminals is unknown in this case, we assume that the final layout will be a square containing all blocks and that the terminals can be everywhere within this square (figure 4.21). Only the area of the circuit blocks (memories, data-paths, controllers, address-generators, etc.) is taken into account. We could also try to take the area occupied by the interconnect into account, but as nowadays this overhead is quite small due to the availability of multiple layers of metal on which the buses can be routed, we have decided to neglect this contribution. Anyhow, the accuracy of the wire length estimation is limited by the fact that the placement is unknown. This leads to the following formula:

$$l = 2\sqrt{A}, \tag{4.14}$$

where:

- $A \ [mm^2]$ is the total area of the designed blocks, excluding the interconnect.

4.4.5.5 Capacitance per mm on-chip interconnect. Table 4.2 shows the interconnect capacitances ($C_{1mm \ on-chip} \ [pF]$) for the first metal layer as projected by Semiconductor Industry Associator's roadmap for semiconductors [344].

Table 4.2.   Interconnect power/performance characteristics.

	1995 0.35 μm	1998 0.25 μm	2001 0.18 μm	2004 0.13 μm	2007 0.10 μm
R (Metal1) (Ω/mm)	150	190	290	820	1340
C (Metal1) (pF/mm)	0.17	0.19	0.21	0.24	0.27

Notice that the capacitance values of the interconnect increases for smaller technologies. The reason being that the wire widths are usually scaled with a smaller factor than the thickness of the insulating layers and the spacing between the wires.

In reality, a bus is often routed on several layers of metal, each with their own characteristics. The capacitance values of the different interconnect layers are not dramatically different, though. For instance, for a minimum width wire (3λ) in a 1.2 μm Mosis technology, the capacitance values are 0.115 pF/mm and 0.098 pF/mm for Metal1 and Metal2 respectively [64]. The capacitance values decrease for higher interconnect layers, due to increasing insulator thickness and wire spacing.

4.4.5.6 Capacitance per mm off-chip interconnect. According to [297] the line capacitance ($C_{1mm off-chip} \ [pF/mm]$) of a printed wiring board is 0.1 pF/mm, which is very similar to the on-chip line capacitance.

Again this is only a crude approximation, as in reality there are several interconnection layers available, each with their own characteristics.

---

[11]Most practical circuits have either two or three terminal nets.

4.4.5.7   ASIC I/O pin capacitance.   Typical values for the ASIC I/O pin capacitance, C_{Pin} [pF], range from 1 to 10 pF depending on the packaging type.

4.4.5.8   Memory input capacitances.

- **On-chip memories**

   The only source of data on input capacitances for on-chip memories available to us, is VTI's data book. Table 4.3 shows the ranges of the capacitance values for the different I/O pins of VTI's 0.6 μm SRAM.

   Table 4.3.   Capacitance values for I/O pins of Asynchronous SRAM (on-chip).

	C_{AddrIn}	C_{DataIn}	C_{WE}	C_{CS}	C_{OE}
minimum (pF)	0.04	0.22	0.17	0.05	0.13
maximum (pF)	3.16	1.77	1.69	2.87	1.43

- **Off-chip memories**

   To get an idea about the input capacitances of off-chip memories, we have included table 4.4 which shows the pin capacitances for a typical DRAM chip. Keep in mind, though, that the pin capacitances depend on the memory's package type.

   Table 4.4.   Pin capacitances of Micron 1Meg x 16bit Fast Page DRAM (off-chip)

Parameter	Max. Capacitance
Address Input Capacitance	5 pF
Data I/O Capacitance	7 pF

4.4.5.9   Driver Capacitance.   For the calculation of the bus driver capacitance, $C_{Driver}(b)$ [pF], we take the approach of [246]. They calculated that the total capacitance in the buffer is about 30% of its total load [185]. This is obtained from the assumption that the size ratio in the inverter chain is 4. Thus, the last inverter that drives the load has a capacitance of 1/4 of its load. The total capacitance ratio of the inverter chain is $1/4 + 1/16 + 1/64 + \cdots \approx 0.3$. This leads to the following formula:

$$C_{Driver}(b) = 0.3 \cdot C_{Load}(b) \qquad (4.15)$$

4.4.6   Consequences for High-Level Power Models

The following conclusions can be drawn for high-level power models:

- Memory power consumption is proportional to the real access frequency of the memory.

- For on-chip memories, the power consumption of a memory increases with the size of the memory. The dependency on size is something between linear and logarithmic, depending on the memory library used.

- For off-chip memories, the power consumption can be assumed to be independent of memory size. This is assumption is quite realistic considering that modern low-power memories are partitioned in lots of smaller banks. The power consumption depends primarily on the size of the banks and less on the number of banks.

- Multi-port memories are very power hungry: a two port memory consumes about twice as much energy per memory access as a single port memory storing the same amount of data.

- Off-chip interconnect is difficult to estimate early in the design flow, but relatively important compared to the internal power consumption of modern low-power memories. A fixed capacitance value could be used to model the off-chip power interconnect in high-level models. A value of 20 to 30 pF per bus line seems appropriate:

6 pF	memory I/O pin
3 pF	30 mm off-chip interconnect
5 pF	ASIC I/O pin
1 pF	5 mm on-chip interconnect
5 pF	bus driver
+ ———	
20 pF	total

A value of 30 pF corresponds to interconnect lengths that are three times longer than the ones used in the previous calculation.

- On-chip interconnect is even more difficult to estimate early in the design flow, but relatively unimportant compared to the internal power consumption of on-chip memories. Therefore we can ignore it in high-level power estimations.

4.4.7 Summary on power models

Only dynamic power consumption due to capacitive loading is considered in this book, as the other power components are negligible provided the circuit design is done properly. The power models used for estimating the power consumption of on-chip memories, off-chip memories, and interconnect are presented. Memory and interconnect power consumption is proportional to the real access frequency of the memories. The power consumption of *on-chip* memories increases with size. The power consumption of *off-chip* memories is considered to be independent of size. Interconnect power is difficult to estimate accurately early in the design flow. For the off-chip interconnect, a fixed capacitance value per bus wire is used. The on-chip interconnect can be ignored in high-level power estimations.

5 GEOMETRICAL PROGRAM MODELING

This chapter describes the geometrical data- and control-flow modeling techniques for data-dominated applications. The literature on this topic is very scattered, and many different formalisms are being used. We present a unified view on these techniques and formalisms. Next, we show how these models can be used and extended to accurately describe the occupation of memories used for storing multi-dimensional data. Moreover, we derive the necessary and sufficient constraints that have to be satisfied by the design parameters (i.e. execution and storage order) in order to arrive at a valid algorithm implementation (either in hardware or in software). The optimization techniques presented in the subsequent chapters are all partly based on the models presented here, wherever they have to manipulate or analyse the loop nest code and the array signals.

5.1 AFFINE MODELS

In this section we present the "classical" affine models that have become well-known in the parallel compiler and regular array synthesis world.[1]

5.1.1 Iteration domain

In the parallel compiler and regular array synthesis communities the main interest lies in the accurate modeling of the execution of program statements and the dependencies between these executions in order to be able to exploit the potential parallelism. This has lead to the definition of the concept of an **iteration domain** [134] (also called *iteration space* [325, 368, 239, 431, 433, 304, 34], *index set* [229, 368], *node space* [419, 421], *computation domain* [94], or *index space* [236, 388]).

An iteration domain is a mathematical description of the executions of a statement in a program. It is a geometrical domain in which each point with integral coordinates represents exactly one execution of the statement. Each statement has its own iteration domain, and the set of points in this domain represents all the executions of this statement. In the sequel, we refer to the execution of a specific statement as an **operation**, as in [134].

---

[1]Note that we will only discuss the *modeling* techniques, and *not* the corresponding *analysis* and *manipulation* techniques.

The different mathematical dimensions of such a domain generally correspond to the different iterators of the loops surrounding the corresponding statement (but this is not strictly required). The domain itself is described by means of a set of constraints corresponding to the boundaries of these iterators. A small example is given next.

Example 5.1.1

```
    for ( i = 0; i < 10; i++ )
        for ( j = 0; j < 10; j++ )
S1:     A[i][j] = f(B[i][j]);
```

The dimensions of the iteration domain of statement S1 are denoted by i and j (which, by coincidence, resemble the names of the iterators). The constraints on these dimensions can be extracted from the boundaries of the two loops: $0 \leq i \leq 9$ and $0 \leq j \leq 9$. Consequently, we can describe the iteration domain of statement S1 (denoted by $\mathbf{D}_1^{iter}$) as follows:

$$\mathbf{D}_1^{iter} = \{ \ [i,j] \ | \ 0 \leq i \leq 9 \ \wedge \ 0 \leq j \leq 9 \ \wedge \ [i,j] \in \mathbb{Z}^2 \ \}$$

So in this case, the iteration domain is a two-dimensional domain with dimensions i and j, consisting of points with integral coordinates and bounded by the given constraints. Each of these points represents exactly one execution of statement S1. A graphical representation of this domain is given in Fig. 5.1.

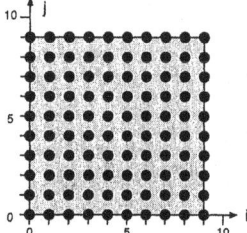

Figure 5.1.   The iteration domain of example 5.1.1.

In general, the loop boundaries are not constant however. Moreover, the program may contain conditional statements. The iteration domain model can easily incorporate these properties though, provided that the loop boundaries and conditions are functions of the iterators of the surrounding loops. This is shown in the next example.

Example 5.1.2

```
    for ( i = 0; i < 10; i++ )
        for ( j = 0; j <= i; j++ )
            if ( i + j >= 10 )
S1:             A[i][j] = f(B[i][j]);
```

In this case, the upper boundary of iterator j is not a constant, but it is a function of i. Moreover, statement S1 is only executed if the condition i + j >= 10 is satisfied. Consequently, we can represent the iteration domain as follows:

$$\mathbf{D}_1^{iter} = \{ \ [i,j] \ | \ 0 \leq i \leq 9 \ \wedge \ 0 \leq j \leq i \ \wedge \ i+j \geq 10 \ \wedge \ [i,j] \in \mathbb{Z}^2 \ \}$$

A graphical representation is shown in Fig. 5.2.

Another important concept is that of **auxiliary dimensions,** which are also called *wildcards* [329]. Sometimes it is not possible to represent an iteration domain only by constraints on its dimensions. In

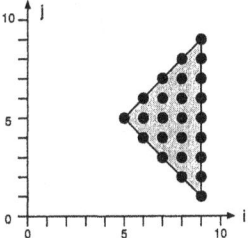

Figure 5.2.   The iteration domain of example 5.1.2.

those cases, it may be necessary to introduce extra dimensions in order to be able to correctly describe the iteration domain.[2] This is for instance the case if the loop iterators are not incremented by 1 after each execution of the loop, as shown in the next example.

Example 5.1.3

```
      for ( i = 0; i < 10; i += 2 )
        for ( j = 0; j <= 10; j += 3 )
          if ( i + j <= 12 )
S1:         A[i][j] = f(B[i][j]);
```

In this case, if we can only express constraints on the 2 dimensions i and j of our iteration domain, we cannot model the iteration domain exactly. Therefore we have to introduce 2 auxiliary dimensions (α and β), which are existentially qualified and which allow us to express more specific constraints. In this particular case, we have to express that i must be a multiple of 2 and j must be a multiple of 3:

$$D_1^{\text{iter}} = \{ \ [i,j] \mid \exists \, \alpha, \beta \text{ s.t. } 0 \leq i \leq 9 \ \wedge \ 0 \leq j \leq 9 \ \wedge \ i + j <= 12 \ \wedge$$
$$i = 2\alpha \ \wedge \ j = 3\beta \ \wedge \ [i,j] \in \mathbb{Z}^2 \ \wedge \ [\alpha,\beta] \in \mathbb{Z}^2 \ \}$$

Note that the real iteration domain only contains the dimensions i and j, i.e. it is the orthogonal projection of the 4-dimensional domain containing i, j, α and β along the α and β dimensions onto the i and j dimensions. A graphical representation of the (projected) iteration domain is shown in Fig. 5.3. Note that the domain is no longer dense.

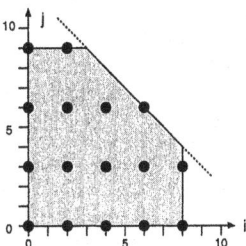

Figure 5.3.   The iteration domain of example 5.1.3.

In general the auxiliary dimensions of a domain are not unique. They only act as a mathematical aid for describing specific constraints. Any combination of auxiliary dimensions and constraints that

---

[2]Note that wildcards can also be used in other contexts, e.g. dependency analysis.

results in the same orthogonal projection onto the real dimensions is therefore equivalent from a modeling point of view.

Of course, from a manipulation or transformation point of view, this is not true. One representation can be more efficient than another one for certain purposes. For instance, certain techniques can only operate efficiently on dense domains or domains that are described as mappings of dense domains. The description of D_1^{iter} in example 5.1.3 does not satisfy this requirement. In [142, 29] techniques are presented for converting general affine domain descriptions to "dense descriptions". A possible outcome for D_1^{iter} in example 5.1.3 could be the following:

$$D_{1,dense}^{iter} = \{ [\alpha, \beta] \mid 0 \leq \alpha \leq 4 \land 0 \leq \beta \leq 3 \land 2\alpha + 3\beta \leq 12 \land [\alpha, \beta] \in \mathbb{Z}^2 \}$$
$$D_1^{iter} = \{ [i, j] \mid \exists [\alpha, \beta] \in D_{1,dense}^{iter} \text{ s.t. } i = 2\alpha \land j = 3\beta \land [i, j] \in \mathbb{Z}^2 \}$$

Note that the inequalities originally constraining i and j have now been converted into constraints on α and β.

Sometimes the reasons for the introduction of auxiliary dimensions are less straightforward. In [142] it is described how one can deal with modulo operations in array index expressions in an efficient way through the introduction of extra (auxiliary) dimensions. We come back to this in section 5.2 on page 90.

Mainly for technical reasons, the iteration domain models being used in the parallel compiler and regular array synthesis were usually restricted to be described only by affine constraints. By doing so, many of the well-known results from linear algebra can be used for analysis and optimization of programs. The class of programs that can be handled by these restricted models is quite large, so one could live with that (at least until recently). However, there is no fundamental reason why the *models* should not be capable of representing arbitrary functions instead of only affine functions. Of course there have been some practical reasons, i.e. if non-affine functions are present, the *techniques* for analyzing and optimizing the corresponding programs may have to be extended and may become computationally more complex. Naturally, the required accuracy of the models heavily depends on the goals one wants to achieve. Einstein's famous quote "Everything should be as simple as possible, but not simpler." is applicable here also.

Before we discuss the model extensions in more detail, we describe the concepts of variable, definition and operand domains next.

5.1.2   Variable, definition and operand domains

As mentioned before, the parallel compiler and regular array synthesis communities are mainly interested in the (parallel) execution of statements. Therefore, it is usually sufficient for them to only model the iteration domains explicitly.

In contrast, in our memory-optimizing context, we are mainly interested in the memory related issues such as memory accesses and storage requirements. That is why we have chosen to model not only the executions of the statements of a program (by means of iteration domains), but also the accesses to the *program variables*, and especially the array variables. This has lead to the concepts of **definition domains, operand domains** and **variable domains**[3] (definition domains and operand domains have also been called *operation space* and *operand space* respectively [419, 421]). Note that in the remainder of this text, we refer to program variables simply as variables, i.e. drop the "program" prefix. There should be no confusion with mathematical variables, which are used in the mathematical descriptions of the different domains.

Typically the variables being written or read during the execution of a program are grouped into sets of similar variables, which are called arrays. These arrays are arranged as multi-dimensional structures, in which each individual variable can be addressed by a unique set of indices. These

---

[3]Although variable domains have never been explicitly described, they have always implicitly been taken into account. Sometimes the term "signal domain" is also used for it in a single-assignment context, but we will not use this term because the ambiguity of the term "signal".

multi-dimensional structures are ideally suited to be modeled by geometrical domains also. Therefore the concept of a **variable domain** was introduced. A variable domain is a mathematical description of an array of variables. Each point with integer coordinates in this domain corresponds to exactly one variable in the array.

During the execution of a program, not every statement accesses each array of variables completely. Typically, the executions of a statement only access part of one or a few arrays. The **definition** and **operand domains** of a statement describe which variables are being accessed during *all* possible executions of that statement. Each point with integer coordinates in these domains corresponds to exactly one variable that is being written (in case of a definition domain) or read (in case of an operand domain). Possibly, a variable is written or read multiple times during the execution of a program, even by the same statement.

The relations between the executions of the statements and the variables that are being written or read, are represented by means of mathematical **mappings** between the dimensions of the iteration domains and the dimensions of the definition or operand domains. We refer to these mappings as the **definition mappings** and **operand mappings** respectively. An example illustrating these concepts is given next.

Example 5.1.4

```
      int A[10][20];
      int B[10];
      for ( i = 0; i < 10; i++ )
          for ( j = 0; j < 10; j++ )
S1:           A[i][2*j] = f(B[i]);
```

First, we can describe the variable domains of arrays A and B, denoted by $\mathbf{D}_A^{var}$ and $\mathbf{D}_B^{var}$ respectively. The boundaries of the domains can be extracted from the declarations as follows (note that a_1, a_2, and b are arbitrarily chosen names):

$$\mathbf{D}_A^{var} = \{ \ [a_1, a_2] \mid 0 \le a_1 \le 9 \ \wedge \ 0 \le a_2 \le 19 \ \wedge \ [a_1, a_2] \in \mathbb{Z}^2 \ \}$$
$$\mathbf{D}_B^{var} = \{ \ b \mid 0 \le b \le 9 \ \wedge \ b \in \mathbb{Z} \ \}$$

Next, given the iteration domain of statement S1 and the index expressions of the array accesses, we can extract the descriptions of the definition and operand mappings (denoted by $\mathbf{M}_{11A}^{def}$ and $\mathbf{M}_{11B}^{oper}$ respectively) and definition and operand domains (denoted by $\mathbf{D}_{11A}^{def}$ and $\mathbf{D}_{11B}^{oper}$ respectively), which are the result of applying the respective mappings to the iteration domain:

$$\mathbf{D}_1^{iter} = \{ \ [i,j] \mid 0 \le i \le 9 \ \wedge \ 0 \le j \le 9 \ \wedge \ [i,j] \in \mathbb{Z}^2 \ \}$$
$$\mathbf{M}_{11A}^{def} = \{ \ [i,j] \to [a_1, a_2] \mid a_1 = i \ \wedge \ a_2 = 2j \ \wedge \ [a_1, a_2] \in \mathbb{Z}^2 \ \}$$
$$\mathbf{D}_{11A}^{def} = \mathbf{M}_{11A}^{def}(\mathbf{D}_1^{iter})$$
$$= \{ \ [a_1, a_2] \mid \exists \ [i,j] \in \mathbf{D}_1^{iter} \text{ s.t. } a_1 = i \ \wedge \ a_2 = 2j \ \wedge \ [a_1, a_2] \in \mathbb{Z}^2 \ \}$$
$$= \{ \ [a_1, a_2] \mid \exists \ [i,j] \in \mathbb{Z}^2 \text{ s.t. } a_1 = i \ \wedge \ a_2 = 2j \ \wedge$$
$$0 \le i \le 9 \ \wedge \ 0 \le j \le 9 \ \wedge \ [a_1, a_2] \in \mathbb{Z}^2 \ \}$$
$$\mathbf{M}_{11B}^{oper} = \{ \ [i,j] \to b \mid b = i \ \wedge \ b \in \mathbb{Z} \ \}$$
$$\mathbf{D}_{11B}^{oper} = \mathbf{M}_{11B}^{oper}(\mathbf{D}_1^{iter})$$
$$= \{ \ b \mid \exists \ [i,j] \in \mathbf{D}_1^{iter} \text{ s.t. } b = i \ \wedge \ b \in \mathbb{Z} \ \}$$
$$= \{ \ b \mid \exists \ [i,j] \in \mathbb{Z}^2 \text{ s.t. } b = i \ \wedge \ 0 \le i \le 9 \ \wedge \ 0 \le j \le 9 \ \wedge \ b \in \mathbb{Z} \ \}$$

The first index in the subscripts of $\mathbf{D}_{11A}^{def}$ and $\mathbf{D}_{11B}^{oper}$ refers to statement S1. The second index refers to the position of the definition or operand respectively (there can be multiple definitions and/or operands in the same statement), while the third index refers to the array to which the domains belongs. A graphical representation of these domains is given in Fig. 5.4.

Note that in general, also the definition or operand domains may require the use of extra auxiliary dimensions, next to the dimensions of the corresponding iteration domain, as shown in the next example:

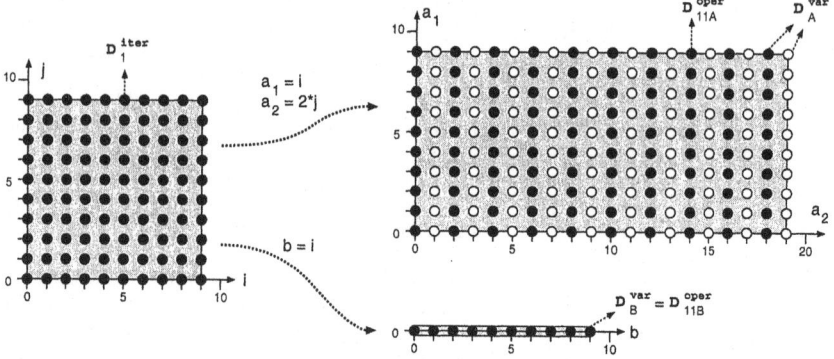

Figure 5.4.   The iteration, variable, definition and operand domains of example 5.1.4.

Example 5.1.5

Consider the following statement, which states that certain elements of array A are function of certain elements of array B (i.e. f is a multiple-input/multiple-output function):

```
S1: A[0,1,2,3,...,9] = f(B[0,2,4,6,...,8],B[20,22,...,28]);
```

For this example we can represent the iteration, definition and operand mappings and domains as follows:

$$\mathbf{D}_1^{iter} = \{\ \mathit{0}\ \}$$
$$\mathbf{M}_{11A}^{def} = \{\ \mathit{0} \to a \mid 0 \leq a \leq 9 \ \wedge \ a \in \mathbb{Z}\ \}$$
$$\mathbf{D}_{11A}^{def} = \{\ a \mid 0 \leq a \leq 9 \ \wedge \ a \in \mathbb{Z}\ \}$$
$$\mathbf{M}_{11B}^{oper} = \{\ \mathit{0} \to b \mid \exists\ \alpha \in \mathbb{Z} \ \text{s.t.}\ b = 2\alpha \ \wedge \ 0 \leq b \leq 8 \ \wedge \ b \in \mathbb{Z}\ \}$$
$$\mathbf{D}_{11B}^{oper} = \{\ b \mid \exists\ \alpha \in \mathbb{Z} \ \text{s.t.}\ b = 2\alpha \ \wedge \ 0 \leq b \leq 8 \ \wedge \ b \in \mathbb{Z}\ \}$$
$$\mathbf{M}_{12B}^{oper} = \{\ \mathit{0} \to b \mid \exists\ \beta \in \mathbb{Z} \ \text{s.t.}\ b = 2\beta \ \wedge \ 20 \leq b \leq 28 \ \wedge \ b \in \mathbb{Z}\ \}$$
$$\mathbf{D}_{12B}^{oper} = \{\ b \mid \exists\ \beta \in \mathbb{Z} \ \text{s.t.}\ b = 2\beta \ \wedge \ 20 \leq b \leq 28 \ \wedge \ b \in \mathbb{Z}\ \}$$

Note that the iteration domain consists of only 1 point in this case, so we don't need any dimensions for it, i.e. the domain only contains the null vector $\mathit{0}$.

5.2   NON-AFFINE MODELS

In this section we extend the affine models from the previous section to include also non-affine and even non-manifest[4] control and data-flow.

5.2.1   Manifest non-affine models

From section 5.1, we know that we can describe programs by means of iteration, variable, definition, and operand domains/mappings, provided that the program is affine and manifest. In general, these domains and mappings have the following form:

$$\mathbf{D}_i^{iter} = \{\ \mathbf{i} \mid \exists\ \alpha \in \mathbb{Z}^{n_{i2}} \ \text{s.t.}\ \mathbf{C}_i^{iter}(\mathbf{i}, \alpha) \geq 0 \ \wedge \ \mathbf{i} \in \mathbb{Z}^{n_{i1}}\ \}$$
$$\mathbf{D}_m^{var} = \{\ \mathbf{s} \mid \exists\ \beta \in \mathbb{Z}^{n_{m2}} \ \text{s.t.}\ \mathbf{C}_m^{var}(\mathbf{s}, \beta) \geq 0 \ \wedge \ \mathbf{s} \in \mathbb{Z}^{n_{m1}}\ \}$$

---

[4]Manifest means that both the control-flow and the data-flow ($\neq$ data values) of the program are known at compile time.

$$\mathbf{M}^{\mathbf{def}}_{ikm} = \{\ \mathbf{i} \to \mathbf{d}\ |\ \exists\,\gamma \in \mathbb{Z}^{nd_{ikm}}\ \text{s.t.}\ \mathbf{C}^{\mathbf{def}}_{ikm}(\mathbf{d},\mathbf{i},\gamma) \geq 0\ \wedge\ \mathbf{d} \in \mathbb{Z}^{n_{m1}}\ \}$$

$$\mathbf{D}^{\mathbf{def}}_{ikm} = \{\ \mathbf{d}\ |\ \exists\,\mathbf{i} \in \mathbf{D}^{\mathbf{iter}}_i, \gamma \in \mathbb{Z}^{nd_{ikm}}\ \text{s.t.}\ \mathbf{C}^{\mathbf{def}}_{ikm}(\mathbf{d},\mathbf{i},\gamma) \geq 0\ \wedge\ \mathbf{d} \in \mathbb{Z}^{n_{m1}}\ \}$$

$$\mathbf{M}^{\mathbf{oper}}_{jlm} = \{\ \mathbf{j} \to \mathbf{o}\ |\ \exists\,\delta \in \mathbb{Z}^{no_{jlm}}\ \text{s.t.}\ \mathbf{C}^{\mathbf{oper}}_{jlm}(\mathbf{o},\mathbf{j},\delta) \geq 0\ \wedge\ \mathbf{o} \in \mathbb{Z}^{n_{m1}}\ \}$$

$$\mathbf{D}^{\mathbf{oper}}_{jlm} = \{\ \mathbf{o}\ |\ \exists\,\mathbf{j} \in \mathbf{D}^{\mathbf{iter}}_j, \delta \in \mathbb{Z}^{no_{jlm}}\ \text{s.t.}\ \mathbf{C}^{\mathbf{oper}}_{jlm}(\mathbf{o},\mathbf{j},\delta) \geq 0\ \wedge\ \mathbf{o} \in \mathbb{Z}^{n_{m1}}\ \}$$

In these equations, $\mathbf{C}^{\mathbf{iter}}_i()$, $\mathbf{C}^{\mathbf{var}}_m()$, $\mathbf{C}^{\mathbf{def}}_{ikm}()$, and $\mathbf{C}^{\mathbf{oper}}_{jlm}()$ represent sets of constraint functions of $\mathbf{D}^{\mathbf{iter}}_i$, $\mathbf{D}^{\mathbf{var}}_m$, $\mathbf{D}^{\mathbf{def}}_{ikm}$, and $\mathbf{D}^{\mathbf{oper}}_{jlm}$ respectively. Note that in practice, many of the inequalities described by these constraint functions can be reduced to equalities, but from a modeling point of view this makes no difference. Therefore, and in order to keep the notation a simple as possible, we continue to use the inequalities notation. Also note that the main dimensions of the iteration domains can be seen as auxiliary dimensions of the corresponding definition and operand domains. In many cases, these dimensions can be eliminated, as they are existentially quantified (just as any other auxiliary dimension).

In the sequel, we do not write down the definition and operand mappings explicitly anymore, unless there is a good reason for it, because the information about these mappings can be readily extracted from the (non-simplified) definition and operand domains descriptions. However, this does not mean that this information is not important! For certain purposes it may be even crucial. For instance, the mappings from the iteration domains to the operand/definition domains indicate what array elements are accessed by what operations and vice versa. The (simplified) domain descriptions alone do not contain the information about these relations.

As already stated in section 5.1, for technical reasons the constraints were required to be affine (at least until recently). However, these mathematical *models* allow to describe *all* practical programs that are manifest, i.e. non-recursive programs that contain for-loops with manifest boundaries, manifest conditions and array accesses with manifest index expressions, *even if these are not affine*. Note that in the worst case one may have to rely on a complete enumeration of the points in the domains, but even an enumeration fits within the models. An (extreme) example is given next:

Example 5.2.1

```
    for ( i = 1; i < 20; i *= 2 )      // multiplication by 2
        for ( j = i; j < i^3; j += i )      // variable increment
S1:      A[i^2+j*i*7+8] = f(B[j-i..j+j*i+20]);   // range
```

The following expressions describe the corresponding domains:

$$\mathbf{D}^{\mathbf{iter}}_1 = \{\ [i,j]\ |\ \exists\,\alpha \in \mathbb{Z}\ \text{s.t.}\ 1 \leq i \leq 19\ \wedge\ i = 2^\alpha\ \wedge$$
$$i \leq j \leq i^3 - 1\ \wedge\ j \bmod i = 0\ \wedge\ [i,j] \in \mathbb{Z}^2\ \}$$

$$\mathbf{D}^{\mathbf{def}}_{11A} = \{\ d\ |\ \exists\,[i,j] \in \mathbf{D}^{\mathbf{iter}}_1\ \text{s.t.}\ d = i^2 + j*i*7 + 8\ \wedge\ d \in \mathbb{Z}\ \}$$

$$\mathbf{D}^{\mathbf{oper}}_{11B} = \{\ o\ |\ \exists\,[i,j] \in \mathbf{D}^{\mathbf{iter}}_1, \beta \in \mathbb{Z}\ \text{s.t.}\ o = j - i + \beta\ \wedge$$
$$0 \leq \beta \leq j*i + 20 + i\ \wedge\ o \in \mathbb{Z}\ \}$$

The drawing of a graphical representation of these domains is left as an exercise to the enthusiastic reader.

Note that it is probably possible to find even more esoteric examples for which the mathematical modeling without enumeration becomes very difficult, if not impossible. But nevertheless, an enumeration is an equally valid description. For instance, $\mathbf{D}^{\mathbf{iter}}_1$ can be described as follows:

$$\mathbf{D}^{\mathbf{iter}}_1 = \{\ [i,j]\ |\ (i = 2\ \wedge\ (j = 2\ \vee\ j = 4\ \vee\ j = 6))\ \vee$$
$$(i = 4\ \wedge\ (j = 4\ \vee\ j = 8\ \vee\ j = 12\ \vee\ j = 16\ \vee\ ...))\ \vee$$
$$...\ \wedge\ [i,j] \in \mathbb{Z}^2\ \}$$

Of course, this kind of descriptions is not encouraged as it can have a disastrous effect on the performance of any software implementation that does not recognize the regularity.

Note that this should not be seen as a model extension. In the past, most practical implementations were limited to contain only affine functions, but the underlying model has always allowed arbitrary manifest functions (although probably only very few people have realized that, as few people seem to understand the difference between a model and an implementation). Moreover, the techniques for analysis and manipulation of these models could only handle affine functions, so there was no practical use or need to extend the implementations, at least not until recently.

As promised in section 5.1 on page 86, we now look in more detail at the techniques for modeling a special class of non-affine manifest index expressions, namely piece-wise linear indexing caused by modulo operations in index expressions. In [142] it is indicated how a loop (nest) can be rewritten in order to get rid of the modulo operation, as shown in the next example.

Example 5.2.2

```
    for ( x = 0; x < 20; x++ )
S1:    A[x] = B[x mod 4];
```

Note that the modulo operation in the index expression of array B causes a piece-wise linear variation of the index. In [142], it is shown that this loop can be rewritten as follows, without altering the behaviour:

```
    for ( x1 = 0; x1 < 5; x1++ )
        for ( x2 = 0; x2 < 4; x2++ )
S1:        A[x1*4 + x2] = B[x2];
```

However, this technique requires that the original loop be transformed *into another one, even only for the purpose of* modeling *the program.[5] Moreover, it quickly becomes rather complex when there are modulo operations present in more than one index expression, with incompatible periods (e.g. the* A[x mod 7] = B[x mod 4] *statement requires already two additional loops, with non-trivial boundaries).*

A better alternative from a modeling point of view is to introduce auxiliary dimensions in the descriptions of the definition and/or operand domains. For instance, for the original example, the straightforward domain descriptions are the following:

$$D_1^{\text{iter}} = \{ \ x \mid 0 \leq x \leq 19 \ \wedge \ x \in \mathbb{Z} \ \}$$
$$D_{11A}^{\text{def}} = \{ \ a \mid \exists \ x \in D_1^{\text{iter}} \ \text{s.t.} \ a = x \ \wedge \ a \in \mathbb{Z} \ \}$$
$$D_{11B}^{\text{oper}} = \{ \ b \mid \exists \ x \in D_1^{\text{iter}} \ \text{s.t.} \ b = x \bmod 4 \ \wedge \ b \in \mathbb{Z} \ \}$$

By introducing auxiliary dimensions, D_{11B}^{oper} *can be rewritten as follows:*

$$D_{11B}^{\text{oper}} = \{ \ b \mid \exists \ x \in D_1^{\text{iter}}, [x_1, x_2] \in \mathbb{Z}^2 \ \text{s.t.} \ b = x_2 \ \wedge$$
$$x = 4x_1 + x_2 \ \wedge \ 0 \leq x_1 \leq 4 \ \wedge \ 0 \leq x_2 \leq 3 \ \wedge \ b \in \mathbb{Z} \ \}$$

This alternative has the advantage that the loops can be left in their original form, and that it can be applied to each operand or definition domain separately.

A potential disadvantage of this technique is that the resulting domain descriptions are no longer "dense" (see the discussion on page 86). Of course, they can always be converted to dense descriptions using the techniques presented in [142, 29].

The main problem left is the modeling of non-manifest programming constructs. This problem is tackled in the next section.

5.2.2 Non-manifest models

The behaviour of many algorithms depends on data values that are not known until the algorithm is executed (e.g. input data). Consequently, the programs implementing these algorithms are not

---

[5]One might even wonder how one is able to perform this transformation if one is not able to model the initial modulo operation
...

manifest, i.e. the data and/or control flow may be different each time the program is executed. The models presented above are not capable of describing this kind of programs (or at least not very accurately) and therefore require further extensions.

One of the oldest and best known extensions is the use of symbolic constants. In scientific computing the problem of optimizing programs for which some of the structure parameters are not yet known is occurring frequently. It is for instance possible to do data-flow analysis in the presence of unknown structure parameters by treating them as symbolic constants [133]. An example of how these symbolic constants can be included in our models is given next.

Example 5.2.3

```
      int A[N][N], B[N*N]; /* N is an unknown constant */
      for ( i = 0; i < N; i++ )
          for ( j = 0; j <= i; j++ )
S1:       A[i][j] = f(B[i*N+j]);
```

The corresponding domain descriptions are the following:

$$\mathbf{D}_1^{\mathbf{iter}} = \{\ [i,j]\ |\ 0 \leq i \leq N-1 \ \wedge\ 0 \leq j \leq i\ \wedge\ [i,j] \in \mathbb{Z}^2\ \}$$
$$\mathbf{D}_A^{\mathbf{var}} = \{\ [a_1,a_2]\ |\ 0 \leq a_1 \leq N-1 \ \wedge\ 0 \leq a_2 \leq N-1 \ \wedge\ [a_1,a_2] \in \mathbb{Z}^2\ \}$$
$$\mathbf{D}_B^{\mathbf{var}} = \{\ b\ |\ 0 \leq b \leq N^2 \ \wedge\ b \in \mathbb{Z}\ \}$$
$$\mathbf{D}_{11A}^{\mathbf{def}} = \{\ [a_1,a_2]\ |\ \exists\ [i,j] \in \mathbf{D}_1^{\mathbf{iter}}\ \text{s.t.}\ a_1 = i\ \wedge\ a_2 = j\ \wedge\ [a_1,a_2] \in \mathbb{Z}^2\ \}$$
$$\mathbf{D}_{11B}^{\mathbf{oper}} = \{\ b\ |\ \exists\ [i,j] \in \mathbf{D}_1^{\mathbf{iter}}\ \text{s.t.}\ b = i*N+j\ \wedge\ b \in \mathbb{Z}\ \}$$

Recently, this concept of symbolic constants and the corresponding analysis techniques have been extended to include *any* value that is not known at compile time [332, 256, 77, 172]. Depending on the author, these unknowns are called *dynamically defined symbolic constants* [256], or *hidden variables* [77] or *placeholder variables* [172]. The main difference between the traditional symbolic constants and the more general hidden variables is that hidden variables may not have a constant value during the execution of the program. But even though the actual values are not known at compile time, the rate at which these values changes, is usually known. Since the information about this rate can be crucial in order to perform optimizations, we should be able to model programs containing hidden variables as accurately as possible, i.e. we should be able to describe the rate at which they change.

Before introducing the general model, we will first look at an example that explains the difference between the classical symbolic constants and hidden variables.

Example 5.2.4

Consider the following programs:

```
for (i=0; i<10; ++i)     for (i=0; i<10; ++i)     for (i=0; i<10; ++i)
    if (i < N)               if (i < input)           if (i < input[i])
S1:  A[i]=f(B[i]);       S2:  A[i]=f(B[i]);       S3:  A[i]=f(B[i]);
```

Although these three programs are syntactically very similar, the conditions in each of them contain a different concept. The left one contains a symbolic constant (assuming that N is an unknown structure parameter), the middle one contains a data dependent variable that remains constant during the execution of the loop, and the right one contains a data dependent variable that varies with each execution of the loop.

The first two cases are very similar: they both contain a value that is not known at compile time, but which remains constant during the execution of the program. Consequently, we can model the corresponding iteration domains in a similar way:

$$\mathbf{D}_1^{\mathbf{iter}} = \{\ i\ |\ 0 \leq i < N \ \wedge\ i \in \mathbb{Z}\ \}$$
$$\mathbf{D}_2^{\mathbf{iter}} = \{\ i\ |\ 0 \leq i < p \ \wedge\ i \in \mathbb{Z}\ \}$$

In the second equation, p represents the unknown variable input. *One can easily see that from a modeling point of view, there is no difference between those two cases. This means that techniques for dealing with symbolic constants can also be used for dealing with data dependent variables, provided that they remain constant during* *the execution of the program. Note that we have not put any constraints on the values that p or N can take. In general, it is not required that these values are integral (e.g. input may be a floating point variable). Of course, some constraints may be imposed by the context of the program, and in that case we can add them to the domain descriptions, but this is not required by the model. In the sequel we assume that dimensions that have not been constrained, can take any **real** value.*

*For the third case, the situation is quite different. In this case, we cannot represent input[i] by means of an unknown variable that remains constant during the execution of the program. Therefore, we must explicitly indicate that the value of this unknown variable may be different for each value of i. In other words, the value of the unknown variable is a **function** of i, although the function is not known at compile time. So we can model the iteration domain as follows:*

$$\mathbf{D}_3^{\texttt{iter}} = \{ \ i \ | \ \exists \ q \ \text{s.t.} \ 0 \leq i < q \ \wedge \ q = F^{\texttt{iter}}(i) \ \wedge \ i \in \mathbb{Z} \ \}$$

*in which $F^{\texttt{iter}}(i)$ represents an **unknown function** of i. Note that this is essentially the same model as the ones used in [332, 256, 77] albeit in a slightly different notation. The functions used to model these unknown variables are also known as **uninterpreted function symbols** [331].*

One may wonder whether the iteration domains of the first and the second program cannot be modeled in a similar way. Not surprisingly, the answer is yes, since the unknowns of the first and second programs can be seen as degenerate cases of varying unknowns, i.e. varying unknowns that do not vary. In other words, they can be modeled by means of unknown functions without input arguments. In this way, we can unify the modeling techniques for symbolic constants, constant unknowns, and varying unknowns. So we can rewrite the expressions above in a uniform way as follows:

$$\mathbf{D}_1^{\texttt{iter}} = \{ \ i \ | \ \exists \ N \ \text{s.t.} \ 0 \leq i < N \ \wedge \ N = F_1^{\texttt{iter}}() \ \wedge \ i \in \mathbb{Z} \ \}$$
$$\mathbf{D}_2^{\texttt{iter}} = \{ \ i \ | \ \exists \ p \ \text{s.t.} \ 0 \leq i < p \ \wedge \ p = F_2^{\texttt{iter}}() \ \wedge \ i \in \mathbb{Z} \ \}$$
$$\mathbf{D}_3^{\texttt{iter}} = \{ \ i \ | \ \exists \ q \ \text{s.t.} \ 0 \leq i < q \ \wedge \ q = F_3^{\texttt{iter}}(i) \ \wedge \ i \in \mathbb{Z} \ \}$$

Note that $F_{1/2}^{\texttt{iter}}()$ are functions without input arguments, i.e. constant functions.

Now that we have indicated how we can model both arbitrary manifest and non-manifest program constructs, we can combine everything into one model. This is done next.

5.2.3 General model

In general, we can describe the geometrical domains associated with programming constructs by means of three kinds of dimensions:

1. **Main dimensions**: these are the real dimensions of the domains. All other dimensions should be eliminated in order to obtain the real domain consisting of points with integer coordinates (although this may be possible only at run-time). Each of these points then corresponds to exactly one program entity (e.g. an operation, a variable, ...). Each main dimension generally corresponds to an iterator in the program (although this iterator may not be explicitly present) or a dimension of a (multi-dimensional) variable.

   For a given semantical interpretation and a given programming construct, the main dimensions of a domain are unique, i.e. all domain descriptions corresponding to a certain programming construct should be mathematically equivalent. The shape of the iteration domain of a statement for instance, is unique for a given loop structure surrounding the statement and a given semantical interpretation of the dimensions. However, through transformations the shapes of domains can be modified, but then the corresponding programming constructs are assumed to be transformed also.

   In the sequel, vectors of main dimensions in formal equations are represented by bold letters, e.g. **i**.

2. **Auxiliary dimensions** (also called wildcard dimensions): these dimensions are not really part of the domains, but are used to be able to express more complex constraints on the domains. In general, auxiliary dimensions cannot be associated with any variables present in the program. An exception are the auxiliary dimensions that correspond to main dimensions of other domains (e.g. the dimensions of the iteration domains also appear in the descriptions of the definition

and operand domains, unless they have been eliminated). Auxiliary dimensions are nothing but a mathematical aid, and are existentially quantified and therefore not unique. Only the result obtained after their elimination matters.

In the sequel, vectors of auxiliary dimensions in formal equations are represented by Greek letters, e.g. α (except when they correspond to the main dimensions of other domains; in that case we leave them in bold to highlight this correspondence).

3. **Hidden dimensions**: these dimensions are also not really part of the domains, but are used to model non-manifest behaviour. Just like auxiliary dimensions, hidden dimensions are not unique either, and have to be eliminated in order to obtain the real domains. The difference is that in general this elimination cannot be done at compile time. Generally, hidden dimensions correspond to symbolic constants or data-dependent variables in the program. Hidden dimensions are always expressed as functions of main dimensions and are also existentially quantified.

In the sequel, vectors of hidden dimensions in formal equations are represented by italic letters, e.g. r.

So in general we can represent the iteration, variable, definition and operand domains as follows:

$$\mathbf{D}_i^{\texttt{iter}} = \{\ \mathbf{i}\ |\ \exists\,\alpha \in \mathbb{Z}^{n_{i2}}, p \ \text{s.t.}$$
$$\mathbf{C}_i^{\texttt{iter}}(\mathbf{i},\alpha,p) \geq 0\ \wedge\ p = \mathbf{F}_i^{\texttt{iter}}(\mathbf{i})\ \wedge\ \mathbf{i} \in \mathbb{Z}^{n_{i1}}\ \}$$
$$\mathbf{D}_m^{\texttt{var}} = \{\ \mathbf{s}\ |\ \exists\,\epsilon \in \mathbb{Z}^{n_{m2}}, r \ \text{s.t.}$$
$$\mathbf{C}_m^{\texttt{var}}(\mathbf{s},\epsilon,r) \geq 0\ \wedge\ r = \mathbf{F}_m^{\texttt{var}}(\phi)\ \wedge\ \mathbf{s} \in \mathbb{Z}^{n_{m1}}\ \}$$
$$\mathbf{M}_{ikm}^{\texttt{def}} = \{\ \mathbf{i} \to \mathbf{d}\ |\ \exists\,\gamma \in \mathbb{Z}^{nd_{ikm}}, u \ \text{s.t.}$$
$$\mathbf{C}_{ikm}^{\texttt{def}}(\mathbf{d},\mathbf{i},\gamma,u) \geq 0\ \wedge\ u = \mathbf{F}_{ikm}^{\texttt{def}}(\mathbf{d})\ \wedge\ \mathbf{d} \in \mathbb{Z}^{n_{m1}}\ \}$$
$$\mathbf{D}_{ikm}^{\texttt{def}} = \{\ \mathbf{d}\ |\ \exists\,\mathbf{i} \in \mathbf{D}_i^{\texttt{iter}}, \gamma \in \mathbb{Z}^{nd_{ikm}}, u \ \text{s.t.}$$
$$\mathbf{C}_{ikm}^{\texttt{def}}(\mathbf{d},\mathbf{i},\gamma,u) \geq 0\ \wedge\ u = \mathbf{F}_{ikm}^{\texttt{def}}(\mathbf{d})\ \wedge\ \mathbf{d} \in \mathbb{Z}^{n_{m1}}\ \}$$
$$\mathbf{M}_{jlm}^{\texttt{oper}} = \{\ \mathbf{j} \to \mathbf{o}\ |\ \exists\,\delta \in \mathbb{Z}^{no_{jlm}}, v \ \text{s.t.}$$
$$\mathbf{C}_{jlm}^{\texttt{oper}}(\mathbf{o},\mathbf{j},\delta,v) \geq 0\ \wedge\ v = \mathbf{F}_{jlm}^{\texttt{oper}}(\mathbf{o})\ \wedge\ \mathbf{o} \in \mathbb{Z}^{n_{m1}}\ \}$$
$$\mathbf{D}_{jlm}^{\texttt{oper}} = \{\ \mathbf{o}\ |\ \exists\,\mathbf{j} \in \mathbf{D}_j^{\texttt{iter}}, \delta \in \mathbb{Z}^{no_{jlm}}, v \ \text{s.t.}$$
$$\mathbf{C}_{jlm}^{\texttt{oper}}(\mathbf{o},\mathbf{j},\delta,v) \geq 0\ \wedge\ v = \mathbf{F}_{jlm}^{\texttt{oper}}(\mathbf{o})\ \wedge\ \mathbf{o} \in \mathbb{Z}^{n_{m1}}\ \}$$

We can make the following remarks:

- The dimensions corresponding to the components of vectors $\mathbf{d}$ and $\mathbf{o}$ of $\mathbf{D}_{ikm}^{\texttt{def}}$ and $\mathbf{D}_{jlm}^{\texttt{oper}}$ are the same as the dimensions of vector $\mathbf{s}$ of $\mathbf{D}_m^{\texttt{var}}$, since $\mathbf{D}_{ikm}^{\texttt{def}}$ and $\mathbf{D}_{jlm}^{\texttt{oper}}$ are always sub-domains of $\mathbf{D}_m^{\texttt{var}}$ (at least for programs where no arrays are ever accessed outside their boundaries, which is an obvious constraint for practical realizations).

- Some of the domain descriptions share names of *mathematical* variables representing the dimensions. This sharing only *suggests* that the descriptions can be constructed in a certain way (e.g. a definition domain description can be constructed by combining and iteration domain description and a definition mapping description). From a mathematical point of view, identically named mathematical variables in independent domain/mapping descriptions are unrelated.

- Unlike the main and auxiliary dimensions, we do not require the hidden dimensions to be integral. As hidden dimensions may correspond to data variables of the program, hidden dimensions may take any value the corresponding data variables can take, even non-integral ones. Depending on the context, an integrality condition may be added (e.g. when the corresponding data variables have an integral type).

- In practice, strict inequalities ($<$ and $>$) originating from program statements such as A[i] > 0 can always be converted to non-strict inequalities because the precision of data types in a program is always limited, even for pseudo-real data types. For instance, if $A[i] > 0$ would result in an inequality $p > 0$, then this inequality be rewritten as $p - \epsilon \geq 0$, in which ϵ is a small enough positive value (depending on the precision of the data type of A). For integral types, $\epsilon = 1$. Therefore, we can avoid all strict inequalities, which results in a simpler *notation*.

- From a mathematical point of view, the distinction between constraint functions and functions for hidden dimensions is somewhat artificial. For instance, $\mathbf{C}_i^{\text{iter}}(\mathbf{i}, \alpha, p) \geq 0 \land p = \mathbf{F}_i^{\text{iter}}(\mathbf{i})$ can be rewritten as $\mathbf{C}_i^{\text{iter}}(\mathbf{i}, \alpha, p) \geq 0 \land p - \mathbf{F}_i^{\text{iter}}(\mathbf{i}) \geq 0 \land \mathbf{F}_i^{\text{iter}}(\mathbf{i}) - p \geq 0$ or $\mathbf{C}_i^{\text{iter}}(\mathbf{i}, \alpha, p)^* \geq 0$ in which $\mathbf{C}_i^{\text{iter}}()^*$ is a combination of the three (vector) functions. Therefore, in the sequel, we assume that these functions have been combined into $\mathbf{C}_i^{\text{iter}}()^*$, but we drop the '*'. The same can be done for the other constraint functions.

This results in the following general model:

$$\mathbf{D}_i^{\text{iter}} = \{ \ \mathbf{i} \mid \exists \ \alpha \in \mathbb{Z}^{n_{i2}}, p \text{ s.t.}$$
$$\mathbf{C}_i^{\text{iter}}(\mathbf{i}, \alpha, p) \geq 0 \land \mathbf{i} \in \mathbb{Z}^{n_{i1}} \} \tag{5.1}$$

$$\mathbf{D}_m^{\text{var}} = \{ \ \mathbf{s} \mid \exists \ \epsilon \in \mathbb{Z}^{n_{m2}}, r \text{ s.t.}$$
$$\mathbf{C}_m^{\text{var}}(\mathbf{s}, \epsilon, r) \geq 0 \land \mathbf{s} \in \mathbb{Z}^{n_{m1}} \} \tag{5.2}$$

$$\mathbf{M}_{ikm}^{\text{def}} = \{ \ \mathbf{i} \to \mathbf{d} \mid \exists \ \gamma \in \mathbb{Z}^{nd_{ikm}}, u \text{ s.t.}$$
$$\mathbf{C}_{ikm}^{\text{def}}(\mathbf{d}, \mathbf{i}, \gamma, u) \geq 0 \land \mathbf{d} \in \mathbb{Z}^{n_{m1}} \} \tag{5.3}$$

$$\mathbf{D}_{ikm}^{\text{def}} = \{ \ \mathbf{d} \mid \exists \ \mathbf{i} \in \mathbf{D}_i^{\text{iter}}, \gamma \in \mathbb{Z}^{nd_{ikm}}, u \text{ s.t.}$$
$$\mathbf{C}_{ikm}^{\text{def}}(\mathbf{d}, \mathbf{i}, \gamma, u) \geq 0 \land \mathbf{d} \in \mathbb{Z}^{n_{m1}} \} \tag{5.4}$$

$$\mathbf{M}_{jlm}^{\text{oper}} = \{ \ \mathbf{j} \to \mathbf{o} \mid \exists \ \delta \in \mathbb{Z}^{no_{jlm}}, v \text{ s.t.}$$
$$\mathbf{C}_{jlm}^{\text{oper}}(\mathbf{o}, \mathbf{j}, \delta, v) \geq 0 \land \mathbf{o} \in \mathbb{Z}^{n_{m1}} \} \tag{5.5}$$

$$\mathbf{D}_{jlm}^{\text{oper}} = \{ \ \mathbf{o} \mid \exists \ \mathbf{j} \in \mathbf{D}_j^{\text{iter}}, \delta \in \mathbb{Z}^{no_{jlm}}, v \text{ s.t.}$$
$$\mathbf{C}_{jlm}^{\text{oper}}(\mathbf{o}, \mathbf{j}, \delta, v) \geq 0 \land \mathbf{o} \in \mathbb{Z}^{n_{m1}} \} \tag{5.6}$$

5.3 ORDER

The primary models described in the previous sections (iteration, variable, definition, and operand domains) are only used to mathematically describe a given program and they do not contain any information about design decisions such as execution order and storage order. Moreover, these models are not even complete, i.e. without additional information, one cannot restore the original program or sometimes not even an equivalent program from the mathematical description. Consider the following example:

Example 5.3.1

```
    for ( i = 0; i < 10; ++i )
S1:    A[i] = ...;

    for ( j = 9; j >= 0 ; --j )
S2:    A[j] = ...;
```

Given this program, we can extract the following mathematical descriptions of the iteration and definition domains:

$$\mathbf{D}_1^{\text{iter}} = \{ \ i \mid 0 \leq i \leq 9 \land i \in \mathbb{Z} \ \}$$
$$\mathbf{D}_{11A}^{\text{def}} = \{ \ a \mid a = i \land i \in \mathbf{D}_1^{\text{iter}} \land a \in \mathbb{Z} \ \}$$
$$\mathbf{D}_2^{\text{iter}} = \{ \ j \mid 0 \leq j \leq 9 \land j \in \mathbb{Z} \ \}$$
$$\mathbf{D}_{21A}^{\text{def}} = \{ \ a \mid a = j \land j \in \mathbf{D}_2^{\text{iter}} \land a \in \mathbb{Z} \ \}$$

Given only these domains, one cannot reconstruct the original program, since the notion of execution order is not present in these domains. For instance, without additional information, one cannot decide whether the statement corresponding to $\mathbf{D}_1^{\text{iter}}$ should be executed before or after the statement corresponding to $\mathbf{D}_2^{\text{iter}}$, although in the original non-single-assignment program, it is clear that this order is very important. Moreover, from the domain descriptions, one cannot even derive in what direction the loops have to be executed, since the domain descriptions are independent of the execution order of the loops.

5.3.1 Execution and storage order

For manifest single-assignment programs, the domain descriptions are sufficient to reconstruct *an equivalent* program, because in that case one can (at least in theory) find all data-dependencies between the operations, and consequently a valid order.

However, if we want to perform optimizations, one of the things we have to do, is to decide on an **execution order** of the statements. Another important decision we have to make is the decision on the **storage order**, i.e. the layout of the arrays in the memory/memories (see also chapters 1 and 12).

We can express these orders by assigning an "**execution date**" to every operation and a "**storage address**" to every variable. The meaning of the terms "execution date" and "storage address" is context dependent. Sometimes they are defined in terms of some absolute unit such as a clock cycle or a memory location, but in many cases, the absolute numbers are not relevant, i.e. only the relative order of operations or storage locations is important. For instance, an execution date can be expressed in terms of a number of operations that have been executed before that date, even though the operations may have different time durations in the final implementation of the program. Also, a *storage* location may not correspond to one physical *memory* location, but possibly to several ones. We can make use of these observations to simplify certain optimization problems, as indicated at the end of this section, on page 96. Nevertheless, when comparing execution dates or storage addresses, we should make sure that their respective scales and offsets are taken into account.

Anyhow, as we are dealing with large sets of operations and large sets of variables, it would be infeasible to assign an execution date to each individual operation and a storage address to each individual variable. Not only would the memory optimization problem become intractable for large applications, but also the implementation cost (controller and address generation overhead) would be prohibitive. Therefore, we have to assign execution dates and storage addresses to *groups* of operations and variables respectively in order to maintain some regularity [105].

One of the basic requirements that the storage order and the execution order have to fulfill, is that each variable can have only one storage address and each operation can be executed at only one execution date (provided that variables and operations can be seen as atomic units, which is true in this context). This requirement is compatible with one of the properties of a mathematical *function*: a function evaluates to only one value for each distinct set of input arguments (although different argument sets may result in the same value). Therefore, it is a straightforward choice to use (deterministic) functions to describe execution and storage order. As arguments of these functions, we can use the coordinates of the operations/variables in the corresponding domains, since these coordinates are unique with respect to all other operations/variables in the same set. So we can describe an execution or storage order with exactly one function per set of operations/variables. A small example is shown next:

Example 5.3.2

```
    int A[10][2]
    for ( i = 0; i < 10; ++i )
      for ( j = 0; j < 2; ++j )
S1:     A[i][j] = ...;
```

For this program, the corresponding iteration and variable domains are the following:

$$\mathbf{D}_1^{\text{iter}} = \{\ [i,j]\ |\ 0 \le i \le 9\ \wedge\ 0 \le j \le 1\ \wedge\ [i,j] \in \mathbb{Z}^2\ \}$$
$$\mathbf{D}_A^{\text{var}} = \{\ [s_1,s_2]\ |\ 0 \le s_1 \le 9\ \wedge\ 0 \le s_2 \le 1\ \wedge\ [s_1,s_2] \in \mathbb{Z}^2\ \}$$

Assuming that this program is executed sequentially, and that we are only interested in the execution order of the operations corresponding to S1, we can express the (relative) execution date of an operation by means of the following function defined over $\mathbf{D}_1^{\mathtt{iter}}$:

$$O_1^{\mathtt{time}}(i, j) = 2 * i + j$$

In other words, the relative date at which the operation corresponding to the point with coordinates $[i, j]$ is executed, equals $2 * i + j$.

Similarly, we can define a storage address function for array A. Assuming that the elements of A are stored in a column-major way, we can describe the storage addresses with the following function defined over $\mathbf{D}_A^{\mathtt{var}}$:

$$O_A^{\mathtt{addr}}(s_1, s_2) = s_1 + 10 * s_2$$

In other words, the storage address at which the variable corresponding to the point with coordinates $[s_1, s_2]$ is stored, equals $s_1 + 10 * s_2$. A graphical representation is shown in Fig. 5.5.

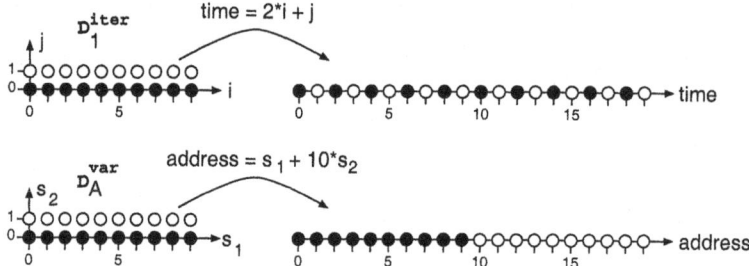

Figure 5.5.   The execution dates and addresses of example 5.3.2.

5.3.2 Differences between execution and storage order

Although execution order and storage order have been treated in a very similar way, there are some differences that have to be taken into account during memory optimizations.

First of all, as stated above, for the purpose of memory related optimizations, it is usually sufficient to know the *relative* execution order of operations and/or memory accesses.[6] As mentioned on page 95, this may allow us (under certain conditions) to use simpler functions for describing the relative execution order than the ones describing the *absolute* execution order. By doing so, the mathematical complexity of the optimization problems may be drastically reduced. A small example of such a simplification is given next:

Example 5.3.3

```
      for ( i = 0; i < 10; ++i )
          for ( j = 0; j <= i; ++j )
S1:       ...
```

For this program, which has a non-rectangular iteration domain for S1, describing the execution order of the operations corresponding to statement S1 in terms of the number of times that this statement is executed, would require a polynomial function of degree 2:

$$O_1^{\mathtt{time}}(i, j) = \frac{i * (i + 1)}{2} + j$$

---

[6]Usually, but not always. When dealing with timing constraints for instance, absolute timing functions should be used.

*However, provided that we are only interested in the **relative** execution order, we can replace this polynomial function by a linear one, which results in the same relative order **over the iteration domain**:*

$$O_1^{\text{time}}(i,j)' = i * 10 + j$$

Using this function instead of the previous one, most likely results in simpler optimization problems. A graphical comparison of both orders is shown in Fig. 5.6.

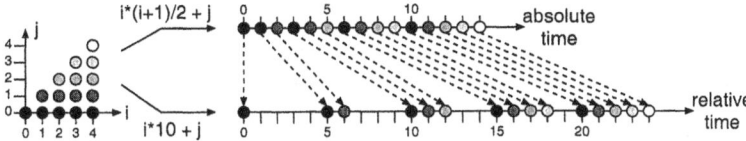

Figure 5.6.   The "absolute" execution order of example 5.3.3 versus an equivalent relative order.

A procedure for extracting linear relative execution order functions from a program is described in [109], Appendix 3.A.

For storage addresses, the situation is somewhat different as one is usually interested in the memory *size* required for a set of variables. In order to accurately relate *memory* sizes to distances between *storage* addresses, it is important to keep the relation between storage addresses and memory addresses as simple as possible, i.e. in practice this relation should be linear (or piece-wise linear, e.g. because one logical storage address range may be divided over several physical memories or because modulo addressing is being used). Therefore, one has less freedom to choose an appropriate ordering function than in the case of the execution order. Using an ordering function with a non-linear relation to the physical memory addresses could result in serious estimation errors.

Another important difference between execution order and storage order is the fact that, in practice, a storage location can be treated as an atomical unit for many purposes (even though it may correspond to several memory locations), while on the other hand operations may be decomposed into several "sub-operations" with different (but related) execution dates and durations. For instance, on a programmable processor, the execution of a statement may take several clock cycles. Moreover, depending on the memory type, the fetching of operands and storing of results may occur in different clock cycles than the actual data-path operation, e.g. operands may be fetched in one cycle, the data-path operation may be executed in the next cycle and the results may be stored during yet another clock cycle. It may even be possible that the memory accesses of different operations are interleaved in time, e.g. due to bandwidth considerations [438].

So, a simple time order model associating an execution date with every operation may not be sufficient to model practical applications accurately. In subsection 5.3.3 we present a more extended timing model, based on the following observations:

- In practice, if the memory accesses corresponding to an operation do not coincide with the actual execution of the operation, the time offset for the accesses with respect to the operations is always the same for operations corresponding to the same statement. For instance, operands may always be fetched one clock cycle before the operation and the results may always be stored one clock cycle after the operation. Moreover, it makes no sense to fully decouple the memory accesses from the corresponding operations, i.e. memory accesses always have to be "synchronized" with their corresponding data operations. For instance, for a statement being executed in a loop, it generally makes no sense to perform all read accesses first, all data operations after that, and finally all write accesses. Doing so would *implicitly* require a (potentially large) extra buffer for storing the operands and/or the definition. In case one really wants to model such a buffering strategy, it should be done explicitly, i.e. by explicitly modeling the transfers to and from the buffers (by means of extra iteration, operand, and/or definition domains).[7]

---

[7]This restriction may be too cumbersome for tasks that need a higher level of abstraction, i.e. the early stages in the ATOMIUM methodology. Therefore these tasks may (temporarily) require different timing models. For tasks that need more accurate cost

Note that this does not mean that *every* storage location always should be modeled explicitly. It is for instance possible that an operand is fetched a few cycles before it is actually needed by an operation, such that it has to be stored temporarily in a register. But as long as the memory accesses are not decoupled from the corresponding data operations, the required number of extra storage locations remains limited (usually at most a few locations, such that they can reside in foreground memory). In that case these (few) extra storage locations usually can be neglected (compared to the large background storage), certainly for memory-intensive applications, and need not to be modeled explicitly (in most contexts).

Based on this reasoning, it seems natural to express the time order of the *accesses* as a *function* of the time order of the corresponding *operations*. For instance, if a set of operations is executed at time $2 * (i + j * 10)$, then the corresponding write accesses could occur at time $(2 * (i + j * 10)) + 1$, i.e. one time unit later than the actual operations. In practice, the time offset between the accesses and the data operations is constant, or can at least assumed to be constant.[8]

This timing model is fully compatible with the one described in [441], where the time offsets for read and write accesses are measured from the start of the body of loops. In that model, it is implicitly assumed that the corresponding data operations are executed at fixed time offsets from the start of the loop bodies, such that the read and write accesses remain synchronized with the data operations.

■ For the domain models presented above, it would be difficult to define timing functions on stand-alone definition or operand domains (which correspond to memory accesses), since in general the mappings between the iteration domains and definition or operand domains can be non-injective, such that one point in a definition or operand domain may correspond to several memory accesses. It would then be very difficult to distinguish between accesses represented by the same point. Therefore, it is easier to use indirect timing functions, such that the timing functions of the memory accesses are defined in terms of the timing functions of the corresponding operations. For operations there is no such ambiguity problem, since each point in an iteration domain corresponds to one operation and vice versa, and consequently we can associate a unique execution date with each memory access.

■ In case one is only interested in the relative execution order of operations and/or memory accesses, one can assume that each operation and/or access takes exactly one time unit to complete, which is very often even true in practice. By doing so, two operations or accesses either completely overlap in time or don't overlap at all, i.e. operations or accesses cannot partially overlap in time. This can make the mathematical modeling and optimization somewhat simpler. In case it is not possible to use such a simplified model, one has to rely on a more detailed one. In this text we assume that we can make this simplification though.

5.3.3 A simple execution and storage order model

Based on the reasoning given above, we can define the following ordering functions:

■ $O_i^{\texttt{time}}()$: the time order function of $\mathbf{D}_i^{\texttt{iter}}$. Evaluating this function for a point in $\mathbf{D}_i^{\texttt{iter}}$ results in the "execution date" of the corresponding operation. The meaning of the term "execution date" is context dependent, as stated before, i.e. it may refer to an absolute execution date (e.g. in terms of a number of clock cycles) or to a relative execution date (e.g. in terms of a number of executions).

■ $O_{ikm}^{\texttt{wtime}}()$: the time *offset* function of $\mathbf{D}_{ikm}^{\texttt{def}}$. The resulting offset is an offset relative to $O_i^{\texttt{time}}()$. A write access corresponding to a point in $\mathbf{D}_{ikm}^{\texttt{def}}$ occurs at this offset in time relative to the execution

---

estimations, i.e. the lower stages in the ATOMIUM methodology, hidden buffers are undesirable though and should be modeled explicitly.

[8]On modern processors it is possible that this time offset is not always constant or is even unpredictable (e.g. because of run-time operation scheduling and out-of-order execution), but at least the processor has to make sure that the relative order is a valid one. So, if we specify a valid relative order in the program (based on constant offsets), the processor is allowed to change this order as long as the I/O behaviour of the program is not altered.

of the corresponding operation. In practice, this offset is always constant (or can at least assumed to be constant). Again, the same remarks apply with respect to the time unit used.

- $O_{jlm}^{\text{rtime}}()$: the time *offset* function of $\mathbf{D}_{jlm}^{\text{oper}}$. The resulting offset is an offset relative to $O_j^{\text{time}}()$. A read access corresponding to a point in $\mathbf{D}_{jlm}^{\text{oper}}$ occurs at this offset in time relative to the execution of the corresponding operation. Again, in practice, this offset is always constant (or can assumed to be constant).

- $O_m^{\text{addr}}()$: the storage order function of $\mathbf{D}_m^{\text{var}}$. Evaluating this function for a point in $\mathbf{D}_m^{\text{var}}$ results in the storage address at which the corresponding variable is stored. In general, there is a (piece-wise) linear relation between storage addresses and absolute memory addresses (although the storage order function may be non-linear).

In general, each of these functions may have (extra) hidden or auxiliary dimensions as arguments and may be accompanied by extra constraints on these extra dimensions (e.g. a symbolic constant), represented by $\mathbf{C}_i^{\text{time}}() \geq 0$, $\mathbf{C}_{ikm}^{\text{wtime}}() \geq 0$, $\mathbf{C}_{jlm}^{\text{rtime}}() \geq 0$, and $\mathbf{C}_m^{\text{addr}}() \geq 0$. In practice, $\mathbf{C}_{ikm}^{\text{wtime}}()$ and $\mathbf{C}_{jlm}^{\text{rtime}}()$ are generally not present, and therefore we do not mention them explicitly any more in the sequel in order to keep our notations a bit simpler.

The next example illustrates the use of these time and address ordering functions:

Example 5.3.4

```
      double A[3][N];
      float B[3][N];
      for ( i = 0; i < 3; ++i )
          for ( j = 0; j < N; ++j )
S1:       A[i][j] = f(B[i][j]);
```

Let us suppose that statement S1 is executed in 3 clock cycles: in the first cycle, an element of B is read; in the second one, function f ( ) is evaluated; and finally, the result is written in an element of A. This would result in the following time order functions (and constraints), assuming that the symbolic constant N is only known to be larger than zero:

$$O_1^{\text{time}}(i, j, N) = 3(N * i + j) + 1$$
$$\mathbf{C}_1^{\text{time}}(N) = N > 0$$
$$O_{11B}^{\text{rtime}}(x) = x - 1$$
$$O_{11A}^{\text{wtime}}(x) = x + 1$$

which lead to the following execution dates for the read and write operations respectively:

$$3(N * i + j)$$
$$3(N * i + j) + 2$$

For arrays A and B we assume the following storage address functions (a_1, a_2 and b_1, b_2 represent the dimensions of the variable domains of A and B respectively):

$$O_A^{\text{addr}}(a_1, a_2, N) = N * a_1 + a_2$$
$$\mathbf{C}_A^{\text{addr}}(N) = N > 0$$
$$O_B^{\text{addr}}(b_1, b_2, N) = N * b_1 + b_2$$
$$\mathbf{C}_B^{\text{addr}}(N) = N > 0$$

If we assume that A and B are stored in the same memory, one after the other, and assuming that a float and a double require 4 and 8 memory locations respectively, the corresponding memory address functions could be the following:

$$8N * a_1 + 8a_2 + [0..7]$$
$$4N * b_1 + 4b_2 + 24N + [0..3]$$

Note that the relation between the storage address functions and their memory address counterparts is linear (assuming N is a known constant).

We could extend these simple models to take into account operations with non-unit durations for example but, for practical purposes, these simple models are usually more than sufficient. Therefore we do not discuss these possible extensions. In Appendix 3.A of [109] we describe a procedure, based on this model, for the extraction of linear relative execution order functions from a program. These functions are accurate enough for use in our storage order optimization strategy.[9] For other tasks in the ATOMIUM methodology, specific timing models may be required though. In contrast, the storage order functions can be used everywhere in the methodology.

5.4   DEPENDENCIES

The domain and order models presented in the previous sections are sufficient to be able to perform data-flow analysis of the program. An important concept in data-flow analysis is that of a data dependency. In literature several types of data dependencies are described [134, 34, 236].

In general a data dependency denotes a kind of precedence constraint between operations. The basic type of dependency is the **value-based flow dependency**. A value-based flow dependency between two operations denotes that the first one produces a data value that is being consumed by the second one, so the first one has to be executed before the second one.

Other types of dependencies (e.g. memory-based flow dependencies, output dependencies, anti-dependencies, ...) also correspond to precedence constraints, but these are purely storage related, i.e. they are due to the sharing of storage locations between different data values. This kind of dependencies is only important for the analysis of procedural non-single-assignment code. Eventually, the goal of dependency analysis is to find all the value-based flow dependencies, as they are the only "real" dependencies. Given the value-based flow dependencies, it is (in theory) possible to convert any procedural non-single-assignment code to single-assignment form, where the only precedence constraints left are the value-based flow-dependencies.

For reasons of simplicity, we assume from now on that the code that we are analyzing or optimizing has been converted to single-assignment form. The theory and techniques presented in the sequel are however also applicable and extensible to non-single-assignment code, provided that the value-based flow dependencies are known. We also use the term *flow dependency* or simply *dependency* to refer to a value-based flow dependency in the sequel.

Just like we do not describe individual operations or individual variables, we also do not describe individual dependencies between operations either. Instead we describe *groups* of dependencies between *groups* of operations. A simple example is shown next.

Example 5.4.1

```
      int A[5][15];
      for ( i = 0; i < 5; ++i )
      {
          for ( j = 0; j < 10; ++j )
S1:         A[i][j] = f1(...);
      ...
          for ( k = 5; k < 15; ++k )
S2:         ... = f2(A[i][k]);
      }
```

The domain descriptions for this example are the following:

$$\mathbf{D}_A^{\mathbf{var}} = \{ \ [a_1, a_2] \ | \ 0 \leq a_1 \leq 4 \ \wedge \ 0 \leq a_2 \leq 14 \ \wedge \ [a_1, a_2] \in \mathbb{Z}^2 \ \}$$
$$\mathbf{D}_1^{\mathbf{iter}} = \{ \ [i, j] \ | \ 0 \leq i \leq 4 \ \wedge \ 0 \leq j \leq 9 \ \wedge \ [i, j] \in \mathbb{Z}^2 \ \}$$

---

[9]Remember Einstein.

$$\mathbf{D}_{11A}^{\mathtt{def}} = \{ \ [a_1, a_2] \ | \ \exists \ [i, j] \in \mathbf{D}_1^{\mathtt{iter}} \ \mathrm{s.t.} \ a_1 = i \ \wedge \ a_2 = j \ \wedge \ [a_1, a_2] \in \mathbb{Z}^2 \ \}$$

$$\mathbf{D}_2^{\mathtt{iter}} = \{ \ [i, k] \ | \ 0 \leq i \leq 4 \ \wedge \ 5 \leq k \leq 14 \ \wedge \ [i, k] \in \mathbb{Z}^2 \ \}$$

$$\mathbf{D}_{21A}^{\mathtt{oper}} = \{ \ [a_1, a_2] \ | \ \exists \ [i, k] \in \mathbf{D}_2^{\mathtt{iter}} \ \mathrm{s.t.} \ a_1 = i \ \wedge \ a_2 = k \ \wedge \ [a_1, a_2] \in \mathbb{Z}^2 \ \}$$

A graphical representation of these domains is shown in Fig. 5.7.

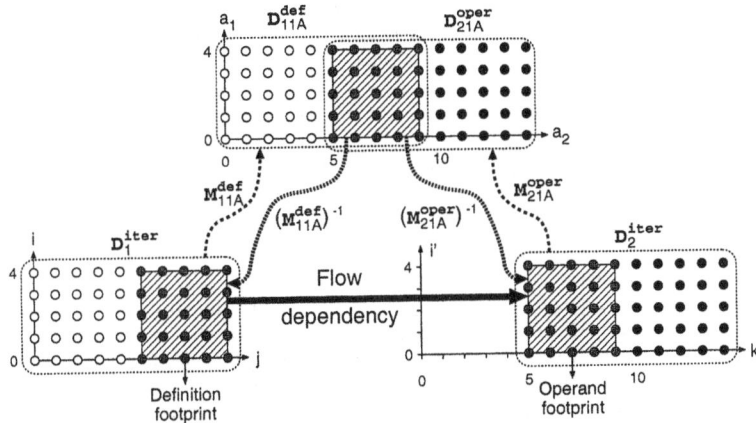

Figure 5.7.   The domains of example 5.4.1, and the corresponding flow dependency.

One can see that some of the elements of array A *that are being produced during the executions of statement* S1 *are also being consumed during the executions of statement* S2. *Consequently there exists a flow dependency between these two statements. First we can find the elements of the array that contribute to the dependency by simply intersecting the definition and operand domain. The points in the intersection correspond to the array elements that are being produced by* S1 *and consumed by* S2. *Given these elements, we can find out which operations (i.e. executions of the statements) correspond to them by applying the inverse definition or operand mapping.*

*We refer to the result of the inverse mappings as the **definition footprint** and **operand footprint** respectively. Note that in general the definition and operand mappings may be non-injective, but this poses no problems in this general model as we impose no restriction on the nature of mappings. Non-injectivity may complicate the analysis of these dependencies though. Techniques for dealing with non-injective mappings have been described in [421].*

*The most general way to describe value-based dependencies is by means of **dependency relations**, introduced in [329], which are mappings from one iteration domain to another one. For our example we obtain the following dependency relation, denoted by* $\mathbf{M}_{1211A}^{\mathtt{flow}}$:

$$\mathbf{M}_{1211A}^{\mathtt{flow}} = \{ \ [i, j] \rightarrow [i', k] \ | \ \exists \ [a_1, a_2] \in \mathbf{D}_A^{\mathtt{var}} \ \mathrm{s.t.}$$
$$\mathbf{M}_{11A}^{\mathtt{def}}(i, j) = [a_1, a_2] = \mathbf{M}_{21A}^{\mathtt{oper}}(i', k) \ \wedge$$
$$[i, j] \in \mathbf{D}_1^{\mathtt{iter}} \ \wedge \ [i', k] \in \mathbf{D}_2^{\mathtt{iter}} \ \}$$
$$= \{ \ [i, j] \rightarrow [i', k] \ | \ i = i' \ \wedge \ j = k \ \wedge$$
$$0 \leq i \leq 4 \ \wedge \ 0 \leq j \leq 9 \ \wedge \ 0 \leq i' \leq 4 \ \wedge \ 5 \leq k \leq 14 \ \wedge$$
$$[i, j] \in \mathbb{Z}^2 \ \wedge \ [i', k] \in \mathbb{Z}^2 \ \}$$
$$= \{ \ [i, j] \rightarrow [i, k] \ | \ 0 \leq i \leq 4 \ \wedge \ 5 \leq j = k \leq 9 \ \wedge \ [i, j, k] \in \mathbb{Z}^3 \ \}$$

When we apply this mapping to $\mathbf{D}_1^{\mathtt{iter}}$, *we obtain the operand footprint. Similarly, when we apply the inverse mapping to* $\mathbf{D}_2^{\mathtt{iter}}$, *we obtain the definition footprint.*

Based on this example, we can extend our general model presented in subsection 5.2.3 to include value-based flow dependency relations. A dependency due to an overlap between a definition domain

$\mathbf{D}_{ikm}^{\text{def}}$, belonging to an iteration domain $\mathbf{D}_i^{\text{iter}}$, and an operand domain $\mathbf{D}_{jlm}^{\text{oper}}$, belonging to an iteration domain $\mathbf{D}_j^{\text{iter}}$, is represented by the following dependency relation:

$$M_{ijklm}^{\text{flow}} = \{ \ \mathbf{i} \to \mathbf{j} \ | \ \exists \ s \in \mathbf{D}_m^{\text{var}} \ \text{s.t.} \ \mathbf{i} \in \mathbf{D}_i^{\text{iter}} \ \wedge \ \mathbf{j} \in \mathbf{D}_j^{\text{iter}} \ \wedge$$
$$\mathbf{M}_{ikm}^{\text{def}}(\mathbf{i}) = s = \mathbf{M}_{jlm}^{\text{oper}}(\mathbf{j}) \ \} \tag{5.7}$$

Of course the definition and operand domains causing a dependency always belong to the same array. In general there may be several dependencies present between two statements, if there are multiple overlapping definition and operand domains.

5.5 MEMORY OCCUPATION MODELS

As stated before, the primary domain models do not contain all information about the programs being modeled. Especially the execution and storage order are missing in these models. If one wants to perform storage optimization, this information about order is crucial. However, the execution and storage order are exactly the subject of optimization, i.e. the result of an optimization is a(n) (new) execution and/or storage order. Hence we must be able to decide whether such orders are valid or not, before we can try to optimize them.

Therefore, we have to derive some necessary and sufficient conditions that the execution and storage order must satisfy. For reasons mentioned earlier, we restrict ourselves to single-assignment programs. The result can be extended to non-single-assignment programs (provided that accurate data-flow analysis is possible), but this would not give extra insights and it would only obscure things. If a less accurate (conservative) data-flow analysis would be used for non-single-assignment programs, the memory occupation models described in this section would still be usable in the sense that they would be conservative too, i.e. they would describe a worst-case memory occupation.

Before we can come up with these conditions, we must first gain some insight in the problem of memory occupation. Therefore, we have further extended the mathematical models presented earlier to also accurately describe memory occupation. These extended models have not yet been described in literature before.

5.5.1 Memory occupation

We can say that a memory location is occupied as long as it contains a data value that *may* still be needed by the program being executed or by its environment. During the execution of a program, a certain memory location may contain several distinct values during disjoint time intervals. It must be clear that an execution and storage order are invalid if they result in two distinct values occupying the same memory location at the same time. But before we can check this, we must first know *when a memory location is occupied.* The following example shows how we can find this out:

Example 5.5.1

```
    int A[15];
    for ( i = 0; i < 10; ++i )
S1:    A[i] = ...;
    ...
    for ( j = 5; j < 15; ++j )
S2:    ... = f(A[j]);
```

The corresponding iteration, definition and operand domains have been graphically represented at the top of Fig. 5.8. For this simple example, one can easily see that there are some elements of A being produced during the executions of statement S1 and also being consumed during the executions of statement S2. This results in partially overlapping definition and operand domains, as indicated at the top of Fig. 5.8. Consequently, there exists a (value-based) flow-dependency between these two statements.

For each array element that is written during the execution of the first statement and read during the execution of the second one, we can easily find out when it is written, when it is read and where it is stored, provided that we know the execution order of both statements and the storage order of the array. This is indicated for one of

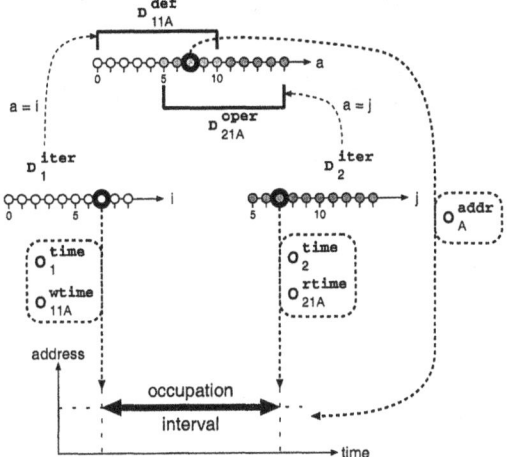

Figure 5.8.   The memory occupation of example 5.5.1.

the common array elements in Fig. 5.8. So, in an address/time diagram, shown at the bottom of the figure, we can see which is the location being occupied and when it is occupied. Note that in practice, some values may be read multiple times, so the corresponding memory location is occupied until it is read for the last time.

From the above example we can see that in principle, given the primary domain descriptions and ordering functions, we can derive the occupation intervals for each memory location. However, just as we don't want to describe individual operations and individual variables, but groups of operations and groups of variables instead, we also don't want to describe the memory occupation for each individual location either. If possible, we should be able to come up with closed mathematical expressions describing the memory occupation of *groups* of memory locations. This is done next.

5.5.2 Binary occupied address/time domain

As indicated above, it is possible to derive the memory occupation interval and address for each variable that is being written during the execution of one statement and that is being read during the execution of another statement, i.e. each variable that contributes to a (value-based) flow-dependency between two statements. Not surprisingly, given two statements, one of which writes elements of an array and one of which reads elements of the same array, it is possible to come up with a closed expression describing the memory occupation for *all* values being written by the first statement *and* being read by the second one.

First of all, given two such statements, we can easily find the mathematical description for the commonly accessed array elements and the addresses they occupy, provided that we know the storage order for that array, by taking the intersection of the corresponding definition and operand domains, and applying the storage order to the intersection:

$$\mathbf{D}_{ijklm}^{\mathrm{addr}} = \{\ a\ |\ \exists\ \mathbf{s} \in \mathbf{D}_m^{\mathrm{var}}, w\ \text{s.t.}\ a = O_m^{\mathrm{addr}}(\mathbf{s}, w)\ \wedge$$
$$\mathbf{C}_m^{\mathrm{addr}}(w) \geq 0\ \wedge\ \mathbf{s} \in \mathbf{D}_{ikm}^{\mathrm{def}} \bigcap \mathbf{D}_{jlm}^{\mathrm{oper}}\ \} \tag{5.8}$$

This expression describes all of the addresses that are (potentially) being occupied during some time due to a flow-dependency between two statements. Note that in theory, the constraint $\mathbf{s} \in \mathbf{D}_m^{\mathrm{var}}$ should be redundant, at least for valid programs, since the definition and operand domains should always be subsets of the variable domains (otherwise an array is being read or written outside its boundaries).

In practice however, this extra constraint may contain some extra information. For instance, if the reads or writes are non-manifest, and no other information is known except that they should not access the array outside its boundaries, then it is not necessary to add this information explicitly to all the definition or operand domains. Instead, it can be added only to the variable domain.

In order to be complete, we also need to know *when* these addresses are occupied. From the execution order, we can derive when the addresses are written and when they are read:

$$\mathbf{D}_{ikm}^{\mathtt{wtime}} = \{\ w \mid \exists\, \mathbf{i} \in \mathbf{D}_i^{\mathtt{iter}}, x \ \text{s.t.}$$
$$w = O_{ikm}^{\mathtt{wtime}}(O_i^{\mathtt{time}}(\mathbf{i}, x)) \ \wedge \ \mathbf{C}_{ikm}^{\mathtt{time}}(x) \geq 0\ \} \tag{5.9}$$
$$\mathbf{D}_{jlm}^{\mathtt{rtime}} = \{\ r \mid \exists\, \mathbf{j} \in \mathbf{D}_j^{\mathtt{iter}}, y \ \text{s.t.}$$
$$r = O_{jlm}^{\mathtt{rtime}}(O_j^{\mathtt{time}}(\mathbf{j}, y)) \ \wedge \ \mathbf{C}_{jlm}^{\mathtt{time}}(y) \geq 0\ \} \tag{5.10}$$

We are now ready to combine these 3 expressions, as we know that each address in equation 5.8 is *possibly*[10] occupied from the corresponding time in equation 5.9 till *at least*[11] the corresponding time in equation 5.10. This results in the following expression:

$$\mathbf{D}_{ijklm}^{\mathtt{BOAT}} = \{\ [a,t] \mid \exists\, \mathbf{s} \in \mathbf{D}_m^{\mathtt{var}}, \mathbf{i} \in \mathbf{D}_i^{\mathtt{iter}}, \mathbf{j} \in \mathbf{D}_j^{\mathtt{iter}}, x, y, w \ \text{s.t.}$$
$$a = O_m^{\mathtt{addr}}(\mathbf{s}, w) \ \wedge \ \mathbf{C}_m^{\mathtt{addr}}(w) \geq 0 \ \wedge$$
$$\mathbf{M}_{ikm}^{\mathtt{def}}(\mathbf{i}) = \mathbf{s} = \mathbf{M}_{jlm}^{\mathtt{oper}}(\mathbf{j}) \ \wedge$$
$$t \geq O_{ikm}^{\mathtt{wtime}}(O_i^{\mathtt{time}}(\mathbf{i}, x)) \ \wedge \ \mathbf{C}_{ikm}^{\mathtt{time}}(x) \geq 0 \ \wedge$$
$$t \leq O_{jlm}^{\mathtt{rtime}}(O_j^{\mathtt{time}}(\mathbf{j}, y)) \ \wedge \ \mathbf{C}_{jlm}^{\mathtt{time}}(y) \geq 0\ \} \tag{5.11}$$

This expression is the description of a two-dimensional geometrical domain, which we call a **binary occupied address/time domain**[12] (BOAT-domain). Each point with integer coordinates in this domain represents an occupied address/time tuple, i.e. an address that is (possibly) being occupied at that time.

For a given execution and storage order, this equation contains *all* information available at compile time about the (potential) memory occupation due to a flow-dependency between two statements. Comparison with equation 5.7 reveals that the mathematical constraints present in a dependency relation are also present in equation 5.11. In fact a BOAT-domain is nothing more than a mapping of a dependency on an address/time space.

In order to get a better feeling of what this equation represents, we have a look at a small example:

Example 5.5.2

```
    int A[5][5];
    for ( i = 0; i < 5; ++i )
      for ( j = 0; j < 5; ++j )
S1:     A[i][4-j] = f(...);

    for ( k = 0; k < 5; k += 2 )
      for ( l = 0; l < 5; ++l )
S2:     if ( l <= k ) ... = g(A[l][k]);
```

For this program, we have the following domain descriptions:

$$\mathbf{D}_A^{\mathtt{var}} = \{\ [s_1, s_2] \mid 0 \leq s_1 \leq 4 \ \wedge \ 0 \leq s_2 \leq 4 \ \wedge \ [s_1, s_2] \in \mathbb{Z}^2\ \}$$

---

[10]Possibly, because the program may be non-manifest.
[11]At least, because the same variables may be read again by other statements.
[12]Binary, because it is always associated with a (value-based) flow-dependency between *two* statements (possibly identical).

$$\mathbf{D}_1^{\mathtt{iter}} = \{ \ [i,j] \mid 0 \le i \le 4 \ \wedge \ 0 \le j \le 4 \ \wedge \ [i,j] \in \mathbb{Z}^2 \ \}$$
$$\mathbf{D}_{11A}^{\mathtt{def}} = \{ \ [d_1,d_2] \mid \exists \ [i,j] \in \mathbf{D}_1^{\mathtt{iter}} \ \text{s.t.} \ d_1 = i \ \wedge \ d_2 = 4 - j \ \}$$
$$\mathbf{D}_2^{\mathtt{iter}} = \{ \ [k,l] \mid \exists \ \alpha \in \mathbb{Z} \ \text{s.t.} \ 0 \le k \le 4 \ \wedge \ 0 \le l \le 4 \ \wedge \ l \le k = 2\alpha \ \wedge \ [k,l] \in \mathbb{Z}^2 \ \}$$
$$\mathbf{D}_{21A}^{\mathtt{oper}} = \{ \ [o_1,o_2] \mid \exists \ [k,l] \in \mathbf{D}_2^{\mathtt{iter}} \ \text{s.t.} \ o_1 = l \ \wedge \ o_2 = k \ \}$$

If we assume that this program is executed sequentially, and that each of the statements S1 and S2 can be executed in one clock cycle, we get the following time order functions:

$$O_1^{\mathtt{time}}(i,j) = 5i + j \qquad\qquad O_2^{\mathtt{time}}(k,l) = 25 + \frac{5k}{2} + l$$
$$O_{11A}^{\mathtt{vtime}}(x) = x \qquad\qquad O_{11B}^{\mathtt{rtime}}(y) = y$$

We also assume a row-major storage order function for array A:

$$O_A^{\mathtt{addr}}(s_1,s_2) = 5s_1 + s_2$$

This results in the following BOAT-domain description:

$$\mathbf{D}_{1211A}^{\mathtt{BOAT}} = \{ \ [a,t] \mid \exists \ [s_1,s_2] \in \mathbb{Z}^2, [i,j] \in \mathbb{Z}^2, [k,l] \in \mathbb{Z}^2, \alpha \in \mathbb{Z} \ \text{s.t.}$$
$$a = 5s_1 + s_2 \ \wedge \ t \ge 5i + j \ \wedge \ t \le 25 + \frac{5k}{2} + l \ \wedge$$
$$s_1 = i \ \wedge \ s_2 = 4 - j \ \wedge \ s_1 = l \ \wedge \ s_2 = k \ \wedge$$
$$0 \le i \le 4 \ \wedge \ 0 \le j \le 4 \ \wedge$$
$$0 \le k \le 4 \ \wedge \ 0 \le l \le 4 \ \wedge \ l \le k \ \wedge \ k = 2\alpha \ \wedge$$
$$0 \le s_1 \le 4 \ \wedge \ 0 \le s_2 \le 4 \ \}$$

This expression can easily be simplified by elimination of several existentially quantified dimensions and removal of redundant (in)equalities:

$$\mathbf{D}_{1211A}^{\mathtt{BOAT}} = \{ \ [a,t] \mid \exists \ s_1 \in \mathbb{Z}, \alpha \in \mathbb{Z} \ \text{s.t.}$$
$$a = 5s_1 + 2\alpha \ \wedge \ t \ge 5s_1 + 4 - 2\alpha \ \wedge \ t \le 25 + 5\alpha + s_1 \ \wedge$$
$$0 \le s_1 \le 4 \ \wedge \ 0 \le \alpha \le 2 \ \wedge \ s_1 \le 2\alpha \ \}$$

A graphical representation of this BOAT-domain is given in Fig. 5.9. At the top of the figure all memory accesses, both writes and reads, are shown in an address/time diagram. The addresses which are both written and read contribute to the BOAT-domain, as indicated by the thick lines. The corresponding address/time tuples are duplicated at the bottom of the figure, where the exact BOAT-domain is shown. Note that the addresses contained in the BOAT-domain are discrete, whereas time is continuous (e.g. when an address is occupied from time $t = 1$ till $t = 2$, it is also occupied at time $t = 1.3763$, but an address with a non-integral value is meaningless in practice). For practical purposes, time can also be considered to be discrete, as we assume that nothing interesting can happen at non-integral points in time.[13]

Note that in this special case, where all constraints and ordering functions are manifest and affine, and all constraints are convex, the resulting BOAT-domain is a linearly bounded lattice (LBL) [391].

Another example shows how non-manifest behaviour can be incorporated in the BOAT-domain descriptions:

Example 5.5.3

```
int A[5][5], B[5];
for ( i = 0; i < 5; ++i )
   for ( j = 0; j < 5; ++j )
```

---

[13] Of course, as someone correctly pointed out, it is always possible that an alpha particle impact toggles a memory bit in between two discrete moments in time. Model extensions to deal with these phenomena are not being considered (yet). ☺

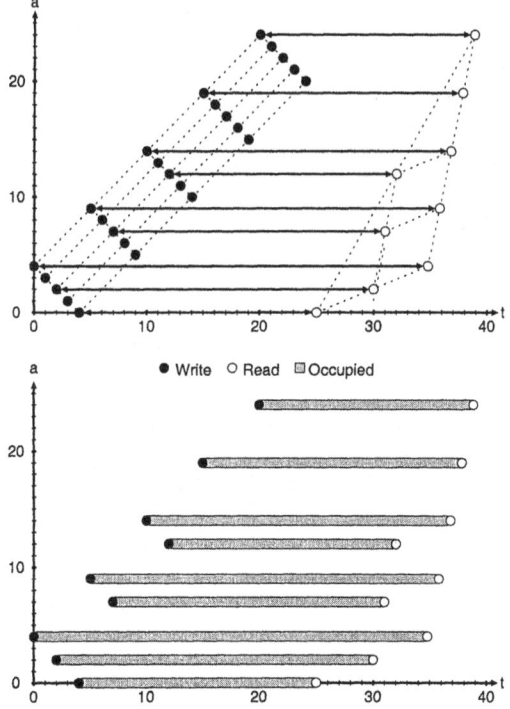

Figure 5.9.   The construction of the BOAT-domain of example 5.5.2.

```
S1:       A[i][j] = f(...);

    for ( k = 0; k < 5; ++k )
        for ( l = 0; l < 5; ++l )
            if ( B[k] >= 0 )          /* non-manifest */
S2:                ... = g(A[k][l]);
```

Again, we can extract the domain descriptions in a straightforward way:

$$\mathbf{D}_A^{\mathtt{var}} = \{ \ [s_1, s_2] \ | \ 0 \le s_1 \le 4 \ \wedge \ 0 \le s_2 \le 4 \ \wedge \ [s_1, s_2] \in \mathbb{Z}^2 \ \}$$
$$\mathbf{D}_1^{\mathtt{iter}} = \{ \ [i, j] \ | \ 0 \le i \le 4 \ \wedge \ 0 \le j \le 4 \ \wedge \ [i, j] \in \mathbb{Z}^2 \ \}$$
$$\mathbf{D}_{11A}^{\mathtt{def}} = \{ \ [d_1, d_2] \ | \ \exists \ [i, j] \in \mathbf{D}_1^{\mathtt{iter}} \ \mathrm{s.t.} \ d_1 = i \ \wedge \ d_2 = j \ \}$$
$$\mathbf{D}_2^{\mathtt{iter}} = \{ \ [k, l] \ | \ \exists \ p \ \mathrm{s.t.} \ 0 \le k \le 4 \ \wedge \ 0 \le l \le 4 \ \wedge \ p \ge 0 \ \wedge$$
$$p = \mathbf{F}_2^{\mathtt{iter}}(k) \ \wedge \ [k, l] \in \mathbb{Z}^2 \ \}$$
$$\mathbf{D}_{21A}^{\mathtt{oper}} = \{ \ [o_1, o_2] \ | \ \exists \ [k, l] \in \mathbf{D}_2^{\mathtt{iter}} \ \mathrm{s.t.} \ o_1 = k \ \wedge \ o_2 = l \ \}$$

Note that $\mathbf{F}_2^{\mathtt{iter}}()$ is an unknown function, which is used to model the non-manifest behaviour of the program: we are not sure whether any of the elements of A is ever read.

If we assume the same ordering functions as in example 5.5.2, we arrive at the following BOAT-domain description (after simplification):

$$\mathbf{D}_{1211A}^{\mathtt{BOAT}} = \{ \ [a, t] \ | \ \exists \ [s_1, s_2] \in \mathbb{Z}^2, p \ \mathrm{s.t.}$$

$$a = 5s_1 + s_2 \ \wedge \ t \geq 5s_1 + s_2 \ \wedge \ t \leq 25 + s_15 + s_2 \ \wedge$$
$$0 \leq s_1 \leq 4 \ \wedge \ 0 \leq s_2 \leq 4 \ \wedge \ p \geq 0 \ \wedge \ p = \mathbf{F}_2^{\mathtt{iter}}(s_1) \ \}$$

A graphical representation of the BOAT-domain is shown in Fig. 5.10. Due to the condition in the program, it is not known until run-time whether any of the elements of A is ever read and whether the corresponding memory locations are therefore ever occupied (according to our definition of occupation, see subsection 5.5.1). We do know however that there are some relations between some of the occupied address/time tuples: if for a given execution of the k-loop, the condition B[k] >= 0 evaluates to true, then all of the corresponding executions of statement S2 are executed. Consequently, if one of the corresponding addresses is read, all of them are read. In other words, the addresses can be divided into groups of addresses whose run-time occupation is related. These groups are indicated by the differently shaded regions in the figure.

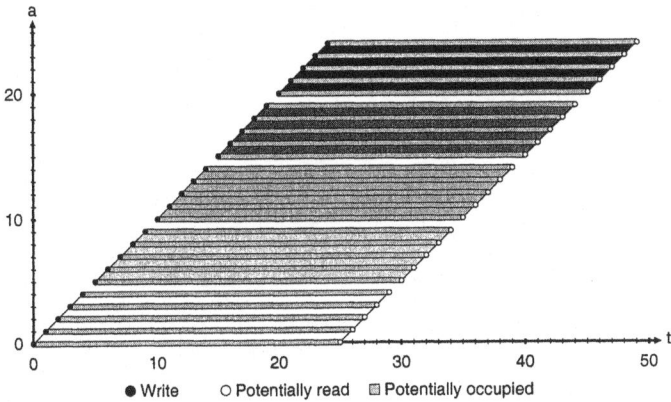

Figure 5.10.   The BOAT-domain of example 5.5.3.

This information is explicitly present in the description of the BOAT-domain due to the presence of $\mathbf{F}_2^{\mathtt{iter}}()$. It must be intuitively clear that this can be very valuable in a memory optimization context. Ignoring it and taking worst-case assumptions can prohibit certain optimizations.

5.5.3 Occupied address/time domain

We have shown that it is possible to accurately model the memory occupation for a set of variables being written by one statement and being read by another one, in other words, for each variable contributing to a value-based flow-dependency. However, we are not only interested in the memory occupation due to flow-dependencies, but rather in the memory occupation of a complete set of variables, i.e. a complete array[14].

We can easily find the complete memory occupation of an array if we know the memory occupation due to *each* value-based flow dependency related to that array. The memory occupation of the array is then simply the *union* of the memory occupation due to each of the flow dependencies. Consequently, we can also model the memory occupation of an array by a geometrical domain, which we call the **occupied address/time domain** (OAT-domain) of that array. This domain is simply the union of the BOAT-domains of all value-based flow-dependencies related to that array:

$$\mathbf{D}_m^{\mathtt{OAT}} = \bigcup_{ijkl} \mathbf{D}_{ijklm}^{\mathtt{BOAT}} \tag{5.12}$$

---

[14]It is assumed that the storage order is common for the complete array. If required, an array can always be divided first into sub-arrays (e.g. basic sets [32]) each having their own storage order.

In general, the different BOAT-domains of an array are not disjoint. For instance, if an array element is written by one statement and read by two other statements, then this element contributes to two flow-dependencies and the corresponding BOAT-domains have certain address/time tuples in common. An example of the memory occupation for a complete array is given next.

Example 5.5.4

```
    int A[4][4];
    for ( i = 0; i < 4; ++i )
        for ( j = 0; j < 4; ++j )
S1:         if ( j <= i ) A[i][j] = f1(...);

    for ( k = 0; k < 4; ++k )
        for ( l = 0; l < 4; ++l )
S2:         if ( l > k ) A[k][l] = f2(...);

    for ( m = 0; m < 4; ++m )
        for ( n = 0; n < 4; ++n )
S3:         if ( n >= m ) ... = g1(A[3-m][n]);

    for ( o = 0; o < 4; ++o )
        for ( p = 0; p < 4; ++p )
S4:         if ( p <= o ) ... = g2(A[3-o][p]);
```

This program contains 2 statements in which elements of A are written and 2 statements in which elements are read. Consequently, 4 value-based flow-dependencies can be present (one from each writing statement to each reading statement), and in fact they are present here. This means that we can describe 4 BOAT-domains.

The variable, iteration, definition and operand domains are the following:

$$\mathbf{D}_A^{\mathtt{var}} = \{\ [s_1, s_2]\ |\ 0 \le s_1 \le 3\ \wedge\ 0 \le s_2 \le 3\ \wedge\ [s_1, s_2] \in \mathbb{Z}^2\ \}$$
$$\mathbf{D}_1^{\mathtt{iter}} = \{\ [i, j]\ |\ 0 \le i \le 3\ \wedge\ 0 \le j \le 3\ \wedge\ j \le i\ \wedge\ [i, j] \in \mathbb{Z}^2\ \}$$
$$\mathbf{D}_{11A}^{\mathtt{def}} = \{\ [s_1, s_2]\ |\ \exists\ [i, j] \in \mathbf{D}_1^{\mathtt{iter}}\ \text{s.t.}\ s_1 = i\ \wedge\ s_2 = j\ \wedge\ [s_1, s_2] \in \mathbb{Z}^2\ \}$$
$$\mathbf{D}_2^{\mathtt{iter}} = \{\ [k, l]\ |\ 0 \le k \le 3\ \wedge\ 0 \le l \le 3\ \wedge\ l > k\ \wedge\ [k, l] \in \mathbb{Z}^2\ \}$$
$$\mathbf{D}_{21A}^{\mathtt{def}} = \{\ [s_1, s_2]\ |\ \exists\ [k, l] \in \mathbf{D}_2^{\mathtt{iter}}\ \text{s.t.}\ s_1 = k\ \wedge\ s_2 = l\ \wedge\ [s_1, s_2] \in \mathbb{Z}^2\ \}$$
$$\mathbf{D}_3^{\mathtt{iter}} = \{\ [m, n]\ |\ 0 \le m \le 3\ \wedge\ 0 \le n \le 3\ \wedge\ n \ge m\ \wedge\ [n, m] \in \mathbb{Z}^2\ \}$$
$$\mathbf{D}_{31A}^{\mathtt{oper}} = \{\ [s_1, s_2]\ |\ \exists\ [n, m] \in \mathbf{D}_3^{\mathtt{iter}}\ \text{s.t.}\ s_1 = 3 - m\ \wedge\ s_2 = n\ \wedge\ [s_1, s_2] \in \mathbb{Z}^2\ \}$$
$$\mathbf{D}_4^{\mathtt{iter}} = \{\ [o, p]\ |\ 0 \le o \le 3\ \wedge\ 0 \le p \le 3\ \wedge\ p \le o\ \wedge\ [o, p] \in \mathbb{Z}^2\ \}$$
$$\mathbf{D}_{41A}^{\mathtt{oper}} = \{\ [s_1, s_2]\ |\ \exists\ [o, p] \in \mathbf{D}_4^{\mathtt{iter}}\ \text{s.t.}\ s_1 = 3 - o\ \wedge\ s_2 = p\ \wedge\ [s_1, s_2] \in \mathbb{Z}^2\ \}$$

Again, we assume a procedural execution order in which each statement takes 1 time unit to execute, and a column-major storage order for A:

$$O_1^{\mathtt{time}}(i, j) = 4i + j \qquad\qquad O_{11A}^{\mathtt{wtime}}(x) = x$$
$$O_2^{\mathtt{time}}(k, l) = 16 + 4k + l \qquad\qquad O_{21A}^{\mathtt{wtime}}(x) = x$$
$$O_3^{\mathtt{time}}(n, m) = 32 + 4m + n \qquad\qquad O_{31A}^{\mathtt{rtime}}(x) = x \qquad\qquad O_A^{\mathtt{addr}}(s_1, s_2) = s_1 + 4s_2$$
$$O_4^{\mathtt{time}}(o, p) = 48 + 4o + p \qquad\qquad O_{41A}^{\mathtt{rtime}}(x) = x$$

This leads us to the following BOAT-domain descriptions, after simplification:

$$\mathbf{D}_{1311A}^{\mathtt{BOAT}} = \{\ [a, t]\ |\ \exists\ [s_1, s_2] \in \mathbb{Z}^2\ \text{s.t.}$$
$$a = s_1 + 4s_2\ \wedge\ t \ge 4s_1 + s_2\ \wedge\ t \le 44 - 4s_1 + s_2\ \wedge$$
$$0 \le s_1 \le 3\ \wedge\ 0 \le s_2 \le 3\ \wedge\ s_2 \le s_1\ \wedge\ s_2 \ge 3 - s_1\ \}$$
$$\mathbf{D}_{1411A}^{\mathtt{BOAT}} = \{\ [a, t]\ |\ \exists\ [s_1, s_2] \in \mathbb{Z}^2\ \text{s.t.}$$
$$a = s_1 + 4s_2\ \wedge\ t \ge 4s_1 + s_2\ \wedge\ t \le 60 - 4s_1 + s_2\ \wedge$$

$$0 \leq s_1 \leq 3 \ \wedge \ 0 \leq s_2 \leq 3 \ \wedge \ s_2 \leq s_1 \ \wedge \ s_2 \leq 3 - s_1 \ \}$$

$$\mathbf{D}^{\text{BOAT}}_{2311A} = \{ \ [a,t] \ | \ \exists \ [s_1, s_2] \in \mathbb{Z}^2 \ \text{s.t.}$$

$$a = s_1 + 4s_2 \ \wedge \ t \geq 16 + 4s_1 + s_2 \ \wedge \ t \leq 44 - 4s_1 + s_2 \ \wedge$$

$$0 \leq s_1 \leq 3 \ \wedge \ 0 \leq s_2 \leq 3 \ \wedge \ s_2 > s_1 \ \wedge \ s_2 \geq 3 - s_1 \ \}$$

$$\mathbf{D}^{\text{BOAT}}_{2411A} = \{ \ [a,t] \ | \ \exists \ [s_1, s_2] \in \mathbb{Z}^2 \ \text{s.t.}$$

$$a = s_1 + 4s_2 \ \wedge \ t \geq 16 + 4s_1 + s_2 \ \wedge \ t \leq 60 - 4s_1 + s_2 \ \wedge$$

$$0 \leq s_1 \leq 3 \ \wedge \ 0 \leq s_2 \leq 3 \ \wedge \ s_2 > s_1 \ \wedge \ s_2 \leq 3 - s_1 \ \}$$

A graphical representation[15] *of these BOAT- and OAT-domains is shown in Fig. 5.11. Note that there is an overlap between* $\mathbf{D}^{\text{BOAT}}_{1311A}$ *and* $\mathbf{D}^{\text{BOAT}}_{1411A}$ *at addresses 3 and 6, and an overlap between* $\mathbf{D}^{\text{BOAT}}_{2311A}$ *and* $\mathbf{D}^{\text{BOAT}}_{2411A}$ *at addresses 9 and 12. This is because these addresses are being read more than once. The dotted triangles represent the read and write accesses during the different loops. Note that write (or read) operations of the same statement may contribute to different BOAT-domains in different ways (e.g. not all writes of the same statement must contribute to the same BOAT-domain).*

5.5.4   Collective occupied address/time domain

Now that we know how to model the memory occupation for a complete array, we can easily derive an expression for the occupation of a memory itself. It is again simply the union of the memory occupation of all arrays *assigned to that memory*. The result is again a geometrical domain, which we call the **collective occupied address/time domain** (COAT-domain) for that memory:

$$\mathbf{D}^{\text{COAT}} = \bigcup_m \mathbf{D}^{\text{OAT}}_m \tag{5.13}$$

It must be clear that one of the preconditions for an execution and storage order to be valid, is that none of the OAT-domains of different arrays assigned to the same memory overlap, because this would mean that 2 arrays are using the same memory location at the same time, which is incorrect.

An example of a COAT-domain for a non-manifest program and a memory containing 3 arrays is given next.

Example 5.5.5

```
    int A[10], B[5], C[10];
    for ( i = 0; i < 10; ++i )
S1:    C[i] = f1(...);

    for ( j = 0; j < 5; ++j )
S2:    A[j] = f2(...);

    for ( k = 5; k < 10; ++k )
S3:    if ( C[k-5] > 0 )
S4:        A[k] = f3(...);
       else
S5:        B[k-5] = f4(...);

    for ( l = 0; l < 10; ++l )
S6:    if ( l >= 5 && C[l] <= 0 )
S7:        ... = f5(B[l-5]);
       else
S8:        ... = f6(A[l]);
```

The variable, iteration, definition and operand domains for this example are the following:

$$\mathbf{D}^{\text{var}}_A = \{ \ s_a \ | \ 0 \leq s_a \leq 9 \ \wedge \ s_a \in \mathbb{Z} \ \}$$

$$\mathbf{D}^{\text{var}}_B = \{ \ s_b \ | \ 0 \leq s_b \leq 4 \ \wedge \ s_b \in \mathbb{Z} \ \}$$

---

[15]Don't try to do this yourself. It can result in a nervous breakdown. ☺

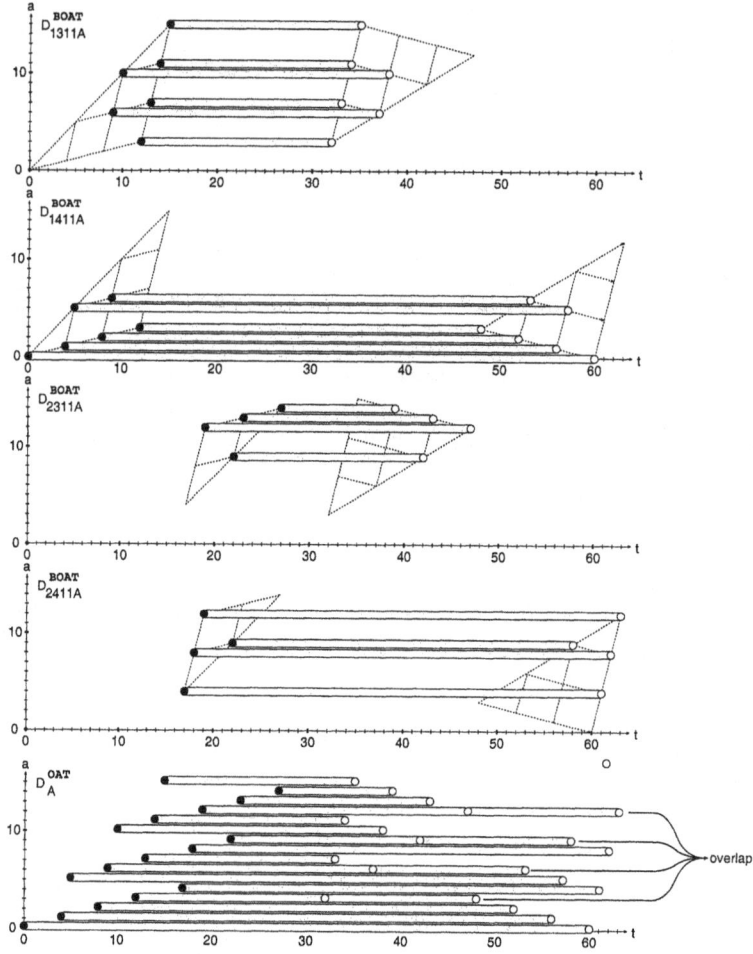

Figure 5.11.   The BOAT- and OAT-domains of example 5.5.4.

$$\mathbf{D}_C^{\mathrm{var}} = \{\ s_c \mid 0 \leq s_c \leq 9 \ \wedge\ s_c \in \mathbb{Z}\ \}$$
$$\mathbf{D}_1^{\mathrm{iter}} = \{\ i \mid 0 \leq i \leq 9 \ \wedge\ i \in \mathbb{Z}\ \}$$
$$\mathbf{D}_2^{\mathrm{iter}} = \{\ j \mid 0 \leq j \leq 4 \ \wedge\ j \in \mathbb{Z}\ \}$$
$$\mathbf{D}_3^{\mathrm{iter}} = \{\ k \mid 5 \leq k \leq 9 \ \wedge\ k \in \mathbb{Z}\ \}$$
$$\mathbf{D}_4^{\mathrm{iter}} = \{\ k \mid 5 \leq k \leq 9 \ \wedge\ p = F(k) \ \wedge\ p > 0 \ \wedge\ k \in \mathbb{Z}\ \}$$
$$\mathbf{D}_5^{\mathrm{iter}} = \{\ k \mid 5 \leq k \leq 9 \ \wedge\ p = F(k) \ \wedge\ p \leq 0 \ \wedge\ k \in \mathbb{Z}\ \}$$
$$\mathbf{D}_6^{\mathrm{iter}} = \{\ l \mid 0 \leq l \leq 9 \ \wedge\ l \in \mathbb{Z}\ \}$$
$$\mathbf{D}_7^{\mathrm{iter}} = \{\ l \mid 0 \leq l \leq 9 \ \wedge\ q = F(l) \ \wedge\ l \geq 5 \ \wedge\ q \leq 0 \ \wedge\ l \in \mathbb{Z}\ \}$$
$$\mathbf{D}_8^{\mathrm{iter}} = \{\ l \mid 0 \leq l \leq 9 \ \wedge\ q = F(l) \ \wedge\ (l < 5 \ \vee\ q > 0) \ \wedge\ l \in \mathbb{Z}\ \}$$
$$\mathbf{D}_{11C}^{\mathrm{def}} = \{\ s_c \mid \exists\, i \in \mathbf{D}_1^{\mathrm{iter}} \ \text{s.t.}\ s_c = i\ \}$$
$$\mathbf{D}_{21A}^{\mathrm{def}} = \{\ s_a \mid \exists\, j \in \mathbf{D}_2^{\mathrm{iter}} \ \text{s.t.}\ s_a = j\ \}$$

$$D_{31C}^{oper} = \{ s_c \mid \exists k \in D_3^{iter} \text{ s.t. } s_c = k - 5 \}$$
$$D_{41A}^{def} = \{ s_a \mid \exists k \in D_4^{iter} \text{ s.t. } s_a = k \}$$
$$D_{51B}^{def} = \{ s_b \mid \exists k \in D_5^{iter} \text{ s.t. } s_b = k - 5 \}$$
$$D_{61C}^{oper} = \{ s_c \mid \exists l \in D_6^{iter} \text{ s.t. } s_c = l \}$$
$$D_{71B}^{oper} = \{ s_b \mid \exists l \in D_7^{iter} \text{ s.t. } s_b = l - 5 \}$$
$$D_{81A}^{oper} = \{ s_a \mid \exists l \in D_8^{iter} \text{ s.t. } s_a = l \}$$

We assume the following execution and storage orders (assuming that arrays A, B and C all have the same element size):

$$O_1^{time}(i) = i \qquad\qquad O_{11C}^{vtime}(x) = x$$
$$O_2^{time}(j) = j + 10 \qquad\qquad O_{21A}^{vtime}(x) = x$$
$$O_3^{time}(i) = 2(k - 5) + 15 \qquad O_{31C}^{rtime}(x) = x$$
$$O_4^{time}(i) = 2(k - 5) + 16 \qquad O_{41A}^{vtime}(x) = x \qquad\qquad O_A^{addr}(s_a) = s_a$$
$$O_5^{time}(i) = 2(k - 5) + 16 \qquad O_{51B}^{vtime}(x) = x \qquad\qquad O_B^{addr}(s_b) = s_b + 5$$
$$O_6^{time}(i) = 2l + 25 \qquad\qquad O_{61C}^{rtime}(x) = x \qquad\qquad O_C^{addr}(s_b) = s_c + 10$$
$$O_7^{time}(i) = 2l + 26 \qquad\qquad O_{71B}^{rtime}(x) = x$$
$$O_8^{time}(i) = 2l + 26 \qquad\qquad O_{81A}^{rtime}(x) = x$$

From this, we can derive the following BOAT-domains (after simplification):

$$D_{1311C}^{BOAT} = \{ [a, t] \mid 10 \le a \le 14 \wedge a - 10 \le t \le 2a - 5 \wedge a \in \mathbb{Z} \}$$
$$D_{1611C}^{BOAT} = \{ [a, t] \mid 10 \le a \le 19 \wedge a - 10 \le t \le 2a + 5 \wedge a \in \mathbb{Z} \}$$
$$D_{2811A}^{BOAT} = \{ [a, t] \mid 0 \le a \le 4 \wedge a + 10 \le t \le 2a + 26 \wedge F(a) > 0 \wedge a \in \mathbb{Z} \}$$
$$D_{4811A}^{BOAT} = \{ [a, t] \mid 5 \le a \le 9 \wedge 2a + 6 \le t \le 2a + 26 \wedge F(a) > 0 \wedge a \in \mathbb{Z} \}$$
$$D_{5711B}^{BOAT} = \{ [a, t] \mid 5 \le a \le 9 \wedge 2a + 6 \le t \le 2a + 26 \wedge F(a) \le 0 \wedge a \in \mathbb{Z} \}$$

The corresponding graphical representations can be found in Fig. 5.12. The resulting OAT-domains and COAT-domain can be found in Fig. 5.13. Note that the OAT-domains of arrays A and B seem to overlap graphically. This is caused by a virtual overlap between D_{4811A}^{BOAT} and D_{5711B}^{BOAT}. One can verify that this overlap is non-existent in reality:

$$D_{4811A}^{BOAT} \bigcap D_{5711B}^{BOAT} = \{ [a, t] \mid 5 \le a \le 9 \wedge 2a + 6 \le t \le 2a + 26 \wedge$$
$$\underline{F(a) > 0 \wedge F(a) \le 0} \wedge a \in \mathbb{Z} \} = \phi$$

So, the memory occupation of arrays A and B is partly conditional, and the conditions for A and B are complementary.

In practice, this means that some of the elements of arrays A and B can share storage locations, simply because they can never occupy those locations at the same time.

If we would not have modeled this exclusive condition accurately, we could not have been sure that this storage order was valid, and we would have to make sure that the address ranges of arrays A and B were disjoint. This would require extra storage locations. So this example already clearly illustrates how a more accurate modeling of non-manifest programs may result in better solutions, compared to the ones obtained through traditional worst-case modeling techniques.

5.6 VALIDITY CONSTRAINTS

In the previous section, we have demonstrated how the memory occupation for a given storage and execution order can be accurately modeled. However, as stated before, the execution and storage order are the subject of optimization. We therefore have to be able to decide whether a given order is valid. We say that a combination of storage and execution order for a program is valid if it satisfies the following constraints:

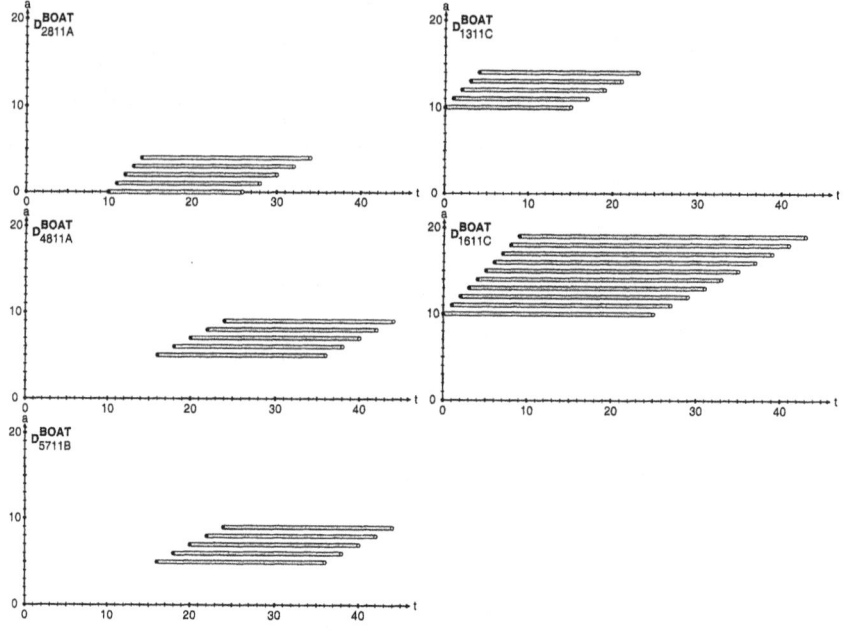

Figure 5.12.   The BOAT-domains of example 5.5.5.

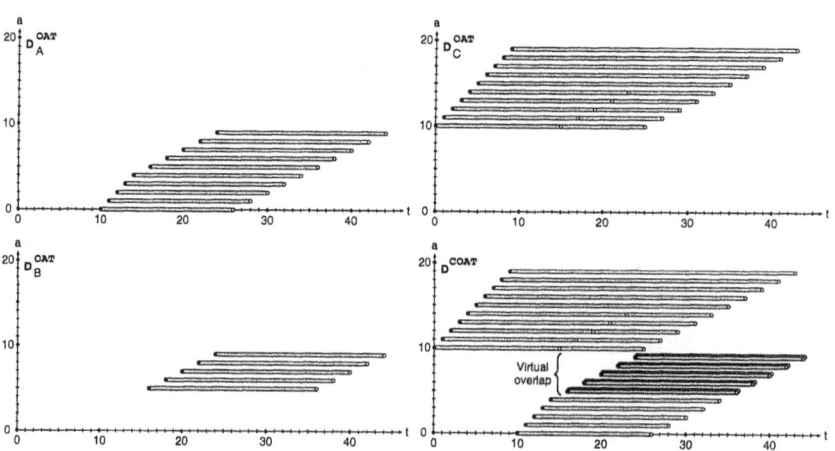

Figure 5.13.   The OAT- and COAT-domains of example 5.5.5.

1. No value of any variable (e.g. array element) is ever read from a memory location before it has been written to that location.

2. No memory location *might*[16] ever be overwritten when it *might* still contain a data value of a variable that *might* still have to be read.

In this section we derive a set of necessary mathematical conditions that the storage and/or execution order have to satisfy. Together, these conditions form a sufficient set for having a valid storage and execution order. Again we limit ourselves to the single-assignment case. Extensions for the non-single-assignment case are possible but require an extended data-flow analysis.

5.6.1 Necessary conditions for the execution order

The first validity constraint only affects the execution order. Given the primary domain models of section 5.2, we can easily express this constraint mathematically:

$$\forall\, i, j, k, l, m: \quad \text{if} \quad \exists\, \mathbf{i} \in \mathbf{D}_i^{\texttt{iter}}, \mathbf{j} \in \mathbf{D}_j^{\texttt{iter}}, \mathbf{s} \in \mathbf{D}_m^{\texttt{var}}, x, y$$
$$\text{s.t.} \quad \mathbf{M}_{ikm}^{\texttt{def}}(\mathbf{i}) = \mathbf{s} = \mathbf{M}_{jlm}^{\texttt{oper}}(\mathbf{j}) \,\wedge$$
$$\mathbf{C}_{ikm}^{\texttt{time}}(x) \geq 0 \,\wedge\, \mathbf{C}_{jlm}^{\texttt{time}}(y) \geq 0$$
$$\text{then} \quad O_{ikm}^{\texttt{wtime}}(O_i^{\texttt{time}}(\mathbf{i}, x)) < O_{jlm}^{\texttt{rtime}}(O_j^{\texttt{time}}(\mathbf{j}, y)) \tag{5.14}$$

This can be read as: *"If a data value is written to memory during the execution of one statement and it is read during the execution of another statement, then the write access corresponding to the first operation should be executed before the read access corresponding to the second operation."*

A closer look at this equation reveals that it corresponds to the precedence constraint due to a flow dependency, which is described by the dependency relation in equation 5.7. In fact equation 5.14 states that an operation that is the source of a flow dependency should be executed before the operation that is the sink.

Alternatively, equation 5.14 can be written as:

$$\forall\, i, j, k, l, m: \quad \{\, t \mid \exists\, \mathbf{i} \in \mathbf{D}_i^{\texttt{iter}}, \mathbf{j} \in \mathbf{D}_j^{\texttt{iter}}, \mathbf{s} \in \mathbf{D}_m^{\texttt{var}}, x, y \text{ s.t.}$$
$$\mathbf{M}_{ikm}^{\texttt{def}}(\mathbf{i}) = \mathbf{s} = \mathbf{M}_{jlm}^{\texttt{oper}}(\mathbf{j}) \,\wedge$$
$$\mathbf{C}_{ikm}^{\texttt{time}}(x) \geq 0 \,\wedge\, \mathbf{C}_{jlm}^{\texttt{time}}(y) \geq 0 \,\wedge$$
$$O_{ikm}^{\texttt{wtime}}(O_i^{\texttt{time}}(\mathbf{i}, x)) \geq t \geq O_{jlm}^{\texttt{rtime}}(O_j^{\texttt{time}}(\mathbf{j}, y)) \,\}$$
$$= \phi \tag{5.15}$$

i.e. for a valid execution order, each of these sets should be empty.

Note that $O_{ikm}^{\texttt{wtime}}()$, $O_i^{\texttt{time}}()$, $O_{jlm}^{\texttt{rtime}}()$ and $O_j^{\texttt{time}}()$ are in general unknown functions, i.e. they are the result of the optimization process. Also note that it does not matter whether we use absolute or relative timing functions, as the condition only states something about the relative order.

This condition is necessary to obtain a valid execution order, but it is not sufficient. In order to have a sufficient set of conditions, we must also take into account the storage order. But we first show an example of a necessary condition for the execution order.

Example 5.6.1
Suppose we have the following domain descriptions available (assuming that they have been extracted from some single-assignment program specification), without any given ordering:

$$\mathbf{D}_1^{\texttt{iter}} = \{\, [i, j] \mid 0 \leq i \leq 4 \,\wedge\, 0 \leq j \leq 4 \,\wedge\, [i, j] \in \mathbb{Z}^2 \,\}$$
$$\mathbf{D}_2^{\texttt{iter}} = \{\, [k, l] \mid 0 \leq k \leq 4 \,\wedge\, 0 \leq l \leq 4 \,\wedge\, [k, l] \in \mathbb{Z}^2 \,\}$$
$$\mathbf{D}_{11A}^{\texttt{def}} = \{\, [s_1, s_2] \mid \exists\, i, j \text{ s.t. } s_1 = i \,\wedge\, s_2 = j \,\wedge\, [i, j] \in \mathbf{D}_1^{\texttt{iter}} \,\}$$
$$\mathbf{D}_{21A}^{\texttt{oper}} = \{\, [s_1, s_2] \mid \exists\, k, l \text{ s.t. } s_1 = l \,\wedge\, s_2 = k \,\wedge\, [k, l] \in \mathbf{D}_2^{\texttt{iter}} \,\}$$

---

[16]"Might", because the program can be non-manifest.

We can now easily check whether a set of execution ordering functions is valid or not (we assume that $O_{11A}^{\text{utime}}(x) = O_{21A}^{\text{rtime}}(x) = x$). For instance, suppose we have the following ordering functions:

$$O_1^{\text{time}}(i,j) = 5i + j$$
$$O_2^{\text{time}}(k,l) = 5k + l + 25$$

We must now check whether this order satisfies equation 5.14. This can be done by checking the set of equation 5.15 for emptiness:

$$\{ t \mid \exists [i,j] \in \mathbb{Z}^2, [k,l] \in \mathbb{Z}^2, [s_1, s_2] \in \mathbb{Z}^2 \text{ s.t.}$$
$$0 \le i \le 4 \wedge 0 \le j \le 4 \wedge 0 \le k \le 4 \wedge 0 \le l \le 4 \wedge$$
$$s_1 = i = l \wedge s_2 = j = k \wedge 5i + j \ge t \ge 5k + l + 25 \}$$

which we can simplify to:

$$\{ t \mid \exists [i,j] \in \mathbb{Z}^2 \text{ s.t. } 0 \le i \le 4 \wedge 0 \le j \le 4 \wedge 4i - 4j - 25 \ge t - 5j - i - 25 \ge 0 \}$$

Based on the well known results of linear algebra, we know that for this case, where all expressions are affine and manifest, it is sufficient to evaluate the expression $4i - 4j - 25 \ge 0$ for each of the extremal points of the domain described by the other inequalities in order to prove that this set is empty. One can easily verify that for $[i,j] = [0,0]$, $[i,j] = [0,4]$, $[i,j] = [4,0]$, and $[i,j] = [4,4]$, the expression $4i - 4j - 25$ always evaluates to a value smaller than zero. Therefore, this execution order satisfies the necessary condition of equation 5.14.

A possible program corresponding to this execution order is the following:

```
      for ( i = 0; i < 5; ++i )
         for ( j = 0; j < 5; ++j )
S1:         A[i][j] = ...;

      for ( k = 0; k < 5; ++k )
         for ( l = 0; l < 5; ++l )
S2:         ... = A[l][k];
```

Alternatively, we could have tried the following execution order:

$$O_1^{\text{time}}(i,j) = 10i + 2j$$
$$O_2^{\text{time}}(k,l) = 10k + 2l + 1$$

Again, we can check whether there is a solution to equation 5.15:

$$\{ t \mid \exists [i,j] \in \mathbb{Z}^2, [k,l] \in \mathbb{Z}^2, [s_1, s_2] \in \mathbb{Z}^2 \text{ s.t.}$$
$$0 \le i \le 4 \wedge 0 \le j \le 4 \wedge 0 \le k \le 4 \wedge 0 \le l \le 4 \wedge$$
$$s_1 = i = l \wedge s_2 = j = k \wedge 10i + 2j \ge t \ge 10k + 2l + 1 \}$$

After simplification:

$$\{ t \mid \exists [i,j] \in \mathbb{Z}^2 \text{ s.t. } 0 \le i \le 4 \wedge 0 \le j \le 4 \wedge 10i + 2j \ge t \ge 10j + 2i + 1 \}$$

One can easily verify that there exist several solutions. For instance, $[i,j] = [1,0]$ is a solution, which means that A[1,0] is read (at date $t = 3$) before it is written (at date $t = 10$). Therefore, this execution order is not valid. An (invalid) program that corresponds to this order would look like this:

```
      for ( i = 0, k = 0; i<5 && k<5; ++i, ++k )
         for ( j = 0, l = 0; j<5 && l<5; ++j, ++l )
S1:         A[i][j] = ...;
```

```
S2:        ... = A[l][k];
```

From this, we can see that a valid order can be obtained by interchanging the k and l loops (without touching the i and j loops). This order can be described as follows:

$$O_1^{\text{time}}(i,j) = 10i + 2j$$
$$O_2^{\text{time}}(k,l) = 10l + 2k + 1$$

The resulting program could look like this:

```
    for ( i = 0, l = 0; i<5 && l<5; ++i, ++l )
        for ( j = 0, k = 0; j<5 && k<5; ++j, ++k )
S1:         A[i][j] = ...;
S2:         ... = A[l][k];
```

Evaluating equation 5.15 for this order yields the following set of solutions (after simplification):

$$\{\, t \mid \exists\, i,j \in \mathbb{Z}^2 \text{ s.t. } 0 \le i \le 4 \ \wedge\ 0 \le j \le 4 \ \wedge\ \underline{0 \ge t - 10i - 2j \ge 1} \,\}$$

which is obviously empty. In other words, this order satisfies the necessary condition of equation 5.14.

5.6.2 Necessary conditions for the execution and storage order

The second validity constraint can be expressed as follows:

$$\forall\, i_1, j_1, k_1, l_1, m_1, i_2, k_2, m_2 :$$

$$\text{if} \quad \exists\, \mathbf{s}_1 \in \mathbf{D}_{m_1}^{\text{var}}, \mathbf{i}_1 \in \mathbf{D}_{i_1}^{\text{iter}}, \mathbf{j}_1 \in \mathbf{D}_{j_1}^{\text{iter}}, x_1, y_1, w_1,$$

$$\mathbf{s}_2 \in \mathbf{D}_{m_2}^{\text{var}}, \mathbf{i}_2 \in \mathbf{D}_{i_2}^{\text{iter}}, x_2, w_2 \text{ s.t.}$$

$$\mathbf{M}_{i_1 k_1 m_1}^{\text{def}}(\mathbf{i}_1) = \mathbf{s}_1 = \mathbf{M}_{j_1 l_1 m_1}^{\text{oper}}(\mathbf{j}_1) \ \wedge$$

$$\mathbf{C}_{i_1 k_1 m_1}^{\text{time}}(x_1) \ge 0 \ \wedge\ \mathbf{C}_{j_1 l_1 m_1}^{\text{time}}(y_1) \ge 0 \ \wedge\ \mathbf{C}_{m_1}^{\text{addr}}(w_1) \ge 0 \ \wedge$$

$$\mathbf{M}_{i_2 k_2 m_2}^{\text{def}}(\mathbf{i}_2) = \mathbf{s}_2 \ \wedge$$

$$\mathbf{C}_{i_2 k_2 m_2}^{\text{time}}(x_2) \ge 0 \ \wedge\ \mathbf{C}_{m_2}^{\text{addr}}(w_2) \ge 0 \ \wedge$$

$$O_{m_1}^{\text{addr}}(\mathbf{s}_1, w_1) = O_{m_2}^{\text{addr}}(\mathbf{s}_2, w_2) \ \wedge\ (\mathbf{s}_1 \ne \mathbf{s}_2 \ \vee\ m_1 \ne m_2)$$

$$\text{then} \quad O_{i_2 k_2 m_2}^{\text{wtime}}(O_{i_2}^{\text{time}}(\mathbf{i}_2, x_2)) < O_{i_1 k_1 m_1}^{\text{wtime}}(O_{i_1}^{\text{time}}(\mathbf{i}_1, x_1)) \ \vee$$

$$O_{i_2 k_2 m_2}^{\text{wtime}}(O_{i_2}^{\text{time}}(\mathbf{i}_2, x_2)) > O_{j_1 l_1 m_1}^{\text{rtime}}(O_{j_1}^{\text{time}}(\mathbf{j}_1, y_1)) \tag{5.16}$$

This expression can be read as: *"If a first (program) variable is written by a statement S_{i_1} and read by a statement S_{j_1}, and a second variable is written by a statement S_{i_2}, and these variables are stored at the same memory address, and they either have different indices in the same array or belong to different arrays, then the write access corresponding to S_{i_2} should occur before the write access corresponding to S_{i_1} or after the read access corresponding to S_{j_1}."*

Together with the conditions imposed by equation 5.14, these conditions make sure that both validity constraints are satisfied and therefore form a sufficient set.

Note that due to the presence of the logical or in the last line of the if part of equation 5.16, we can distinguish 2 slightly different sets of conditions: conditions that are related to the intra-array order (i.e. between elements of the same arrays) and conditions that are related to the inter-array order (i.e. between elements of different arrays).

Nevertheless, it must be clear that this expression is potentially hard to evaluate, due to the possibly large number of (mathematical) variables, and especially due to the presence of inequalities and logical or's, which can easily lead to an explosion of the number of scalar (in)equalities to be checked.

In practice however, it is usually possible to simplify equation 5.16 by enforcing more stringent but simpler constraints, without sacrificing too much optimality. Examples of this will be shown later on, but for instance, if one makes sure that the address ranges of two arrays are not overlapping, then these conditions are always satisfied for each pair of variables, one of which belongs to the first array and one of which belongs to the second array.

It is also important to note that equation 5.16 is again strongly related to the presence of value-based flow-dependencies, as we can recognize the constraints that are also present in equation 5.7. In other words, equation 5.16 only has to be evaluated in the presence of flow-dependencies. So, performing a dependency analysis first may already drastically reduce the number of conditions to be checked, although even then this number equals the number of flow-dependencies times the number of statements writing to array elements present in the program (assuming that each statement writes to only one array, but the extension to tuple writes is straightforward). This also indicates that the constraints can be reformulated for the non-single-assignment case, provided that an accurate data-flow analysis is available.

We can again rewrite equation 5.16 as follows:

$$
\begin{aligned}
\forall i_1, j_1, k_1, l_1, m_1, i_2, k_2, m_2 : \\
\{\ t\ |\ \exists\ s_1 \in D_{m_1}^{var}, i_1 \in D_{i_1}^{iter}, j_1 \in D_{j_1}^{iter}, x_1, y_1, w_1, \\
s_2 \in D_{m_2}^{var}, i_2 \in D_{i_2}^{iter}, x_2, w_2 \text{ s.t.} \\
M_{i_1 k_1 m_1}^{def}(i_1) = s_1 = M_{j_1 l_1 m_1}^{oper}(j_1)\ \wedge \\
C_{i_1 k_1 m_1}^{time}(x_1) \geq 0\ \wedge\ C_{j_1 l_1 m_1}^{time}(y_1) \geq 0\ \wedge\ C_{m_1}^{addr}(w_1) \geq 0\ \wedge \\
M_{i_2 k_2 m_2}^{def}(i_2) = s_2\ \wedge \\
C_{i_2 k_2 m_2}^{time}(x_2) \geq 0\ \wedge\ C_{m_2}^{addr}(w_2) \geq 0\ \wedge \\
O_{m_1}^{addr}(s_1, w_1) = O_{m_2}^{addr}(s_2, w_2)\ \wedge\ (s_1 \neq s_2\ \vee\ m_1 \neq m_2)\ \wedge \\
t = O_{i_2 k_2 m_2}^{wtime}(O_{i_2}^{time}(i_2, x_2))\ \wedge \\
O_{i_1 k_1 m_1}^{wtime}(O_{i_1}^{time}(i_1, x_1)) \leq t \leq O_{j_1 l_1 m_1}^{rtime}(O_{j_1}^{time}(j_1, y_1))\ \} \\
= \phi
\end{aligned}
\tag{5.17}
$$

In other words, for a valid execution and storage order, each of these sets should be empty. Let us again have a look at some examples.

Example 5.6.2

```
    int A[10];
    for ( i = 0; i < 5; ++i )
S1:    A[i] = f1(...);

    for ( j = 0; j < 5; ++j )
S2:    ... = f2(A[j]);

    for ( k = 0; k < 5; ++k )
S3:    A[k+5] = f3(...);

    for ( l = 0; l < 5; ++l )
S4:    ... = f4(A[l+5]);
```

The corresponding domains are the following:

$$
\begin{aligned}
D_A^{var} &= \{\ s\ |\ 0 \leq s \leq 9\ \wedge\ s \in \mathbb{Z}\ \} \\
D_1^{iter} &= \{\ i\ |\ 0 \leq i \leq 4\ \wedge\ i \in \mathbb{Z}\ \} \\
D_2^{iter} &= \{\ j\ |\ 0 \leq j \leq 4\ \wedge\ j \in \mathbb{Z}\ \} \\
D_3^{iter} &= \{\ k\ |\ 0 \leq k \leq 4\ \wedge\ k \in \mathbb{Z}\ \}
\end{aligned}
$$

$$\mathbf{D}_4^{\text{iter}} = \{\, l \mid 0 \leq l \leq 4 \ \wedge \ l \in \mathbb{Z} \,\}$$
$$\mathbf{D}_{11A}^{\text{def}} = \{\, s \mid \exists\, i \in \mathbf{D}_1^{\text{iter}} \text{ s.t. } s = i \,\}$$
$$\mathbf{D}_{21A}^{\text{oper}} = \{\, s \mid \exists\, j \in \mathbf{D}_2^{\text{iter}} \text{ s.t. } s = j \,\}$$
$$\mathbf{D}_{31A}^{\text{def}} = \{\, s \mid \exists\, k \in \mathbf{D}_3^{\text{iter}} \text{ s.t. } s = k + 5 \,\}$$
$$\mathbf{D}_{41A}^{\text{oper}} = \{\, s \mid \exists\, l \in \mathbf{D}_4^{\text{iter}} \text{ s.t. } s = l + 5 \,\}$$

One can easily verify that the following execution order satisfies equation 5.14:

$$O_1^{\text{time}}(i) = i \qquad\qquad\qquad O_{11A}^{\text{rtime}}(x) = x$$
$$O_2^{\text{time}}(j) = j + 5 \qquad\qquad\quad O_{21A}^{\text{rtime}}(x) = x$$
$$O_3^{\text{time}}(k) = k + 10 \qquad\qquad O_{31A}^{\text{rtime}}(x) = x$$
$$O_4^{\text{time}}(l) = l + 15 \qquad\qquad O_{41A}^{\text{rtime}}(x) = x$$

Let us assume the following storage order:

$$O_A^{\text{addr}}(s) = s \bmod 5$$

We can now check whether this order, combined with the execution order above, is valid. But first of all, we should do some dependency analysis in order to find the non-empty value-based flow-dependencies. This reduces the number of conditions to be checked. In this case, dependency analysis tells us that there are only two data-dependencies: one from S1 to S2 and one from S3 to S4. Also, for this program fragment there are 2 statements writing elements to an array: S1 and S3. So, there are 4 possible combinations of flow-dependencies and writing statements that have to be checked: (S1→S2, S1), (S1→S2, S3), (S3→S4, S1) and (S3→S4, S3). Consequently, we have 4 sets that have to be checked for emptiness:

$$\{\, t \mid \exists\, [s_1, i_1, j, s_2, i_2] \in \mathbb{Z}^5 \text{ s.t.}$$
$$s_1 = i_1 = j \ \wedge \ 0 \leq i_1 \leq 4 \ \wedge \ 0 \leq j \leq 4 \ \wedge \ s_2 = i_2 \ \wedge \ 0 \leq i_2 \leq 4 \ \wedge$$
$$s_1 \bmod 5 = s_2 \bmod 5 \ \wedge \ s_1 \neq s_2 \ \wedge \ t = i_2 \ \wedge \ i_1 \leq t \leq j + 5 \,\} = \phi$$

$$\{\, t \mid \exists\, [s_1, i, j, s_2, k] \in \mathbb{Z}^5 \text{ s.t.}$$
$$s_1 = i = j \ \wedge \ 0 \leq i \leq 4 \ \wedge \ 0 \leq j \leq 4 \ \wedge \ s_2 = k + 5 \ \wedge \ 0 \leq k \leq 4 \ \wedge$$
$$s_1 \bmod 5 = s_2 \bmod 5 \ \wedge \ s_1 \neq s_2 \ \wedge \ t = k + 10 \ \wedge \ i \leq t \leq j + 5 \,\} = \phi$$

$$\{\, t \mid \exists\, [s_1, k_1, l, s_2, k_2] \in \mathbb{Z}^5 \text{ s.t.}$$
$$s_1 = k_1 + 5 = l + 5 \ \wedge \ 0 \leq k_1 \leq 4 \ \wedge \ 0 \leq l \leq 4 \ \wedge \ s_2 = k_2 + 5 \ \wedge$$
$$0 \leq k_2 \leq 4 \ \wedge \ s_1 \bmod 5 = s_2 \bmod 5 \ \wedge \ s_1 \neq s_2 \ \wedge$$
$$t = k_2 + 10 \ \wedge \ k_1 + 10 \leq t \leq l + 15 \,\} = \phi$$

$$\{\, t \mid \exists\, [s_1, k, l, s_2, i] \in \mathbb{Z}^5 \text{ s.t.}$$
$$s_1 = k + 5 = l + 5 \ \wedge \ 0 \leq k \leq 4 \ \wedge \ 0 \leq l \leq 4 \ \wedge \ s_2 = i \ \wedge \ 0 \leq i \leq 4 \ \wedge$$
$$s_1 \bmod 5 = s_2 \bmod 5 \ \wedge \ s_1 \neq s_2 \ \wedge \ t = i \ \wedge \ k + 10 \leq t \leq l + 15 \,\} = \phi$$

After simplification we get the following constraints:

$$\{\, t \mid \exists\, [i_1, i_2] \in \mathbb{Z}^2 \text{ s.t. } 0 \leq i_1 \leq 4 \ \wedge \ 0 \leq i_2 \leq 4 \ \wedge$$
$$i_1 \bmod 5 = i_2 \bmod 5 \ \wedge \ i_1 \neq i_2 \ \wedge \ t = i_2 \ \wedge \ i_1 \leq t \leq i_1 + 5 \,\} = \phi$$

$$\{\, t \mid \exists\, [i, k] \in \mathbb{Z}^2 \text{ s.t. } 0 \leq i \leq 4 \ \wedge \ 0 \leq k \leq 4 \ \wedge$$
$$i \bmod 5 = (k + 5) \bmod 5 \ \wedge \ i \neq k \ \wedge \ t = k + 10 \ \wedge \ i \leq t \leq i + 5 \,\} = \phi$$

$$\{\, t \mid \exists\, [k_1, k_2] \in \mathbb{Z}^2 \text{ s.t. } 0 \leq k_1 \leq 4 \ \wedge \ 0 \leq k_2 \leq 4 \ \wedge$$
$$k_1 \bmod 5 = k_2 \bmod 5 \ \wedge \ k_1 \neq k_2 \ \wedge \ t = k_2 + 10 \ \wedge \ k_1 + 10 \leq t \leq k_1 + 15 \,\} = \phi$$

$$\{\, t \mid \exists\, [k, i] \in \mathbb{Z}^2 \text{ s.t. } 0 \leq k \leq 4 \ \wedge \ 0 \leq i \leq 4 \ \wedge$$
$$k \bmod 5 = i \bmod 5 \ \wedge \ k \neq i \ \wedge \ t = i \ \wedge \ k + 10 \leq t \leq k + 15 \,\} = \phi$$

One can easily verify that each of these sets is indeed empty: for the first one, the condition $i_1 \bmod 5 = i_2 \bmod 5 \;\wedge\; i_1 \neq i_2$ can never be satisfied within the given ranges of i_1 and i_2. This means that none of the variables contributing to the first flow-dependency is stored at the same location, such that there cannot be a conflict. For the second one, the condition $t = k + 10 \;\wedge\; i \leq t \leq i + 5$ can never be satisfied for the given ranges of i and k, which means that the memory locations being occupied due to the presence of the first flow-dependency are no longer occupied when the memory writes due to statement S3 occur. For the third and fourth equations, similar reasonings hold. This can also be seen in Fig. 5.14.

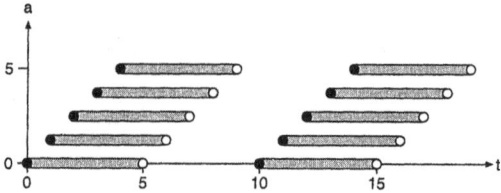

Figure 5.14.   The memory occupation of example 5.6.2.

A slightly different example, this time involving 2 different arrays, is given next.

Example 5.6.3

```
    int A[10];
    for ( i = 0; i < 5; ++i )
S1:    A[i] = f1(...);

    for ( j = 0; j < 5; ++j )
S2:    B[j] = f2(A[j]);

    for ( k = 0; k < 5; ++k )
S3:    ...  = f3(B[k]);
```

We can again readily extract the domain descriptions:

$$\mathbf{D}_A^{\text{var}} = \{\; s_a \mid 0 \leq s_a \leq 4 \;\wedge\; s \in \mathbb{Z} \;\}$$
$$\mathbf{D}_B^{\text{var}} = \{\; s_b \mid 0 \leq s_b \leq 4 \;\wedge\; s \in \mathbb{Z} \;\}$$
$$\mathbf{D}_1^{\text{iter}} = \{\; i \mid 0 \leq i \leq 4 \;\wedge\; i \in \mathbb{Z} \;\}$$
$$\mathbf{D}_2^{\text{iter}} = \{\; j \mid 0 \leq j \leq 4 \;\wedge\; j \in \mathbb{Z} \;\}$$
$$\mathbf{D}_3^{\text{iter}} = \{\; k \mid 0 \leq k \leq 4 \;\wedge\; k \in \mathbb{Z} \;\}$$
$$\mathbf{D}_{11A}^{\text{def}} = \{\; s_a \mid \exists\, i \in \mathbf{D}_1^{\text{iter}} \text{ s.t. } s_a = i \;\}$$
$$\mathbf{D}_{21A}^{\text{oper}} = \{\; s_a \mid \exists\, j \in \mathbf{D}_2^{\text{iter}} \text{ s.t. } s_a = j \;\}$$
$$\mathbf{D}_{21B}^{\text{def}} = \{\; s_b \mid \exists\, j \in \mathbf{D}_2^{\text{iter}} \text{ s.t. } s_b = j \;\}$$
$$\mathbf{D}_{31B}^{\text{oper}} = \{\; s_b \mid \exists\, k \in \mathbf{D}_3^{\text{iter}} \text{ s.t. } s_b = k \;\}$$

and verify that the following execution order satisfies equation 5.14:

$$O_1^{\text{time}}(i) = i \qquad\qquad O_{11A}^{\text{vtime}}(x) = x$$
$$O_2^{\text{time}}(j) = 2j + 5 \qquad O_{21A}^{\text{rtime}}(x) = x \qquad O_{21B}^{\text{vtime}}(x) = x + 1$$
$$O_4^{\text{time}}(k) = k + 15 \qquad O_{31B}^{\text{rtime}}(x) = x$$

Let us assume the following storage order:

$$O_A^{\text{addr}}(s_a) = s_a \quad and \quad O_B^{\text{addr}}(s_b) = s_b$$

Again, we have to check for the emptiness of all sets of the kind of equation 5.17. Dependency analysis reveals 2 non-empty value-based flow-dependencies, such that we have to check only 4 sets:

$$\{ \; t \;|\; \exists \; [s_{a1}, i_1, j, s_{a2}, i_2] \in \mathbb{Z}^5 \;\text{s.t.}\; s_{a1} = i_1 = j \;\wedge\; 0 \leq i_1 \leq 4 \;\wedge$$
$$0 \leq j \leq 4 \;\wedge\; s_{a2} = i_2 \;\wedge\; 0 \leq i_2 \leq 4 \;\wedge\; s_{a1} = s_{a2} \;\wedge\; s_{a1} \neq s_{a2} \;\wedge$$
$$t = i_2 \;\wedge\; i_1 \leq t \leq 2j + 5 \; \} = \phi$$
$$\{ \; t \;|\; \exists \; [s_a, i, j_1, s_b, j_2] \in \mathbb{Z}^5 \;\text{s.t.}\; s_a = i = j_1 \;\wedge\; 0 \leq i \leq 4 \;\wedge$$
$$0 \leq j_1 \leq 4 \;\wedge\; s_b = j_2 \;\wedge\; 0 \leq j_2 \leq 4 \;\wedge\; s_a = s_b \;\wedge$$
$$t = 2j_2 + 6 \;\wedge\; i \leq t \leq 2j_1 + 5 \; \} = \phi$$
$$\{ \; t \;|\; \exists \; [s_{b1}, j_1, k, s_{b2}, j_2] \in \mathbb{Z}^5 \;\text{s.t.}\; s_{b1} = j_1 = k \;\wedge\; 0 \leq j_1 \leq 4 \;\wedge$$
$$0 \leq k \leq 4 \;\wedge\; s_{b2} = j_2 \;\wedge\; 0 \leq j_2 \leq 4 \;\wedge\; s_{b1} = s_{b2} \;\wedge\; s_{b1} \neq s_{b2} \;\wedge$$
$$t = 2j_2 + 6 \;\wedge\; 2j_1 + 6 \leq t \leq k + 15 \; \} = \phi$$
$$\{ \; t \;|\; \exists \; [s_b, i, j, s_a, k] \in \mathbb{Z}^5 \;\text{s.t.}\; s_b = j = k \;\wedge\; 0 \leq j \leq 4 \;\wedge$$
$$0 \leq k \leq 4 \;\wedge\; s_a = i \;\wedge\; 0 \leq i \leq 4 \;\wedge\; s_b = s_a \;\wedge$$
$$t = i \;\wedge\; 2j + 6 \leq t \leq k + 15 \; \} = \phi$$

Note that this time, we have 2 slightly different kinds of conditions. The first one and the third one are again related to intra-array storage and execution order, while the second one and the fourth one are related to the order for two different arrays. We can immediately see that the first and the third set are empty due to the conditions $s_{a1} = s_{a2} \;\wedge\; s_{a1} \neq s_{a2}$ and $s_{b1} = s_{b2} \;\wedge\; s_{b1} \neq s_{b2}$ respectively. After simplification, the second equation becomes the following:

$$\{ \; t \;|\; \exists \; s_a \in \mathbb{Z} \;\text{s.t.}\; 0 \leq s_a \leq 4 \;\wedge\; t = 2s_a + 6 \;\wedge\; s_a \leq t \leq 2s_a + 5 \; \} = \phi$$

in which the condition $t = 2s_a + 6 \;\wedge\; t \leq 2s_a + 5$ can never be satisfied, so this set is also empty, which means that there is no conflict in time between the memory occupation of the different arrays. A similar reasoning holds for the fourth set. A graphical representation can be found in Fig. 5.15.

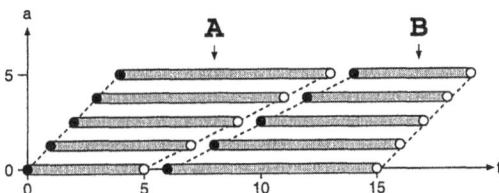

Figure 5.15.   The memory occupation of example 5.6.3.

5.7 CONCLUSIONS

In this chapter we have presented an overview of the geometrical modeling techniques for data-processing programs that are described in previous work, and "glued" everything together in a uniform and powerful model. Moreover, we have extended these models towards an accurate description of the occupation of memories by data, and derived a set of necessary and sufficient constraints for the most important design parameters, namely the execution and storage order of data. As illustrated by several examples, these models and constraints are very powerful and are able to deal efficiently with complex situations. Therefore they are crucial for the execution and storage order optimizations and they underly most of the techniques investigated in the following chapters.

6 POLYHEDRAL DATA-FLOW ANALYSIS FOR DATA STORAGE AND TRANSFER EXPLORATION

This chapter describes a theoretical polyhedral basis for data-flow analysis of multi-dimensional signal processing algorithms, especially in the context of DTSE. As mentioned in chapter 1, a data-flow analysis operating with groups of scalars is necessary in order to handle realistic RMSP applications. Formally, these groups are represented by images of polyhedra and are noted as basic sets further on. They can be used to identify useful signal partitions for almost all other DTSE steps but the actual decision on which partitions to withhold depends on the step and on the actual application at hand. The results also have to be carefully evaluated because in some cases the partitioning can lead to an explosion into small basic sets which are too fine grain for an acceptable solution complexity of our steps.

Section 6.1 introduces the basic definitions and concepts of this chapter. Section 6.2 describes an analytical decomposition of indexed signals in basic sets. Section 6.3 presents the construction of data-flow graphs at the level of basic sets of signals. The material in this chapter is based on [31], chapters 2 and 3 (see also [33]).

6.1 BASIC DEFINITIONS AND CONCEPTS

Definitions A *polyhedron* is a set of points $P \subset \Re^n$ satisfying a finite set of linear inequalities: $P = \{ \mathbf{x} \in \Re^n \mid \mathbf{A} \cdot \mathbf{x} \geq \mathbf{b} \}$, where $\mathbf{A} \in \Re^{m \times n}$ and $\mathbf{b} \in \Re^m$. If P is a bounded set, then P is called a *polytope*. If $\mathbf{x} \in \mathbf{Z}^n$, then P is called an *integral* polyhedron/polytope. The set $\{ \mathbf{y} \in \Re^m \mid \mathbf{y} = \mathbf{A}\mathbf{x} , \mathbf{x} \in \mathbf{Z}^n \}$ is called the *lattice* generated by the columns of matrix $\mathbf{A}$.

Each array reference $A[x_1(i_1, \ldots, i_n)] \cdots [x_m(i_1, \ldots, i_n)]$ of an m-dimensional signal A, in the scope of a nest of n loops having the iterators $i_1, \ldots, i_n$, is characterized by an *iterator space* and an *index space*. The iterator space signifies the set of all iterator vectors $\mathbf{i} = (i_1, \ldots, i_n) \in \mathbf{Z}^n$ in the scope of the array reference. The index space is the set of all index vectors $\mathbf{x} = (x_1 \ldots , x_m) \in \Re^m$ of the array reference.

If the loop boundaries are affine mappings with integer coefficients of the surrounding loop iterators, the increment steps of the loops[1] are ± 1, and the conditions in the scope of the array reference are

---

[1] This condition can be relaxed, as shown further.

relational and/or logical operations between affine mappings of the loop iterators[2], then the iterator space can be represented by one or several disjoint integral (iterator) polytopes $\mathbf{A} \cdot \mathbf{i} \geq \mathbf{b}$, where $\mathbf{A} \in \mathbf{Z}^{(2n+c) \times n}$ and $\mathbf{b} \in \mathbf{Z}^{2n+c}$. The first $2n$ linear inequalities are derived from the loop boundaries and the last c inequalities are derived from the control-flow conditions[3].

If, in addition, the indices of an array reference are affine mappings with integer coefficients of the loop iterators, the index space consists of one or several *linearly bounded*[4] *lattices* (LBL) [391] – the image of a vectorial affine function over the iterator polytope(s):

$$\{ \mathbf{x} = \mathbf{T} \cdot \mathbf{i} + \mathbf{u} \mid \mathbf{A} \cdot \mathbf{i} \geq \mathbf{b}, \mathbf{i} \in \mathbf{Z}^n \} \tag{6.1}$$

where $\mathbf{x} \in \mathbf{Z}^m$ is the index (coordinate) vector of the m-dimensional signal. The affine function is characterized by $\mathbf{T} \in \mathbf{Z}^{m \times n}$ and $\mathbf{u} \in \mathbf{Z}^m$.

The references to indexed signals from an RMSP algorithm are called *definition* (the result of an assignment) and *operand* (an assignment argument) *domains* [422].

For instance, the index space of the operand domain $Delta[i][j][(2*m+1)*(2*n+1)]$ in line (4) from the Silage code in figure 4.12 is represented by:

$$\left\{ \begin{bmatrix} x \\ y \\ z \end{bmatrix} = \begin{bmatrix} 1 & 0 \\ 0 & 1 \\ 0 & 0 \end{bmatrix} \begin{bmatrix} i \\ j \end{bmatrix} + \begin{bmatrix} 0 \\ 0 \\ (2m+1)(2n+1) \end{bmatrix} \right|$$
$$\left. \begin{bmatrix} 1 & 0 \\ -1 & 0 \\ 0 & 1 \\ 0 & -1 \end{bmatrix} \begin{bmatrix} i \\ j \end{bmatrix} \geq \begin{bmatrix} m \\ -M \\ n \\ -N \end{bmatrix} \right\}.$$

The index space of the operand domain $A[j][n+i]$ in line (4) from the Silage code in figure 6.6 is represented by:

$$\left\{ \begin{bmatrix} x \\ y \end{bmatrix} = \begin{bmatrix} 0 & 1 \\ 1 & 0 \end{bmatrix} \begin{bmatrix} i \\ j \end{bmatrix} + \begin{bmatrix} 0 \\ n \end{bmatrix} \right| \left. \begin{bmatrix} 1 & 0 \\ -1 & 0 \\ 0 & 1 \\ 0 & -1 \\ 1 & -1 \end{bmatrix} \begin{bmatrix} i \\ j \end{bmatrix} \geq \begin{bmatrix} 0 \\ -n+1 \\ 0 \\ -n+1 \\ 1 \end{bmatrix} \right\}.$$

In general, the index space may be a *collection* of linearly bounded lattices. E.g., a conditional instruction $if\ (\ i \neq j\ )$ determines two LBL's for the domains of signals within the scope of the condition – one corresponding to $i \geq j+1$, and another corresponding to $i \leq j-1$. Without any decrease in generality, it is assumed in the sequel that each index space is represented by a single linearly bounded lattice.

Definition The signals common to a given set of domains and only to them constitute a *basic set* [27, 31].

Example The signals common to the definition domain $A[j][n+i+1]$ from line (4) and to the operand domain $A[j][n+i]$ $(j \geq i)$ from line (5) in figure 6.6 have the index space characterized by the LBL (in non-matrix format): $\{ x = j, y = i+n \mid n-1 \geq j \geq i \geq 1 \}$. Representing the definition/operand domains from the code, symbolically, by ellipses (see figure 6.1), this LBL corresponds to the (cross-) hatched area. But the LBL above partially covers the index space of the operand $A[i][n+i]$ from line (3) (the cross-hatched region). Therefore, the signals common to $A[j][n+i+1]$ and $A[j][n+i]$ $(j \geq i)$ *and only to them* (the horizontally hatched region) have a different index space: $\{ x = j, y = i+n \mid n-1 \geq j, j \geq i+1, i \geq 1 \}$, which is the basic set of the given two domains. In figure 6.1, the intersection of the ellipses determines 6 regions, each one corresponding to a basic set.

---

[2]The polyhedral representation of the iterator space is still valid if the array reference scope also contains data-dependent (but iterator independent) conditions.

[3]This representation of the polytope $\mathbf{A} \cdot \mathbf{i} \geq \mathbf{b}$ is usually not minimal. The elimination of redundant inequalities will be presented in subsection 6.2.2.2.

[4]As the definition domain of the mappings is not $\mathbf{Z}^n$ but a polytope – which is *bounded* by hyperplanes (characterized by *linear* equations).

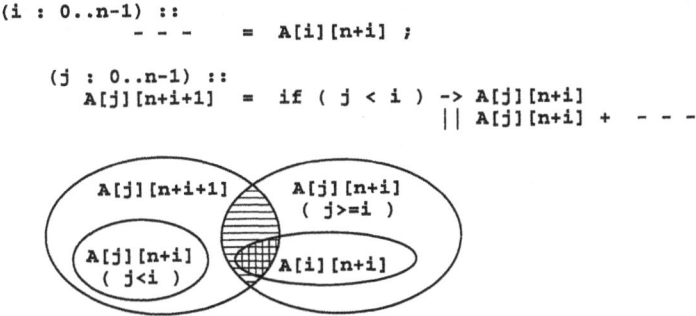

```
(i : 0..n-1) ::
      - - -     = A[i][n+i] ;

   (j : 0..n-1) ::
      A[j][n+i+1] = if ( j < i ) -> A[j][n+i]
                              || A[j][n+i] +  - - -
```

Figure 6.1.   Domains of signals extracted from an RMSP algorithm

A collection of 2 domains results in at most 3 basic sets; 3 overlapping domains result in maximally 7 sets. More generally, assume the number of definition/operand domains of a multi-dimensional signal is denoted by d. As the basic sets belonging to only one domain are at most C_d^1 – the number of domains, the basic sets belonging to two domains are at most C_d^2 (all the possible combination of two domains) a.s.o, it follows that the number of basic sets is upper-bounded by $C_d^1 + C_d^2 + \cdots + C_d^d = 2^d - 1$. However, this evaluation proved to be extremely pessimistic in practical cases. In the illustrative example in figure 6.6, for instance, the number of domains for signal A is $d = 9$, and the resulting number of basic sets is only 10 (see subsection 6.2.1).

Section 6.2 describes thoroughly the signal decomposition into basic sets.

6.2 ANALYTICAL PARTITIONING OF INDEXED SIGNALS

Once the collections of definition/operand domains are extracted from the RMSP algorithm, a partitioning process into non-overlapping "pieces" – the *basic sets* – is performed for each collection. The aim of the decomposition process is to determine which parts of the operand domains are not needed any more after the computation of a definition domain. The last question is directly related to the evaluation of the storage requirements, as it allows to compute exactly how many memory locations are needed and how many can be freed when a certain group of signals is produced.

In the example in figure 6.6, there are three collections of domains: $\{B[i]\}$, $\{alpha[i]\}$, $\{A[j][0], A[j][i+1], A[j][i], A[j][n+i+1]\ (j < i), A[j][n+i]\ (j < i), A[j][n+i+1]\ (j \geq i),$ $A[j][n+i]\ (j \geq i), A[i][n+i], A[i][2n]\}$. Only the collection of signal A needs partitioning.

The main benefit of the analytical decomposition into groups of signals – called *basic sets* – is that it makes possible the estimation of the memory size by means of a data-flow analysis *at basic set level* (rather than scalar level). Simultaneously, the partitioning can be employed to check whether the program complies with the single-assignment requirement, and to provide a diagnosis regarding the eventual multiple assignments. Moreover, the useless signals are detected and eliminated, avoiding hence unnecessary memory allocation.

6.2.1 The partitioning algorithm

The partitioning process – described in the sequel – yields a separate inclusion graph for each of the collections of linearly bounded lattices (describing the index spaces of the domains of an M-D signal). The direct inclusion relation is denoted by "$\subset$", while its transitive closure (the existence of a path in the graph) is denoted by "$\prec$". An inclusion graph is constructed as follows ([31], chapter 2):

Construction of the inclusion graphs

void ConstructInclusionGraph(*collection of LBL's – the index spaces of signal domains*) {
    initialize inclusion graph with the def/opd index spaces as nodes ;
    while *new nodes can still be appended to the graph*

> **for** *each pair of LBL's* (Lbl_1, Lbl_2) *in the collection*
>   *such that* $(Lbl_1 \not\prec Lbl_2)$ && $(Lbl_2 \not\prec Lbl_1)$ {
>   compute $Lbl = Lbl_1 \cap Lbl_2$;
>   **if** $(Lbl \neq \emptyset)$

(1)   **if** $(Lbl == Lbl_1)$ AddInclusion(Lbl_1, Lbl_2) ;
>   **else if** $(Lbl == Lbl_2)$ AddInclusion(Lbl_2, Lbl_1) ;

(2)   **else if** *there exists already* $lbl \equiv Lbl$
>   **if** $(lbl \not\prec Lbl_1)$ AddInclusion(lbl, Lbl_1) ;
>   **if** $(lbl \not\prec Lbl_2)$ AddInclusion(lbl, Lbl_2) ;

(3)   **else** add Lbl to the collection ;
>   set $Lbl \subset Lbl_1$;   set $Lbl \subset Lbl_2$;
>   }
> }

At the beginning, the inclusion graph of an indexed signal contains only the index spaces of the corresponding array references as nodes. Every pair of LBL's – between which no inclusion relation is known so far[5] – is intersected. If the intersection produces a non-empty Lbl, there are basically three possibilities: (1) the resulting Lbl is one of the intersection operands: in this case, inclusion arcs between the corresponding nodes must be added to the graph; (2) an *equivalent* linearly bounded lattice[6] exists already in the collection: in this case, arcs must be added only if necessary between the nodes corresponding to the equivalent LBL and to the operands; (3) the resulting Lbl is a new element of the collection: a new node is appended to the graph, along with two arcs towards the operands of intersection. The construction of the inclusion graph ends when no more elements – nodes or arcs – can be created.

The procedure *AddInclusion* creates new inclusion relations (arcs) between groups of signals, but deletes the resulting transitive arcs: keeping a strict hierarchy for the LBL's is essential for the partitioning phase. The intersection "$\cap$" of two linearly bounded lattices is also a linearly bounded lattice, which is computed as described in subsection 6.2.2.

Example The partitioning algorithm is exemplified for the simple Silage code in figure 6.6. Initially, the inclusion graph of signal A contains only the nodes labeled from **a** to **i** , corresponding to the operand/definition domains (indicated as $A[\ ][\ ]$) extracted from the source code. The first execution of the *while* loop in the algorithm adds to the graph the new vertices $\mathbf{j}, \ldots, \mathbf{r}$, along with the sequence of arcs: $(\mathbf{j}, \mathbf{b}), (\mathbf{j}, \mathbf{c}), (\mathbf{k}, \mathbf{b}), (\mathbf{k}, \mathbf{d}), (\mathbf{l}, \mathbf{b}), (\mathbf{l}, \mathbf{h}), (\mathbf{m}, \mathbf{d}), (\mathbf{m}, \mathbf{g}), (\mathbf{n}, \mathbf{e}), (\mathbf{n}, \mathbf{f}),$ $(\mathbf{o}, \mathbf{e}), (\mathbf{o}, \mathbf{i}), (\mathbf{p}, \mathbf{f}), (\mathbf{p}, \mathbf{g}), (\mathbf{q}, \mathbf{g}), (\mathbf{q}, \mathbf{h}), (\mathbf{r}, \mathbf{g}), (\mathbf{r}, \mathbf{i})$. Also the inclusion arcs $(\mathbf{d}, \mathbf{h})$, and $(\mathbf{a}, \mathbf{c})$ are added. The second execution of the *while* loop creates the new arcs $(\mathbf{m}, \mathbf{q})$ and $(\mathbf{k}, \mathbf{l})$, eliminating $(\mathbf{m}, \mathbf{g})$ and $(\mathbf{k}, \mathbf{b})$ at the same time, as they become transitive. The latter is done in the procedure *AddInclusion*, which creates new inclusion relations, deleting the transitive arcs. E.g., node $\mathbf{l}$ represents the LBL

$$\left\{ \begin{bmatrix} x \\ y \end{bmatrix} = \begin{bmatrix} 0 & 1 \\ 0 & 0 \end{bmatrix} \begin{bmatrix} i \\ j \end{bmatrix} + \begin{bmatrix} 0 \\ n \end{bmatrix} \ \middle| \ \begin{bmatrix} 0 & 1 \\ 0 & -1 \end{bmatrix} \begin{bmatrix} i \\ j \end{bmatrix} \geq \begin{bmatrix} 0 \\ -n+1 \end{bmatrix} \right\} \quad \text{resulting}$$

from the intersection of the linearly bounded lattices corresponding to nodes $\mathbf{h}$ and $\mathbf{b}$:

$$\left\{ \begin{bmatrix} x \\ y \end{bmatrix} = \begin{bmatrix} 0 & 1 \\ 1 & 0 \end{bmatrix} \begin{bmatrix} i \\ j \end{bmatrix} + \begin{bmatrix} 0 \\ n \end{bmatrix} \ \middle| \ \begin{bmatrix} -1 & 1 \\ 1 & 0 \\ 0 & -1 \end{bmatrix} \begin{bmatrix} i \\ j \end{bmatrix} \geq \begin{bmatrix} 0 \\ 0 \\ -n+1 \end{bmatrix} \right\} \text{, and}$$

$$\left\{ \begin{bmatrix} x \\ y \end{bmatrix} = \begin{bmatrix} 0 & 1 \\ 1 & 0 \end{bmatrix} \begin{bmatrix} i \\ j \end{bmatrix} + \begin{bmatrix} 0 \\ 1 \end{bmatrix} \ \middle| \ \begin{bmatrix} 0 & 1 \\ 0 & -1 \\ 1 & 0 \\ -1 & 0 \end{bmatrix} \begin{bmatrix} i \\ j \end{bmatrix} \geq \begin{bmatrix} 0 \\ -n+1 \\ 0 \\ -n+1 \end{bmatrix} \right\}.$$

The final inclusion graph is shown in figure 6.2 .

---

[5]i.e., between which there is currently no path in the graph.
[6]Two linearly bounded lattices of the same indexed signal are equivalent if they represent the same set of indices. E.g., $\{x = i + j \mid 0 \leq i \leq 2, \ 0 \leq j \leq 2\}$ and $\{x = i \mid 0 \leq i \leq 4\}$ are equivalent. Testing LBL's equivalence can be done employing LBL intersection (subsection 6.2.2) and size determination (subsection 6.2.3).

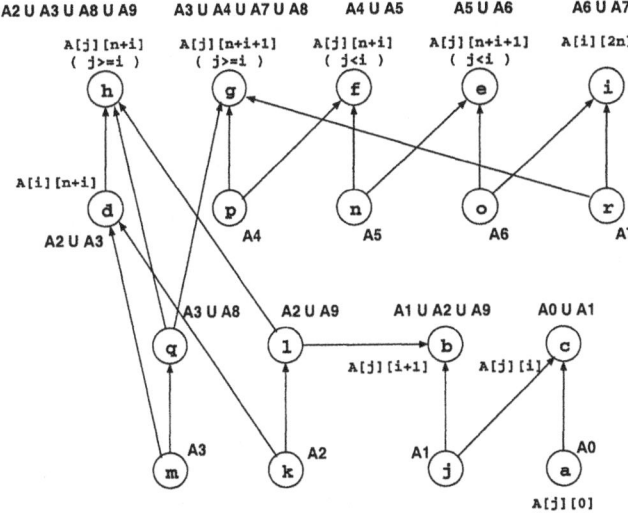

Figure 6.2. The inclusion graph corresponding to the indexed signal A from the Silage example in figure 6.6

Afterwards, the basic sets of a signal are derived from the inclusion graph with a simple bottom-up technique. If a node has no components, a new basic set – equal to the corresponding linearly bounded lattice – is introduced; otherwise, if all its components have already been partitioned, the union of the component partitions and (potentially) a new basic set will constitute the current node partitioning. In the latter case, the new basic set appears only if there is a difference between the size of the node and the total size of its components.

The computation of LBL sizes is described in subsection 6.2.3. The efficiency of this operation and of the intersection procedure are crucial for the practical time complexity of the whole algorithm.

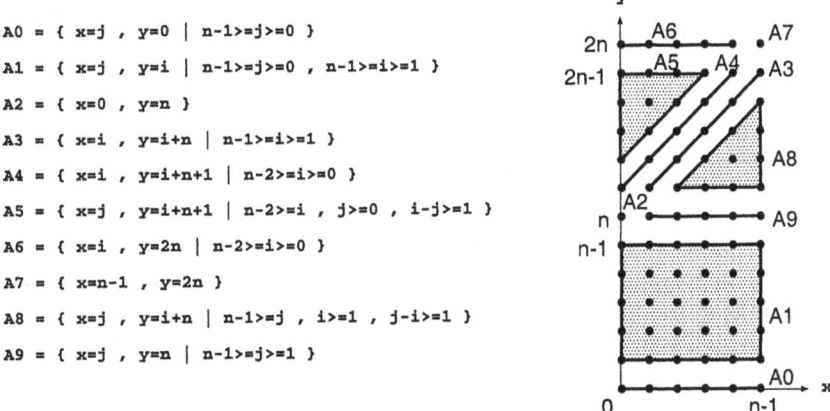

Figure 6.3. The basic sets of signal A represented in non-matrix format and graphically

Determination of the basic sets

```
void CreateBasicSets(inclusion graph) {
    for every LBL (node in the inclusion graph) having no arc incident to it
        create a new basic set P, equal to that LBL ;
    while there are still non-partitioned LBL's {
        select a non-partitioned LBL (node in the inclusion graph) such that
            all arcs incident to it emerge from partitioned LBL's ( LBLᵢ = ∪ⱼPᵢⱼ );
        if ( size(LBL) > size(∪ᵢLBLᵢ) )
            create a new basic set P equal to LBL − ∪ᵢLBLᵢ ;
            partition the LBL into ( ∪ᵢ ∪ⱼ Pᵢⱼ ; P );
        else  // size(LBL) = size(∪ᵢLBLᵢ)
            partition the LBL into ( ∪ᵢ ∪ⱼ Pᵢⱼ );
    }
}
```

Example (cont'd) The LBL's in the inclusion graph are partitioned in the order: first **a, j, k, m, p, n, o, r** creating the basic sets A0÷A7; afterwards, **c, d, e, f, i, l, q, b, g, h**. When the LBL's **l** and **q** are processed, two more basic sets are created, as $size(l) > size(k)$ and $size(q) > size(m)$. The final partitioning for our example is indicated in figure 6.2, and represented in figure 6.3. Each domain has an index space which consists of one (e.g., $A[j][0] \rightsquigarrow$ A0) or several basic sets (e.g., $A[j][i] \rightsquigarrow$ A0 ∪ A1).

Remark Basic sets cannot always be represented as linearly bounded lattices (as it is the case in this illustrative example). In general, they can be *decomposed*, or they can be expressed as differences of affine mappings of polytopes. E.g., signal A from the Silage code in figure 4.12 has two basic sets; one of them, $A1$, cannot be represented as a linearly bounded lattice (see figure 6.9b). Although the novel concept of *basic sets* relies on *linearly bounded lattices*, the latter concept cannot replace the former one.

6.2.2 Intersection of linearly bounded lattices

This section describes thoroughly the procedure for intersecting two linearly bounded lattices ([31], chapter 2). As already mentioned, the efficiency of this operation together with the computation of an LBL size are essential for the practical time complexity of the whole partitioning algorithm.

Let $\{x = T_1 i_1 + u_1 \mid A_1 i_1 \geq b_1\}$, $\{x = T_2 i_2 + u_2 \mid A_2 i_2 \geq b_2\}$ be two LBLs derived from the same indexed signal, where T_1 and T_2 have obviously the same number of rows (the signal dimension). Intersecting the two linearly bounded lattices means, first of all, solving a linear Diophantine system[7] [363] (chapter 5) $T_1 i_1 - T_2 i_2 = u_2 - u_1$ having the elements of i_1 and i_2 as unknowns. If the system has no solution, the intersection is empty. Otherwise, let

$$\begin{bmatrix} i_1 \\ i_2 \end{bmatrix} = \begin{bmatrix} V_1 \\ V_2 \end{bmatrix} i + \begin{bmatrix} v_1 \\ v_2 \end{bmatrix}$$

be the solution of the Diophantine system[8]. If the set of coalesced constraints

$$A_1 V_1 \cdot i \geq b_1 - A_1 v_1 \tag{6.2}$$
$$A_2 V_2 \cdot i \geq b_2 - A_2 v_2$$

has at least one integer solution (see subsection 6.2.3), than the intersection is a new linearly bounded lattice defined by $\{x = T \cdot i + u \mid A \cdot i \geq b\}$, where

$$T = T_1 V_1 \quad , \quad u = T_1 v_1 + u_1 \tag{6.3}$$
$$A = \begin{bmatrix} A_1 V_1 \\ A_2 V_2 \end{bmatrix} \quad , \quad b = \begin{bmatrix} b_1 - A_1 v_1 \\ b_2 - A_2 v_2 \end{bmatrix}$$

---

[7] Finding the integer solutions of a system of linear equations with integer coefficients.
[8] The solution of the linear Diophantine system has always this form, as shown in subsection 6.2.2.1.

The problem of solving a linear Diophantine system was proven to be of polynomial complexity [363] (section 5.3). All the known methods are based on bringing the system matrix to the Hermite Normal Form [363] (chapter 4), [296] (section I.7.4). They only differ from each other basically by the modality in which this representation is obtained. Numerous researchers proposed algorithms for computing Hermite and Smith Normal Forms, as well as the related problem of solving systems of linear Diophantine equations. After 1977, more recent results have been obtained, overcoming the main shortcoming of the classical methods, namely the "intermediate expression swell" [179]. This undesirable effect signifies the quick and uncontrollable increase of elements of the intermediate matrices (see subsection 6.2.2.1). Therefore, several algorithms with provable polynomial worst-case complexity have been proposed (e.g., [145, 208, 363, 179]).

It must be emphasized that the size of Diophantine systems in most of our practical cases does not justify the overhead (in programming and practical computation effort) implied by the use of one of those more sophisticated algorithms with provable polynomial worst-case complexity. Therefore, we have searched for a simpler technique which works well for smaller problem sizes, taking into account the dimensions of the indexed signals, as well as the depth of the loop nests, in our current RMSP applications. The technique employed for solving linear Diophantine systems ([31], chapter 2) is explained thoroughly in subsection 6.2.2.1.

An undesirable side-effect of intersection is the rapid size increase of the polytope description for the resulting linearly bounded lattice (eq. 6.3), due to the coalescing of the two constraint sets (eq. 6.2). Therefore, minimizing the set of constraints proved to be a necessity in order to restrict the computational effort of, e.g., counting the lattice points of the resulting LBL's (see subsection 6.2.3). An exhaustive theory of the reducibility of linear inequalities is presented, for instance, in [250] (chapter 5). However, the practical technique employed for minimizing the set of linear constraints is based on the double description of polyhedra [285]. The implementation in use stems from the polyhedral library developed at IRISA [427]. The main ideas of this approach are briefly presented in subsection 6.2.2.2.

6.2.2.1 The Hermite normal form and linear Diophantine systems.

Let the system of m linear equations with n unknowns, having integer coefficients, be

$$
\begin{aligned}
a_{11}x_1 + \cdots + a_{1n}x_n &= a_1 \\
&\cdots \\
a_{m1}x_1 + \cdots + a_{mn}x_n &= a_m
\end{aligned}
\tag{6.4}
$$

Two questions arise: (1) how can it be decided whether the system (6.3) has integer solutions (compatible) or not, and (2) if the system is compatible, how can the general solution be derived.

In this subsection, an algorithm answering simultaneously at both questions will be described ([31], chapter 2). The algorithm builds the Hermite Normal Form [363] (section 4.1) of the system matrix.

Definition An $m \times n$ integer matrix $\mathbf{H}_{m,n}$ ($m \leq n$) is in the Hermite Normal Form if:

1. $\mathbf{H}_{m,n}$ is lower-triangular ($h_{ij} = 0$, if $i < j$);

2. $h_{ii} > 0$, $i = \overline{1,m}$;

3. $h_{ij} \geq 0$ (or, equivalently, ≤ 0 [[296], (p. 190)]) , and $|h_{ij}| < h_{ii}$, if $i > j$.

If an $m \times n$ integer matrix $\mathbf{A}_{m,n}$ ($m \leq n$) has linearly independent rows, then there exists a unimodular matrix[9] $\mathbf{S}_{n,n}$ transforming the matrix $\mathbf{A}_{m,n}$ into a Hermite Normal Form $\mathbf{H}_{m,n}$:

$$
\mathbf{A}_{m,n} \cdot \mathbf{S}_{n,n} = \mathbf{H}_{m,n}
$$

---

[9] A square matrix with integer coefficients, having the determinant value ± 1.

Furthermore, it can be proven [363] (section 4.2) that the Hermite Normal Form $H_{m,n}$ is unique. However, there can be several unimodular transformations that bring matrix $A_{m,n}$ to the Hermite Normal Form.

The algorithm for solving linear Diophantine systems results in a straightforward way from the following two lemma's:

Lemma 1   There is a unimodular transformation changing the linear equation

$$a_{11}x_1 + \cdots + a_{1n}x_n = a_1 \in \mathbf{Z} \qquad , \text{where } a_{1i} \in \mathbf{Z} - \{0\} \tag{6.5}$$

in $c_{11}X_{11} = a_1$, where $c_{11} = gcd(a_{11}, \ldots, a_{1n})$.

Proof: As (6.5) can be written $c_{11}(a'_{11}x_1 + \cdots + a'_{1n}x_n) = a_1$, it can be assumed that the greatest common divisor $gcd(a_{11}, \ldots, a_{1n}) = 1$.

Without loss in generality, let a_{11} be one of the coefficients such that

$$|a_{11}| = M_1 \overset{def}{=} \min \{|a_{11}|, \ldots, |a_{1n}|\}$$

The unimodular substitution $x_1 = X_1 + c_2 x_2 + \cdots + c_n x_n$ transforms equation (6.5) in

$$a_{11}X_1 + (a_{11}c_2 + a_{12})x_2 + \cdots + (a_{11}c_n + a_{1n})x_n = a_1$$

The coefficients c_k can be chosen as integers satisfying

$$|a_{11}c_k + a_{1k}| \leq \frac{1}{2}M_1 \qquad k = \overline{2,n}$$

Indeed, writing $a_{1k} = q_k|a_{11}| + r_k$, with $0 \leq r_k < |a_{11}|$, it is easy to see that c_k can be chosen as:

$$c_k = \begin{cases} -sgn(a_{11})q_k & \text{, if } \quad 0 \leq r_k \leq |a_{11}|/2 \\ -sgn(a_{11})(q_k + 1) & \text{, if } \quad |a_{11}|/2 < r_k < |a_{11}| \end{cases}$$

Therefore, a unimodular transformation was constructed such that the form of equation (6.5) is preserved, but the absolute values of the coefficients (less a_{11}) are at most half of the smallest absolute value of the initial coefficients. Continuing this process iteratively, after at most $\lfloor \log_2 M_1 \rfloor$ transformations, an equation of the form $X_i + \cdots = a_1$ is obtained. Finally, one more unimodular substitution $X_i = X_{11} - \cdots$ leads to $X_{11} = a_1$. $\square$

Remark   Equation (6.5) has solutions iff c_{11} divides a_1. Assuming this condition is fulfilled, the general solution of (6.5) can be expressed as an affine function of $n - 1$ variables. This solution is obtained applying the unimodular transformations backwards.

Lemma 2   If $rank\ A_{m,n} = m \leq n$ (where $A_{m,n}$ is the coefficient matrix of system (eq. 6.3)), there is a unimodular substitution changing (eq. 6.3) in:

$$\begin{aligned} c_{11}X_{11} &= a_1 \\ c_{21}X_{11} + c_{22}X_{22} &= a_2 \\ \cdots & \\ c_{m1}X_{11} + c_{m2}X_{22} + \cdots + c_{mm}X_{mm} &= a_m \end{aligned} \tag{6.6}$$

Proof: As in the proof of *Lemma 1*, there is a sequence of unimodular transformations changing the first equation from the system (eq. 6.3) in $c_{11}X_{11} = a_1$. This sequence simultaneously modifies the second equation in $c_{21}X_{11} + (a'_{22}X_2 + \cdots + a'_{2n}X_n) = a_2$. But the second term can be changed, as in the previous proof in $c_{22}X_{22}$. Employing the same strategy, the canonical form (eq. 6.6) will eventually be obtained, if $rank\ A_{m,n} = m$. $\square$

Remark   If $rank\ A_{m,n} < m$, let i be the the first equation in (eq. 6.3) such that its left side is a linear combination of the previous ones. Applying the previous approach, the sequence of unimodular transformations leads the i-th equation to $c_{i1}X_{11} + \cdots + c_{i,i-1}X_{i-1,i-1} = a_i$, thus the expected extra term $a'_{ii}X_i + \cdots + a'_{in}X_n$ will vanish.

The previous proof allows to derive the general solution of the linear Diophantine system (eq. 6.3): denoting $r \stackrel{def}{=} rank\ \mathbf{A}_{m,n}$, any r linearly independent equations of the initial system are brought to the *canonical form* (eq. 6.6); then the new system, having $X_{11}, \ldots, X_{rr}$ as unknowns, is solved by a simple forward substitution. Substituting the values of $X_{11}, \ldots, X_{rr}$ in the unimodular transformations applied backwards, the solution of the initial system (eq. 6.3) is obtained:

$$
\begin{aligned}
x_1 &= x_1(t_1, \ldots, t_{n-r}) \\
x_2 &= x_2(t_1, \ldots, t_{n-r}) \\
&\cdots \\
x_n &= x_n(t_1, \ldots, t_{n-r})
\end{aligned}
$$

The unknowns are affine functions of $n-r$ variables t_i. The solution can be expressed in matrix format as $\mathbf{x} = \mathbf{V}_{n,n-r} \cdot \mathbf{t} + \mathbf{v}$, where $\mathbf{x}$ and $\mathbf{v}$ are n-dimensional vectors, and $\mathbf{t}$ is an $(n-r)$-dimensional vector.

When the system has no integer solution, this will become apparent at some equation i while solving the canonical system (6.6): either it follows that X_{ii} cannot be an integer (if the left hand side of the equation is linearly independent from the previous ones), or the identity which should be obtained is not produced.

It can be noticed that the matrix of the canonical system (eq. 6.6) is in the Hermite Normal Form if $c_{ij} \geq 0$ and $|c_{ij}| < c_{ii}$. This supplementary requirement can be achieved by applying eventually one extra unimodular transformation – as described in *Lemma 1* – after each sequence of transformations which processes an equation. Concluding, the unimodular transformation $\mathbf{S}_{n,n}$ which brings the system matrix $\mathbf{A}_{m,n}$ to the Hermite Normal Form can be expressed as a factorization:

$$
\mathbf{S}_{n,n} = \mathbf{U}^{11} \cdot \mathbf{U}^{12} \cdots \mathbf{\Pi}^{1j_1} \cdots \mathbf{U}^{m1} \cdot \mathbf{U}^{m2} \cdots \mathbf{\Pi}^{mj_m}
$$

where $\mathbf{U}^{ij}$ are unimodular transformations like the one constructed in the proof of *Lemma 1*

$$
\mathbf{U}^{ij} = \begin{bmatrix} 1 & c_2 & \cdots & c_n \\ 0 & 1 & \cdots & 0 \\ & & \ddots & \\ 0 & 0 & \cdots & 1 \end{bmatrix}
$$

Bringing the non-zero coefficients X_{ii}, obtained after processing equation i, on the main diagonal necessitates usually a column permutation $\mathbf{\Pi}^{ij_i}$, which is a unimodular transformation as well. The maximum number of factors $\mathbf{U}$ and $\mathbf{\Pi}$ is $\sum_{i=1}^{m} \lfloor \log_2 M_i \rfloor + 2m - 1$, where $M_i = \min\{|a_{ii}^{(i)}|, \ldots, |a_{i,n}^{(i)}|\}$ (the superscript signifies the transformed coefficients of equation i).

Example /* the loop boundaries are not relevant here */
    $(i: \ldots)::$
        $(j: \ldots)::$
            $A[2*i + 3*j][5*i + 3*j] = \cdots$
                $\cdots = \cdots A[-4*i - 5*j + 1][2 - i - j] \cdots$

The two LBL's corresponding to the signal domains are

$$
\left\{ \begin{bmatrix} x \\ y \end{bmatrix} = \begin{bmatrix} 2 & 3 \\ 5 & 3 \end{bmatrix} \begin{bmatrix} i \\ j \end{bmatrix} + \begin{bmatrix} 0 \\ 0 \end{bmatrix} \middle| \cdots \right\}
$$

and

$$
\left\{ \begin{bmatrix} x \\ y \end{bmatrix} = \begin{bmatrix} -4 & -5 \\ -1 & -1 \end{bmatrix} \begin{bmatrix} i \\ j \end{bmatrix} + \begin{bmatrix} 1 \\ 2 \end{bmatrix} \middle| \cdots \right\}.
$$

In order to find the intersection of the two LBL's, the following linear Diophantine system has to be solved:

$$2i_1 + 3j_1 + 4i_2 + 5j_2 = 1$$
$$5i_1 + 3j_1 + i_2 + j_2 = 2$$

The sequence of unimodular transformations determined as described in the proof of *Lemma 1*

$$i_1 = I_1 - j_1 - 2i_2 - 2j_2$$
$$j_1 = X_{11} - 2I_1 - j_2$$

applied to the first equation leads to $X_{11} = 1$. The same sequence applied to the second one, transforms it in:

$$-2X_{11} + (9I_1 - 9i_2 - 7j_2) = 2$$

The next sequence:

$$j_2 = J_2 + I_1 - i_2$$
$$I_1 = I_1' + i_2 + 4J_2$$
$$J_2 = X_{22} - 2I_1'$$

leads the second equation to $-2X_{11} + X_{22} = 2$. Thus, the system is compatible, as its canonical system has the solution $X_{11} = 1$, $X_{22} = 4$. Applying the transformations backwards, the general solution results:

$$
\begin{aligned}
i_1 &= -X_{11} &+7X_{22} &+t_1 &-12t_2 \\
j_1 &= +X_{11} &-13X_{22} &-2t_1 &+23t_2 \\
i_2 &= & &+t_2 & \\
j_2 &= &+5X_{22} & &-9t_2
\end{aligned}
\tag{6.7}
$$

where t_1 and t_2 are two parameters renaming i_2 and I_1'.

In matrix format, the solution of the system is

$$
\begin{bmatrix} i_1 \\ j_1 \\ \cdots \\ i_2 \\ j_2 \end{bmatrix} =
\begin{bmatrix} 1 & -12 \\ -2 & 23 \\ \cdots & \cdots \\ 1 & 0 \\ 0 & -9 \end{bmatrix}
\begin{bmatrix} t_1 \\ t_2 \end{bmatrix} +
\begin{bmatrix} 27 \\ -51 \\ \cdots \\ 0 \\ 20 \end{bmatrix}
$$

The intersection of the two LBL's is a new linearly bounded lattice defined by (eq. 6.3) $\{ \mathbf{x} = \mathbf{T} \cdot \mathbf{i} + \mathbf{u} \mid \cdots \}$, where

$$
\mathbf{T} = \mathbf{T}_1 \mathbf{V}_1 = \begin{bmatrix} 2 & 3 \\ 5 & 3 \end{bmatrix} \begin{bmatrix} 1 & -12 \\ -2 & 23 \end{bmatrix} = \begin{bmatrix} -4 & 45 \\ -1 & 9 \end{bmatrix}
$$

$$
\mathbf{u} = \mathbf{T}_1 \mathbf{v}_1 + \mathbf{u}_1 = \begin{bmatrix} 2 & 3 \\ 5 & 3 \end{bmatrix} \begin{bmatrix} 27 \\ -51 \end{bmatrix} + \begin{bmatrix} 0 \\ 0 \end{bmatrix} = \begin{bmatrix} -99 \\ -18 \end{bmatrix}
$$

With the supplementary transformation[10] $X_{22} = X_{22}' + 2X_{11}$, the unimodular transformation that brings the system matrix to the Hermite Normal Form results from (6.7):

$$
\mathbf{S}_{n,n} = \begin{bmatrix} 13 & 7 & 1 & -12 \\ -25 & -13 & -2 & 23 \\ 0 & 0 & 1 & 0 \\ 10 & 5 & 0 & -9 \end{bmatrix}
$$

As each unimodular transformation reduces the absolute values of the coefficients of an equation (except the smallest one) to at most 1/2 of the smallest coefficient, this method employs usually a shorter sequence of unimodular transformations than the approaches in [282] (chapter 5), [268] (section A.3)] – where the reduction factor is 1, or in [329] – where the reduction factor is 2/3. The explicit computation of the greatest common divisor – as in [296] (section I.7.4) – is also avoided.

---

[10]In practice, this extra transformation is embedded in the previous U-type transformation, as well as the column permutations Π.

6.2.2.2 The double description of polyhedra. The repeated intersections between the LBL's during the partitioning algorithm (see subsection 6.2.1) generate new LBL's with larger and larger constraint sets $A \cdot i \geq b$. Keeping permanently a minimal representation of the polytopes in the LBL's is a necessary operation in order to prevent the explosive increase of the inequality sets. The procedure employed stems from the polyhedral library developed at IRISA, France [427]. The approach is based on the *double description of polyhedra*.

In 1896, Minkowski showed that any polyhedron defined implicitly

$$\mathcal{P} = \{\, x \mid Ax = b \,,\ Cx \geq d \,\}$$

has an equivalent dual parametric representation [363] (section 8.9):

$$\mathcal{P} = \{\, x \mid x = V\nu + L\lambda + R\mu \,,\ \mu, \nu \geq 0 \,,\ \sum \nu_i = 1 \,\}$$

characterized by a convex combination of vertices (the columns of matrix V), a linear combination of lines (the columns of matrix L), and a positive combination of rays (the columns of matrix R).

In 1936, Motzkin proved the *Decomposition Theorem for Polyhedra*, according to which any polyhedron can be *uniquely* decomposed into a polytope $\mathcal{V} = ConvexHull\{V_i\}$ generated by a convex combination of extreme vertices, and a *polyhedral cone*. In its turn, the cone can be partitioned into a *lineality space* $\mathcal{L}$ – the largest linear space contained in the cone, and a *pointed cone* $\mathcal{R}$ – represented by a positive combination of extreme rays [Schrijver 86 (chapter 8), Nemhauser 88 (chapter I.4)].

An algorithm to compute the dual representations of a polyhedron $\mathcal{P}$, that is, given A, b, C, d, compute V, L, R and reciprocally has been proposed in [285]. A similar algorithm has been also implemented in the polyhedral library of IRISA [428].

The elimination of redundant inequalities in (6.1) and (6.2) can be achieved by computing the double representation of the inequality set $A \cdot i \geq b$:
(1) by selecting a basis of the lineality space, the redundant equalities can be detected and eliminated (as $A \cdot i \geq b$ may contain also equalities embedded, and these are generating the lineality space);
(2) the non-redundant inequalities must be saturated[11] by at least n extreme vertices/rays, where n is the space dimension.

Example The index space of the operand domain $A[j][n + i]$ in line (4) from the Silage code in figure 6.6 is represented by the LBL:

$$\left\{ \begin{bmatrix} x \\ y \end{bmatrix} = \begin{bmatrix} 0 & 1 \\ 1 & 0 \end{bmatrix} \begin{bmatrix} i \\ j \end{bmatrix} + \begin{bmatrix} 0 \\ n \end{bmatrix} \quad \middle| \quad \begin{bmatrix} 1 & 0 \\ -1 & 0 \\ 0 & 1 \\ 0 & -1 \\ 1 & -1 \end{bmatrix} \begin{bmatrix} i \\ j \end{bmatrix} \geq \begin{bmatrix} 0 \\ -n+1 \\ 0 \\ -n+1 \\ 1 \end{bmatrix} \right\}.$$

The goal is to eliminate the redundant inequalities (if any) from the polytope of this LBL.

A preprocessing step is required in order to transform the initial inhomogeneous system of dimension n into a homogeneous system of dimension $(n + 1)$: $\xi A \cdot i - \xi b \geq 0$. This homogeneous system of inequalities represents a polyhedral cone. By means of the above transformation, both the vertices and the rays of the initial inhomogeneous polyhedron have a unified representation as rays in the polyhedral cone.

Applying the double description algorithm [285] to the homogeneous set of constraints, it results that the lineality space is the null space, and the ray space is generated by the vectors:

$$\mathcal{R} = \{\, r_1 = [n-1 \ \ 0 \ \ 1] \,,\ r_2 = [1 \ \ 0 \ \ 1] \,,\ r_3 = [n-1 \ \ n-2 \ \ 1] \,\}$$

As the resulting three rays have the general form $(\xi v_x, \xi v_y, \xi)$, with $\xi > 0$, they all represent vertices of coordinates (v_x, v_y) in the initial polytope (see figure 6.4). As none of the rays saturates

---

[11] A ray r saturates the inequality $a^T x \geq 0$ when $a^T r = 0$.

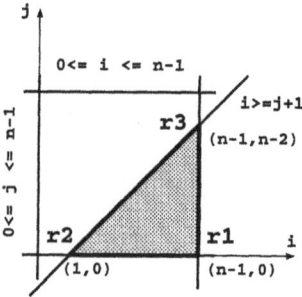

Figure 6.4.   Elimination of redundant inequalities

the 1st and the 4th constraint, while the other inequalities are saturated by exactly two rays (*vertices* in inhomogeneous coordinates) each, it follows that the 1st and 4th constraints are redundant. The given LBL must be replaced by

$$\left\{ \begin{bmatrix} x \\ y \end{bmatrix} = \begin{bmatrix} 0 & 1 \\ 1 & 0 \end{bmatrix} \begin{bmatrix} i \\ j \end{bmatrix} + \begin{bmatrix} 0 \\ n \end{bmatrix} \middle| \begin{bmatrix} -1 & 0 \\ 0 & 1 \\ 1 & -1 \end{bmatrix} \begin{bmatrix} i \\ j \end{bmatrix} \geq \begin{bmatrix} -n+1 \\ 0 \\ 1 \end{bmatrix} \right\}.$$

6.2.3 Computing the number of scalars addressed by an array reference

The polytope size means, in this context, the number of distinct n-dimensional points, having integer coordinates, inside a polytope: $Card \{ \mathbf{i} \in \mathbf{Z}^n \mid \mathbf{A} \cdot \mathbf{i} \geq \mathbf{b} \}$.

The LBL size (or image size of a polytope) means the number of distinct m-dimensional points of integer coordinates belonging to the respective lattice (see also (6.1)):

$$Card \{ \mathbf{x} \in \mathbf{Z}^m \mid \mathbf{x} = \mathbf{T} \cdot \mathbf{i} + \mathbf{u}, \ \mathbf{A} \cdot \mathbf{i} \geq \mathbf{b}, \mathbf{i} \in \mathbf{Z}^n \}$$

The LBL size means, in fact, the number of scalars addressed by an array reference.

The importance of a correct and efficient method for the computation of index space sizes [31] is manifold: (1) this is needed during the partitioning of the indexed signals (see subsection 6.2.1), and (2) this allows to determine the size of the basic sets, as well as the number of dependencies between these groups of signals (see subsection 6.3.1). The latter essential information can be exploited in order to estimate the storage requirements for applicative algorithmic specifications, as explained intuitively in subsection 6.2 .

Remark Most of the memory-related research work assumes that all loops in the specifications have constant boundaries; if this is not the case, approximations by taking the extreme values of the boundaries should yield fairly good results (e.g., [405]). According to such an assumption, the number of scalars addressed by an array reference is approximated by the size of the hypercube encompassing the iterator space of that array reference (instead of computing the size of the index space). This approximation is good only if, in addition, the loops have unitary steps, if there are no iterator-dependent conditions in the array scope, and if the affine mapping **Ti+u** in (6.1) is injective.

Example   (i : 8 .. 15) ::
                    (j : $i - 8 .. i + 8 .. 2$) :: $\cdots A[i+j] \cdots$

As iterator j takes values between 0 and 23, the hypercube encompassing the iterator space is [8,15]×[0,23], and it contains 192 points. The size of the LBL corresponding to the signal domain $A[i + j]$ is however 16 (therefore, 12 times less than the approximation). The error is due both to the incorrect evaluation of the iterator space size – which contains 72 points rather than 192, and to the fact that the affine mapping $t(i, j) = i + j$ is not injective (e.g., $A[12]$ is addressed by several pairs of iterators (i, j): (8,4), (9,3), (10,2)).

It must be mentioned also that loops having increment steps different from 1

$$(\, i : \ m \, .. \, M \, .. \, Step \,) :: \ \cdots A[f(i)] \ \cdots$$

can be easily "normalized" with the affine transformation $i = i' \cdot Step + m$, thus being equivalent to

$$(\, i' : \ 0 \, .. \, \left\lfloor \frac{M - m}{Step} \right\rfloor) :: \ \cdots A[f(i')] \ \cdots$$

For instance, the nest of loop in the example above is equivalent to

$$(\, i : \ 8 \, .. \, 15 \,) ::$$
$$(\, j : \ 0 \, .. \, 8 \,) :: \qquad \cdots A[2 * i + 2 * j - 8] \ \cdots$$

Concluding, for the general affine case, the approximation mentioned above may be very crude, being improper to use for memory estimation/allocation as it can lead to an exaggerated number of storage locations. Section 6.2.3.2 and 6.2.3.3 will present correct solutions to the problem of index space size determination. It is assumed along this subsection that all loops have been "normalized" in a preprocessing phase[12].

It must be emphasized also that enumerative techniques can always be applied to compute the number of scalars in an array reference. These approaches are obviously simple and extremely efficient for array references with "small" iterator spaces. In image and video processing applications most of the array references are characterized by huge iterator spaces: an enumerative technique, although very simple, can be too computationally expensive for such applications (see table 6.1).

6.2.3.1 Computing the size of polytopes. The computation of the number of points inside a polytope, having integer coordinates, has been tackled long ago for the 3-D case [281]. Then, the importance of Dedekind sums was revealed in this context:

$$D_p^{q,r} \stackrel{def}{=} \sum_{x=1}^{p-1} \left(\frac{qx}{p} - \left\lfloor \frac{qx}{p} \right\rfloor \right) \left(\frac{rx}{p} - \left\lfloor \frac{rx}{p} \right\rfloor \right)$$

Recently, it was proven that Dedekind sums can be computed in polynomial time, employing an algorithm similar to Euclid's [123]. Therefore, also our problem could be solved in polynomial time. Unfortunately, Dyer's method involves a preprocessing step where a decomposition of a special form is needed:

$$\{ \mathbf{x} \in \Re^d : \mathbf{Ax} \le \mathbf{b} \} \quad \longrightarrow \quad \left\{ \mathbf{x} \in \Re^d : 1 > \frac{x_1}{a_1} > \cdots > \frac{x_d}{a_d} > 0 \right\}$$

Although this splitting process can be achieved in polynomial time too [81], the above method appears to be of rather theoretical interest. First of all, it is not clear how to extend these results to higher dimensions than 4 (it was proven that case 4-D can be reduced to 3-D in polynomial time). In addition, there is no polynomial-time algorithm available for evaluating Dedekind sums of higher order $D_p^{q_1,\cdots,q_{d-1}}$.

Recent work in the field of data-flow analysis and code parallelization also focused on polytope size determination. [387] introduced an algorithm for counting the number of iterations in nested loops with parametrizable bounds, but the polytopes have restricted representations in this case. [126] presents a "fast" counting method for general polytopes employing rational fractions of polynomials. Unfortunately, the practical efficiency of the approach degrades quickly as the polytope dimension increases.

As an accurate and general solution – able to handle signals of any dimension – was needed, a novel technique based on the Fourier-Motzkin elimination [88] has been developed.

Research on linear inequalities systems prior to 1950 consisted in isolated efforts by a few investigators. One of the oldest result is the elimination technique for reducing the number of variables

---

[12]Loop boundaries containing the *floor* $\lfloor \ \rfloor$ and *ceiling* $\lceil \ \rceil$ functions create no special problem, as the routines of our system are designed to take into account only the points with integer coordinates.

in the system. The first description of the method can be found in one of the works of Fourier from 1826. Later, in his thesis from 1936, Motzkin brings important contributions, applying the method in the context of his theory on linear inequalities and the double description of polyhedra [284] (see subsection 6.2.2.2).

The theoretical importance of the Fourier-Motzkin elimination as revealed more recently, when employing it, Kuhn gave a remarkably simple proof of the *"Feasibility Theorem"* – which states the necessary and sufficient condition that ensures the solvability of a system of linear inequalities [219]:

$$\sum_{j=1}^{n} a_{ij} x_j \geq b_i , \qquad i = \overline{1, m} \tag{6.8}$$

Afterwards, the Fourier-Motzkin elimination has been successfully applied for proving the *Farkas Lemma*, or the *Fundamental Duality Theorem* of linear programming [122]. Dantzig and Eaves formulate the dual problem of the Fourier-Motzkin elimination, employing it to generate the positive integer solutions of a system of linear equalities with integer coefficients [88].

The basic idea of the Fourier-Motzkin elimination is extremely simple: the system (6.8) may be partitioned into three sets of inequalities according to whether the coefficients of x_n are positive, negative, or zero. The initial system (6.8) may be rewritten in the form:

$$\begin{cases} x_n \geq D_1(\overline{x}) \\ \vdots \\ x_n \geq D_p(\overline{x}) \end{cases} \qquad \begin{cases} x_n \leq E_1(\overline{x}) \\ \vdots \\ x_n \leq E_q(\overline{x}) \end{cases} \qquad \begin{cases} 0 \leq F_1(\overline{x}) \\ \vdots \\ 0 \leq F_r(\overline{x}) \end{cases} \tag{6.9}$$

where $D_i(\overline{x})$, $E_j(\overline{x})$, $F_k(\overline{x})$ are affine functions of $\overline{x} = (x_1, \ldots, x_{n-1})$. The given system (6.8) may be solved by first solving the *reduced* system

$$\begin{cases} D_i(\overline{x}) \leq E_j(\overline{x}) \\ 0 \leq F_k(\overline{x}) \end{cases} \qquad i = \overline{1, p} \quad j = \overline{1, q} \quad k = \overline{1, r} \tag{6.10}$$

and then finding the values of x_n satisfying

$$\max_i D_i(\overline{x}) \leq x_n \leq \min_j E_j(\overline{x})$$

It can be proven that such an x_n always exists providing there exists an $\overline{x}$ satisfying (eq. 6.10). Furthermore, if $p = 0$ or $q = 0$ then the range of x_n is unbounded, and the existence of an $\overline{x}$ satisfying (eq. 6.10) entails an infinite number of solutions for system (eq. 6.8).

As (6.10) is also a system of linear inequalities (with one variable less), the elimination procedure can be applied to x_{n-1}, a.s.o., until all but a single variable, say x_1, are gone. The original system (6.8) is solvable if and only if the final system (obtained by the successive elimination of $n - 1$ variables)

$$x_1 \geq \alpha_i \ , \quad x_1 \leq \beta_j \ , \quad 0 \leq \gamma_k$$

is consistent, that is $\beta_j - \alpha_i \geq 0$ and $\gamma_k \geq 0$ for all i, j, k.

Recently, the Fourier-Motzkin elimination has been employed for testing the data-dependence between array references [329]. This is equivalent to the problem of checking the emptiness[13] of a polytope. The method is claimed to be "fast and practical for performing data dependence analysis" [329], although [36] claims to be "quite time consuming" for data dependence problems in many variables.

The routine for counting the lattice points inside a given polytope is described below. In the sequel, the columns of a matrix are denoted by subscripted vectors: e.g., $\mathbf{A} = [\mathbf{a}_1 \ \mathbf{a}_2 \ \cdots]$; the number of columns is denoted by $\mathbf{A}.nCol$. The main idea of the algorithm is the following: the number of

---

[13]in terms of points having integer coordinates.

points of integer coordinates inside the given n-dimensional polytope, and having as first coordinate z_1, is equal to the number of points inside the polytope $a_2z_2 + a_3z_3 + \cdots \geq b - a_1z_1$ (which has one dimension less). The required result is obtained by accumulating this number over the entire discrete range of z_1 (determined by the Fourier-Motzkin elimination).

Counting the points with integer coordinates inside a polytope

```
int CountPolytopePoints(A,b) {           // the given polytope is Az≥b
   if (A.nCol==1) return Range(A,b);      // handles the trivial case az≥b
   if ( FourierMotzkinElim(A,b) ≤ 0 ) return error_code;
   // special cases, when Az≥b is an unbounded polyhedron or an empty set,
   // are detected; otherwise, the range of z₁ – the first element of z –
   // i.e., [z₁ᵐⁱⁿ, z₁ᵐᵃˣ] is determined
   N = 0 ;
   for ( int z₁ = ⌈z₁ᵐⁱⁿ⌉ ; z₁ ≤ ⌊z₁ᵐᵃˣ⌋ ; z₁ + +)
      N += CountPolytopePoints( [a₂ a₃ ···] , b − a₁z₁ );
   return N ;
}
```

The routine *Range*, handling the case when A is a vector ($az \geq b$), is called also from *FourierMotzkinElim*. It checks whether the range of z is unbounded (in which case it returns the $error\_code = -1$), or whether the range of z is empty (in which case it it returns the $error\_code = 0$). Otherwise, it returns the number of integers in the range of z, that is

$$\min_{\{i \,|\, a_i < 0\}} \left\lfloor \frac{b_i}{a_i} \right\rfloor - \max_{\{i \,|\, a_i > 0\}} \left\lceil \frac{b_i}{a_i} \right\rceil + 1$$

The worst-case time complexity of the algorithm is exponential. The influence of this negative aspect is attenuated by eliminating the resulting identities and, partially, the redundant inequalities. In addition, the initial number of columns of matrix A is the number of the surrounding loops, which is usually small. Moreover, the method is not usually applied on the constraint set $A \cdot i \geq b$, but on *reduced* constraint systems.

6.2.3.2 Computing the size of linearly bounded lattices. When the affine function $t : Z^n \to Z^m$, defined by $t(i) = Ti + u$ is injective, the number of m-dimensional points with integer coordinates belonging to the *image* of a polytope is equal to the number of n-dimensional points with integer coordinates inside the polytope (the LBL size is equal to the polytope size).

Two questions arise: (1) how can it be decided whether function t is injective or not, and (2) what is to be done if the injectivity condition is not fulfilled. In this context, it has to be emphasized that generating all lattice points in the polytope and collecting their images in a set is very inefficient for "large" polytopes (see table 6.1).

For any matrix $T \in Z^{m \times n}$ there exists a unimodular matrix $S \in Z^{n \times n}$ such that $P \cdot T \cdot S = \begin{bmatrix} H_{11} & 0 \\ H_{21} & 0 \end{bmatrix}$, where H_{11} is a nonsingular lower-triangular matrix, and P is a row permutation [268] (section A.3). Such a unimodular matrix can be computed with the technique described in subsection 6.2.2.1 for solving linear Diophantine systems. Denoting $rank \, H_{11} = r$, and $S^{-1}i = j$, the following situations may be encountered:

(1) $r = n$. The affine mapping $x = Ti + u = \begin{bmatrix} H_{11} \\ H_{21} \end{bmatrix} j + u$ is injective, as $x_1 = x_2$ implies $j_1 = j_2$ (as H_{11} is lower-triangular), hence $i_1 = i_2$.

(2) $r < n$. Then $x_1 = x_2$ implies that only the first r components of j_1 and j_2 are resp. equal. This means that all vectors j satisfying $AS \cdot j \geq b$, and having the same prefix $j_1 \ldots j_r$, contribute to the image set with one single distinct value because they are mapped to the same point. A prefix $j_1 \ldots j_r$ is called *valid* if there exist vectors j satisfying $AS \cdot j \geq b$, and having that prefix. Consequently, the affine image size of the polytope is the number of all *valid* prefixes $j_1 \ldots j_r$. If the number of valid prefixes is equal to the polytope size, the affine mapping is injective (see the examples in the sequel).

Therefore, the complete algorithm for counting the image size of the polytope $\mathbf{Az} \geq \mathbf{b}$ becomes the following:

Counting the points with integer coordinates inside an LBL

int CountImagePolytope(*Lbl*) {   // $Lbl = \{\, \mathbf{x} = \mathbf{T} \cdot \mathbf{i} + \mathbf{u} \mid \mathbf{A} \cdot \mathbf{i} \geq \mathbf{b} \,\}$

    compute unimodular matrix **S** such that $\mathbf{P} \cdot \mathbf{T} \cdot \mathbf{S} = \begin{bmatrix} \mathbf{H}_{11} & 0 \\ \mathbf{H}_{21} & 0 \end{bmatrix}$, with $\mathbf{H}_{11}$
        lower triangular, and **P** a permutation matrix;
    let $r = rank\ \mathbf{H}_{11}$;
    if (r=T.nCol) **return** CountPolytopePoints(**A**,**b**) ;   // case (1): mapping t is injective
    else return CountPrefixes(**AS**, **b**, r) ;   // case (2)
}

int CountPrefixes(**A**,**b**,r) {// Generates potential prefixes of length r for the vectors
        // in the polytope $\mathbf{Az} \geq \mathbf{b}$. Only valid prefixes contribute to the image size
    if $(r = 0)$ // If the whole prefix has been generated, it is checked whether
        if (nonEmptyPolytope(**A**,**b**)) **return** 1 ; **else return** 0 ;
    // the prefix is valid or not
    FourierMotzkinElim(**A**,**b**) ;
    // $\mathbf{Az} \geq \mathbf{b}$ is assumed to be a non-empty bounded polyhedron
    // *FourierMotzkinElim* returns the range of z_1 – the first element of **z**
    N = 0 ;
    // for every possible first component of the prefix (of length r)
    //   the rest of the prefix (of length $r - 1$) is generated
    for (int $z_1 = \lceil z_1^{min} \rceil$; $z_1 \leq \lfloor z_1^{max} \rfloor$; $z_1 + +$)
        N += CountPrefixes([$\mathbf{a}_2\ \mathbf{a}_3 \cdots$] , $\mathbf{b} - \mathbf{a}_1 z_1$, $r - 1$);
    return N ;
}

bool nonEmptyPolytope(**A**,**b**) {
    // Returns **true** ($\neq 0$) if the integral polyhedron $\mathbf{Az} \geq \mathbf{b}$ is not empty
    if (A.nCol==1) **return** Range(**A**,**b**); // handles $\mathbf{az} \geq \mathbf{b}$;
    if (FourierMotzkinElim(**A**,**b**) ≤ 0) **return** *error_code*;
    // *FourierMotzkinElim* returns the range of z_1 – the first element of **z**
    for (int $z_1 = \lceil z_1^{min} \rceil$; $z_1 \leq \lfloor z_1^{max} \rfloor$; $z_1 + +$)
        if (nonEmptyPolytope([$\mathbf{a}_2\ \mathbf{a}_3 \cdots$] , $\mathbf{b} - \mathbf{a}_1 z_1$)) **return** 1 ;
    return 0 ;
}

Remark The routine *nonEmptyPolytope* is equivalent to the data dependence checking employed in the Omega test [329].

Two illustrative examples are briefly discussed in the sequel.

Example 1 $(i : 0 .. 255) ::$
$$(j : 0 .. 255) :: \quad \cdots M[2*i+3*j+1][5*i+j+2][4*i+6*j+3] \cdots$$

$$\{ \mathbf{x} = \mathbf{Ti} + \mathbf{u} | \mathbf{Ai} \geq \mathbf{b}\} = \left\{ \begin{bmatrix} x \\ y \\ z \end{bmatrix} = \begin{bmatrix} 2 & 3 \\ 5 & 1 \\ 4 & 6 \end{bmatrix} \begin{bmatrix} i \\ j \end{bmatrix} + \begin{bmatrix} 1 \\ 2 \\ 3 \end{bmatrix} \right\}$$

$$\left\{ \begin{bmatrix} 1 & 0 \\ -1 & 0 \\ 0 & 1 \\ 0 & -1 \end{bmatrix} \begin{bmatrix} i \\ j \end{bmatrix} \geq \begin{bmatrix} 0 \\ -255 \\ 0 \\ -255 \end{bmatrix} \right\}$$

As there is a unimodular matrix $\mathbf{S} = \begin{bmatrix} -1 & 3 \\ 1 & -2 \end{bmatrix}$ such that $\mathbf{TS} = \begin{bmatrix} \mathbf{H}_{11} \\ \cdots \\ \mathbf{H}_{21} \end{bmatrix} = \begin{bmatrix} 1 & 0 \\ -4 & 13 \\ \cdots & \cdots \\ 2 & 0 \end{bmatrix}$, it results that $rank\ \mathbf{H}_{11} = 2 = \mathbf{T}.nCol$, hence the mapping is injective. Therefore, the LBL size is equal to $size(\mathbf{Ai} \geq \mathbf{b}) = 256 \times 256 = 65536$. The size of the polytope can be computed either with the routine $CountPolytopePoints(\mathbf{A}, \mathbf{b})$ or, more efficiently in this case, by taking into account that both iterators have constant bounds with range 256.

Example 2 $(\ i:\ 0\ ..\ 7\ )::$
$$(\ j:\ 0\ ..\ 7\ )::$$
$$(\ k:\ 0\ ..\ 7\ )::\qquad \cdots M[i + k][j + k] \cdots$$

With the unimodular matrix $\mathbf{S} = \begin{bmatrix} 1 & 0 & -1 \\ 0 & 1 & -1 \\ 0 & 0 & 1 \end{bmatrix}$, $\mathbf{TS} = \begin{bmatrix} 1 & 0 & 0 \\ 0 & 1 & 0 \end{bmatrix}$. As $r < n\ (2 < 3)$, the prefixes $(j_1\ j_2)$ of all vectors $\mathbf{j}$ satisfying $\mathbf{AS} \cdot \mathbf{j} \geq \mathbf{b}$, that is $\{7 \geq j_1 - j_3 \geq 0, 7 \geq j_2 - j_3 \geq 0, 7 \geq j_3 \geq 0\}$, have to be checked for validity. Eliminating j_3 from the previous set of inequalities, the prefixes $(j_1\ j_2)$ result to have the ranges $j_1 \in [0, 14]$ and $j_2 \in [\max\{0, j_1 - 7\}, \min\{14, j_1 + 7\}]$.

There are 169 pairs $(j_1\ j_2)$ having these ranges. All of them prove to be valid prefixes: for instance, the prefix (1 6) is valid as there are two vectors $- [1\ 6\ 0]^T$ and $[1\ 6\ 1]^T$ – satisfying the system of inequalities $\mathbf{AS} \cdot \mathbf{j} \geq \mathbf{b}$. It follows that the size of the index space of $M[i + k][j + k]$ is 169. As this value is inferior to the size of the iterator space – which is 512 – it follows that the affine mapping $\mathbf{t}(i, j, k) = \mathbf{T} \cdot [i\ j\ k]^T$ is not injective. Indeed, this can be easily seen noticing that the triplets (i, j, k) equal to (0,1,1) and (1,2,0) are mapped to the same point (1,2) in the index space.

Table 6.1.   Experimental results for *Example 2*. The computation of the size of linearly bounded lattices.

Loop bounds	Size polytope (iterator sp.)	Size LBL (index sp.)	LBL/Pol. ratio [%]	CPU	
				Enum. Polytope	*CountImage*
3	64	37	57	0.0004s	0.0005s
7	512	169	33	0.012s	0.015s
15	4096	721	17	0.087s	0.050s
31	32768	2977	9.0	1.05s	0.18s
63	262,144	12097	4.6	17.48s	0.70s
127	2,097,152	48769	2.3	5m23s	2.75s
255	16,777,216	195841	1.1	1h55m36s	10.86s
511	134,217,728	784897	0.5	?	43.25s

Table 6.1 shows the CPU times (column 6) for the computation of the LBL size when the upper-boundaries of the three loops (column 1) are increased gradually. It can be noticed that the ratio (column 4) between the LBL size (column 3) and the iterator space size (column 2) is decreasing dramatically if the affine mapping is not injective: therefore, the approximation of the index space size by the iterator space size must be avoided!

The CPU times obtained employing an enumerative technique are displayed in column 5. They may be lower than those corresponding to the routine *CountImagePolytope* for small iterator spaces, but they grow dramatically with the LBL size (as the enumerative techniques have to keep track of the index values). Although the run times for the routine are also increasing, the method is still effective even for huge sizes of the iterator space.

6.2.3.3 Index space determination for array references. Let $\{\mathbf{x} = \mathbf{T} \cdot \mathbf{i} + \mathbf{u} \mid \mathbf{A} \cdot \mathbf{i} \geq \mathbf{b}\}$ be the LBL of a given array reference in a program. If matrix $\mathbf{T}$ is square and nonsingular, the index space is included into an *image polytope* which can be easily determined: noticing that the iterator vector $\mathbf{i} = \mathbf{T}^{-1} \cdot (\mathbf{x} - \mathbf{u})$ must represent a point inside the iterator polytope $\mathbf{A} \cdot \mathbf{i} \geq \mathbf{b}$, the image polytope results to be $\mathbf{A} \cdot \mathbf{T}^{-1} \cdot (\mathbf{x} - \mathbf{u}) \geq \mathbf{b}$.

Even in this case when matrix $\mathbf{T}$ is invertible, not all the points in the image polytope belong to the index space. For instance, if $(i : 0..9) :: \cdots A[5 * i] \cdots$ the image polytope of the array reference is $0 \le x \le 45$. But the indices of A take only 10 from these 46 values.

This subsection will present a more direct method for computing the LBL size ([31], chapter 2), based on the effective construction of the index space of an array reference (see figure 6.5), even when matrix $\mathbf{T}$ is singular, or non-square[14].

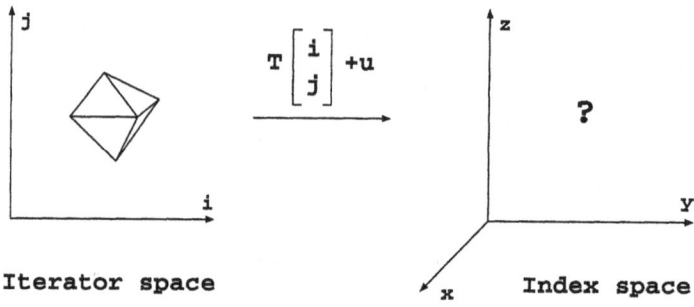

Figure 6.5.   Affine transformation of a polytope from the iterator space to the index space

As shown in subsection 6.2.3.2, for any matrix $\mathbf{T} \in \mathbf{Z}^{m \times n}$ with $rank\ \mathbf{T} = r$, and assuming the first r rows of $\mathbf{T}$ are linearly independent, there exists a unimodular matrix $\mathbf{S} \in \mathbf{Z}^{n \times n}$ such that $\mathbf{T} \cdot \mathbf{S} = \begin{bmatrix} \mathbf{H}_{11} & 0 \\ \mathbf{H}_{21} & 0 \end{bmatrix}$, where $\mathbf{H}_{11} \in \mathbf{Z}^{r \times r}$ is a lower-triangular matrix of rank r, and $\mathbf{H}_{21} \in \mathbf{Z}^{(m-r) \times r}$ [268] (section A.3). The block matrix is called the reduced Hermite form of matrix $\mathbf{T}$.

Let $\mathbf{S}^{-1}\mathbf{i} \stackrel{def}{=} \mathbf{j} \equiv \begin{bmatrix} \mathbf{j}_1 \\ \mathbf{j}_2 \end{bmatrix}$, where $\mathbf{j}_1, \mathbf{j}_2$ are r-, respectively $(n-r)$-, dimensional vectors. Then

$$\mathbf{x} = \mathbf{T}\mathbf{i} + \mathbf{u} = \mathbf{T}\mathbf{S}\mathbf{j} + \mathbf{u} = \begin{bmatrix} \mathbf{H}_{11} \\ \mathbf{H}_{21} \end{bmatrix} \mathbf{j}_1 + \mathbf{u} \qquad (6.11)$$

Denoting $\mathbf{x} = \begin{bmatrix} \mathbf{x}_1 \\ \mathbf{x}_2 \end{bmatrix}$, and $\mathbf{u} = \begin{bmatrix} \mathbf{u}_1 \\ \mathbf{u}_2 \end{bmatrix}$ (where $\mathbf{x}_1$, $\mathbf{u}_1$ are r-dimensional vectors), it follows that $\mathbf{x}_1 = \mathbf{H}_{11}\mathbf{j}_1 + \mathbf{u}_1$. As $\mathbf{H}_{11}$ is nonsingular (being lower-triangular of rank r), $\mathbf{j}_1$ can be obtained explicitly:

$$\mathbf{j}_1 = \mathbf{H}_{11}^{-1}(\mathbf{x}_1 - \mathbf{u}_1) \qquad (6.12)$$

The iterator vector $\mathbf{i}$ results with a simple substitution:

$$\mathbf{i} = \mathbf{S}\begin{bmatrix} \mathbf{j}_1 \\ \mathbf{j}_2 \end{bmatrix} = \begin{bmatrix} \mathbf{S}_1 & \mathbf{S}_2 \end{bmatrix} \begin{bmatrix} \mathbf{H}_{11}^{-1}(\mathbf{x}_1 - \mathbf{u}_1) \\ \mathbf{j}_2 \end{bmatrix} = \mathbf{S}_1\mathbf{H}_{11}^{-1}(\mathbf{x}_1 - \mathbf{u}_1) + \mathbf{S}_2\mathbf{j}_2$$

where $\mathbf{S}_1$ and $\mathbf{S}_2$ are the submatrices of $\mathbf{S}$ containing the first r, respectively the last $n - r$, columns of $\mathbf{S}$. As the iterator vector must represent a point inside the iterator polytope $\mathbf{A} \cdot \mathbf{i} \ge \mathbf{b}$, it follows that:

$$\mathbf{A}\mathbf{S}_1\mathbf{H}_{11}^{-1}\mathbf{x}_1 + \mathbf{A}\mathbf{S}_2\mathbf{j}_2 \ge \mathbf{b} + \mathbf{A}\mathbf{S}_1\mathbf{H}_{11}^{-1}\mathbf{u}_1 \qquad (6.13)$$

---

[14]i.e., the dimension of the index space differs from the dimension of the iterator space.

As the rows of matrix H_{11} are r linearly independent r-dimensional vectors, each row of H_{21} is a linear combination of the rows of H_{11}. Then from (6.11), it results that there exists a matrix $C \in \Re^{(m-r) \times r}$ such that[15]

$$x_2 - u_2 = C \cdot (x_1 - u_1) \tag{6.14}$$

The system of inequalities (6.13) and the system of equations (6.14) characterize the index space of the given array reference.

If $n > r$, the *image polytope* of the index space can be obtained by taking the projection of the n-dimensional polytope (6.13) on the r-dimensional subspace defined by the first r coordinates[16]. This image polytope is usually not dense (it contains "holes"), as not all its points represent valid prefixes of length r in the polytope (6.13) (see subsection 6.2.3.2).

Even if $n = r$ (therefore, no projection is needed), the image polytope is dense if and only if matrix H_{11} is unimodular, as it can be seen from (6.12). Indeed, assuming that $|det\ H_{11}| \neq 1$, and taking into account that the elements of j_1 must be integers, it follows (by dividing and multiplying the right member of (6.12) with $|det\ H_{11}|$) that the points x inside the LBL must satisfy the supplementary constraints

$$|det\ H_{11}|\ |\ h_i^T(x_1 - u_1) \qquad \forall i = \overline{1, r} \tag{6.15}$$

where h_i^T are the rows of the matrix with integer coefficients $|det\ H_{11}|\ H_{11}^{-1}$, and $a|b$ means "a divides b". If H_{11} is unimodular, the constraints (6.15) are obsolete.

In conclusion, the size of the LBL can be computed, taking into account (6.13), with the routine $CountPrefixes\left(\left[AS_1 H_{11}^{-1}\ AS_2\right],\ b + AS_1 H_{11}^{-1} u_1,\ r\right)$ described in subsection 6.2.3.2, with a slight modification: if H_{11} is not unimodular, a valid prefix must satisfy also the constraints (6.15).

The two examples in subsection 6.2.3.2 will constitute again benchmark tests for this method.

Example 1 (revisited)

As $m = 3$ and $r = n = 2$, we have $x_1 = [x\ y]^T$ and $x_2 = [z]$. Taking also into account the results obtained already in subsection 6.2.3.2,

$$S_1 \equiv S = \begin{bmatrix} -1 & 3 \\ 1 & -2 \end{bmatrix}, \quad H_{11} = \begin{bmatrix} 1 & 0 \\ -4 & 13 \end{bmatrix}, \quad H_{11}^{-1} = \frac{1}{13}\begin{bmatrix} 13 & 0 \\ 4 & 1 \end{bmatrix}, \quad H_{21} = \begin{bmatrix} 2 & 0 \end{bmatrix}.$$

Condition (6.13) yields

$$\begin{array}{ccccc} 3320 & \geq & -x & +3y & \geq & 5 \\ 3316 & \geq & 5x & -2y & \geq & 1 \end{array}$$

As the only row in H_{21} is two times the first row in H_{11}, the image polytope is completed with the equality (6.14): $z - 3 = 2(x - 1)$.

As $det\ H_{11} = 13$, the image polytope is not dense. Condition (6.15) yields $13\ |\ 4(x - 1) + (y - 2)$ (as $13\ |\ 13(x - 1)$ is always satisfied).

The number of points in the image polytope, satisfying also the divisibility condition above, are 65536 – the size of the LBL (which is also here the size of the iterator space).

Example 2 (revisited)

As $m = r = 2$ and $n = 3$, we denote the $(n - r)$-dimensional vector $j_2 = [\lambda]$. As $H_{11} = \begin{bmatrix} 1 & 0 \\ 0 & 1 \end{bmatrix}$ (thus unimodular), and H_{21} does not exist (as $m = r$), with $S_1 = \begin{bmatrix} 1 & 0 \\ 0 & 1 \\ 0 & 0 \end{bmatrix}$, $S_2 =$

---

[15] The coefficients of matrix C are determined by backward substitutions from the equations: $H_{21}.row(i) = \sum_{j=1}^{r} c_{ij} \cdot H_{11}.row(j)$ for any $i = 1, \ldots, m - r$.

[16] In practical point of view, eliminating with the Fourier-Motzkin technique the $n - r$ variables of j_2.

$\begin{bmatrix} -1 \\ -1 \\ 1 \end{bmatrix}$, only condition (6.13) yields

$$
\begin{array}{rclcl}
7 & \geq & x & -\lambda \geq 0 \\
7 & \geq & y & -\lambda \geq 0 \\
7 & \geq & & \lambda \geq 0
\end{array}
$$

The number of points $\mathbf{x} = [x\ y]^T$ in the image polytope is 169 – the valid prefixes of length 2 (see subsection 6.2.3.2).

The experimental run times – displayed in table 6.2 for the same upper-bounds of loops as in table 6.1 – are similar, only slightly higher due to the matrix multiplications.

Table 6.2.   Experimental results for *Example 2*. The computation of the LBL size by index space determination

Loop bounds	3	7	15	31	63	127	255	511
CPU [sec]	0.0005	0.020	0.060	0.20	0.78	2.88	11.02	43.40

When matrix $\mathbf{H}_{11}$ is not unimodular, the CPU times can be significantly higher than those yielded by the routine *CountImagePolytope*. The reason is that the method of index space determination operates on the *image* polytope (which contains "holes"), while *CountImagePolytope* operates on the *iterator* polytope – which can be significantly "smaller" when $\mathbf{H}_{11}$ is not unimodular: as shown in table 6.3, the "holes" in the image polytope represent more than 92%. Therefore, the former method can never be better.

Table 6.3.   Experimental results for *Example 1*. Comparison between the two methods for computing the size of an index space

Loop bounds	Size image polytope	Size LBL (index sp.)	Ratio [%] LBL/Image Polytope	CPU [sec] *CountImage Polytope*	Index sp. detection
255	845,836	65,536	7.74	0.0080	2.02
511	3,395,596	262,144	7.72	0.0157	8.08
767	7,649,292	589,824	7.71	0.0235	18.28
1023	13,606,924	1,048,576	7.70	0.0312	32.25

Remark In order to carry out a better comparison between the two methods, the CPU times for *CountImagePolytope* were determined here without exploiting the fact that the iterators in *Example 1* have constant ranges. Taking into account also this aspect (as it usually happens), the CPU times for *CountImagePolytope* are even lower: 0.00017 seconds for all the table entries – as the computation of the LBL size is reduced to a multiplication of iterator ranges in all the four cases.

6.3   THE POLYHEDRAL DATA-FLOW GRAPH

After the partitioning process described in section 6.2, a data-flow graph (DFG) with exact dependence relations is produced (figure 4.7). However, unlike the classic case of data-flow analysis, the nodes in the graph do not correspond to individual variables/signals, but to *groups* of signals (covered by the basic sets derived in section 6.2), and the arcs correspond to the dependence relations between these groups. The nodes in the data-flow graph are weighted with the size of their corresponding basic sets, and the arcs between nodes are weighted with the exact number of dependences between the basic sets corresponding to the nodes. Based on the polyhedral graphs, a data-flow analysis is done in order to provide accurate estimation of storage requirements for RMSP algorithms [27, 31] when a *partial computation ordering* is imposed (as explained in chapter 4).

Subsection 6.3.1 presents the computation of dependences between the basic sets of signals, while subsection 6.3.2 explains the modifications of a data-flow graph in order to deal with delayed signals.

6.3.1 Computation of dependence relations

As the nodes in the DFG and their weights – the basic sets of signals and their sizes – are known from section 6.2, the construction of the graph is completed by determining the arcs between the nodes – the *dependence relations* between the basic sets (in particular, array references), and their weights – the *number of dependences*.

```
#define n 6
#define W fix<16,4>

func main (in : W)  B : W[]=
begin
        (j : 0 .. n-1) ::
        begin
(1)       A[j][0] = in;
          (i : 0 .. n-1)::
(2)           A[j][i+1] = A[j][i] + W(1);
        end;
        (i : 0 .. n-1) ::
        begin
(3)       alpha[i] = A[i][n+i];
          (j : 0 .. n-1) ::
(4)           A[j][n+i+1] = if (j < i) -> A[j][n+i]
(5)                            || A[j][n+i] + alpha[i]
                          fi;
        end;
(6)     (i : 0 .. n-1) :: B[i] = A[i][2*n];
end;
```

Figure 6.6.   Illustrative Silage code

Suppose for the moment, without decrease in generality, that two basic sets are represented as LBL's:

$$ S_1 = \{ \mathbf{x} = \mathbf{T}_1 \lambda + \mathbf{u}_1 \mid \mathbf{A}_1 \lambda \geq \mathbf{b}_1 \} \ , \quad S_2 = \{ \mathbf{x} = \mathbf{T}_2 \mu + \mathbf{u}_2 \mid \mathbf{A}_2 \mu \geq \mathbf{b}_2 \} $$

and the two basic sets belong respectively to the index spaces of a definition domain and of an operand domain within the same instruction scope:

$$ D_1 = \{ \mathbf{x} = \mathbf{C}_1 \mathbf{I}_1 + \mathbf{d}_1 \mid \mathbf{A} \mathbf{I}_1 \geq \mathbf{b} \} \ , \quad D_2 = \{ \mathbf{x} = \mathbf{C}_2 \mathbf{I}_2 + \mathbf{d}_2 \mid \mathbf{A} \mathbf{I}_2 \geq \mathbf{b} \} $$

Solving the linear Diophantine system in $(\mathbf{I}_1, \lambda)$ as variables: $\mathbf{C}_1 \mathbf{I}_1 + \mathbf{d}_1 = \mathbf{T}_1 \lambda + \mathbf{u}_1$ and substituting the solution[17] in the sets of constraints $\mathbf{A}_1 \lambda \geq \mathbf{b}_1$ and $\mathbf{A} \mathbf{I}_1 \geq \mathbf{b}$, the expression of the iterator vector corresponding to the basic set S_1 is obtained:

$$ \{ \mathbf{I}_1 = \overline{\mathbf{T}}_1 \alpha + \overline{\mathbf{u}}_1 \mid \overline{\mathbf{A}}_1 \alpha \geq \overline{\mathbf{b}}_1 \} \tag{6.16} $$

---

[17]The system has always solution as basic set S_1 is included in D_1.

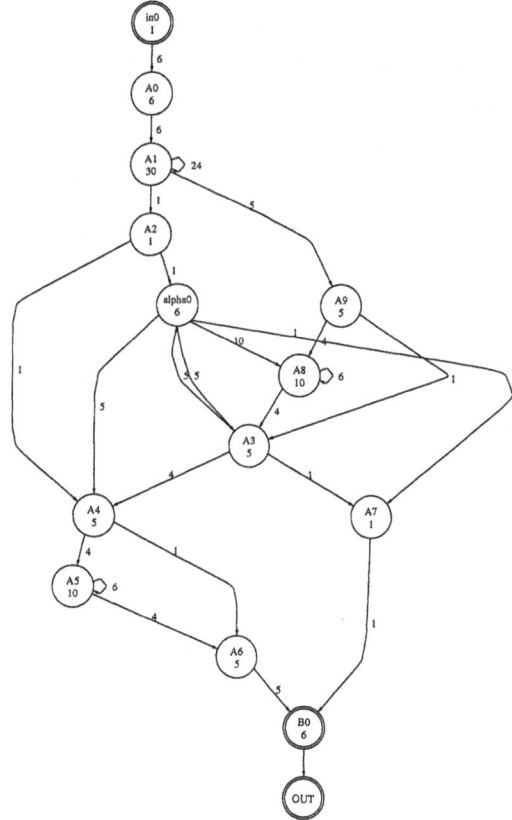

Figure 6.7.   The data-flow graph for the illustrative example in figure 6.6 (n=6)

The expression of the iterator vector corresponding to the basic set S_2 is obtained similarly:

$$\{ \mathbf{I}_2 = \overline{\mathbf{T}}_2 \beta + \overline{\mathbf{u}}_2 \mid \overline{\mathbf{A}}_2 \beta \geq \overline{\mathbf{b}}_2 \} \tag{6.17}$$

There is a dependence relation between S_1 and S_2 if there is at least one iterator vector corresponding to both of them. The number of iterator vectors yields, in this case, the number of dependences. The problem is solved by intersecting the linearly bounded lattices (6.16) and (6.17). If the intersection is empty, there is no dependence relation between S_1 and S_2. Otherwise, the size of the intersection (see subsection 6.2.3) represents the number of dependences.

Example In the Silage code of figure 6.6, basic set $A8$ belongs to the index space $D8$ of the definition domain $A[j][n+i+1]$ ($j \geq i$) (node **g** in figure 6.2), and basic set $A9$ belongs to the index space $D9$ of the operand domain $A[j][n+i]$ ($j \geq i$) (node **h** in figure 6.2). Employing a non-matrix notation, the linearly bounded lattices of the basic sets and of the index spaces are (see figure 6.3):

$$
\begin{aligned}
A8 &= \{ x = \lambda_1 , y = \lambda_2 + n \mid n - 1 \geq \lambda_1 , \lambda_2 \geq 1 , \lambda_1 - \lambda_2 \geq 1 \} \\
A9 &= \{ x = \mu , y = n \mid n - 1 \geq \mu \geq 1 \} \\
D8 &= \{ x = j , y = i + n + 1 \mid n - 1 \geq j \geq 0 , n - 1 \geq i \geq 0 , j \geq i \}
\end{aligned}
$$

$$D9 = \{ x = j, y = i + n \mid n - 1 \geq j \geq 0, n - 1 \geq i \geq 0, j \geq i \}$$

The set of iterators corresponding to $A8$ in the index space $D8$ is described by $\{ i_1 = \alpha', j_1 = \alpha'' \mid n - 1 \geq \alpha'', \alpha' \geq 0, \alpha'' - \alpha' \geq 2 \}$. The set of iterators corresponding to $A9$ in the index space $D9$ is: $\{ i_2 = 0, j_2 = \beta \mid n - 1 \geq \beta \geq 1 \}$.

The intersection of the two LBL's is represented as: $\{ i = 0, j = \gamma \mid n - 1 \geq \gamma \geq 2 \}$. Hence, the number of dependences between $A9$ and $A8$ is $n - 2$, i.e. the size of the intersection. Therefore, the arc between the nodes $A9$ and $A8$ in figure 4.7 has the weight 4 (as $n = 6$).

It has been assumed so far that a basic set was represented by a single LBL. This does not always happen as a basic set may result from the routine $CreateBasicSets$ (section 6.2.1) to be a difference of the form $LBL - (\cup_i LBL_i)$ (e.g., the basic set A1 in figure 6.9b). In this latter case, the basic set is decomposed into a collection of mutually disjoint LBL's. The formal approach for this is discussed in section 6.6.

More generally, given two basic sets represented as unions of mutually disjoint linearly bounded lattices – $\cup_i S_1^i$ and, respectively, $\cup_j S_2^j$ – the number of dependences between the two basic sets equals $\sum_{i,j} nr\_dependences (S_1^i, S_2^j)$. The number of dependences between each pair of LBL's is computed as shown at the beginning of the subsection.

The data-flow graph of the example in figure 6.6 is shown in figure 4.7. The nodes are labeled with the signal name, basic set number, and size; the arcs are labeled with the number of dependences. OUT is a dummy node, necessary for handling delayed signals.

6.3.2 Handling the delayed signals

RMSP algorithms describe the processing of streams of data samples. The source code of these algorithms can be imagined as surrounded by an implicit loop having $time$ as iterator. Consequently, each signal in the algorithm has an *implicit* extra dimension corresponding to the $time$ axis. RMSP algorithms contain usually *delayed* signals, i.e. signals produced or inputs in previous data-sample processings, which are consumed during the current sample processing. The delay operator "@" refers relatively to the past samples. The delayed signals must be kept "alive" during several time iterations, i.e. they must be stored in the background memory during several data-sample processings.

In order to handle the delays in an RMSP algorithm, an appealing solution because of its simplicity would be to add an *explicit* extra dimension – corresponding to the $time$ axis – to all the array references in the behavioral specification. The code transformation for the motion detection kernel (figure 4.12) is presented in figure 6.8. Choosing a maximum value for the number of data-samples, the equivalent code could be processed in the same way as presented so far.

However, this "simple" approach presents a major shortcoming: introducing one extra dimension increases substantially the computational effort for realistic RMSP applications. This effect is especially noticeable in applications with "deep" nested loops, as the motion detection kernel, where the explicit introduction of the $time$ loop (see figure 6.8) causes an increase of the processing time from 37 seconds to 56 seconds.

In order to simulate the effect of the delayed signals in terms of memory requirements, a more effective method for handling delays is introduced in the sequel ([31], chapter 2).

First, the delayed operand domains take part in the partitioning process (section 6.2) as any other signal domain. Afterwards, the construction of the data-flow graph needs the following preprocessing step:

DFG preprocessing for delayed signal handling

create a dummy node OUT;
for *each basic set* b
    let $D(b)$ be its highest delay value (when belonging to an operand domain);
    create $D(b)$ copies of the basic set node, each one labeled from 1 to $D(b)$;
    for *every copy labeled* $1, \ldots, D(b) - 1$

```
(t: 0 .. ∞ )::
 begin
    optDelta[t][0] = W(0);

    (i: m .. M)::
       (j: n .. N)::
          begin
             Delta[t][i][j][0] = W(0);

             (k: i-m .. i+m)::
                (l: j-n .. j+n)::
                   Delta[t][i][j][(k-i+m)*(2*n+1)+1-j+n+1] = A[t][i][j]-A[t-1][k][l]
                                                          + Delta[t][i][j][(k-i+m)*(2*n+1)+1-j+n];

             optDelta[t][(i-m)*(N-n+1)+j-n+1] = Delta[t][i][j][(2*m+1)*(2*n+1)]
                                              + optDelta[t][(i-m)*(N-n+1)+j-n];
          end;

    opt[t] = optDelta[t][(M-m+1)*(N-n+1)];
 end;
```

Figure 6.8.   Explicit extension with the *time* dimension for the motion detection kernel

   create a dependence from the copy node to OUT;
for *every basic set* b *belonging to an output signal, or having a max delay* $D(b) > 0$
   create a dependence arc from its corresponding node to OUT;

The basic idea is to create bogus dependence relations towards an "output" node from the groups of signals produced in the current sample processing and consumed in a future one, or from the groups produced in the past and still necessary in the future. E.g., if there is an operand $sig@3$, the signal sig produced two sample processing ago, will be consumed only in the next sample processing. These groups of signals must be kept "alive" during the entire current sample processing.

Remark In practice, all copies of a basic set having a unique dependence relation (only towards the node OUT) are treated as a single item, so handling signals with high delay values does not cause computational problems.

This modification of the data-flow graph allows to take into account the effect of delays, "translating" the basic sets which affect the memory requirements from previous data-sample processings into the current one. When the delay values are constant, the extension of all domains with one extra dimension – corresponding to the implicit time loop – is avoided, hence reducing the computational effort.

Remark Although our model allows the handling of signals with non-constant delay values (affine functions of loop iterators), the necessity of an explicit time dimension cannot be avoided in this case. However, it could be possible to find out (by solving an ILP) the maximum delay values inside the scope of those signals, and to replace the affine delays by constant values. The shortcoming of such a strategy is a possible excessive storage requirement, due to unnecessary basic set copies and false dependences introduced in the data-flow graph.

Figure 6.9a shows the data-flow graph for the Silage code in figure 4.12. The basic sets corresponding to delayed signals are labeled with "$@delay\_value$".

6.4   EXTENDED DATA-FLOW ANALYSIS FOR NONLINEAR INDICES

As shown in section 6.2, qualitative and quantitative information concerning the dependence relations between groups of signals – called *basic sets* – can be derived. This information is essential to compute accurate estimations of storage requirements for applicative algorithms containing large amounts of scalars (as explained in chapter 4). The polyhedral model of the data-flow analysis presented in section 6.2 is based on the assumptions that all signal indices and loop boundaries are *affine* functions of the loop iterators. This covers the majority of the RMSP application domain.

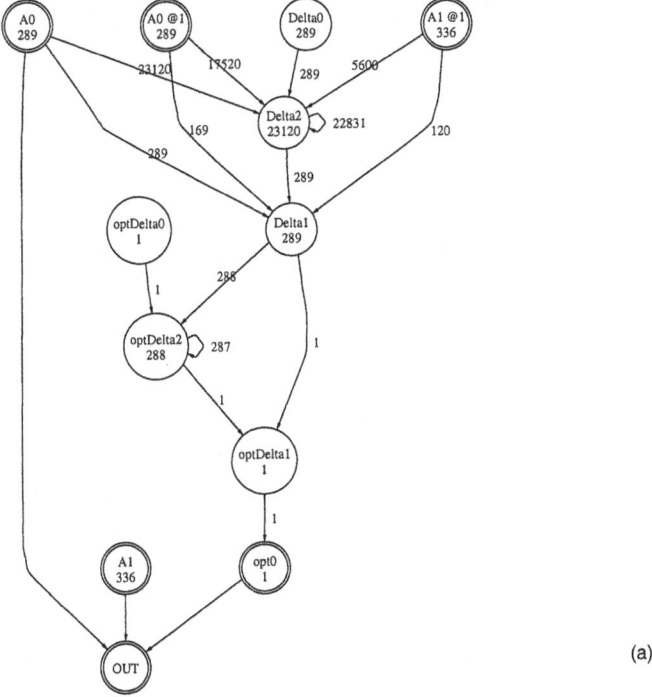

(a)

Delta0 = { x=i , y=j , z=0 | M>=i>=m , N>=j>=n }
Delta1 = { x=i , y=j , z=(2m+1)(2n+1) | M>=i>=m , N>=j>=n }
Delta2 = { x=i , y=j , z=-(2n+1)i-j+(2n+1)k+l+2mn+m+n |
                      M>=i>=m , N>=j>=n , m>=k-i>=-m , n>=l-j>=-n }   (b)

optDelta0 = { x=0 }
optDelta1 = { x=(M-m+1)(N-n+1) }
optDelta2 = { x=(N-n+1)i+j-m(N-n+1)-n | M>=i>=m , N>=j>=n , (N-n+1)i+j >= (N-n+1)m+n+1 }

A0 = { x=i , y=j | M>=i>=m , N>=j>=n }
A1 = { x=i , y=j | M+m>=i>=0 , N+n>=j>=0 } - A0

Figure 6.9.   (a) Data-flow graph with delays for the example in figure 4.12 (M=N=20 , m=n=4) (b) the basic sets (nodes) represented in non-matrix form

However, in practice, some signals may be referenced by index functions which are *nonlinear*, or *data dependent* (see, e.g., [303, 337]). Examples of systems with such signals are frequently occurring in speech and image processing, and in numerical algorithms. The automated handling of more general index functions is also required in order to gain *more independence* from the way an algorithm has been specified initially by the system designer [422]. Unfortunately, no state-of-the-art automated methods can deal with such index expressions.

The most important practical subclass of the nonlinear extensions consist of *modulo expressions of affine index functions* [143]. These are present when, for instance, images are scanned by moving neighbourhoods of pixels in order to take into account boundary effects.

This chapter presents novel results in the automated handling of M-D signals with nonlinear indices ([29], chapter 3). A possible strategy is the transformation of nested loops to higher dimensional loop nests, in order to obtain only affine recurrences [143]. A detailed algorithm to perform this transformation when the indices are *affine functions modulo constant* is presented in section 6.5.

In the field of regular and systolic array synthesis, loop transformations are employed for scheduling and mapping techniques of loop nests with uniform dependences [333], and more recently also piecewise linear scheduling and mapping of affine dependences [94, 388, 420]. Application of sequences of unimodular loop nest transformations is also encountered within parallelizing compilers for massively parallel machines, for example in [431, 35]. In this field, loop transformations aim at maximizing the parallelism in the code. A good overview of the state-of-the-art is presented in [36]. The applicability of these synthesis and compiler techniques can be extended by the method described in this chapter.

The algorithm which eliminates the modulo indices – presented in section 6.5 – has the important benefit that the vast amount of loop oriented transformation techniques which are currently in use can be retained *without* any changes. Furthermore, if this re-indexing approach is used as a preprocessing step in our context, the data-flow analysis presented in section 6.2 can directly handle RMSP algorithms with modulo index functions.

Remark In fact, the class of nonlinear index expressions which can be automatically processed is even larger. Due to the identities $a \bmod b = a - b \cdot div(a, b)$, $\lfloor a/b \rfloor = div(a, b)$, $\lceil a/b \rceil = -div(-a, b)$ (where div means integer division), the indices containing also div , $floor$, and $ceiling$ functions can be handled as well.

Section 6.6 presents an extension of the polyhedral data-flow model – presented in section 6.2 – for handling continuous piecewise-affine index functions and loop boundaries.

6.5 MODULO INDEX FUNCTIONS

Given a nest of n loops – defined by the iterators $\mathbf{i}_n = (i_1, \ldots, i_n)$ – having affine, manifest boundaries, and containing m index functions of constant *modulo* type:

$(i_1 : l_{11} \; .. \; u_{11}) ::$
$\quad (i_2 : l_{21}i_1 + l_{22} \; .. \; u_{21}i_1 + u_{22}) ::$
$\quad \quad \cdots \cdots \cdots$
$\quad \quad \quad (i_n : l_{n1}i_1 + l_{n2}i_2 + \cdots + l_{nn} \; .. \; u_{n1}i_1 + u_{n2}i_2 + \cdots + u_{nn}) ::$
$\quad \quad \quad \quad \textbf{begin} \; \cdots \; A[\cdots][(a_{p1}i_1 + \cdots + a_{pn}i_n + b_p) \bmod C_p][\cdots] \cdots \; \textbf{end};$

with $p = \overline{1, m}$, $C_p \in \mathbf{N}^*$; the goal is to obtain an equivalent nest of loops with only affine index expressions (thus without any *modulo* index).

Section 6.5.1 presents a transformation technique which eliminates modulo index functions from nested loops [29]. The novel approach is illustrated in subsection 6.5.2. The complexity of the loop transformation is discussed in subsection 6.5.3.

6.5.1 *Transformation of nested loops with modulo indexing to affine recurrences*

The main idea of the method is to obtain a new set of $m + n$ variables, replacing the n loop iterators, such that: (1) all index expressions become affine functions; (2) there are $m + n$ linearly independent expressions of the new variables that can be lower- and upper-bounded; (3) the matrix of these expressions is lower-triangular. The last requirement strongly suggests the application of the Hermite Normal Form [296]. Because of requirements (2) and (3), an equivalent nest of loops – having as iterators the new variables – will result.

In order to achieve our goal, the m affine index functions have to be transformed as follows:

$$a_{p1}i_1 + \cdots + a_{pn}i_n + b_p \stackrel{def}{=} q_p C_p + r_p \; ,$$

where q_p, r_p are, respectively, the quotient and the rest of the division of the left member by C_p . This requirement leads to the following linear Diophantine system – having as unknowns $i_j, j = \overline{1, n}$

and q_k, r_k , $k = \overline{1,m}$:

$$
\begin{aligned}
r_1 - b_1 &= a_{11}i_1 + \cdots + a_{1n}i_n & -C_1q_1 \\
r_2 - b_2 &= a_{21}i_1 + \cdots + a_{2n}i_n & -C_2q_2 \\
&\quad\cdots\cdots\cdots \\
r_m - b_m &= a_{m1}i_1 + \cdots + a_{mn}i_n & -C_mq_m
\end{aligned}
$$

or, with a matrix notation (after column permutation):

$$
\begin{bmatrix} r_1 - b_1 \\ \vdots \\ r_m - b_m \end{bmatrix} = \mathbf{T}_{m,m+n} \cdot \begin{bmatrix} q_1 \\ \vdots \\ q_m \\ i_1 \\ \vdots \\ i_n \end{bmatrix} \tag{6.18}
$$

As $rank\ \mathbf{T}_{m,m+n} = m$ (the columns of q_k are linearly independent), $\mathbf{T}_{m,m+n}$ can be brought to the form $[\,\mathbf{H}_m \vdots \mathbf{0}_{m,n}\,]$. Here, $\mathbf{H}_m$ is the Hermite Normal Form, obtained by a unimodular transformation $\mathbf{S}_{m+n}$ of order $m + n$, determined as shown in subsection 6.2.2.1:

$$
\mathbf{T}_{m,m+n} \cdot \mathbf{S}_{m+n} = [\,\mathbf{H}_m \vdots \mathbf{0}_{m,n}\,]
$$

If $(\mathbf{X}_m^T, \overline{\mathbf{X}}_n^T)$ are the new variables resulted after the transformation, the initial system (eq. 6.17) becomes:

$$
\begin{bmatrix} r_1 - b_1 \\ \vdots \\ r_m - b_m \end{bmatrix} = [\,\mathbf{H}_m \vdots \mathbf{0}_{m,n}\,] \begin{bmatrix} X_{j_1} \\ \vdots \\ X_{j_m} \\ \overline{X}_{j_1} \\ \vdots \\ \overline{X}_{j_n} \end{bmatrix} \tag{6.19}
$$

As $\mathbf{T}_{m,m+n} \cdot \mathbf{S}_{m+n} = [\,\mathbf{H}_m \vdots \mathbf{0}_{m,n}\,]$, from (6.17) and (6.19) it results (with a self explanatory notation) that:

$$
\begin{bmatrix} q_1 \\ \vdots \\ q_m \\ i_1 \\ \vdots \\ i_n \end{bmatrix} = \mathbf{S}_{m+n} \cdot \begin{bmatrix} X_{j_1} \\ \vdots \\ X_{j_m} \\ \overline{X}_{j_1} \\ \vdots \\ \overline{X}_{j_n} \end{bmatrix} \equiv \begin{bmatrix} \mathbf{S}_{m,m} & \mathbf{S}_{m,n} \\ \mathbf{S}_{n,m} & \mathbf{S}_{n,n} \end{bmatrix} \begin{bmatrix} \mathbf{X}_m \\ \overline{\mathbf{X}}_n \end{bmatrix}
$$

From the previous equations, only the last n are relevant for our problem as they relate to the initial iterators:

$$
\mathbf{i}_n = \mathbf{S}_{n,m} \cdot \mathbf{X}_m + \mathbf{S}_{n,n} \cdot \overline{\mathbf{X}}_n \tag{6.20}
$$

It is assumed in the sequel, without loss of generality (see the remark at the end of the subsection), that $rank\ \mathbf{S}_{n,n} = n$. Applying once again the same Hermite Normal Form transformation to the term $\mathbf{S}_{n,n} \cdot \overline{\mathbf{X}}_n$ instead of $\mathbf{T}_{m,m+n} \cdot [\mathbf{q}_m\ \mathbf{i}_n]^T$, equation (6.20) becomes:

$$
\mathbf{i}_n = \mathbf{S}_{n,m} \cdot \mathbf{X}_m + \mathbf{H}_n^{'} \cdot \mathbf{Y}_n \tag{6.21}
$$

with $\mathbf{H}_n'$ – the Hermite normal form of $\mathbf{S}_{n,n}$.

In conclusion, the initial nest of n loops, having as iterators $i_1, \ldots, i_n$, is transformed into an equivalent nest of $m + n$ loops – having as iterators the new variables $X_{j_1}, \ldots X_{j_m}, Y_{j_1}, \ldots, Y_{j_n}$. The outer m loops have boundaries derived from the *rest* conditions $0 \leq r_k \leq C_k - 1$, $k = \overline{1, m}$. The loops corresponding to the rows of $\mathbf{H}_m$ which exhibit a single non-zero element h_{kk} on the diagonal have constant boundaries:

$$\left\lceil \frac{-b_k}{h_{kk}} \right\rceil \leq X_{j_k} \leq \left\lfloor \frac{-b_k + C_k - 1}{h_{kk}} \right\rfloor \tag{6.22}$$

while the boundaries of the loops corresponding to the other rows of $\mathbf{H}_m$ have more complex expressions – given by

$$\left\lceil \frac{L_k(X_{j_1}, \ldots, X_{j_{k-1}})}{h_{kk}} \right\rceil \leq X_{j_k} \leq \left\lfloor \frac{U_k(X_{j_1}, \ldots, X_{j_{k-1}})}{h_{kk}} \right\rfloor \tag{6.23}$$

where L_k and U_k are affine functions that can be easily determined (see, e.g., subsection 6.5.2).

The boundaries of the innermost n loops can be derived from (6.21), taking into account the boundaries of the initial nest of loops. After trivial computations, it follows that $\mathbf{l}_n \leq \mathbf{i}_n \leq \mathbf{u}_n$

$$\text{where} \quad \mathbf{l}_n = \begin{bmatrix} 1 & & & \\ -l_{21} & 1 & & \\ \vdots & & \ddots & \\ -l_{n1} & -l_{n2} & \cdots & 1 \end{bmatrix}^{-1} \begin{bmatrix} l_{11} \\ l_{22} \\ \vdots \\ l_{nn} \end{bmatrix}$$

$$\text{and} \quad \mathbf{u}_n = \begin{bmatrix} 1 & & & \\ -u_{21} & 1 & & \\ \vdots & & \ddots & \\ -u_{n1} & -u_{n2} & \cdots & 1 \end{bmatrix}^{-1} \begin{bmatrix} u_{11} \\ u_{22} \\ \vdots \\ u_{nn} \end{bmatrix}$$

The loops corresponding to the rows of $\mathbf{H}_n'$ – having a single non-zero element on the diagonal equal to 1 – have as boundaries affine functions of $X_{j_1}, \ldots, X_{j_m}$ without the need for ceiling and floor functions :

$$L_k'(X_{j_1}, \ldots, X_{j_m}) \leq Y_{j_k} \leq U_k'(X_{j_1}, \ldots, X_{j_m}) \tag{6.24}$$

The boundaries of the loops corresponding to the other rows k of $\mathbf{H}_n'$ have in general more complex expressions, of the form:

$$\left\lceil \frac{L_k'(\mathbf{X}_m, Y_{j_1}, \ldots, Y_{j_{k-1}})}{h_{kk}'} \right\rceil \leq Y_{j_k} \leq \left\lfloor \frac{U_k'(\mathbf{X}_m, Y_{j_1}, \ldots, Y_{j_{k-1}})}{h_{kk}'} \right\rfloor \tag{6.25}$$

where L_k' and U_k' are linear functions that can be easily determined (see, e.g., subsection 6.5.2). If $\mathbf{S}_{n,n}$ results to be unimodular, the second phase introduces only loops with boundaries of type (6.24) (as all the diagonal elements of $\mathbf{H}_n'$ are equal to 1).

Inside the new nest of loops, any appearance of the former loop iterator i_k is replaced by its corresponding linear expression of $X_{j_1}, \ldots, X_{j_m}, Y_{j_1}, \ldots, Y_{j_n}$, derived from (6.21). In particular, the *modulo* indices become linear expressions of the new iterators, as desired:

$$(a_{p1}i_1 + \cdots + a_{pn}i_n + b_p) \bmod C_p \equiv r_p = \sum_{k=1}^{p} h_{pk} X_{j_k} + b_p$$

The elimination of *modulo* indices increases, unfortunately, the number of loops. Although the method presented above is not unique, it must be emphasized that in order to make the initial nest of loops fully affine, the emergence of extra loops cannot be avoided.

Remark If $rank\ \mathbf{S}_{n,n} < n$, the algorithm still works with a slight modification. Suppose, without loss of generality, that $rank\ \mathbf{S}_{n,n} = n - 1$, and the last of its rows is a linear combination of the previous ones. Then, while computing $\mathbf{H}'_n$, it results also that the last row of $\mathbf{H}'_n$ is a linear combination of the previous ones, and hence $h'_{n,n} = 0$. Therefore, the initial iterator i_n (besides the other ones) is a linear combination of $X_{j_1}, \ldots, X_{j_m}, Y_{j_1}, \ldots, Y_{j_{n-1}}$ (i.e. Y_{j_n} is no longer necessary). This means that the introduction of a new loop iterator, corresponding to Y_{j_n} is no longer needed. More generally, if $rank\ \mathbf{S}_{n,n} = r < n$, $n-r$ of the last n loops, corresponding to the new iterators $Y_{j_1}, \ldots, Y_{j_n}$, are no longer needed, and consequently, the final result will be a nest of $m + r$ loops (rather than $m + n$).

6.5.2 Example of modulo indexing elimination

The following illustrative example is extracted from an image processing application:

$$(i :\ 0 .. 574\ )::$$
$$\quad (j :\ 0 .. 719\ )::$$
$$\quad\quad (k :\ 0 .. 7\ )::$$
$$\quad\quad\quad (l :\ 0 .. 15\ )::\quad \cdots A[k + (575 * j + i) \bmod 11][l + (720 * i + j) \bmod 7] \cdots$$

The initial Diophantine linear system is:

$$\begin{aligned}
r_1 &= 720i + j - 7q_1 \\
r_2 &= i + 575j - 11q_2
\end{aligned}$$

Applying successively the unimodular transformations:

$$\begin{aligned}
j &= j^1 - 720i + 7q_1 \\
q_2 &= q_2^1 - 37636i + 366q_1 \\
q_1 &= -q_1^1 - 3i - 11q_2^1 + 575j^1
\end{aligned}$$

the first Hermite Normal Form results: $\mathbf{H}_m = \mathbf{I}_2$ with $\mathbf{X}_m = [j^1\ q_1^1]^T$ and $\overline{\mathbf{X}}_n = [i\ q_2^1]^T$.

This is clearly a result where only diagonal elements remain, so the first $m = 2$ new iterator boundaries should become constant. Indeed, from the $rest$ conditions $0 \leq r_1 \leq 6$, respectively $0 \leq r_2 \leq 10$, the two outer loops – having as iterators j^1 and q_1^1 - are fully determined:

$$0 \leq j^1 \leq 6 \quad \text{and} \quad 0 \leq q_1^1 \leq 10$$

Computing the values of the initial iterators as functions of the new variables $\mathbf{X}_m$ and $\overline{\mathbf{X}}_n$, it results that: $\begin{bmatrix} i \\ j \end{bmatrix} = \begin{bmatrix} 0 & 0 \\ 4026 & 7 \end{bmatrix} \begin{bmatrix} j^1 \\ q_1^1 \end{bmatrix} + \begin{bmatrix} 1 & 0 \\ -741 & -77 \end{bmatrix} \begin{bmatrix} i \\ q_2^1 \end{bmatrix}$. With the unimodular substitution $q_2^1 = -q_2^2 - 9i$, the Hermite Normal Form corresponding to the matrix $\mathbf{S}_{n,n} = \begin{bmatrix} 1 & 0 \\ -741 & -77 \end{bmatrix}$ is obtained: $\mathbf{H}'_n = \begin{bmatrix} 1 & 0 \\ -48 & 77 \end{bmatrix}$. The two inner loops have the iterators $\mathbf{Y}_n = [i\ q_2^2]^T$, their boundaries being determined from the initial nest of loops:

$$0 \leq i \leq 574$$

$$\left\lceil \frac{-4026j^1 - 7q_1^1 + 48i}{77} \right\rceil \leq q_2^2 \leq \left\lfloor \frac{-4026j^1 - 7q_1^1 + 48i + 719}{77} \right\rfloor$$

The final solution can now be handled by methods dealing with transformations on purely affine index functions.

$$(j^1 :\ 0 .. 6\ )::$$
$$\quad (q_1^1 :\ 0 .. 10\ )::$$
$$\quad\quad (i :\ 0 .. 574\ )::$$
$$\quad\quad\quad (q_2^2 :\ \left\lceil \frac{-4026j^1 - 7q_1^1 + 48i}{77} \right\rceil .. \left\lfloor \frac{-4026j^1 - 7q_1^1 + 48i + 719}{77} \right\rfloor\ )::$$

$$(k : \ 0 \ .. \ 7) ::$$
$$(l : \ 0 \ .. \ 15) :: \quad \cdots A[k \ + \ q_i^1][l \ + \ j^1] \cdots$$

As mentioned already in subsection 6.2.3, loop boundaries containing the *floor* $\lfloor \ \rfloor$ and *ceiling* $\lceil \ \rceil$ functions create no special problem, hence the LBL's of the array references can be easily derived.

6.5.3 Complexity of the method

The elimination of *modulo* indices in nested loops is a transformation which can be carried out in polynomial time, applying only integer arithmetic.

A few considerations on the computational effort for the normal form algorithm in the general case will be presented first. Afterwards, we focus on the complexity of the problem. Suppose a given matrix $\mathbf{T}$ has r rows and c columns. Each unimodular transformation modifies the r equations and the c unknowns (as it is in fact a change of variables). Consequently, the number of computations per iteration is $(r+c)c$. As shown already in subsection 6.2.2.1, the necessary number of transformations is upper bounded by $\sum_{i=1}^{r} \log_2 M_i + 2r - 1$, where $M_i = \max\{|t_{ii}^{(i)}|, \ldots, |t_{ic}^{(i)}|\}$, and $|t_{ij}^{(i)}|$ are the values of the coefficients after processing the transformations on the previous rows.

The algorithm was applied twice in our case, first for matrix $\mathbf{T}_{m,m+n}$. Exploiting the special form of the Diophantine system from which this matrix $\mathbf{T}_{m,m+n}$ is derived, the first phase complexity can be shown to be polynomial. First, the necessary number of iterations is upper bounded by $m(\log_2 \mathcal{C} + 2)$ where $\mathcal{C} \overset{def}{=} \max\{C_i\}$: when processing row i, as the coefficient of q_i (that is C_i) is unmodified by the previous transformations, the number of iterations for this row is at most $\lfloor \log_2 C_i \rfloor + 2$. Second, the number of unknowns involved in the change of variables is at most $n+1$ at each transformation (rather than $n + m$ in the general case). With these two remarks, the worst-case complexity for the first phase is $O(mn(m + n) \log \mathcal{C})$.

The undesirable effect of large integer entries (as M_i cannot be upper-bounded any more) affects the run-time complexity of the algorithm when applied to matrix $\mathbf{S}_{n,n}$ (second phase). Due to this undesirable effect, the "classical" Hermite Normal Form algorithms cannot be considered of polynomial complexity. A pathological example is shown in [179]: reducing to the lower-triangular form a 20×20 matrix, having integer elements between 0 and 10, an entry higher than 10^{5011} has been obtained in an intermediate matrix. Applying different classical algorithms on random matrices of the same order, it was observed that the emergence of entries $\sim 10^{500}$ is quite frequent. After 1977, several algorithms able to avoid the uncontrolled increase of some elements in the intermediate matrices have been proposed (e.g., [145, 208] and [363] (section 5.3)).

It must be emphasized that the size of the practical examples in our case does not justify the overhead implied by the use of one of those more sophisticated algorithms: keeping the entries of intermediate matrices in a controlled range necessitates supplementary computations, which will affect the run times although they have a provable polynomial complexity. However, because such algorithms are known and can be embedded in the method, the transformation of nested loops with modulo indexing has a worst-case polynomial complexity.

The computation of the matrices $\mathbf{l}_n$ and $\mathbf{u}_n$ requires $O(n^3)$ time: the inverses of the matrices in the expressions of $\mathbf{l}_n$ and $\mathbf{u}_n$ can be simply computed, employing a scheme similar to the one in the LU factorization [163] (section 3.2).

6.6 CONTINUOUS PIECEWISE-AFFINE INDEX FUNCTIONS

The polyhedral data-flow model presented in section 6.2 can handle in a straightforward way also *continuous* piecewise-affine index functions[18]. This property could, for example, be exploited to carry out the data-flow analysis of RMSP algorithms containing M-D signals with nonlinear indices, as any continuous mapping can be approximated by a continuous piecewise-affine mapping.

---

[18]In fact, the extensions of the functions in $\Re^n$.

The basic idea is to reduce the index spaces of piecewise-affine index functions to *unions* of disjoint "standard" linearly bounded lattices (eq. 6.1). Chua's "canonical" piecewise-affine (explicit) representation [73] can be employed for modifying the affine mapping $\mathbf{T} \cdot \mathbf{i} + \mathbf{u}$ in (6.1) into a *continuous piecewise-affine mapping*:

$$\mathbf{x} = \mathbf{T} \cdot \mathbf{i} + \mathbf{u} + \sum_{k=1}^{p} \mathbf{c}_k \left| \mathbf{a}_k^T \cdot \mathbf{i} - b_k \right| \tag{6.26}$$

where $\mathbf{c}_k \in \mathbf{Z}^m$, $\mathbf{a}_k \in \mathbf{Z}^n$, and $b_k \in \mathbf{Z}$. A good overview of explicit and implicit piecewise-affine representations in circuit analysis is presented in [406]. There is, however, a basic difference between the problems encountered in piecewise-affine circuit analysis and data-flow analysis: the former requires the *real* solutions of equations, while the latter requires only the *integer* solutions. This is, perhaps, one of the few examples where the fact of working with integers makes things easier than if working with reals[19].

The outcome of piecewise-affine mappings as indices is that array references can have an index space represented by (eq. 6.26), which is a union of at most 2^p mutually disjoint linearly bounded lattices[20]:

$$\bigcup_K \{ \mathbf{x} = \mathbf{T}' \cdot \mathbf{i} + \mathbf{u}' \mid \mathbf{A}' \cdot \mathbf{i} \geq \mathbf{b}' \} \tag{6.27}$$

where $K \subseteq \{1, \ldots, p\}$ is a subset of indices. Each LBL in the union (eq. 6.27) corresponds to a region of the iterator space where $\mathbf{a}_k^T \cdot \mathbf{i} - b_k \geq 0$ for any $k \in K$, and $\mathbf{a}_k^T \cdot \mathbf{i} - b_k < 0$ for any $k \in \overline{K}$. Therefore, each affine mapping is characterized by

$$\mathbf{T}' = \mathbf{T} + \sum_{k \in K} \mathbf{c}_k \mathbf{a}_k^T - \sum_{k \in \overline{K}} \mathbf{c}_k \mathbf{a}_k^T$$

$$\mathbf{u}' = \mathbf{u} - \sum_{k \in K} b_k \mathbf{c}_k + \sum_{k \in \overline{K}} b_k \mathbf{c}_k$$

and each iterator region is defined by:

$$\{ \mathbf{i} \in \mathbf{Z}^n \mid \mathbf{A}' \cdot \mathbf{i} \geq \mathbf{b}' \} = \{ \mathbf{i} \in \mathbf{Z}^n \mid \mathbf{A} \cdot \mathbf{i} \geq \mathbf{b} \}$$
$$\bigcap \{ \mathbf{i} \in \mathbf{Z}^n \mid + \mathbf{a}_k^T \cdot \mathbf{i} - b_k \geq 0 , \ \forall k \in K \}$$
$$\bigcap \{ \mathbf{i} \in \mathbf{Z}^n \mid - \mathbf{a}_k^T \cdot \mathbf{i} + b_k \geq 1 , \ \forall k \in \overline{K} \}$$

Remark Also the loop boundaries can be continuous piecewise-affine functions of the surrounding loop iterators: the iterator polytopes $\mathbf{A} \cdot \mathbf{i} \geq \mathbf{b}$ can be similarly modified.

Example The index space of signal A in

$$(i : 0 .. 7) :: \ \cdots A[2 * i - 1 + |3 * i - 7|][i + 2 * |i - 4|] \ \cdots$$

can be written according to (6.26) :

$$\left\{ \begin{bmatrix} x \\ y \end{bmatrix} = \begin{bmatrix} 2 \\ 1 \end{bmatrix} i + \begin{bmatrix} -1 \\ 0 \end{bmatrix} + \begin{bmatrix} 1 \\ 0 \end{bmatrix} |3i - 7| + \begin{bmatrix} 0 \\ 2 \end{bmatrix} |i - 4| \ \middle| \ 7 \geq i \geq 0 \right\}$$

The index space can be decomposed in a union of three disjoint LBL's (as the fourth one is empty):

$$\left\{ \begin{bmatrix} x \\ y \end{bmatrix} = \begin{bmatrix} 5 \\ 3 \end{bmatrix} i + \begin{bmatrix} -8 \\ -8 \end{bmatrix} \ \middle| \ 7 \geq i \geq 4 \right\} \quad \text{for } 3i - 7 \geq 0, \, i - 4 \geq 0$$

$$\left\{ \begin{bmatrix} x \\ y \end{bmatrix} = \begin{bmatrix} 7 \\ 5 \end{bmatrix} \right\} \quad \text{for } 3i - 7 \geq 0, \, i - 4 < 0$$

---

[19] Finding all the real solutions in a bounded domain of the equation $\mathbf{f}(\mathbf{x}) = 0$, where $\mathbf{f}$ is a mapping of the form (6.26), is usually more computationally expensive (even if the mapping is separable) than a trivial enumerative technique for finding only the integer solutions.

[20] In section 6.2, the index space of each array reference could be represented by a single LBL, as the indices were affine mappings.

$$\left\{ \begin{bmatrix} x \\ y \end{bmatrix} = \begin{bmatrix} -1 \\ -1 \end{bmatrix} i + \begin{bmatrix} 6 \\ 8 \end{bmatrix} \,\middle|\, 2 \geq i \geq 0 \right\} \quad \text{for } 3i - 7 < 0, \, i - 4 < 0$$

The unions of disjoint linearly bounded lattices are already employed in the polyhedral data-flow model. For instance, when evaluating the dependence relations between basic sets of signals (subsection 6.3.1), it has been assumed – for the sake of simplicity – that a basic set is represented by a single LBL. This does not always happen as a basic set may result from the routine $CreateBasicSets$ (subsection 6.2.1) to be a difference of the form $LBL - (\cup_i LBL_i)$ (e.g., the basic set A1 in figure 6.9b). In this latter case, the basic set can be decomposed into a collection of mutually disjoint LBL's, as shown in the sequel.

Let a basic set be represented as the difference $LBL - (\cup_i LBL_i)$, or, $LBL \cap (\cap_i \overline{LBL_i})$. The problem is reduced to decomposing the complement $\overline{LBL_i}$ of a given linearly bounded lattice

$$LBL_i = \{ \, \mathbf{x} = \mathbf{Tz} + \mathbf{u} \mid \mathbf{a}_1 \mathbf{z} \geq b_1 \, , \, \cdots \, , \, \mathbf{a}_{r_i} \mathbf{z} \geq b_{r_i} \, \}$$

into a collection of mutually disjoint lattices. From the set identity

$$\overline{C_1 \cap \cdots \cap C_n} \equiv \overline{C_1} \cup \cdots \cup \overline{C_n} = \overline{C_1} \cup (C_1 \cap \overline{C_2}) \cup \cdots \cup (C_1 \cap \cdots \cap C_{n-1} \cap \overline{C_n})$$

where $\overline{C_1}$, $C_1 \cap \overline{C_2}$, $\ldots$, $C_1 \cap \cdots \cap C_{n-1} \cap \overline{C_n}$ are clearly disjoint sets, it follows that $\overline{LBL_i} = \cup_{j=1}^{r_i} \overline{LBL_{ij}}$, where the lattices

$$\overline{LBL_{ij}} = \{ \, \mathbf{x} = \mathbf{Tz} + \mathbf{u} \mid \mathbf{a}_1 \mathbf{z} \geq b_1 \, , \, \cdots \, , \, \mathbf{a}_{j-1} \mathbf{z} \geq b_{j-1} \, , \, -\mathbf{a}_j \mathbf{z} \geq -b_j + 1 \, \}$$

are mutually disjoint[21].

In conclusion, the given basic set can be represented as a union of at most $\Pi_i \, r_i$ disjoint linearly bounded lattices, namely: $\bigcup_{\{j_1, j_2, \cdots\}} (LBL \cap \cap_i \overline{LBL_{ij_i}})$. The complexity of the decomposition method is proportional to this product.

Example The algorithm $CreateBasicSets$ (subsection 6.2.1) applied to the motion detection kernel in figure 4.12 yields a basic set A1 (see figure 6.9b) which is the difference between two linearly bounded lattices.

$$A1 = \{ \, x = i, \, y = j \mid M + m \geq i \geq 0, \, N + n \geq j \geq 0 \, \} - A0$$
$$A0 = \{ \, x = i, \, y = j \mid M \geq i \geq m, \, N \geq j \geq n \, \}$$

After the removal of redundant inequalities, the decomposition of A1 into disjoint LBL's results to be $A1 = A1a \cup A1b \cup A1c \cup A1d$, where:

$$A1a = \{ \, x = i, \, y = j \mid m - 1 \geq i \geq 0, \, N + n \geq j \geq 0 \, \}$$
$$A1b = \{ \, x = i, \, y = j \mid M + m \geq i \geq M + 1, \, N + n \geq j \geq 0 \, \}$$
$$A1c = \{ \, x = i, \, y = j \mid M \geq i \geq m, \, n - 1 \geq j \geq 0 \, \}$$
$$A1d = \{ \, x = i, \, y = j \mid M \geq i \geq m, \, N + n \geq j \geq N + 1 \, \}$$

The decomposition of the basic set A1 has been carried out in 0.01 seconds. The data-flow graph of the motion detection kernel, where now all the nodes represent linearly bounded lattices, is shown in figure 6.10.

Note that different from the partitioning of the indexed signals into basic sets, the decomposition into disjoint linearly bounded lattices is not unique. Another possible decomposition is, for instance:

$$A1a' = \{ \, x = i, \, y = j \mid M + m \geq i \geq 0, \, n - 1 \geq j \geq 0 \, \}$$
$$A1b' = \{ \, x = i, \, y = j \mid M + m \geq i \geq 0, \, N + n \geq j \geq N + 1 \, \}$$
$$A1c' = \{ \, x = i, \, y = j \mid m - 1 \geq i \geq 0, \, N \geq j \geq n \, \}$$
$$A1d' = \{ \, x = i, \, y = j \mid M + m \geq i \geq M + 1, \, N \geq j \geq n \, \}$$

---

[21] The last inequality $-\mathbf{a}_j \mathbf{z} \geq -b_j + 1$ is equivalent to $\mathbf{a}_j \mathbf{z} < b_j$. It is also assumed, without loss of generality, that the mappings of LBL and LBL_i are identical (as LBL_i is included in LBL).

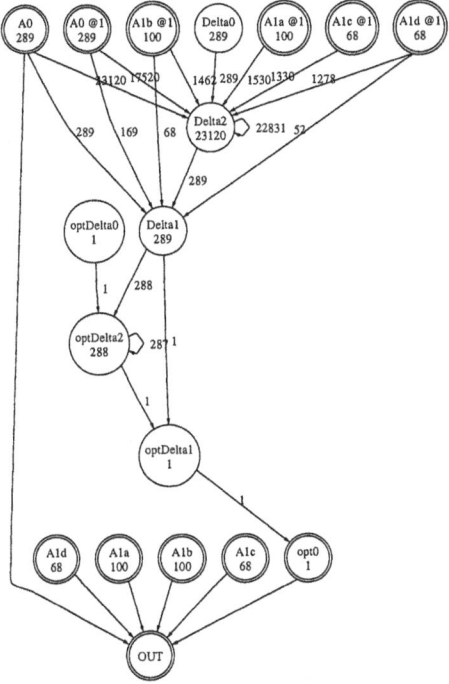

Figure 6.10.   Data-flow graph for the example in figure 4.12 (M=N=20 , m=n=4)   when all the nodes represent linearly bounded lattices

Different decomposition solutions can be obtained modifying the order of LBL's in the intersection $\cap_i \overline{LBL_i}$, and/or interchanging the inequalities in each LBL_i.

Remark Also the *union* of two LBL's can be decomposed with a similar technique in a collection of disjoint LBL's, as $LBL_1 \cup LBL_2 = LBL_1 \cup (LBL_2 - LBL_1)$.

More generally than in subsection 6.3.1, given two array references with piecewise-affine indices, having the index spaces represented as unions of mutually disjoint linearly bounded lattices – $\cup_i A_1^i$ and, respectively, $\cup_j A_2^j$ – the number of dependencies between the two array references is: $\sum_{i,j} nr\_dependences\left(A_1^i, A_2^j\right)$.
The number of dependences between each pair of LBL's belonging to the index spaces can be computed as shown in subsection 6.3.1.

6.7 CONCLUSIONS

In this chapter, the input specification analysis required to deal with the geometric modelling introduced in the previous chapter has been discussed. The main issues to solve are exact array data-flow analysis and signal partitioning, for which novel techniques have been presented which are particularly suited for our DTSE context.

7 SPECIFICATION ISSUES AND PRUNING STRATEGIES

In this chapter, several issues concerning input specification representation and preprocessing steps for the actual DTSE methodology will be briefly discussed. These issues become much more complex when applied to practical designs, but in the scope of the book they cannot be presented in more detail.

7.1 INPUT SPECIFICATION ISSUES

As already mentioned in section 3.2, the single-definition form of the specification code should be stimulated for all critical signal statements. In addition, the use of hierarchy should be heavily encouraged in the application specification. Function hierarchy should be used to model the separation of the DTSE related part and the arithmetic/control related part of the specification. In particular, this will allow to split the application into 3 main layers:

1. a procedural process control flow top layer which is not of real interest for DTSE, except for the potential sequence/timing constraints which are imposed on the memory transfers

2. a middle layer which contains all the relevant information for the DTSE steps, and which can be further divided up in several sublayers (at least initially)

3. a bottom layer of local arithmetic/logic/local control functions which contain (and "hides") all operations to "scalar" data and local control flow. Moreover, it will be used to hide some of the undesired constructs which are not of real interest to the DTSE tasks (see section 7.2).

An example of this for the RMSP domain is shown below:

```
/* LAYER 1 */
/* procedural process control flow in C description */
int main() {
  if (mode1)
    tmp = fnc1(in);
  else
    tmp = fnc2(in);

  while (mode2)
```

```
    tmp= fnc3(tmp);
  ...
}

/* LAYER 2 */
/* applicative data flow in DFL description */
func fnc1(in: W[][][]) out:W[] =
/* M-D data flow to be stored/transferred in background memory */
begin
(i : 0 .. 511 ) ::
  begin
  (k : i .. 511 ) ::
    B[i][k]= g(in[i+1][k-i+5][k]);
  ...
  (j : 0 .. (i mod 255)) ::
    A[i][2*j] = if (i+j)<511 -> f(A[i-1][2*j-i-5],
                                  B[i div 2][2*j-1]) fi;
  ...
  end;
end;

/* LAYER 3 */
/* description of all arithmetic, logic, data-dependent operations
   to be mapped on data-path, foreground memory and controller */
func f(in1,in2: W) out:W =
begin
  ...
end;
```

For non-RMSP applications such as the data-dominated protocol layers in network components (e.g. ATM layer 3-4), the splitting in 3 layers also applies, but here the semantics of the middle and bottom layer cannot be purely data-flow any longer. Due to the run-time data-dependent table accessing, an extension to procedural single-definition models and analysis is required. As the data-flow between these data types now resembles the scalar data-flow in many ways, the analysis is not as complex any more as in the case of M-D arrays. So the problems mentioned in section 3.2 in terms of allowing procedural specifications do not apply here in the same way.

In summary, all arithmetic, constants/scalars and all undesired non-linear and/or data-dependent constructs are hidden in the lowest level functions which are called from higher levels in the middle layer where the (M-D) data flow is expressed by means of loops and indexed signals. Only these middle layer functions should be passed to DTSE because only they contain the relevant information. Usually, it is also better to expand all the function hierarchy in between the top level (expressing the system level processes) and the bottom layer of "scalar" functions. In this way, the (M-D) data flow functions in LAYER 2 are expanded to 1 flattened layer to create more freedom for transformation and optimisation of the storage and system communication. Indeed, if the hierarchy is maintained it means that a function which is called twice (or more) will have to be treated in exactly the same way independent of the position of its call in the code (context). As a result, it cannot be optimized in a fully context-dependent way any more because that would lead to different versions of the optimized function specification. As motivated earlier, the context-dependent optimisation is essential in most applications to obtain really good results. If a function is called only once, this restriction is not valid but then also the advantage of leaving the hierarchy in this exploration context is less clear.

The expansion of the hierarchy in the middle layer can lead to complex code, especially when combined with the extra signal array partitioning introduced in subsection 1.5.7. Fortunately, a number of good ways to deal with this complexity are available. These will be discussed in section 7.2.

When all DTSE tasks have been completed, the end result (new intermediate layer) has to be hooked up again with the "scalar" functions describing the originally hidden code in LAYER 3. This is not trivial to implement but feasible technically.

7.2   PRUNING AND RELATED PREPROCESSING STAGES

First, it should be clear that pruning is essential in order to reduce the complexity, both during manual exploration and within tools. If e.g. only 1 page of code could be handled, then either the initial code has to be heavily pruned or it has to be partitioned in "functions" where the subdivision in functions is procedurally interpreted, i.e. all calls to a given function use the same DTSE based organisation for the content of the function. The latter option is inherently easier to support but it leads to a large reduction in the potential search space as all the mismatches between production and consumption of data are then fixed, leading to buffers in between the functions executions later on. Practice has shown that the biggest gains are situated in these buffers and they can only be optimized by a very global view [53]. This means that code up to several dozen pages should be dealt with in the worst case.

Unfortunately, experiments show that the "exact/optimal" pruning techniques do not allow more than a factor of about 2 in reduction. Therefore also "inexact" (potentially suboptimal) pruning heuristics are needed [101]). These are not leading to incorrect or "unsafe" solutions though, if they are carefully integrated in the methodology. A useful approach is also to perform the optimisation iteratively, so first look at the global problem, and locally optimize further by applying the script on memory-critical partitions with less pruning.

First it should be noted that the term "pruning" should be used with care because it encompasses several "preprocessing" substeps. In particular these substeps are:

1. hierarchical rewriting (see example in section 7.1) which reduces the search space by interpreting the "hidden" functions as already fixed in terms of DTSE decisions.

2. hiding of undesired constructs (data-dependent conditions, scalar and logic operations) in lower level functions.

3. code expansion of functions containing M-D data flow, to increase the optimisation freedom.

4. pruning to reduce the graph complexity for chains, recursions and conditions [101]. Many of these are "exact", i.e. they retain all the freedom in the background memory exploration.

5. weight-based removal of less important parts of the code during the most time-critical optimisations which do not have to be based on the knowledge of all the M-D data flow dependencies. These remove information and are thus "inexact".

6. partitioning of graphs to exploit divide-and-conquer principles, but this should be performed with care in order not to affect the global view too much. When subgraphs are almost not communicating, most DTSE transformation steps can operate on separate subgraphs, without loosing the near-optimal view. Memory organisation related steps however (SCBD, memory assignment and in-place optimisation) are affected even by interference between non-communicating subgraphs which are executed in parallel. So for these steps some suboptimality can be introduced.

In the systematic methodology, these subtasks can be performed in the order given above but in practice iterations will be needed (see [101]). All these substeps have to be "undone" (depruning) after finishing the DTSE stage before continuing to the subsequent stages in the system design trajectory.

8 GLOBAL TRANSFORMATION STRATEGIES FOR POWER AND STORAGE SIZE REDUCTION

In this chapter, the main focus will be on global transformations which improve the original code in terms of DTSE related cost factors. First, data-flow transformations will be considered in section 8.1 to section 8.3. They allow to remove redundant access in the data-flow and they serve as enabler for the subsequent transformation steps by removing data-flow bottle-necks wherever it is cost effective to do so.

Then, global loop (and other control-flow) transformations are treated (in section 8.4 and following). We review the classical system organization and identify the problem of synchronization as the cause of the typical area and power cost inefficiencies of this organization. These inefficiencies are mainly located in the buffers between large sub-systems. Our strategy to reduce these buffers and to improve the access locality is based on loop transformations and is able to deal with stringent synchronization constraints. The strategy is applicable when multi-media applications are mapped on custom processors but also towards programmable processors or embedded processor cores. To illustrate the effectiveness and feasibility of this strategy we use a relevant class of multimedia applications as a red-thread example throughout this chapter.

8.1 GLOBAL DATA-FLOW TRANSFORMATIONS CONTEXT

In this section and the following two, we will exploit the data-dominated characteristics of our target domain to propose a formalized data-flow transformation methodology, which can achieve large savings in the system power without having to worry about the detailed data-path, foreground registers, and controller architecture. The main reason why this step is the first real optimisation stage in our overall DTSE methodology, is that it heavily affects the impact of all other steps too much to postpone it. Indeed, when global data-flow changes are introduced on the main data type instances in the algorithm, all costs are heavily modified so that the trade-offs for the other steps become totally different. The data-flow methodology on its own is however also usable as a stand-alone optimization step in other contexts.

In chapter 2 it has been shown that very little related work is available for such system-level transformations, especially for an embedded application context where power and area cost are important. Here, we will introduce a number of system-level data-flow transformations which are intended for data-dominant applications and which modify the algorithmic data-flow to achieve a better trade-off in terms of data transfers and/or storage. One class also servers as enabling transformations for other tasks in our global DTSE methodology because they break data-flow bottle-necks. Our approach will be illustrated on a real-life driver application, namely the H.263 video conferencing decoder standard.

8.2 H.263 VIDEO DECODING

H.263 is a draft recommendation for video coding for narrow telecommunication channels at < 64kbit/s [178]. The coding/decoding is a block based algorithm that exploits spatial and temporal redundancy. From the intra frames (I), inter (P) frames are predicted by using motion compensation. Interpolation between I and P frames then yields bidirectional (B) frames. Several extensions are orthogonal to P and PB mode, including the overlapped block motion compensation (OBMC) mode which is discussed later. The data-flow for the continuous P mode is shown in figure 8.1.

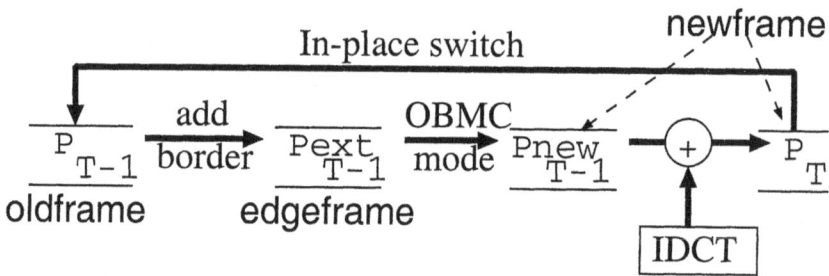

Figure 8.1.   Data-flow for continuous P frame decoding.

Given this application and the power model of chapter 4, we have performed a detailed power exploration of the frame related storage and transfers [291]. We have started from the original public domain C code, available from Telenor Research [178] and have derived a relevant reference memory organisation from this (figure 8.2).

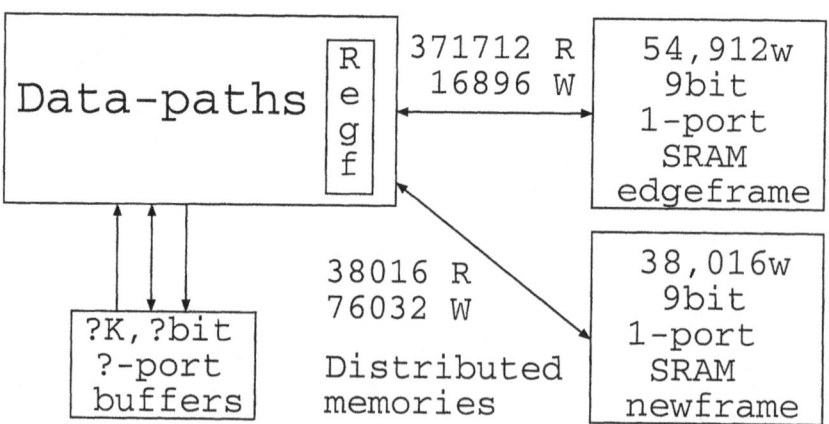

Figure 8.2.   Memory organisation of reference design for H.263 decoder when consecutive P frames are decoded. Frame transfers (Read and Write) are indicated on the arrows.

8.3 DATA-FLOW TRANSFORMATION METHODOLOGY

Our global data-flow transformation methodology involves several stages. The order in which they are applied is very important and has been extracted by us from the dependencies between the stages and the best way to break the unavoidable cycles in these dependencies. This enabling/disabling effect of the transformations is crucial to investigate in-depth before deciding on the most promising order in a methodology, as described in the following subsections. The end results is the following sequence:

1. advanced signal substitution including moving conditional scopes.

2. modifying computation order in associative chains

3. shifting of "delay lines" through the algorithm.

4. recomputation issues.

These four stages will now be applied to modules in the H.263 test-vehicle (and similar applications) to illustrate them and to demonstrate their effect. The order in which they are executed will also be motivated. In addition, the potential for design support and techniques to (partly) automate the crucial tasks will be discussed.

A fifth stage can be introduced after these four optimizing stages in order to break bottle-necks in the global data-flow by means of so-called look-ahead transformations [139, 312, 314], but then generalized to multi-dimensional data-flow. They are sometimes needed to create more freedom during the loop transformation, data reuse decision or SCBD steps. These enabling data-flow transformations have been applied by us for instance on a cavity detection application [89] but will not be further discussed here.

8.3.1 Advanced signal substitution and propagation

This class of data-flow transformations involves signal substitution and propagation over global condition and loop scopes. It is for instance important to remove dependencies on signals that are "copies" of other signals and which would normally have to be stored separately. This optimisation is usually applied automatically by the algorithm designer when writing simple chains of signals, but it is usually not performed over conditional scope boundaries where it can have an even larger but also more complex effect. The overall goal of the propagation is to remove pure copies (not found during pruning) and to remove any redundant array accesses by putting them under the most restricted conditional scope which is really required for the functionality of the algorithm. Consider e.g. the borders of images in algorithms like H.263 which are not just filled with zero's but with real data copies in a non-trivial way [178]. To simplify the control-flow in the original C, these data are duplicated in the frame signals (see $edge\,frame$ in figure 8.1) prior to the actual image manipulations, resulting in storage and transfer overhead both for reading and writing. To reduce this overhead, the dependences on the border data can be checked by (manifest) conditions on the position of the pixels to be read. Now, instead of storing and accessing duplicate data, the original pixels are read at the boundary row/columns of the image frame. These conditions have to be implemented in the controller and will steer the data-path. Usually, also some local buffering is necessary then. Several degrees of optimisation are possible here, starting from a simple context-independent caching of the border data (which is apparently selected in most industrial designs) up to a heavily optimized context-dependent checking and reduced local buffering. All of these alternatives make the storage for the extra borders superfluous but only the latter option allows to remove all redundant frame accesses.

In the H.263 design, the $edge\,frame$ memory area is reduced by about a factor 13/9: from 54,912 words (combined Y and C) of 9 bit to 38,016 words. In terms of power consumption, our detailed models show that we obtain a data-dependent saving of between 24% and 27% by the combined effect of less transfers and a reduced frame size. The memory organisation corresponding to this for the P mode is illustrated in figure 8.3. Note that writing the $old\,frame$ is not needed here due to the in-place switch with $new\,frame$ at the end of each frame period.

In terms of potential tool support, the main bottle-neck here is the application of context-dependent transformations. In the H.263 application, this is required to introduce all the necessary boundary check conditions. A quite generally applicable (within video/image processing) parametrizable template to

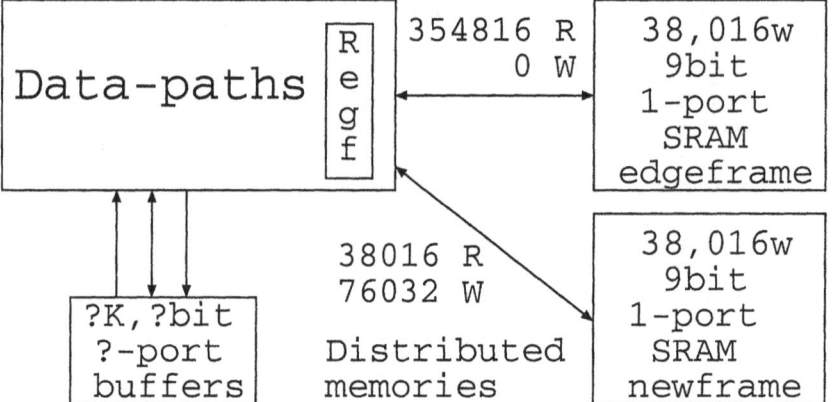

Figure 8.3.   Memory organisation and frame transfers for P mode of H.263 decoder after border reduction

perform this could be described in a transformation library. This transformation can then be applied in an interactive way to selected code regions. In addition, there is a need to have a fast estimation routine for the memory size and the number of accesses in both cases. Accurately counting these manually for the complex code is very error-prone and tedious.

Because this class of data-flow transformations reduces the array signals present in the algorithm and the choice on its application is not directly affected by the subsequent stages, it should be performed as the first stage in our methodology.

8.3.2 Optimized associative chains

Examples of such chains are the accumulation of signal data. The recursive nature of the corresponding loops normally restricts the potential for transforming these to obtain more locality of access. The power of combined loop/data-flow transformations is however much larger because the freedom created by exploiting the associative chains acts as a powerful enabler on the scope of the loop transformations. Compared to applying only "standard" loop transformations [35], more locality can be achieved in this way. Another example of using these associative chains is to decrease the word-length of the intermediate data storage at the places in the flow graph where the most data "pass" or where the storage cut is best made when different graph partitions are assigned to different processing units. This leads to a reduced memory width and data transfer cost, on top of the obtained memory size reduction.

These associativity-based transformations can affect the bit-true behaviour so in general the conditions under which they are allowed have to be checked [201].

In the H.263 code, an example of the effect of this stage is present in the overlapped-block motion compensation ($OBMC$) mode (see figure 8.4) of the forward motion compensation for the luminances Y. In particular, each new Y pixel is the weighted sum of three prediction values, which are obtained by using three out of the five available motion vectors for that particular macro-block, in conjunction with the motion vectors of the adjacent macro-blocks. The other 2 intermediate variables are just copies. Which variables contain relevant data and which are copies depends on the location of the subblock. Three contributions are accumulated for each subblock of the $new frame$ block, coming from different regions with different motion vectors. Hence the term overlapped block motion compensation. The 8×8 blocks are divided in 4×4 subblocks of which every pixel is produced as a combination of three contributions: all pixels of block 0 in figure 8.4 are constructed by the accumulation of a contribution of every block 0 of the $old frame$, every pixel of block 1 is constructed by contributions of every block 1 in the $old frame$, and so on.

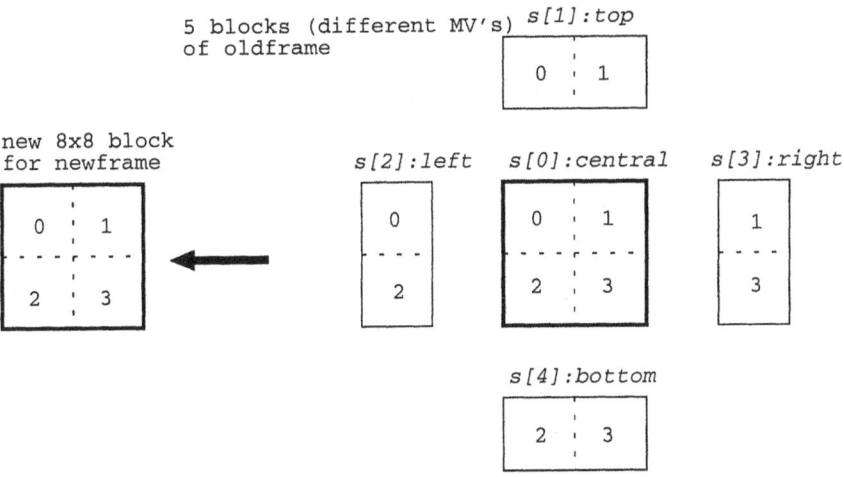

Figure 8.4.   Principle of Y oriented reconstruction operation in $OBMC$ mode.

Here, every 8 × 8 block within a macro-block is produced by a combination of the contributions of 5 regions in the previous P frame, which are accumulated. Instead of performing this in separate steps (which requires some buffering in between), it is possible to produce the four 4 × 4 subblocks consecutively, where all contributions are accumulated directly for each subblock. The actual locality of access is however only obtained after a combination with a loop transformation and the introduction of a memory hierarchy level [291]. The effect on power is relatively local and the contribution in the overall power budget for H.263 is not very large (a few %) but still relevant.

Optimizing associative chains allows to introduce more locality of access which heavily influences the estimated storage and transfer cost afterwards (by high-level in-place estimation). As accurate estimates of these costs are required to drive the 2 remaining categories of transformations, and as the dependence on the outcome of these other categories towards the associative ordering is limited, this transformation should be performed as the second stage in our methodology.

8.3.3 Shifting of "delay lines"

The effect of this can be illustrated by the move of a delay present in the initial data-flow to an edge with a smaller word-length or to an edge with less signal words. In principle, this requires the move of the delays through the graph by retiming, but then generalized to handle delays on M-D array signals. In this way, the total associated storage can be heavily reduced and also the corresponding area and power cost is decreased. Also these generalized M-D retiming transformations can potentially be automated with our algebraic and delay optimisation framework [201, 202].

An example of this is related to the delay of the *oldframe* which is present in the original motion compensation loop in the H.263 video decoder top layer, shown at the right-hand side in figure 8.5. This delay of an (array) signal over the time axis – symbolized with @ in the data-flow language (DFL) of Mentor – can be shifted over the loop by retiming, as depicted in figure 8.6. It means however that the IDCT data would have to be buffered intermediately. This extra cost has to be taken into account when considering the global trade-off. For the H.263 related reconstruction, this transformation is not effective because it is unlikely that the "delay" inside the motion compensation component can be put at a position where the cut-set through the flow-graph contains less bits to be stored than in the original position in figure 8.5. However, in general it could well be that the reconstruction does not act on the full old frame but that some form of decimation takes place first. If the delay is then moved, only the decimated frame should be stored, leading to a significant saving.

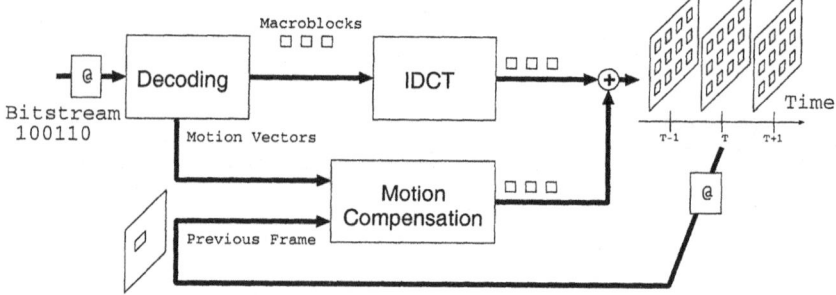

Figure 8.5.   Retiming illustration: original FG

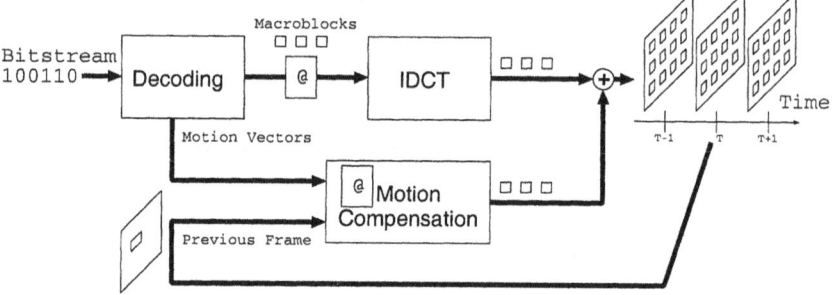

Figure 8.6.   Retiming illustration: alternative

Shifting of "delay lines" modifies the position of the main storage related signal accesses so it should be performed in front of the last stage, namely recomputation, which is usually costly and very sensitive to these positions.

8.3.4 Recomputation issues

A typical example of this class of transformations is the recomputation of a large amount of data from a limited amount of "key date", if the large amount of data is used multiple times. This leads to less long-living data and less global transfers to recover these data from the higher layers in the memory hierarchy to the data-path. Because of the typically costly data-path overhead involved, this stage is only useful in a context where long life-times would otherwise lead to much extra storage. It usually involves a power-area trade-off too. Independent research on this issue within the context of fixed storage/transfer costs on flow-graph edges has been published very recently [152].

This transformation can be useful within video decoders like the H.263 to modify the original macro-block based traversal during motion-compensation to a block-based one, which has some advantages under certain conditions [291]. The trade-off between these alternatives is however difficult to make manually and requires a detailed analysis of all the side effects. Design support is clearly needed here.

This last stage of transformations is very sensitive to the estimated storage and access cost at different positions in the algorithm. Hence, it is very dependent on the outcome of all the previous stages. Moreover, the area overhead of applying it is in general relatively high so it should only be used as a last recourse to reduce the power cost. Hence, it should be performed as the last stage.

8.4   GLOBAL LOOP AND CONTROL-FLOW TRANSFORMATIONS

In the rest of this chapter, the main focus will be on the steering of the global loop (and other control-flow) transformations that will improve the data access locality and regularity. The main illustration will be on instruction-set processor architectures to illustrate that our global DTSE methodology is indeed also for a large part extensible towards programmable target architectures.

As indicated in chapter 1, an important property that distinguishes embedded multimedia applications from many other signal processing applications is the presence of *real-time constraints*. At first sight it may be tempting to think that this lack of the cycle count dimension in the design optimization search space simplifies the design steps such as the loop transformation steering considerably. However, the optimization of the remaining major cost factors that have been identified in chapter 1, namely chip area, power consumption, and design time, becomes much more difficult when real-time constraints have to be met. Especially when the system has to be built using predefined processors or processor cores, timing constraints are hard to take into account.

As a result, designers usually fail to make a real trade-off between the remaining cost factors in those cases and opt to focus on the minimization of the design time, while the reduction of area and power cost become secondary goals that are often neglected. Methodology support for making a better global trade-off is therefore certainly worthwhile and even necessary to arrive at cost effective solutions.

A methodology that is able to deal effectively with all instances of the wide variety of multimedia applications is unlikely to exist. In contrast, methodologies that are targeted towards certain classes of applications are much more feasible.

One of the most important classes is that of pseudo-regular algorithms, i.e. algorithms that are mainly consisting of regular loop nests and that contain only a limited amount of data-dependent behaviour. Many of the current algorithm standards (including the ones which are still under development like H26x and object-oriented video coders [43]) are based on a hierarchy of sub-systems which are called from a system layer. The main data-dependent behaviour and irregularities are situated in the higher layers, i.e. in the global control-flow. Typical sub-systems called from there are motion estimation and compensation, wavelet schemes, contour extraction, segmentation, and discrete cosine transforms. Inside these usually complex sub-systems, the algorithms are heavily based on pseudo-regular loop nests, occasionally mixed with data-dependent constructs. It is exactly with these sub-systems, and not with the higher layers, that the main data storage and transfer related costs are associated.

These observations have motivated us to come up with a global loop and control flow transformation based methodology that heavily optimizes memory related power and area and that is also able to take into account real-time constraints. The target domain is the large class of (pseudo-)regular algorithms and algorithm kernels.

Our methodology, which is presented in this chapter, is powerful enough to deal with entire sub-systems, even when they contain irregularly nested loops and partly data-dependent behaviour, because in real-time systems we can take worst-case assumptions to size the buffers and to compute the maximal power consumption. Hence, we believe that our approach is valid for most of the sub-systems that have a crucial impact on the storage related system cost, and even that it remains valid for advanced compression schemes.

Our main target architecture in the rest of this chapter is based on programmable video or multimedia processors like the Texas Instruments TMS32060, the Philips TriMedia, or the Chromatic Mpact, because it is one of the current trends to use these processors in multimedia systems [180], but most principles remain valid for architectures based on custom processors.

As reviewed in chapter 2, none of the earlier loop transformation related approaches determine the best global steering strategy over the entire specification for a given application and only few address the power cost.

After this introduction, we review the classical system organization in section 8.5 and identify synchronization constraints as the cause of area and power inefficiencies. Next, in section 8.6, we present a sub-class of algorithms that we use as a red-thread example throughout this chapter. The overall methodology consists of several major steps: synchronisation (especially for programmable processors) serves as an enabler of global loop transformations (section 8.7), and is followed by buffer size reduction (section 8.8) to reduce power and area by means of the loop transformations themselves. The other steps needed to arrive at the full memory organisation are the topic of the subsequent chapters.

8.5   CLASSICAL SYSTEM ORGANIZATION

Many real-time signal processing systems are organized in a globally pipelined way, as depicted in Fig. 8.7. They get their new input data from some source (e.g. a video camera) at a fixed rate and store it in an input buffer. Similarly, the results of the processing are stored in an output buffer, from which they are read by another system (e.g. another processing stage) later on, also at a fixed rate.

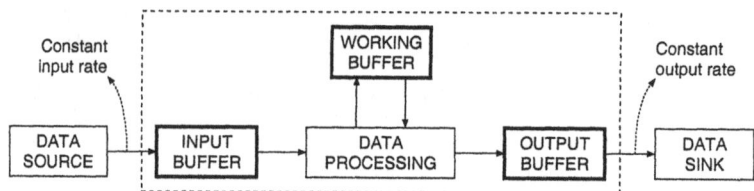

Figure 8.7.   Typical signal processing (sub-)system setup.

This setup has the advantage that it allows the actual processing to take place on the data in the buffers without imposing stringent synchronization constraints. Only at relatively large distances in time (e.g. at the start of each new image frame) some synchronization is necessary between the processing and the I/O. This is the main reason why this setup is so popular among system designers: one can optimize the different sub-systems without having to deal with fine-grain synchronization issues. A typical mode of operation is shown in Fig. 8.8. The input and output buffers are usually divided into several parts with alternating functionality ("ping-pong" buffers), as shown in the figure.

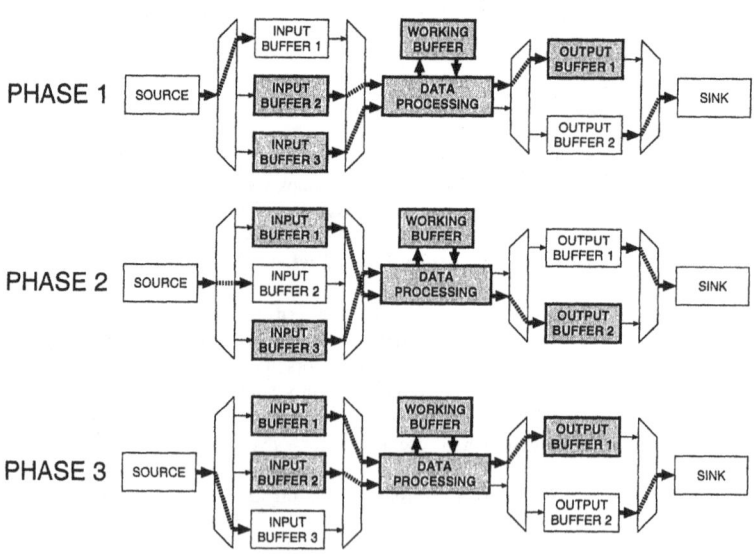

Figure 8.8.   Typical execution cycle.

However, this organization can be very costly in terms of area and power consumption. Especially in video and image processing applications, the buffers between sub-systems can become very large (e.g. several image frames) and consequently consume a lot of area. Moreover, for this kind of applications, the number of memory accesses is usually quite large, which results in a large power consumption due to memory transfers. This effect is further amplified by the large sizes of the buffers, as the capacitive load increases with the memory size.

Techniques for reducing buffer sizes and power consumption are commonly known in different contexts. For instance, the buffer sizes can be reduced by enforcing synchronization at (much) shorter distances in time. Power consumption on the other hand can be reduced by introducing (multi-level) memory hierarchy (caches), which reduce the number of expensive accesses to the larger memories.

However, even though designers of real-time multimedia systems may realize that these techniques can reduce their system cost, they rarely ever apply them in practice, because they are reluctant to increase the complexity of their design by decreasing the synchronization interval. Especially when the system includes programmable components like DSP processor cores, fine-grain synchronization is considered to be practically impossible for high-throughput applications such as video.

In the following sections, we show that this "common sense" approach is too pessimistic and we present a systematic methodology for reducing both the buffer sizes and the power consumption that is able to take into account stringent synchronization constraints, but first we present our demonstrator class of algorithms.

8.6 DEMONSTRATOR CLASS OF ALGORITHMS

As a demonstrator we use the class of motion-estimation-like algorithms, a variant of which is used in the MPEG video compression standard [235] for instance. The presented principles in the following sections are nevertheless very general and can be applied to a very large class of signal processing algorithms or algorithm kernels, provided that they exhibit a relatively large degree of regularity, and that data-dependent behaviour is mainly localized between different (image) frames (see section 8.4).

The class of motion-estimation-like algorithms contains a large number of variants (e.g. [216]) but they are all characterized by the following properties:

- data of multiple frames have to be available at the same time;

- a block-based way of operation: data blocks in different frames, but around the same location, are combined;

- a huge number of arithmetic operations and memory accesses per second is required for real-time operation (e.g. 2.6×10^9 operations/second and 5.3×10^9 memory accesses/second for motion estimation on standard television).

Typically, an image frame is subdivided into small blocks (e.g. 8×8 pixels) and each of these blocks (which we refer to as *current blocks* or CBs) is compared with blocks in a region around the same center, but in a previous frame (which we refer to as *reference windows* or RWs) as shown in Fig. 8.9. The position of the block in this RW that matches the CB best is then considered to be the previous version of the CB. The distance between the CB and the best match in the RW is represented by a motion vector. These motion vectors can then be used to compress the image stream. Typically only the motion vectors and some correction information have to be transmitted, which allows to obtain very high compression ratios.

A C-like description of the typical structure of such an algorithm is shown in Fig. 8.10. We only show the relevant portions of the algorithm, such as the loop nests and the buffer read and write instructions. Note that we explicitly included the input and output sections, which we assume to operate in parallel with the processing section (see also section 8.5). The actual implementation of the INPUT and OUTPUT statements is processor-dependent. They can be implemented through direct-memory-access (DMA) or through real input and output instructions for instance.

This description is compatible with the typical system setup presented in Fig. 8.7. We assume that synchronization only occurs at each execution of the time loop, which corresponds to the start of a new frame. Realistically we can also assume that the order in which the input and output data arrive/leave is specified and cannot be changed (e.g. due to standardized camera input or a raster scan

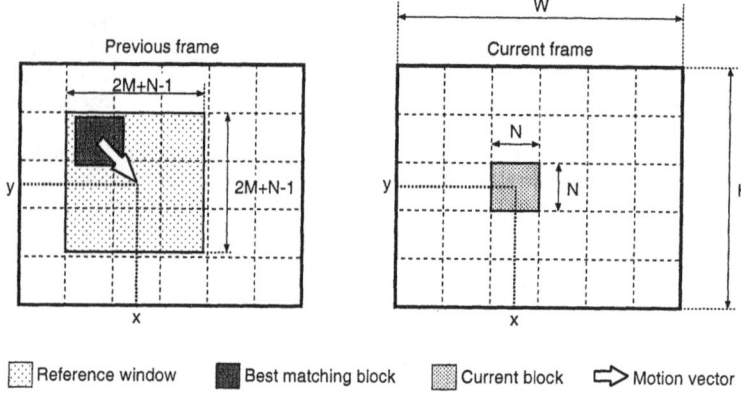

Figure 8.9.   The motion estimation algorithm and its parameters.

display output). In this case, the input pixels arrive in a row-wise manner, from left to right and from top to bottom. A similar order is assumed for the results, but a higher level of granularity, i.e. one result per block of N^2 pixels.

Although our demonstrator contains only one processing loop nest, our methodology also applies to the far more general case where processing is not embedded in one perfect loop nest. We have selected this demonstrator because it is one of the most demanding classes of algorithms in terms of memory bandwidth, and yet is still simple enough to be used as a red-thread example.

8.7   SYNCHRONIZATION ISSUES

One of the things that we need to do in order to be able to decrease the buffer sizes, is to decrease the lengths of the synchronization intervals. This also means that we have to match the processing speed with the I/O rates. In order to obtain the largest possible buffer size reduction, perfect synchronization is required, i.e. synchronization at the highest sample rate. In practice however, one may opt to relax this constraint by increasing the synchronization interval to create somewhat more freedom (at the cost of an increased buffer size). In the remainder of this chapter, we assume that perfect synchronization is desired, and we demonstrate how it can be obtained in a systematic way.

If we assume that the algorithm is to be executed on a programmable DSP or DSP core, there are basically two alternatives to achieve this: we can either rely on hardware interrupts or make sure that the time intervals spent on processing have a constant length, and that this length equals the length of the smallest I/O sample intervals (in that case, either DMA or explicit I/O instructions can be used). However, in a video processing context, where sample rates can be very high, an interrupt driven approach is not feasible due to the non-negligible overhead, especially on programmable processors. In contrast, in a speech processing context this alternative is usually feasible. But whatever alternative we choose, we always have to distribute the processing over the different sample intervals such that I/O streams and processing stay in perfect synchronization. In the remainder of this section we further investigate how we can implement the processing time equalization, but most of the conclusions remain valid for the interrupt alternative.

In practice, there are two kinds of programming constructs that can prevent us from achieving constant processing times: conditional branches and imperfectly nested loops. These are discussed in more detail in the following subsections.

8.7.1   Conditional branches

The problem of equalizing the duration of the different branches of a condition is well known and rather easy to solve. A simple padding with dummy code (e.g. NOP instructions) can do the job. In

```
for ( T = -infinity; T < +infinity; ++T )      ⇨ Infinite time loop
{
 ┌ ···································································································
I│   for ( Yin = 0; Yin < H; ++Yin )          ⇨ For each pixel of frame
N│     for ( Xin = 0; Xin < W; ++Xin )
P│        INPUT(image[Xin,Yin,T+1]);          ⇨ Read new pixels in input buffer
U│ ···································································································
T│  || ⇨ Do in parallel
 │ {
 │   for ( Xproc = 0; Xproc < W/N; ++Xproc )
 │     for ( Yproc = 0; Yproc < H/N; ++Yproc )    ⇨ For each CB
 │     {
 │        Init1();
 │
 │        for ( i = -M; i < M; ++i )              ⇨ For each position of CB in RW
 │          for ( j = -M; j < M; ++j )
 │          {
 │             Init2();
P│
R│             for ( k = 0; k < N; ++k )          ⇨ For each pixel of CB
O│               for ( l = 0; l < N; ++l )
C│               {
E│                 ┌ ···································································
S│                 │ xCB = Xproc*N + k;
S│                 │ yCB = Yproc*N + l;
I│                 │ READ(image[xCB,yCB,T]);      ⇨ Read pixel from current frame
N│                 ···································································
G│                 │ xRW = xCB + i;
 │                 │ yRW = yCB + j;
 │                 │ if (WithinBoundaries(xRW,yRW))
 │                 │    READ(image[xRW,yRW,T-1]); ⇨ Read pixel from previous frame
 │                 ···································································
 │                 Process();
 │               }
 │
 │             PostProcess();
 │          }
 │        WRITE(result[Xproc,Yproc,T]);       ⇨ Write new result to output buffer
 │     }
 │ }
 │  || ⇨ Do in parallel
O│ {
U│   for ( Yout = 0; Yout < H/N; ++Yout )     ⇨ For each CB
T│     for ( Xout = 0; Xout < W/N; ++Xout )
 │        OUTPUT(result[Xout,Yout,T-1]);      ⇨ Produce old result at output
 │ }
}
```

Figure 8.10.   The original demonstrator program.

case of nested conditions, this approach is easily extendible in a hierarchical way. Therefore we do not describe this in more detail.

8.7.2 Imperfectly nested loops

A problem that is less well known and not described in literature in a formal way, is the problem of imperfectly nested loops. Imperfectly nested loops containing I/O instructions result in an irregular I/O rate. In practice, loops are almost never perfectly nested, especially when implemented on programmable processors, because there is usually at least some kind of initialization required (even for so-called zero-overhead loops).

This problem also can be tackled by inserting dummy code, but it is not as straightforward as in the case of conditional branches. Fig. 8.11a contains a graphical representation of the execution of a single loop containing an I/O instruction. Before the loop is entered, some entry code is executed (e.g. the initialization of the loop counter), and afterwards some exit code is executed (e.g. some post-processing). Assuming that the I/O instruction is the only one accessing the I/O stream, a constant I/O rate is obtained once the loop is started. This situation changes, however, in case we nest this loop in another one, as shown in Fig. 8.11b (where we have only shown the resulting execution pattern of the inner loop). The resulting I/O rate is no longer constant now, due to the execution of the entry and exit code at a lower frequency. In order to avoid this irregularity, we can insert some dummy code in the inner loop. This dummy code is to be executed only when the entry and exit code is not executed. This is shown in Fig. 8.11c.

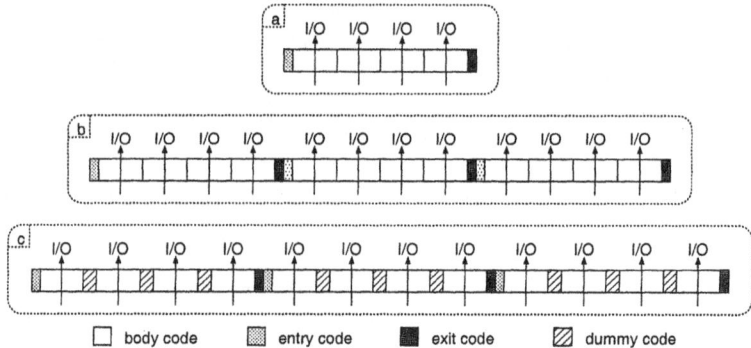

Figure 8.11.   Imperfectly nested loop synchronization issues.

A possible pseudo-code template for a perfectly synchronized doubly nested loop is the following:

```
          entry1();
loop1:    entry2();
          goto loop2b;   // skip dummy code the first time
loop2:    dummy2();
loop2b:   body2();       // body includes I/O operation
          if <loop 2 not finished> goto loop2;
          exit2();
          if <loop 1 not finished> goto loop1;
          exit1();
```

From Fig. 8.11c one can derive that for a doubly nested loop, the duration of the dummy code in the inner loop has to be equal to the sum of the durations of the entry and exit code of that loop. If we extend this to an arbitrary number of nested loops in which the inner loop contains an I/O operation, we obtain the following formula:

$$[dummy\ code]_l = \sum_{k=2}^{l} [exit\ code]_k + [entry\ code]_k \qquad \text{for } l = 2 \text{ to } N \qquad (8.1)$$

where the $[\ldots]$ represents the duration of a code fragment, and N is the number of nested loops (the outer loop has number 1). In other words, the duration of the dummy code to be inserted in a certain loop not only depends on the durations of its own entry and exit code, but also on those of the surrounding loops. So, in order to obtain perfect synchronization, it is very important to minimize the durations of the entry and exit code for *all* loops, even the outer ones, as this has direct influence on the amount of dummy code to be inserted in the inner loops. In contrast, in case perfect synchronization is not required, the code of the inner loop is the most important one to optimize in terms of speed (this is what happens conventionally). The following small example illustrates the difference (we assume for reasons of simplicity that the loops only have entry code).

Example 8.7.1

```
for i = 1 .. N1
  [E2 cycles]          // entry code j-loop
  for j = 1 .. N2
    [E3 cycles]        // entry code k-loop
```

```
for k = 1 .. N3
  [E4 cycles]            // entry code l-loop
  for l = 1 .. N4
    [E5 cycles]          // body code inner loop
```

For the loop nest shown above the number of cycles for the case where no synchronization is required, is given by the following formula:

$$(((E5 \times N4 + E4) \times N3 + E3) \times N2 + E2) \times N1$$

If we assume that $E[2..5] = 5$ and $N[1..4] = 10$, we end up with 55550 cycles. If we increase E2 to 20, the cycle count becomes 55700, which is an increase of only 0.3% compared to the original code.

To obtain perfect synchronization on the other hand, we have to insert dummy code as follows:

```
for i = 1 .. N1
  if (i > 1) [D1 cycles]     // dummy code i-loop
  [E2 cycles]                // entry code j-loop
  for j = 1 .. N2
    if (j > 1) [D2 cycles]   // dummy code j-loop
    [E3 cycles]              // entry code k-loop
    for k = 1 .. N3
      if (k > 1) [D3 cycles] // dummy code k-loop
      [E4 cycles]            // entry code l-loop
      for l = 1 .. N4
        if (l > 1) [D4 cycles] // dummy code l-loop
        [E5 cycles]            // body code inner loop
```

The sizes of the dummy code are given by equation 8.1:

$$D4 = E2 + E3 + E4 \qquad D3 = E2 + E3 \qquad D2 = E2 \qquad D1 = 0$$

The resulting duration is now given by the following formula (after simplification):

$$(E2 + E3 + E4 + E5) \times N4 \times N3 \times N2 \times N1$$

Evaluating this formula for the original parameters results in a cycle count of 200000. If we again increase E2 to 20, the cycle count now becomes 350000, which is an increase of 75%!

This example clearly illustrates the fact that we have to try to obtain a perfectly nested loop around I/O instructions (or synchronization points in general), i.e. *we should avoid as much code between loop boundaries as possible if we want to achieve perfect synchronization in an efficient way.* This is even true for regular applications such as motion estimation, and it is therefore one of the main driving forces behind our optimization strategy presented in the following sections.

In case more than one I/O instruction is accessing the same stream (possibly in a different loop nest), our synchronization techniques are extendible in a straightforward way.

As mentioned before, in practice one can of course relax the synchronization constraint somewhat to avoid a too large cycle overhead, at the cost of a small increase in buffer size. In this way also algorithms with a reasonable amount of irregularities can be optimized in an efficient way. In the subsequent sections we assume that pixel-level synchronization is required, however, and illustrate that it is feasible to achieve this even for a high-throughput algorithm such as full-search motion estimation.

8.8   BUFFER SIZE REDUCTION

In this section we present our strategy for buffer size reduction and apply it to our demonstrator algorithm. As stated before, this can be achieved by increasing the synchronization rate and carefully distributing the processing over the synchronization intervals.

8.8.1 Aligning of I/O and processing

We start by identifying the different I/O streams. In this case there are two streams: the input pixels and the results. From Fig. 8.10 one can see that the output stream has the lowest rate. In fact, the input rate is N^2 times higher in this case. Therefore, as a first step, we use the output stream as our base stream and we try to interleave the input stream and the processing with it. In practice this means that we have to try to obtain a loop structure that is compatible (i.e. same number of iterations) with the loop structure of the output section, both for the input section and the processing section. Moreover, we should try to "align" the I/O and the processing direction, because failing to do so would result in unavoidable buffers.

We can achieve these goals by means of a set of global loop transformations [54, 53, 420] on the input and processing sections. For the input section, the original loops are both a factor of N larger than their counterparts in the output section. The sequence of transformations applied to the input section in order to obtain a compatible loop structure is shown in Fig. 8.12. This kind of transformations is also commonly used in the parallel compiler world [304, 431], but for different purposes. Note that we have *not* altered the order in which the pixels are being read, so the global I/O behaviour is not changed.

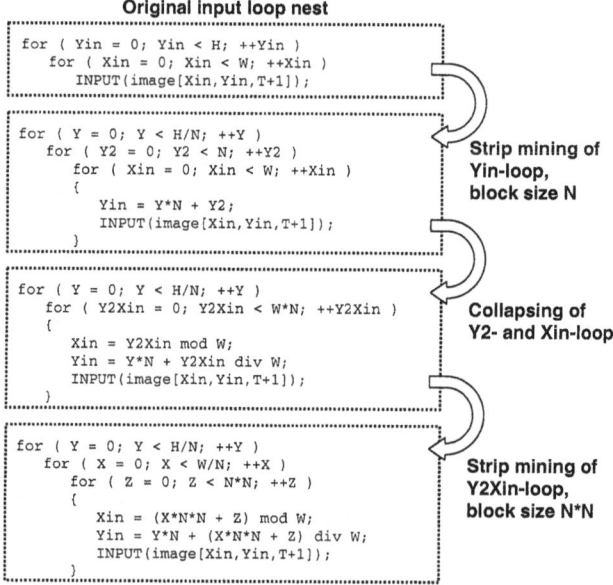

Figure 8.12.   Transformations of the input loop nest.

For the processing section, the problem is much simpler. One can intuitively see that the original column-wise processing order is not compatible with the row-wise I/O order, and that it would result in unavoidable buffering. Therefore, we have to interchange the outer 2 loops in order to align them with the output loop nest. This is shown in Fig. 8.13.

Note that in this case, where the processing section consists of only a single loop nest, the alignment is relatively simple. In general, if the processing section contains several loop nests, the choices can become less straightforward. For instance, one could transform the processing section into a single loop nest with many conditions, or one could transform the input and output loop structures such that they match the processing loop structure. Intermediate solutions are also possible. The choice of the best alternative can be based on code complexity considerations for instance, but also on the target processor. How to deal with this choice effectively in a formal way is a topic of future research.

Original processing loop nest

```
for ( Xproc = 0; Xproc < W/N; ++Xproc )
    for ( Yproc = 0; Yproc < H/N; ++Yproc )
    {
        . . .
    }
```

```
for ( Yproc = 0; Yproc < H/N; ++Yproc )
    for ( Xproc = 0; Xproc < W/N; ++Xproc )
    {
        . . .
    }
```

**Interchange of
Xproc- and
Yproc-loops**

Figure 8.13.   Loop interchange of the processing loop nest.

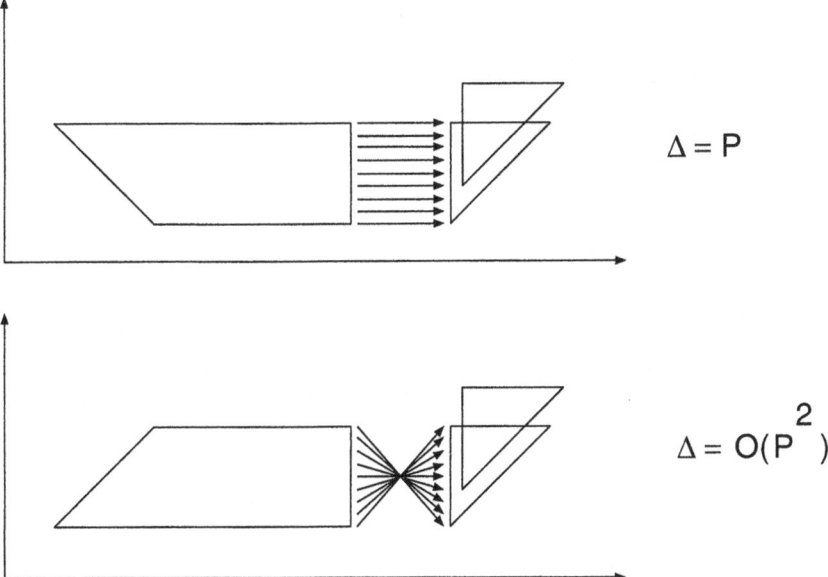

$$\Delta = P$$

$$\Delta = O(P^2)$$

Figure 8.14.   A simple example of removal of dependence crossing to allow more locality of access and hence also better in-place optimisation possibilities.

In addition, cross data dependencies can exist which cause costly intermediate buffers. An example is shown in figure 8.14. These can be locally removed by an appropriate loop reverse, which will however typically move the cross dependency to another node in the data-flow. If it is compensated there by another symmetric cross dependency then they eliminate each other. In general this will not fully happen however. So in order to have a global cost reduction, the effect on all nodes should be considered globally in this substep. A systematic global approach to steer this is a topic of research still.

8.8.2 Bringing I/O and processing closer together

The next step that we have to perform, is to bring the I/O streams and the processing closer together in (relative) time. By doing so, the time that data need to be stored in the buffers is reduced, which is likely to result in smaller buffers.

A graphical representation of how we can do this for the input buffer is shown in Fig. 8.15. In Fig. 8.15a, a snapshot of the original situation is presented, in which we need a buffer size of 3 frames. We can see however that the new pixels enter the buffer a long time (i.e. 1 frame cycle) before they are actually being processed. By delaying the input stream by about 1 frame cycle, we can make sure that the new pixels enter the buffer just before they are needed. This is shown in Fig. 8.15b. Once we have done this, we can see that continuously almost 2/3 of the input buffer does not contain useful data. By organizing the input buffer in a circular way, we can reduce its size to little more than one frame. This is shown in Fig. 8.15c.

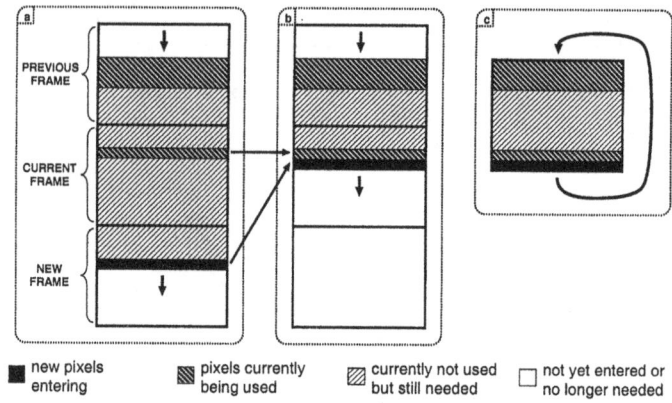

Figure 8.15.   Organization of frame buffer.

For the output buffer, which has typically a much smaller size than the input buffer, we proceed in a similar way. In the original situation, there is a relatively large distance in time between the moment of production of a result and the moment it is sent to the output (also 1 frame cycle), as shown in Fig. 8.16a. By sending the result to the output much earlier, i.e. right after it has been produced, and organizing the buffer in a circular way, we can also drastically reduce its size. This is shown in Fig. 8.16b.

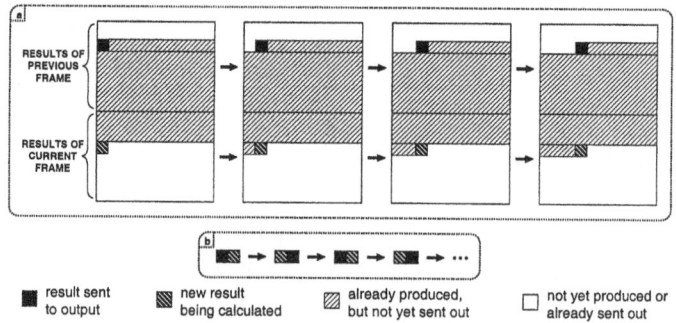

Figure 8.16.   Organization of output buffer.

In order to reflect these modifications in the code, we again need to apply some loop transformations to the input and output sections. In this case, we need to delay the input stream and move the output stream forward in time. This can be done by means of a sequence of loop collapsing, translation, and strip mining transformations. The result of these transformations is shown in Fig. 8.17.

Note that the symbolic index expressions can become relatively complex, containing several div and mod operations. For practical instantiations of the parameters however, many algebraic simpli-

```
for ( Y = 0; Y < H/N; ++Y )
  for ( X = 0; X < W/N; ++X )
    for ( Z = 0; Z < N*N; ++Z )
    {
      Xin = (X*N*N + Z) mod W;
      Yin = ((Y+1) mod (H/N))*N + (X*N*N + Z) div W;   ➪ Next row of blocks
      Tin = (Y+1) div (H/N) + T;                        ➪ Current or next frame
      INPUT(image[Xin,Yin,Tin]);
    }

...

for ( Y = 0; Y < H/N; ++Y )
  for ( X = 0; X < W/N; ++X )
  {
    Xout = (X-1) mod (W/N);                              ➪ Previous block
    Yout = (Y + (X-1) div (W/N)) mod (H/N);             ➪ Current or previous row
    Tout = T + (Y + (X-1) div (W/N)) div (H/N);          ➪ Current or previous frame
    OUTPUT(result[Xout,Yout,Tout]);
  }
```

Figure 8.17.   Time translation of input and output loop nests.

fications can usually be made, resulting in much simpler expressions[1]. In our experiments, we have always been able to convert these expressions to simple incremental pointer arithmetic, which matches the addressing capabilities of most DSPs.

8.8.3 Interleaving

Now that the I/O streams and processing have been aligned and brought closer together in time, we can perform the actual interleaving via another loop transformation: a loop fusion of the outer loops of each section. The result of this is shown in Fig. 8.18.

We have now balanced the processing and the handling of the input stream over the different sample intervals of the output stream, i.e. the lowest rate stream. In the mean time, we have been able to reduce the buffer sizes drastically.

8.8.4 Recursive application

Our next step is to recursively apply the same strategy again, but this time to the next lowest rate I/O stream. In this case, there is only one stream left, namely the input stream. Again, in order to obtain an efficient implementation, we should carefully balance the processing over the input sample intervals and avoid code between loop boundaries.

We only have to concentrate on the inner loops now as the outer loops have already been balanced during the previous steps. We are not able to gain very much anymore concerning the buffer sizes, except for some second order effects. The reason for this is that the outer loops have a much lower execution rate, and consequently their organization has a much higher impact on the buffer sizes. The possible reduction of maybe a few tens of bytes probably does not justify the increased complexity of the algorithm. So we will not try to reduce the buffer sizes any further and only try to balance the input stream and the processing. Again, this can be done by making the non-common outer loops of the input section and the processing section compatible.

In this case, we have applied a strip mining transformation to the input loop and a tiling transformation to the processing loop nest, as shown in Fig. 8.19. Note that the strip mining transformation has not altered the input order, while the tiling transformation has changed the internal processing order. This processing order is not dictated by the external environment, so we have opted for a more regular loop structure.

Now we are ready to interleave the input stream with the processing. This can again be done by means of a loop fusion transformation, as shown in Fig. 8.20.

---

[1]For a systematic approach to achieve this, supported by prototype tools, see [201, 202]

```
for ( T = -infinity; T < +infinity; ++T )
  for ( Y = 0; Y < H/N; ++Y )
    for ( X = 0; X < W/N; ++X )
    {
        for ( Z = 0; Z < N*N; ++Z )
        {
            Xin = (X*N*N + Z) mod W;
            Yin = ((Y+1) mod (H/N))*N + (X*N*N + Z) div W;
            Tin = (Y+1) div (H/N) + T;
            INPUT(image[Xin,Yin,Tin]);
        }
    }
    ||
    {
        Init1();
        for ( i = -M; i < M; ++i )
          for ( j = -M; j < M; ++j )
          {
              Init2();
              for ( k = 0; k < N; ++k )
                for ( l = 0; l < N; ++l )
                {
                    xCB = [X]N + k;
                    yCB = [Y]N + l;
                    READ(image[xCB,yCB,T]);
                    xRW = xCB + i;
                    yRW = yCB + j;
                    if (WithinBoundaries(xRW,yRW))
                        READ(image[xRW,yRW,T-1]);
                    Process();
                }
              PostProcess();
          }
        WRITE(result[X,Y,T]);
    }
    ||
    {
        Xout = (X-1) mod (W/N);
        Yout = (Y + (X-1) div (W/N)) mod (H/N);
        Tout = T + (Y + (X-1) div (W/N)) div (H/N);
        OUTPUT(result[Xout,Yout,Tout]);
    }
```

Figure 8.18.   Fusion of block-traversing loops.

8.8.5 Summary

The result of our strategy is a carefully balanced algorithm implementation, with drastically reduced buffer size requirements. The sizes before and after our optimizations are shown in Table 8.1. Typical values are shown between square brackets. Note that, although the absolute reduction of the size of the output buffer is much smaller, this reduction is still relevant because this buffer is typically stored in the local RAM of the processor, where the cost per bit is larger. This buffer size reduction has significant positive impact on area and power.

Table 8.1.   I/O buffer sizes before and after optimization.

	Before optimization	After optimization
Input buffer	$3(WH)$ pixels [1.2MB]	$W(H + N)$ [0.42MB]
Output buffer	$WH(N^2)$ results [13KB]	2 results [4B]

We can summarize our buffer size reduction strategy as follows:

1. Identify the lowest rate I/O stream.

```
for ( i = 0; i < N; ++i )
  for ( j = 0; j < N; ++j )
  {
    Xin = (X*N*N + i*N+j) mod W;
    Yin = ((Y+1) mod (H/N))*N + (X*N*N + i*N+j) div W;
    Tin = (Y+1) div (H/N) + T;
    INPUT(image[Xin,Yin,Tin]);
  }

  ...

for ( i = 0; i < N; ++i )
  for ( j = 0; j < N; ++j )
    for ( i1 = -M/N; i1 < M/N; ++i1 )
      for ( j1 = -M/N; j1 < M/N; ++j1 )
      {
        Init2();
        for ( k = 0; k < N; ++k )
          for ( l = 0; l < N; ++l )
          {
            xCB = X*N + k;
            yCB = Y*N + l;
            READ(image[xCB,yCB,T]);
            xRW = xCB + i + i1*N;
            yRW = yCB + j + j1*N;
            if (WithinBoundaries(xRW,yRW))
              READ(image[xRW,yRW,T-1]);
            Process();
          }
        PostProcess();
      }

  ...
```

Figure 8.19.   Strip mining of input loop nest and tiling of processing loop nest.

2. Align remaining streams and processing with this stream through loop transformations (typically strip mining, loop collapsing and loop interchanges). Potentially, also cross data dependencies should be removed by means of loop reverse.

3. If possible, shift I/O streams in time to bring the I/O as close as possible to the processing, also via loop transformations (typically loop translation). This step can reduce the sizes of the I/O buffers.

4. Interleave the I/O streams and the processing (typically by means of loop fusions), taking into account the effects of non-perfectly nested loops.

5. Identify the next lowest rate I/O stream and apply the same steps recursively again, without altering the loop structure surrounding the previous I/O streams.

In order to obtain a perfectly synchronous final implementation, we still have to perform some low-level synchronization through the insertion of dummy code as outlined in section 8.7. However, this step is processor-dependent. Moreover, the additional power reduction techniques presented in the next chapters, partly also alter the loop structure. Therefore the low-level synchronization step has to be executed as the last step in the mapping process.

8.9 CONCLUSIONS

In this chapter, we have shown that we can significantly improve the power and area cost in data-dominant applications by applying data-flow transformations. These operate at the system-level to reduce the multi-dimensional data transfers and storage. We have applied this systematic methodology on complex realistic test-vehicles in the video processing domain. Our main demonstrator has been the H.263 teleconferencing video decoding standard. We believe that these results clearly substantiate the validity of the proposed methodology for data-dominated applications. They also show the very promising results on power reduction that can be obtained by system level exploration.

Next, we have outlined a global loop transformation based buffer size and power consumption reduction strategy which is able to deal with stringent I/O constraints and which can be applied to a

```
for ( T = -infinity; T < +infinity; ++T )
  for ( Y = 0; Y < H/N; ++Y )
    for ( X = 0; X < W/N; ++X )
    {
        Init1();
        for ( i = 0; i < N; ++i )
          for ( j = 0; j < N; ++j )
          {
              Xin = (X*N*N + i*N+j) mod W;
              Yin = ((Y+1) mod (H/N))*N + (X*N*N + i*N+j) div W;
              Tin = (Y+1) div (H/N) + T;
              INPUT(image[Xin,Yin,Tin]);

              for ( i1 = -M/N; i1 < M/N; ++i1 )
                for ( j1 = -M/N; j1 < M/N; ++j1 )
                {
                    Init2();
                    for ( k = 0; k < N; ++k )
                      for ( l = 0; l < N; ++l )
                      {
                          xCB = X*N + k;
                          yCB = Y*N + l;
                          READ(image[xCB,yCB,T]);
                          xRW = xCB + i + i1*N;
                          yRW = yCB + j + j1*N;
                          if (WithinBoundaries(xRW,yRW))
                              READ(image[xRW,yRW,T-1]);
                          Process();
                      }
                    PostProcess();
                }
          }
        WRITE(result[X,Y,T]);
        Xout = (X-1) mod (W/N);
        Yout = (Y + (X-1) div (W/N)) mod (H/N);
        Tout = T + (Y + (X-1) div (W/N)) div (H/N);
        OUTPUT(result[Xout,Yout,Tout]);
    }
```

Figure 8.20.   Fusion of i and j loops.

broad class of signal processing applications. It is even possible to (partly) take into account limitations on the memory architecture, such as a limited cache size. The feasibility of this approach has been demonstrated by applying it to a class of video coding algorithms that are among the most memory intensive ones, and are frequently encountered in a multimedia context.

9 MEMORY HIERARCHY AND DATA REUSE DECISION EXPLORATION

Efficient use of an optimized memory hierarchy to exploit temporal locality in the data accesses can have a very large impact on the power consumption in data dominated applications. Effective formalized techniques to deal with this specific task have not been addressed up to now. In this chapter, the design freedom available for the basic problem is explored in-depth and the outline of a systematic solution methodology is proposed. The efficiency of the methodology is illustrated on a real-life motion estimation application. The results obtained for this application show power reductions of about 85% for the memory sub-system compared to the case without memory hierarchy.

9.1 INTRODUCTION

A large part of the power dissipation in data dominated applications is due to data transfers and data storage. This power component can often be reduced by introducing an optimized custom memory hierarchy that exploits the temporal locality in data accesses in the application. The impact of this can be very large, as has been demonstrated by us on an H.263 video decoder [291] and a motion estimation application [437].

The idea of using a custom memory hierarchy to minimize the power consumption is based on the fact that memory power consumption depends primarily on the access frequency and the size of the memory. Power savings can be obtained by accessing heavily used data from smaller memories instead of from large background memories. Such an optimization requires architectural transformations that consist of adding layers of smaller and smaller memories to which frequently used data can be copied [435]. Memory hierarchy optimization introduces copies of data from larger to smaller memories in the data flow graph. This means that there is a trade-off involved here: on the one hand, power consumption is decreased because data is now read mostly from smaller memories, while on the other hand, power consumption is increased because extra memory transfers are introduced. Moreover, adding another layer of hierarchy can also have a negative effect on the area and interconnect cost, and as a consequence also on the power because of the larger capacitances involved. The data reuse decision step has to find the best solution for this trade-off. Note that a memory level related decision is performed in this step too. The actual allocation of memories within each level is however only performed during the memory allocation and assignment step, after the bandwidth and timing constraints have been analyzed and optimized during the intermediate SCBD step.

Memory hierarchy design for power optimization is basically different from caching for performance optimization [130, 220, 254, 204]. The latter determines how to fill the cache such that data has been loaded from main memory before it is needed. Instead of minimizing the number of transfers, the number of transfers is often increased to maximize the chance of a cache hit, leading to wasted power by prefetching data that may never be needed.

As reviewed in chapter 2, none of the earlier memory hierarchy related approaches determine the best memory hierarchy organization for a given (set of) applications and only few address the power cost. In this chapter, we present a formalized methodology for this decision and give an indication of how large the search space really is. This search space is much larger than conventionally exploited in state-of-the-art designs.

The rest of the chapter is organized as follows. Section 9.2 defines the memory hierarchy problem. Section 9.3 defines the data reuse problem and points out important issues towards a methodology. In section 9.4 we propose a methodology for solving the data reuse exploration and decision problem. Section 9.5 discusses the results of the data reuse exploration experiment for the motion estimation application. Section 9.6 briefly discusses the memory hierarchy level assignment substep. Section 9.7 concludes the chapter.

9.2 MEMORY HIERARCHY DESIGN

This section defines the memory hierarchy design task. First, it shows that the main issue is exploiting temporal locality. Then, it shows that memory hierarchy design consists of two steps: data reuse decision, which is the topic of this chapter, and memory hierarchy layer assignment, which is a substep in SCBD, and which is largely still a topic of future research.

9.2.1 Exploiting temporal locality

Memory hierarchy design exploits data reuse local in time to save power by copying data that is reused often in a short time period to a smaller memory, from which the data can then be accessed. Figure 9.1 illustrates this for all read operations to a given array.

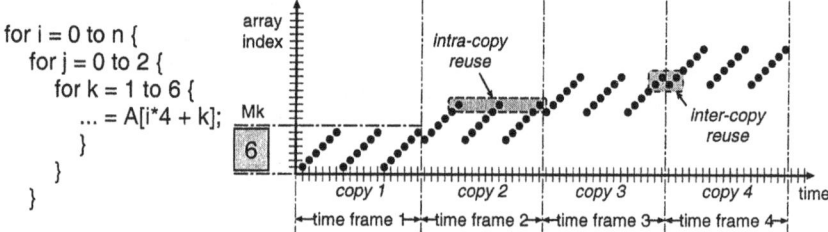

Figure 9.1.   Exploiting data reuse local in time to save power.

The horizontal axis is the time axis. It shows how the data accesses are ordered relatively to each other in time. The vertical axis shows the index of the data element (scalar) that is accessed from the array. Every dot represents a memory read operation, scheduled at a certain time and accessing a given data element. It can be seen that, in this example, most values are read multiple times. Assuming that the data is still needed later on, all data values of the array have to be stored in a large background memory. However, when we look at smaller time-frames (indicated by the vertical dashed lines), we see that only part of the data is needed in each time-frame, so this part of the data would fit in a smaller, less power consuming memory. If there is sufficient reuse of the data in that time-frame, it can be advantageous to copy the data that is used frequently in this time-frame to a smaller memory. Consequently, the second time a data element has to be read, it can be read from the smaller memory instead of the larger memory.

Definition: *time-frame*

The execution time of an application can be subdivided into a number of non-overlapping time-intervals, called *level 1 time-frames*. Each level i time-frame can be subdivided further into non-overlapping level $i + 1$ time-frames.

Definition: *copy-candidate*

A copy-candidate corresponding to an array A and time-frame TF_i is a set of data elements of A that are read in TF_i.

A copy-candidate is a group of data that is a potential candidate to be copied to a smaller memory at a lower level in the hierarchy. During the data reuse task it will be decided which copy-candidates are really worth to be copied in order to save power. Once this decision is made, the transfers for making the copies will be added to the application code and the copy-candidates become real (partial) copies of their corresponding arrays.

Data reuse is the result of **intra**-*copy reuse* and **inter**-*copy reuse* (see figure 9.1). *Intra-Copy* reuse means that each data element is read several times from its memory during one time-frame. *Inter-Copy* reuse means that advantage is taken from the fact that part of the data needed in the next time-frame could already be available in the memory from the previous time-frame, and therefore does not have to be copied again.

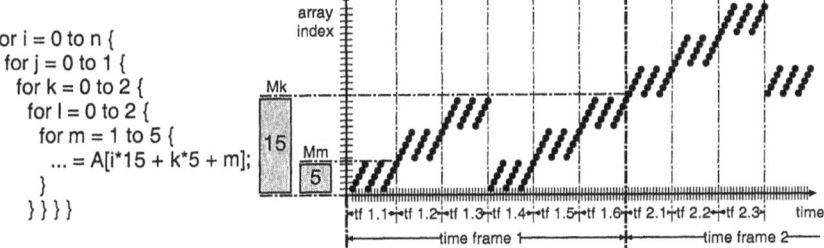

```
for i = 0 to n {
  for j = 0 to 1 {
    for k = 0 to 2 {
      for l = 0 to 2 {
        for m = 1 to 5 {
          ... = A[i*15 + k*5 + m];
        }
}}}}
```

Figure 9.2.   Possibility for multi-level hierarchy.

Taking full advantage of temporal locality for power, usually requires architectural transformations that consist of adding several layers of memories (each corresponding to its own time-frame level) between the large background memories and the small foreground memories (registers in the data-path). Every layer in the memory hierarchy contains smaller memories than the ones used in the layer above it. An example of this is shown in figure 9.2. It shows time-frames that are subdivided into smaller time-frames. Each level of time-frames potentially corresponds to a memory layer in the memory hierarchy.

9.2.2 Substeps in memory hierarchy design

Two major substeps can be identified in memory hierarchy design:

1. *Data reuse decision and exploration.*

   The data reuse decision and exploration step decides *which* intermediate copies have to made for accessing the data in a power efficient way. Possibly the same data has to be copied several times to get the optimal result. Figure 9.2 shows some examples of this. For instance, the data copied in time frame 1.1 and 1.4 is exactly the same, but has to be copied twice to keep memory Mm smaller than memory Mk.

2. *Memory hierarchy layer assignment.*

   The memory layer assignment step decides for each array and copy of an array on *which layer* in the common custom memory hierarchy it will be stored.

After memory layer assignment, an optimal memory architecture has to be derived for each layer. This is done by the subsequent memory allocation and array-to-memory assignment tasks in our DTSE methodology [58, 290]. These tasks are described in [376].

This chapter mainly focuses on the data reuse decision step. The memory layer assignment step is only discussed in section 9.6.

9.3 DATA REUSE DECISION

This section discusses the data reuse exploration and decision step of memory hierarchy design. It points out important elements for a systematic methodology described in the next section.

9.3.1 Search space

The following two assumptions allow to focus the data reuse exploration task, without really restricting the search space in practice:

Only read operations have to be considered   Data reuse exploration and decision has to focus on read operations only. The reason being that repeated reading of the same data *value* makes sense, whereas writing *the same* data *value* usually does not make sense. So there is no need for creating a memory hierarchy for repeatedly written data. The only thing that has to be decided for write operations is in which layer a certain (temporary) array will be written. This is decided in the memory layer assignment step *after* data reuse decisions are made.

Only one array has to be considered at a time   Data reuse *exploration* can be tackled for each array separately. The main reason for this is that each copy in the memory hierarchy has its own root array, i.e., the copies form a tree where the root is the original array that is being accessed. A copy cannot be obtained as the mix of different arrays. We will also assume that the data reuse *decision* can be taken for each array separately. As explained later on, this limits the search space to some extent.

9.3.2 Data reuse factor

The usefulness of memory hierarchy for saving power is strongly related to the array's data reusability, because this is what determines the ratio between the number of read operations from a copy of an array in a smaller memory, and the number of read operations from the array in the larger memory on the next hierarchical level.

The reuse factordata reuse factor of a group of data D stored in a memory on layer i relative to layer $i - 1$, where layer 0 is the memory furthest from the data paths, is defined as:

$$F_R(i, D) = \frac{N_R(M_i, D)}{N_R(M_{i-1}, D)} \tag{9.1}$$

where $N_R(M_i, D)$ is the total number of times an element from D is read from a memory at layer i.

A reuse factor larger than 1 is the result of **intra**-*copy reuse* and **inter**-*copy reuse* (see figure 9.1).

If $F_R(1, D) = 1$, a memory hierarchy is useless and would even lead to an increase of area and power, because the number of read operations from M_0 would remain unchanged while the data also has to be stored and read from M_1.

9.3.3 Classification of data reuse

Figure 9.3 presents a classification of four cases in which data reuse can be exploited by means of memory hierarchy. These four cases are:

(a) *No loops*

In case there are no loops, there is no structured use of arrays. In fact, each array element is treated independently from the others similar to scalars. Depending on the temporal locality, it may be useful to copy a scalar to a foreground register and keep it there, such that it can be reread from the register a number of times. In principle multiple time-frames have to be considered in order to limit the amount of foreground registers. In other words, a scalar can be copied several times from the background memory to a register, from where it can be read multiple times before it is overwritten by another scalar. Only intra-copy reuse is possible here, because at the end of each time-frame,

```
                                                                    for ...
                                                                      for ...
{                                           for ...                    {
  ... = f1(A);        for ...                 for ...                   ... = f1(A[][]);
                        for ...               {                        }
  ...                   {                       ... = f1(A[][]);
  ... = f2(A);            ... = f1(A[][]);      ...                   ...
                        }                       ... = f2(A[][]);      for ...
  ...                                         }                        for ...
  ... = f3(A);                                                         {
}                                                                        ... = f2(A[][]);
                                                                       }

      (a)                   (b)                    (c)                    (d)
```

Figure 9.3.   Classification of data reuse opportunities.

the register will be overwritten by another scalar, such that inter-copy reuse is not possible. This case is *not* considered in our methodology. It is left for scalar methodologies which are more suited for this.

(b) *One read instruction in a (nested) loop*

This is the basic case on which our methodology is based. Both intra-copy and inter-copy reuse are possible here, when the loop nesting (i.e., the order of the different nested loops, and the direction in which the loops are traversed) is fixed. Indeed, in this case, the ordering of the different copies is known and every two consecutive copies can be examined for overlap (i.e., inter-copy reuse).

(c) *Multiple read instructions in a (nested) loop*

Here it is assumed that each read instruction has a different index expression, because otherwise they can be reduced to the previous case by reading once from background memory and storing the result in a temporary foreground register. When the different read instructions are accessing different parts of the array, a different memory hierarchy can be constructed for each of them. However, in practice, these memory hierarchies will contain partly the same data, and are therefore best combined. Determining which part of the memory hierarchy can be shared, can be done with a geometrical data flow analysis.

(d) *Read instructions in different loop nests*

In this case, a memory hierarchy can be derived for each of the loop nests separately. These memory hierarchies can be very different from each other. Depending on the temporal locality (which is only known when the ordering of all loop nests is already fixed), it may be useful to copy the data that is common to these loops to an extra memory layer. Again a geometrical analysis can be used to determine which part of the array is used in common. Remark that when more than two loops are involved, part of the array can be common to only a few of the loops involved, making things much more complex. Also here, as in case (a), multiple time-frames have to be considered: the data can be copied several times to a lower memory layer. In this case, there is however even more freedom than in case (a), because here the data to be copied to the lower memory layer can be different from time-frame to time-frame.

9.4 DATA REUSE METHODOLOGY

In this section, we propose a methodology for data reuse exploration and decision based on a number of assumptions to make the solution feasible for real-life applications.

9.4.1 Assumptions

The following is assumed in our methodology:

- *Single assignment code*

  Single assignment code, where every array value can only be written once but read several times, makes the analysis much easier. It is not fundamentally required, though.

- *Order and direction of loop iterators of nested loops is fixed*

  The fixed nesting order is required to determine the time-frames and copy-candidates corresponding to each read instruction in the loop nest. The fixed iteration direction is required to determine the overlap for estimating the *inter-copy* reuse. This assumption is valid for the position of this step in our global methodology [58].

- *Time-frames are determined by loop boundaries*

  Finding an optimal time-frame hierarchy is a very complex problem. However, we believe that the optimal time-frame boundaries are likely to coincide with the loop boundaries of loop nests. Therefore, we use as a heuristic that the loop boundaries correspond to time-frame boundaries, instead of trying to find globally optimal time-frame boundaries.

- *A copy-candidate contains all data accessed in its time-frame*

  It is assumed that *all* data being accessed by a certain read instruction in a certain time-frame will be copied to a copy-candidate, such that all data required by the read instruction can be found inside the copy-candidate. Ideally, only part of the data that is accessed more than once should be copied. The loss due to this restriction is small in practice.

- *All copy-candidates of a time-frame level are stored in-place*

  It is assumed that at the end of a time-frame, the data copied into the intermediate memory is not needed anymore, and will therefore be overwritten by the data needed in the next time-frame. Therefore the size of the copy-candidate corresponding to a certain time-frame level is determined by the time-frame leading to the largest copy-candidate.

- *Copy-candidates will be stored in perfectly fitting memories*

  It is assumed that a copy-candidate will be stored in a memory with word depth and word-length equal to those of the copy-candidate. In general, this assumption leads to an underestimation of the power cost, as usually several copy-candidates will have to share a memory. This real memory size is however only known until after array-to-memory assignment and can therefore not be taken into account.

- *Inter-copy reuse is fully exploited*

  It is assumed that inter-copy reuse will be fully exploited. This means that data that is already present in the smaller memory from the previous copy will not be copied again. Only the data that is not already present has to be copied over the data of the previous copy that is not needed anymore. This affects the number of accesses to each of the copy-candidates. The size of the copy-candidates is unaffected.

9.4.2 Data reuse decision exploration

Based on the assumptions listed in the previous subsection, we propose a systematic data reuse exploration methodology. Figure 9.4 illustrates the different steps for a small example.

Copy-candidate chains   For each read instruction inside a loop nest, a *copy-candidate chain* can be determined in the following way. The first copy-candidate in the chain contains *all* data elements accessed by the given read instruction during the execution of the loop nest. The second copy-candidate in the chain is associated with the iterations of the outer loop. Each iteration has a corresponding time frame. All data elements that are accessed by the read instruction during a given time frame are

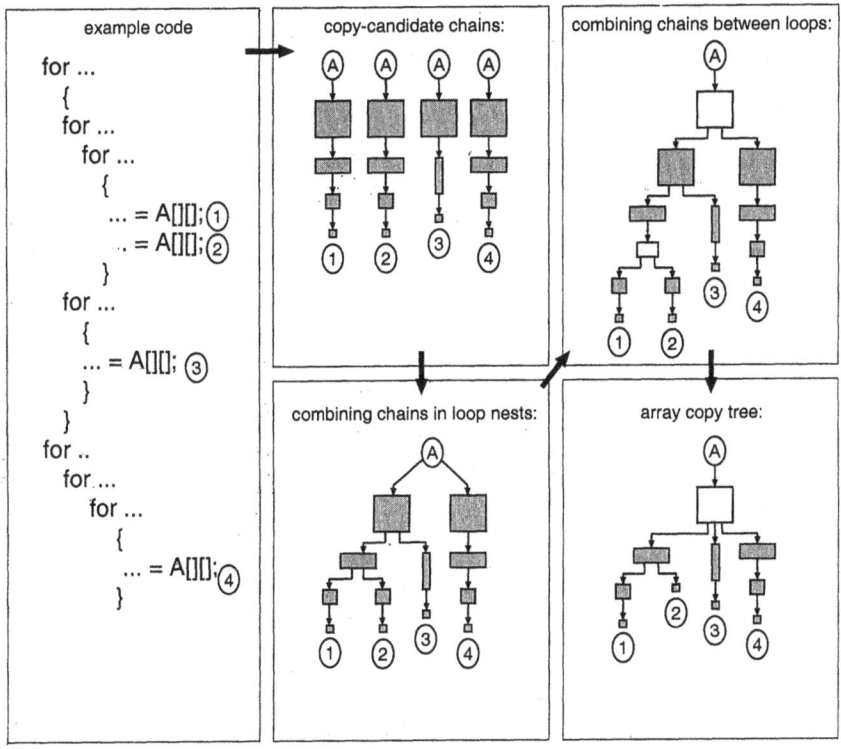

Figure 9.4.   Data reuse exploration methodology: example.

stored in the corresponding copy-candidate. The storage space of the copy-candidate is shared among the different iterations of the loop. In case the number of elements accessed by the read instruction varies between iterations, the size of the copy-candidate is determined by the iteration that accesses the most data elements. This can be repeated for each of the remaining loop nest levels. Together these copy-candidates form the *copy-candidate chain* for the considered read instruction. Such a chain represents the maximal exploitation of data reuse of type (b) in figure 9.3.

Copy-candidate trees   Copy-candidate chains of different read instructions accessing the same array can be combined into a *copy-candidate tree* for that array:

1. *Read instructions belonging to the same loop nest*

   Often, the copy-candidates of read instructions belonging to the same loop nest contain the same data. If this is the case for all iterations, the copy-candidate can be shared between the read instructions. Moreover, if a copy-candidate can be shared, also the copy-candidates before it in the chain can be shared. This, combined with the fact that the first copy-candidate in a chain, which is the array itself, can always be shared between read instructions of the same loop nest, means that the copy-candidate chains of a loop nest can always be combined into a *copy-candidate tree*. Such a tree represents the maximal exploitation of data reuse of type (b) and (c) in figure 9.3.

2. *Read instructions belonging to different loop nests*

   Copy-candidates that cannot be combined in the previous way because they belong to different loop nests (see figure 9.3(d)), can still share the same data. In this case, an extra intermediate copy-candidate can be inserted in the copy-candidate tree to exploit this data reuse opportunity. When more than two copy-candidates share data in this way, the search space grows quickly: for every possible subset of them an extra copy-candidate could be introduced. Which one(s) should be selected? Ideally, only data that would be read several times from the copy-candidate should be copied into it. So, if more than two copy-candidates are partly overlapping, which common data elements should be copied into the intermediate copy-candidate? Because of this significant extra complexity, this freedom will be explored in future research and will not be considered further in this chapter. So we will treat the read instructions as operating on independent arrays.

Two simple rules can be applied to prune copy-candidates from copy-candidate trees because they will never occur in an optimal memory hierarchy: the size of the copy-candidates must decrease from one layer to the next, and the reuse ratio (Eq. 9.1) of each layer must be larger than one.

Copy-candidate graphs From the previous step we can conclude that for every array, a copy-candidate tree can be determined. Each node in the tree represents a copy-candidate. A copy-candidate can be characterized by its required memory size, number of write operations to copy data into it, and the total number of read operations to copies on lower layers in the copy tree. Also its data reuse-factor can be calculated. From this tree other valid copy trees can be derived because it is allowed to copy data from any ancestor node in the tree, not only the parent node. Therefore, we propose to extend the copy-candidate tree to a *copy-candidate graph* in the following way: for every node in the tree, we add edges starting from all its ancestor nodes towards the node itself. All possible trees that can be derived by selecting a single path from the root node to every leave node, represents a valid copy tree. The *array copy tree* is the tree obtained this way with lowest cost. Selecting the array copy tree from all possible copy trees is called data reuse *decision* (see subsection 9.4.4). The cost function we are using for selecting the array copy tree is defined in the next subsection.

9.4.3 Cost function

The cost function for selecting the optimal copy tree is a weighted sum of a power and area estimate for the copy tree CT. The cost function is given by:

$$
\begin{aligned}
cost(CT) = \ \alpha \cdot \ & \sum_{c \in CT} \ [P_r(N_{bits}(c), N_{words}(c), f_{read}(c)) \\
& \qquad + P_w(N_{bits}(c), N_{words}(c), f_{write}(c))] \\
+ \ \beta \cdot \ & \sum_{c \in CT} \ A(N_{bits}(c), N_{word}(c))
\end{aligned}
\tag{9.2}
$$

where

- c is a copy-candidate of the considered copy tree CT,

- $P_{r/w}(N_{bits}, N_{words}, f_{access})$ is the power estimate for read/write operations of a memory with word-length N_{bits}, word depth N_{words}, and that is accessed with a *real* access frequency f_{access},

- $A(N_{bits}, N_{words})$ is the area estimate for a memory with specified parameters, *and*

- α and β are weighting factors for area/power trade-offs.

The *real* access frequency f_{access} of a memory is obtained by multiplying the number of memory accesses per frame with the frame rate of the application (**not** the clock frequency).

9.4.4 Data reuse decision

The end result of the data reuse decision task is an optimal *array copy tree* for each array in the application. Here, it is assumed that the optimal copy tree can be derived for each array independently from the other arrays. This is not completely true because the optimal copy trees depend partly on how well the different copies can share memory space (inter-array in-place [108]). However, because the data reuse decision is best taken early in the design flow, not enough data about inter-array in-place is available to include it in the decision process.

9.5   IMPACT ON AREA AND POWER: DEMONSTRATOR RESULTS

The motion estimation (ME) kernel described in section 8.6 has been analyzed with parameters from the QCIF format (W=176, H=144, m=n=8) with a frame rate of 30 frames/s. The area and power values of the memories are obtained with the proprietary model of Texas Instruments discussed in section 4.4.2.2. Therefore only relative values are provided. First, we illustrate the different steps on the motion estimation application. Then, we discuss the results of applying the methodology on this application.

9.5.1   Data reuse decision exploration for ME application

Copy-candidate chains. For the motion estimation application only the frame arrays *Old (O)* and *New (N)* will be considered here, because the other arrays can easily be stored in a foreground register. Figure 9.5 shows the copy-candidate chains for the two read instructions accessing array O and N respectively. Copy-candidates N_4 and N_5 can be pruned because they are not smaller than N_3.

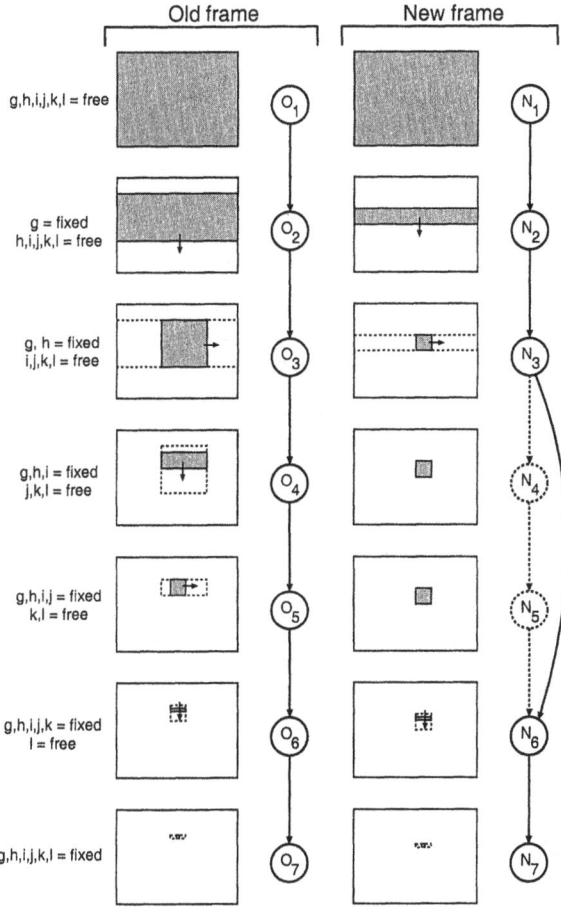

Figure 9.5.   Copy-candidate chains for ME application.

Copy-candidate trees. Since there is only one read instruction from the *Old* frame and one from the *New* frame, there is nothing to combine in this example: the copy-candidate chains for the frame arrays *are* the copy-candidate trees. The nodes O_6 and N_6 can now be pruned, though, because they have a reuse ratio of one (i.e., no reuse).

Copy-candidate graphs. The copy-candidate graphs for the two frame arrays are shown in figure 9.6.

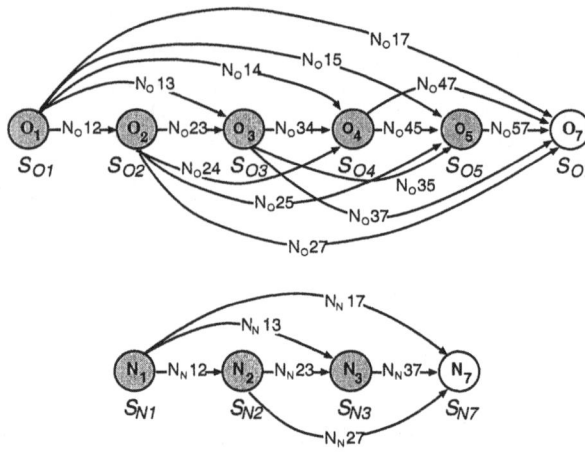

Figure 9.6.   Copy-candidate graphs for ME application.

Array copy trees. In this case the copy-candidate trees are in fact chains.[1] Because of the chains, the search process for finding the array copy tree can be represented as a simple search tree. Figure 9.7 shows these search trees for our motion estimation example. The (optimal) array copy trees are indicated in grey. The dashed lines divide the search trees in a number of layers. Each layer corresponds to a copy-candidate (or time-frame level). The solutions can either include this copy-candidate (copy is shown in search tree) or skip it (copy is not shown). Every node in the tree represents a solution: the copy-candidate on that node is the lowest layer in the hierarchy, and all copy-candidates on the path between the root node and the node itself are intermediate layers in the hierarchy. For each solution, the area (A) and power (P) of the complete array copy tree for that solution relative to the solution without hierarchy (root node) are indicated.

9.5.2 Discussion

A surprising result is that the total memory area can *decrease* by adding extra memory layers (see figure 9.7). The reason for this is that the maximum access frequency of the memories is taken into account in our estimations. If a certain memory would be accessed above its maximum access frequency, this memory will be split into two memories of half the size to increase the memory bandwidth. This splitting introduces overhead. By adding extra memory layers with small memories, the bandwidth requirements of the large background memories can be reduced, and therefore splitting can be avoided for the large memories. This area gain can be larger than the area lost by adding a few small memories.

The optimal memory hierarchy for power is :

- for the *Old* frame, a 3-level hierarchy that leads to a power saving of 83% compared to the solution without memory hierarchy;

---

[1]In general this is not true as demonstrated by us before for the H.263 video conferencing decoder test vehicle in [291].

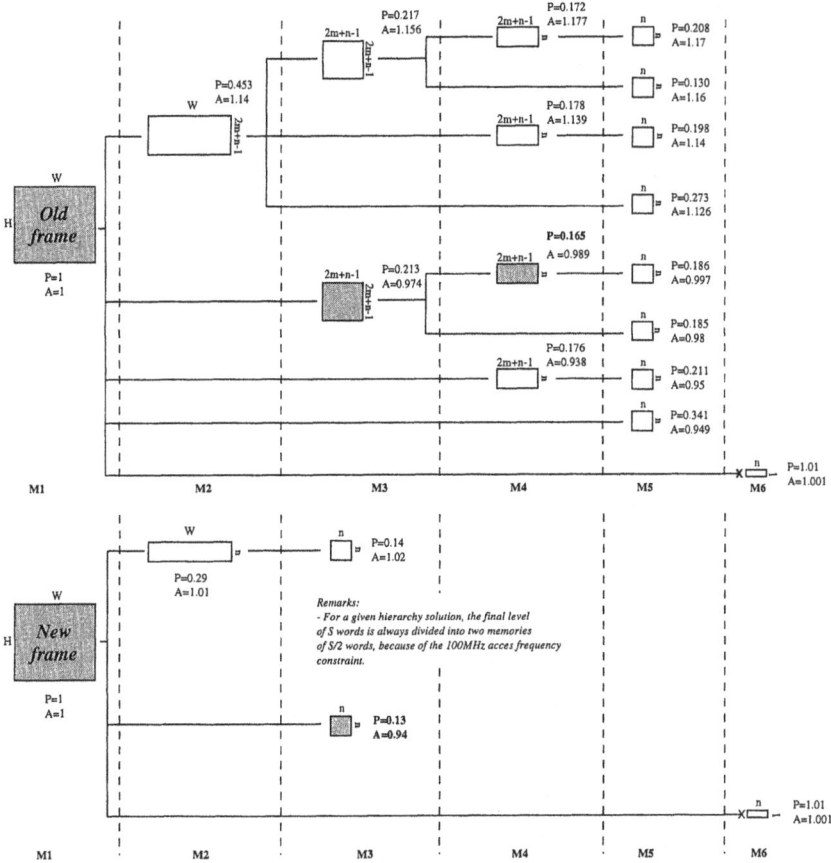

Figure 9.7.   Search trees for data reuse decision.

- for the *New* frame, a 2-level hierarchy that leads to a power saving of 87% compared to the solution without memory hierarchy.

These figures do not include the power dissipation in the interconnect. Taking this into account will result in even larger gains, as off-chip communication dissipates much more power than on-chip communication (figure 9.8). Without memory hierarchy all data transfers are off-chip. With memory hierarchy, most of these are replaced by less power consuming on-chip transfers.

If we compare the result with the one we proposed in an ad hoc way in an earlier paper [437], we note that there a different memory hierarchy with only two levels was selected which results in a higher power consumption. This clearly shows that by using a more systematic design space exploration methodology which exploits the full search space available, as proposed in this chapter, better results can be obtained.

The large power gains obtainable by introducing an optimized memory hierarchy justify that the memory hierarchy should be decided early in the design flow.

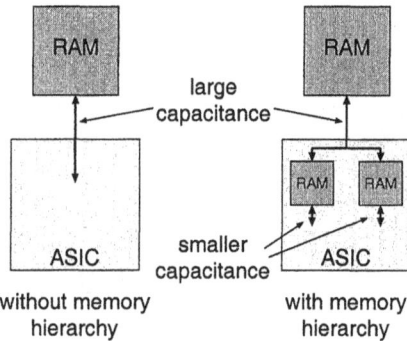

Figure 9.8.   Effect of memory hierarchy on interconnect power.

9.6   GLOBAL MEMORY HIERARCHY LEVEL ASSIGNMENT

As noted in section 9.2, a second substep is necessary to define the full memory hierarchy. We need to assign levels to the different sub-memories carried out for each signal hierarchy.

9.6.1   Relative position in methodology

This assignment is necessary in order to accurately estimate potential conflict accesses during the SCBD step (see chapter 3). So from that view point, it should precede SCBD. On the other hand, in order to make an global decision on the memory hierarchy level assignment (MHLA), accurate information on in-place cost estimation is required and that can only be provided after the loop organisation is fully fixed. Given that during SCBD this loop organisation can change quite a bit still, the order should be the reverse. So, two different ways are possible. We can define a hierarchy organization and then, under the cycle budget and other timing constraints, derive a partial access order to minimize the conflict accesses and to increase the inter-signal data reuse. Or, on the other hand, one can decide a partial order and then build a hierarchy that minimizes the conflicts. We have choosen for the first solution because this is the most promising scheme. Indeed, the potential of conflict reduction is mainly provided by the partial order, compared to level-assignment decisions. The second solution would also lead to situations where the actual in-place would be inefficient because of size mismatches of signals assigned to the same level in the memory hierarchy.

However, in practice, SCBD involves several substeps, some of which are not clearly formalized yet (see chapter 10). Only the last substep requires the full knowledge about the MHLA substep. So in order to partly solve the dilemma, MHLA should be only done just before the last SCBD substep, namely storage-bandwidth optimisation, where the loop organisation is largely fixed already. In that way, the decisions of MHLA which are based on size matching metrics can use quite accurate high-level in-place estimations.

It should also be noted that when several signals are involved, the term memory hierarchy is not so clear-cut any more because it could be that a signal A is assigned to a level 3 memory, but that at the same time another signal B for which the access is not decomposed into intermediate transfers, is assigned also to this level 3 memory (making in a sense a level 1 memory). Therefore, in general it is better to talk about memory partitions in the hierarchy, each with a unique label (e.g. P1, P2, P3 for the hierarchy in figure 3.2).

9.6.2   Illustrative example

Consider, for instance two signals A and B with their own ad hoc hierarchy, defined in figure 9.9 in a parallel loop context. We assume that the one-branch memory hierarchy proposed for A is derived from the operand spaces of loops 1-2-6 (M1a,M2a,M6a). The indices of signal B only depend on the two first loops. The memory hierarchy of B is M1b-M2b where M1b is a little larger than M1a and M2b is smaller than M2a. So, we first have to choose one level-assignment among three possible candidate solutions: *Assign-1,2* or *3*.

Based on the size-matching heuristic, *Assign-3* will be selected as the best solution. In this example, we assume that the inter-signal in-place possibilities are well improved by the partial ordering step, further enabling the merging of the complete hierarchy.

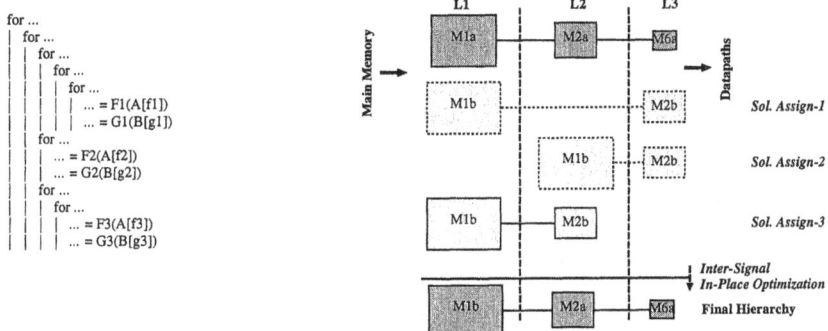

Figure 9.9.   Memory hierarchy level assignment: illustrative example.

9.6.3   How can the result be embedded in the code?

An important task after this substep is to perform transformations which introduce extra transfers between the different memory partitions of a signal which are situated at different levels and which are mainly reducing the power cost. In particular, these involve adding temporary values — to be assigned to a "lower level" — wherever a signal in a "higher" level is read more than once. The more costly read to the "higher" level is then done only once, and the duplicate reads are performed on the lower-level temporary signal. The same can happen in the other direction for writes. If a signal assigned to a higher level is composed of several contributions (e.g. a 32-bit word composed of 4 bytes), it does not make sense to update the final result directly in the higher level memory. Instead, it is usually better to perform the composition from the contributions consecutively (or in a close ordering) in a lower level (or several levels in more complex situations) and then directly transfer the final result to the higher level.

9.7   CONCLUSIONS

Exploiting temporal locality in the memory accesses by means of an optimized memory hierarchy can effectively reduce the power dissipation of data-dominated applications. The memory hierarchy design task can be split into two subtasks: data reuse decision and memory hierarchy layer assignment.

A systematic methodology for the data reuse decision step has been proposed based on a number of realistic assumptions. The feasibility and the large impact of the proposed techniques have been shown on a real-life video application. The results obtained for the motion estimation application show power reductions of about 85% for the memory sub-system compared to the case without memory hierarchy. Similar results have been obtained for other applications not presented in this chapter. These figures do not include the power dissipation in the interconnect which is neglected (see section 4.4). Taking this into account would result in even larger gains however. These large power gains justify that decisions related to the memory hierarchy and especially about the data reuse chains should be decided early in the overall DTSE flow.

10 STORAGE CYCLE BUDGET DISTRIBUTION

In our design methodology, the memory architecture is optimized as a first step before doing the detailed scheduling, and data-path and controller synthesis. Therefore, it has to provide sufficient storage bandwidth (parallel memory ports) such that the application can be scheduled within the cycle budget afterwards. Determining optimal constraints for the subsequent memory allocation and assignment tasks such that the cycle budget can still be met, is the task of storage-bandwidth optimization. This chapter explains what Storage Cycle Budget Distribution is about, with a main focus on the storage bandwidth optimization step. It illustrates its effect and importance on an example. It shows that it is important to take into account which data is being accessed in parallel, instead of only considering the number of simultaneous data accesses. This leads to a problem formulation in terms of the optimization of a conflict graph, for which a cost function is derived at the end of the chapter. The last sections explain how the storage-bandwidth optimization task can be implemented, especially for flat flow graphs as found mainly in network component applications. Extensions needed for hierarchical flow graphs found in RMSP applications are discussed only very briefly.

10.1 INTRODUCTION

As motivated in chapter 1, we believe that the organization of the global communication and data storage are the dominating factors in system level design for many applications and should be defined prior to the data-path and control synthesis stages.. However, this means that the memory architecture has to be defined *before* doing the detailed scheduling. Hence, it has to provide sufficient memory bandwidth (parallel memory ports) such that the application can be scheduled within the cycle budget afterwards. That is where the Storage Cycle Budget Distribution (SCBD) step, with as main substep Storage-Bandwidth Optimization (SBO), comes into play. This is illustrated in figure 10.1. Given a (maximally parallel) control/data-flow graph (CDFG) representing the behaviour of the application to be implemented and the cycle budget in which the application has to be scheduled, it determines for which data parallel access capabilities should be provided such that the cycle budget can still be met with minimum bandwidth requirements on the memory architecture. These requirements are expressed as conflicts in a conflict graph. If the subsequent memory allocation and assignment tasks [27, 28, 245, 339] find a memory architecture that satisfies all conflicts in this graph, it is guaranteed that there exists a valid schedule that meets the cycle budget. The task of SBO is to come up with an optimized conflict graph, allowing the memory allocation and assignment tasks to come up with a cheaper memory architecture with fewer memories and ports. This imposes a partial ordering on the accesses. For hierarchical loop nest based specifications, also the cycle budget distribution over the loop hierarchy is provided as a result. In the context of our DTSE script, a number of ordering constraints have also been added on the input due to preceding steps. Moreover, data reuse tree information has been derived during the data reuse decision step.

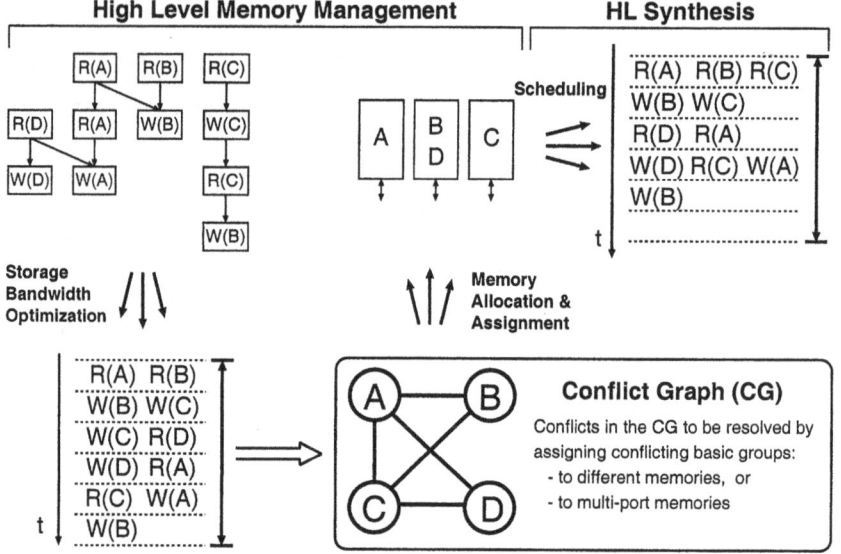

Figure 10.1.   Storage-Bandwidth Optimization in DTSE.

10.1.1 Detailed target architecture

A further detailed instantiation of the target architecture is shown in figure 10.2 as illustration. It consists of four parts: a hierarchical distributed memory architecture, the data paths, a global controller, and address generators for the memories.

The system consists of a set of *processing units* interacting with a shared, distributed *background memory architecture*. Each processing unit executes part of the application under its own thread of control. They can be subdivided into data-paths with a local communication infrastructure, but steered by a single controller. They read the signals they work on from background memory, and write intermediary signals or results back to it. They can share part of the background memory architecture with other processing units. In addition to the background memory, they have some local foreground memory, usually in the form of register files. They use this to store scalars or intermediate signals.

The distributed background memory architecture consists of a number of memories, divided over a number of memory hierarchy layers. Data can be transferred from one hierarchy layer to another (e.g. filling a cache), or from any layer to the foreground memory of a processing unit. Our DTSE stage currently only considers the *background* memory architecture, and only for a *single* processor architecture. The foreground memory will be optimized together with the processing units after our DTSE stage is completed.

The system pipeline between the memory architecture and the data paths contains pipeline registers that temporarily buffer data being transferred between the data paths and the memory. The clocking frequency of the registers in the system pipeline defines the duration of a *storage cycle*.

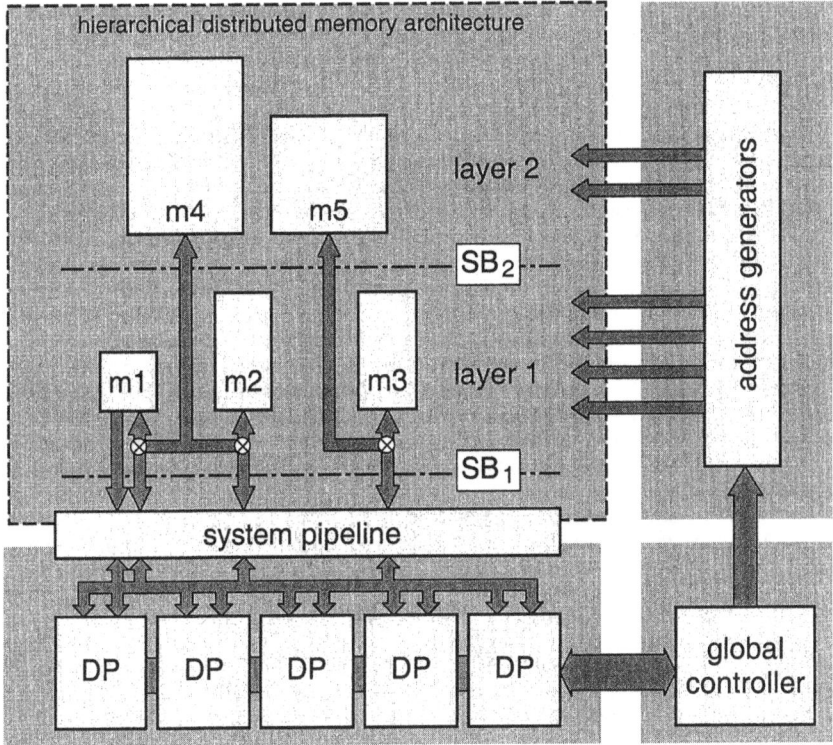

Figure 10.2.   Instantiation of target architecture with system pipeline.

10.1.2 Definitions and problem formulation

Definition: *storage cycle*

A storage cycle is the unit of time used to schedule transfers between the data paths and the memories, and between memories on different memory layers. It defines the granularity at which such transfers can be positioned in time.

There is some freedom for choosing the duration of a storage cycle for a given application. Here are a few points that have to be taken into account when choosing an appropriate duration for the storage cycles for a given application:

- The duration of a storage cycle should not be much larger than the cycle time of the fastest memory one wants to use for the given appliation. Doing so would unnecessarily limit the access frequency of this memory because the time between two consecutive memory accesses would be much larger than required by the memory leading to a waste of resources.

- The duration of the storage cycles should be equal to or a multiple of the data path clock cycles because the data transfers have to remain synchronized with the data path operations.

- Accesses to memories with a cycle time smaller than the duration of the storage cycles can be modeled as single cycle operations, which are simpler and result in shorter run-times for the SBO task. Whereas accesses to memories with a cycle time larger than the duration of the storage cycles have to be modeled as multi cycle operations, which are more complex and result in longer run-times for the SBO task.

From this we can conclude that a good choice for the duration of a storage cycle is to take it equal or slightly larger than the cycle time of the type of memories one wants to use on the first memory layer (which are usually the fastest memories in the whole memory architecture). Then the majority of the accesses can be modeled as single cycle operations. If this choice would limit the access frequency of the memories on higher layers too much, there are two options: increase the duration a bit such that also these transfers can be modeled as single cycle operations, or divide the duration by a small integer number (e.g., two) to increase the scheduling precision. The former approach limits the access frequency of the fastest memories somewhat. The latter approach does not affect the access frequency of the fastest memories, but forces multi-cycle modeling for all data transfers in the application.

We define *storage-bandwidth* (SB) as the *number of ports* in the memory architecture. The SBO task *optimizes* the storage-bandwidth requirements of an application. It *does not minimize* them! Indeed, not all ports are equally costly in a memory architecture: ports on a multi-port memory for instance are more costly because storing data in a multi-port memory is expensive both in terms of area and power. Minimizing the storage-bandwidth corresponds to minimize the maximum number of simultaneous data transfers. Often this requires one huge and very inefficient multi-port memory with a number of ports equal to the maximal number of simultaneous data transfers. This is unacceptable in most cases, especially when much cheaper solutions exist, which is usually the case. Therefore, the SBO task tries to come up with optimal bandwidth constraints such that the final memory architecture can be made as cheap as possible. To this end we define a cost function in Section 10.5 which the SBO task tries to minimize.

The SBO task heavily influences the quality of the results of the subsequent DTSE tasks. Figure 10.3 illustrates this. It shows that a small change in the ordering of data accesses can have a large

Data Access Ordering	Conflict Graph	A valid Memory Configuration
A.B. B.C. C.D. D.A. C.A. t↓ B.....	(A)—(B) (C)—(D) ╳ Chromatic Number = 3	A B D C ↓ ↓ ↓
A.B. B.C. C.D. D.A. C.A.⤳ t↓ B.A.	(A)—(B) (C)—(D) ╳ Chromatic Number = 2	A B C D ↓ ↓

Figure 10.3.   Effect of SBO on resulting memory architecture.

effect on the required memory bandwidth. The ordering shown in the top half of the figure results in a conflict graph with a chromatic number[1] of 3, which means (assuming only single-port memories are available) that a valid memory configuration requires at least 3 memories with this ordering. A small change to this ordering is shown in the bottom half of the figure. This change results in a conflict graph with a chromatic number of only 2, meaning that 2 single-port memories are sufficient for this ordering. This large effect on the resulting memory architecture clearly shows that SBO is in general very useful as a preprocessing step for memory allocation tasks. Remark that, even though a complete ordering is shown for the data accesses in this example, SBO *only* imposes a *partial*

[1] A *c-coloring* of a graph G is a partitioning of G's nodes in c partition classes $V = X_1 + X_2 + \cdots + X_c$ such that every two adjacent nodes belong to a different partition class. In this case, when the members of partition X_i are colored with color i, adjacent nodes will receive different colors. The chromatic number $\chi(G)$ is the smallest number c for which there exists a c-coloring of G. Obviously, $\chi(G) \geq \omega(G)$, where $\omega(G)$ is the size of the maximum clique of G, since every node of a maximum clique must be contained in a different partition class in any minimum coloring of G. For a perfect graph G, $\chi(G_s) = \omega(G_s)$ for all subgraphs G_s of G.

ordering. The subsequent scheduling steps at lower abstraction levels still have a lot of freedom left which can be used to optimize data-path and controller related costs. This can for instance be seen in figure 10.1, where the final schedule and the ordering obtained during SBO are completely different.

The SBO task focuses on the data transfers and does not take the data path operations into account. In many data-dominated applications, the data path operations are very simple and can easily be performed in one cycle. Retrieving the operands from and storing the results in memory are the most time consuming operations in this case. In case there are complex data path operations that take (much) more than one cycle to complete, this can be dealt with in two ways:

1. by *explicitly modelling* the duration of the data path operation by means of extra timing constraints between the data transfers of the operands and the transfer of the results;

2. or, in case of loops, by applying loop pipelining during the data path synthesis task to increase the time between reading the operands and writing the results (see figure. 10.4).

From this we can conclude that it is a reasonable assumption to focus on the data transfers during the DTSE stage of the design flow, leaving the arithmetic and logic parts to the data path and controller synthesis tasks. Since the dominance of the data storage and transfer on the overall system cost is usually very big, suboptimal solutions for these remaining tasks are likely to have only a secondary effect on the overall cost.

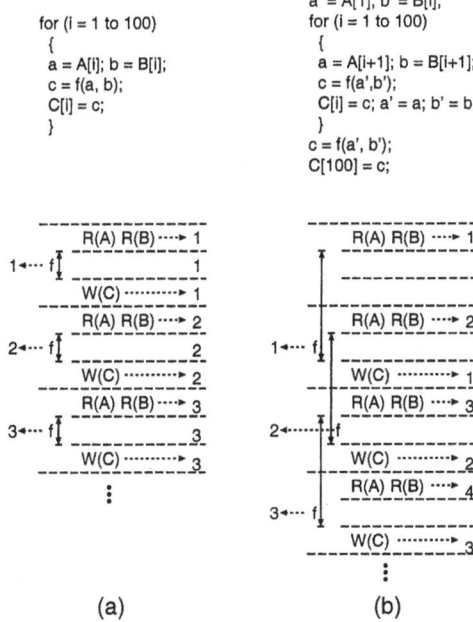

Figure 10.4. Applying loop pipelining to increase freedom for data path synthesis: (a) without loop pipelining, (b) with loop pipelining.

10.1.3 Breakdown in substeps

Sections 9.2 and 9.6 have already motivated that the memory hierarchy layer assignment (MHLA) substep should be part of the SCBD step, prior to the main SBO substep. In addition, additional loop reorganisation should be performed, as illustrated in section 10.4. How these preprocessing substeps involving some loop transformation steering should be formalized is a topic of current research.

The rest of the chapter is organized as follows. Section 10.2 presents the related work. Section 10.3 defines *what* SBO is. After that, Section 10.4 illustrates its effect on an example. From this example, we learn what is important to optimize by the SBO task. This knowledge is then formalized in a cost function in Section 10.5. Section 10.6 explains how memory hierarchy affects SBO. Section 10.12 concludes this chapter. Section 10.7 and following then explains *how* SBO can be done for network component applications applications. The extensions for RMSP applications are only discussed briefly. Section 10.8 explains a custom data-flow analysis for partitioning the dynamically allocated background data in network components into basic groups. Section 10.9 presents the core of the methodology: the Conflict Directed Ordering algorithm. Section 10.10 shows the results of applying the methodology to Alcatel's Segment Protocol Processor application. Section 10.12 concludes the chapter.

10.2 DETAILED RELATED WORK

Several problems are related to storage-bandwidth optimization.

In the register allocation domain, the allocation techniques start from a fully scheduled flow graph and are scalar oriented. A nice literature overview of this domain, which is fairly well understood by now, can be found in [382]. Many of these techniques construct a scalar conflict or compatibility graph and solve the problem using graph coloring or clique partitioning. This conflict graph is fully determined by the schedule which is fixed before. This means that no effort is spent to come up with an optimal conflict graph.

In the less explored domain of memory allocation and assignment for hardware systems, the current techniques start from a given schedule [245, 339, 366], or perform first a bandwidth estimation step [28] which is a kind of crude ordering that does not really optimize the conflict graph either. These techniques have to operate on groups of signals instead of on scalars to keep the complexity acceptable, e.g., the *stream model* of Phideo [245] or the *basic sets* in the ATOMIUM environment [28].

In the parallel compiler domain, [11] proposes a technique to partition arrays into groups of data that have to be assigned to different memories such that they can be accessed simultaneously for an SIMD architecture. They combine the constraints of a number of given access patterns into a single linear address transformation that calculates for every data element the memory in which it should be stored to minimize the total access time. This technique is complementary to SBO. It allows to avoid the allocation of multi-port memories for storing data with self-conflicts, by explicitly splitting arrays into smaller arrays that can be assigned to single port memories.

In the scheduling domain, the techniques optimizing for the number of resources given the cycle budget are of interest to us. Also here most techniques operate on the scalar level, e.g. [320, 409]. The only exceptions currently are the Phideo stream scheduler [410] and the Notre-Dame rotation scheduler [316]. Many of these techniques try to reduce the memory related cost by estimating the required number of registers for a given schedule. Only few of them try to reduce the required memory bandwidth, which they do by minimizing the *number* of simultaneous data accesses [410, 409]. They do not take into account *which* data is being accessed simultaneously. Also no real effort is spent to optimize the data access conflict graphs such that subsequent register/memory allocation tasks can do a better job.

The idea of optimizing a conflict graph is not new in itself. E.g., [323] optimizes a conflict graph in the context of scalar register allocation by removing weighted edges in a coloring problem prior to scheduling. However, the conflicts in their initial conflict graph are determined by the sequential ordering of the input code, whereas in our approach, a partial ordering is determined while optimizing the conflict graph. Also, to the best of our knowledge, this idea was not previously applied to groups of scalars. This is an important difference, because it affects the meaning of the conflicts. In the scalar case, the conflicts are storage conflicts: there is a conflict when the lifetimes of two values overlap; accessibility is usually not taken into account. While, in our case, the conflicts are accessibility conflicts: there is a conflict when two groups of scalars are accessed simultaneously; the life times of the data is not taken into account.

The main difference between our SBO and the related work discussed here is that we try to minimize the required memory bandwidth in advance by optimizing the access conflict graph for groups of scalars within a given cycle budget. We do this by putting ordering constraints on the flow graph, taking into account *which* data accesses are being put in parallel (i.e., will show up as a conflict in the access conflict graph).

10.3 STORAGE-BANDWIDTH OPTIMIZATION

This section defines *what* storage-bandwidth optimization is.

10.3.1 Input: basic groups, control data flow graph, and timing constraints

SBO requires three major inputs: a characterization of the groups of data to be stored in background memory, a CDFG expressing the functionality of the algorithm, and timing constraints including the cycle budget in which the algorithm has to be scheduled. The following subsections describe these inputs.

10.3.1.1 Basic groups and their characteristics. To deal with realistic applications, the memory assignment task should assign groups of scalars to memories instead of individual scalars. We call these groups of scalars *basic groups* (BGs). They form a partitioning of all data that has to be stored in background memory. This partitioning is decided earlier in our methodology, and is done in such a way that for every data access (read or write) in the flow graph it is known which basic group is being accessed. Basic groups are defined by the following properties:

1. *The set of all basic groups forms a partitioning of the data that has to be stored in background memory, i.e., every data item belongs to one and only one basic group.* During the memory assignment substep, each basic group will be assigned to an allocated memory.

2. *The partitioning into basic groups is manifest, i.e., it is decided at compile time.* This means that the data to memory assignment can also be decided at compile time.

3. *Every Read/Write instruction in the algorithm accesses one and only one basic group.* Thus, after basic group to memory assignment, every Read-Write instruction in the algorithm is associated with exactly one physical memory. This is an extremely important property that has a number of important consequences: firstly, a basic group is a set of *full* data words, because a read or write instruction always accesses a complete data word; secondly, this property determines which data can be stored in different basic groups and which data can not. This will be clarified further in Sections 10.8.

4. *The basic groups are as small as possible, i.e., they contain as few data items as possible.* This property is not really necessary, but maximizes the possibility for parallel data accesses and the freedom for the memory allocation and assignment tasks. It also makes the basic group partitioning unique for the given algorithm description. The size of each *basic group* is lower bounded by the previous property.

In the case of multi-dimensional signal processing applications, the basic groups are (parts of) multi-dimensional arrays. One important way of deciding on these partitions is the basic set analysis proposed in section 6.2. In the case of dynamically allocated data as in network applications, the basic groups are (parts of) virtual memory segments.

The set of basic groups is determined early in our design flow as it is used by several tasks in our DTSE methodology (e.g., SBO, memory allocation [28, 32], and memory assignment [376]). Hence, the SBO task takes the set of basic groups as input, together with a characterization of each basic group in terms of:

- word depth, i.e., the number of words in the BG,

- word-length, i.e., the number of bits in one word of the BG,

- average number of read accesses during 1 iteration of the algorithm,

- average number of write accesses during 1 iteration of the algorithm, *and*

- storage level in the memory hierarchy.

10.3.1.2 Control/data-flow graph. A second major input for the SBO task is the control/data-flow graph (CDFG) describing the algorithm to be implemented. As the SBO task focuses on the data storage, only the data accesses and the dependencies between them together with the control-flow (i.e., loops and conditions) are of interest, here. The arithmetic part of the algorithm is considered unimportant at this stage in the optimization process and will be considered after the memory organization has been optimized.

The CDFG must be at the basic group level. With this we mean that, for every data access (i.e., read or write instruction) in the CDFG, it must be indicated which basic group is being accessed.

10.3.1.3 Timing constraints including the cycle budget. The third major input of the SBO task, is the cycle budget in which the algorithm described in the CDFG has to be ordered. The SBO task decides which basic groups should be made accessible in parallel to meet this cycle budget with minimum bandwidth requirements. Next to the cycle budget, other timing constraints that put limitations on the relative and possibly even absolute ordering can be specified and have to be taken into account during SBO.

10.3.2 IO-profiles

IO-profiles play an important role in SBO, as they determine the minimum number of cycles that have to be scheduled between a data access and an other data access that depends on it. The next subsection defines what is meant by an IO-profile. The subsection after that proposes a simple IO-profile assignment strategy.

10.3.2.1 IO-profiles: Definition. An IO-profile specifies the input-output timing characteristics of a memory. It contains IO characteristics for different access modes (normal read/write, fast page mode read/write, burst mode read/write, etc.). Figure 10.5 gives an example of an IO-profile with two modes. Usually the *AddressOffset* is equal to zero. It is also possible that different offsets have the same value (e.g., *AddressOffset* and *DataInOffset* are usually equal in IO-profiles of SRAMs).

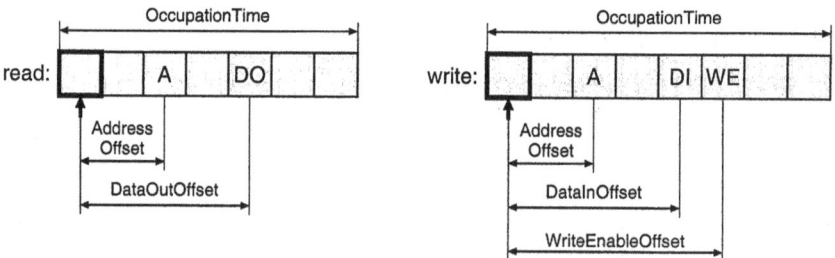

Figure 10.5.   Example of an IO-profile with two modes: read and write .

The same information can also be represented in tabular form:

Mode	Input			Output	Occupation
	Address	Data In	Write Enable	Data Out	Time
Read	*Offset*	×	×	*Offset*	*Delay*
Write	*Offset*	*Offset*	*Offset*	×	*Delay*

where:

- *AddressOffset*, *DataInOffset*, and *WriteEnableOffset* specify the difference between the moment the respective input must be supplied and the beginning of the cycle in which the data access is scheduled.

- *DataOutOffset* specifies the difference between the moment the DataOut output will be ready and the beginning of the cycle in which the data access is scheduled.

- *OccupationTime* specifies how many cycles the data access occupies the memory resource starting from the cycle in which the data access is scheduled. During this OccupationTime no other operations can be scheduled that access the same memory.

  All offsets and occupation times are expressed as a number of cycles.

10.3.2.2 IO-profile assignment. Assuming that all memories on the same layer in the memory hierarchy have the same IO-profile, and that these IO-profiles are specified beforehand, this step is trivial because the storage level for each basic group is part of the input. Otherwise, an optimization decision is involved. Indeed, in order to determine these basic group characteristics, a memory class decision should be made for each basic group, prior to the actual SBO substep. A distinction should be made between Pointer-Addressed Memory (PAM), RAM and ROM based on signal access properties. In addition, the storage in SRAM, conventional DRAM or other DRAM "flavours" should be decided. The I/O profile decision decision will be based mainly on the size and the access frequency of the basic group. It will also depend on the memory partition assigned for the basic group involved, and especially on the on-chip/off-chip choice. After SBO, this I/O profile can be refined.

10.3.3 Conflicts and conflict graphs

As mentioned before, SBO is about minimizing conflicts between basic groups. *Basic group conflicts* are caused by *data access conflicts*. Therefore, this subsection first defines *data access conflicts*, from which, subsequently, the definition of *basic group conflicts* is derived. Next, it defines a *conflict graph* that collects all basic group conflicts, which is then extended towards an *extended conflict graph* that can be used to optimize memory architectures containing multi-port memories.

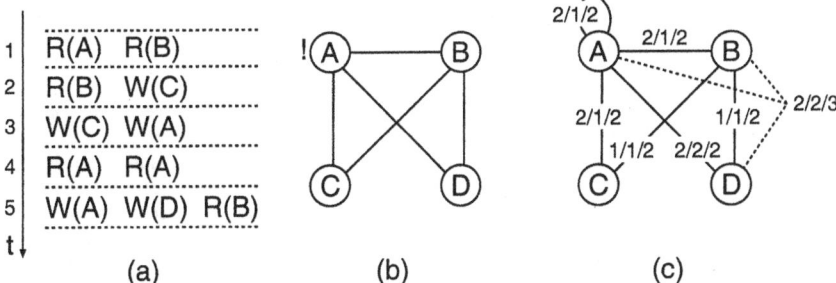

Figure 10.6. Data Access Conflicts: (a) Ordering, (b) Conflict Graph, (c) Extended Conflict Graph.

10.3.3.1 Data access conflicts. When two data accesses in the CDFG are scheduled in the same cycle (due to stringent cycle budget constraints), we say that these data accesses are in conflict. In this case, the basic groups they are accessing should be stored in such a way that they are accessible in parallel. This can be done by storing them either in two different memories, or in a memory with at least two ports.

Example: Looking at the schedule shown in figure 10.6, we see that the data access that reads a value from basic group *A* in *Cycle 1*, is in conflict with the data access that reads a value from basic group *B* also scheduled in *Cycle 1*.

10.3.3.2 Basic group conflicts. We say that two basic groups are in conflict, if and only if there exists a cycle in which two data accesses, one accessing the first basic group and the other one accessing the second basic group, are in conflict.

Example: Looking at the schedule shown in figure 10.6, we see that basic groups B and C are in conflict because they are accessed in parallel in *Cycle 2*. Therefore, they can not be stored together in a 1-port memory. Basic groups C and D are *not* in conflict. Therefore, there is no assignment constraint regarding these two basic groups: if desired, they can be assigned to the same 1-port memory during the memory assignment task.

10.3.3.3 Conflict graph. All basic group conflicts are collected in a conflict graph where the nodes correspond to basic groups and there is an edge between two nodes whenever the corresponding basic groups are in conflict.

Example: An example of a conflict graph is shown in figure 10.6b for the schedule shown in figure 10.6a.

Remark the exclamation mark next to basic group A in figure 10.6b. It indicates that the basic group A is in conflict with itself and can therefore not be stored in a 1-port memory. The conflict graphs discussed here can be used only for memory architectures containing 1-port memories. It contains not enough information for correctly and optimally allocating multi-port memories. When also multi-port memories are allowed in the memory architecture, extended conflict graphs are needed as explained in the next subsection.

As will be shown, conflict graphs are crucial to SBO. They are well known from register [382] and other scalar oriented assignment problems. However, in our case, the nodes correspond to groups of data instead of scalars. The more conflicts there are between basic groups, the less freedom there is for the memory allocation and assignment tasks. Experiments have shown that this typically results in a higher cost of the memory architecture. Therefore, we will define a cost function for conflict graphs (see section 10.5) reflecting this. The idea of SBO is then to come up with a conflict graph with minimal cost such that it is still possible later on (after DTSE) to schedule the CDFG within the cycle budget.

10.3.3.4 Extended conflict graph. When multi-port memories are allowed in the memory architecture, it becomes useful to extend the conflict graph with more information to decide on memory types.

Firstly, this annotation includes the type of conflicts that can occur. More specifically, one has to know for every conflict, the maximum number of reads (R), writes (W), and total number of data accesses (i.e., read or write) that can occur (RW) simultaneously. In the figures this information is shown next to the conflict edges in the form R/W/RW. This information allows to decide which type of ports (Read, Write, or Read-Write) are required on a multi-port memories when certain basic groups are assigned to it (see subsection 10.3.4).

Secondly, when more than two data accesses are scheduled in the same cycle, this results in a conflict between more than two basic groups which should be accurately represented to guarantee valid solutions in later steps of the design trajectory. This type of conflict can be represented in the conflict graph by hyper edges, i.e., edges between more than two nodes.

Finally, it is also possible that a basic group is accessed several times in the same cycle, which results in a self-conflict, represented by a self-loop on the corresponding node. Such a conflict forces a multi-port memory to be allocated for that basic group.

All these extensions lead to the definition of the extended conflict graph which is the main output of the SBO task:

Definition: *extended conflict graph (ECG)*

An Extended Conflict Graph $G(V, S, E, H)$ is an undirected hyper graph, in which the nodes (V) represent basic groups, and the self-edges (S), binary edges (E), and hyper edges (H) represent access conflicts between the basic groups. Every edge $t \in S \cup E \cup H$ is labeled with three numbers called

R, W, and RW. Where R, W, and RW are respectively: the maximum number of simultaneous read operations, the maximum number of simultaneous write operations, and the maximum number of simultaneous data accesses (i.e., read and write operations) that can occur for the given conflict during the execution of the algorithm.

Example: An example of an extended conflict graph is shown in figure 10.6c for the schedule shown in figure 10.6a.

The annotations on the edges quantify the type of conflicts that can occur. For instance, we see that basic groups B and C are in conflict (*Cycle 2*), but that there is at most one simultaneous read and one simultaneous write to these basic groups. Therefore, if they are assigned to the same memory, a memory with one read port and one write port is sufficient.

Basic group A has a self-conflict (*Cycle 4*). Therefore, it has to be assigned to a multi-port memory. Because there are two simultaneous accesses to A, the memory should have at least 2 ports. As there are two simultaneous read accesses to A (*Cycle 4*), at least two ports should have read capabilities. However, as there is only one simultaneous write access to A (*Cycle 3* and *Cycle 5*), only one port requires write capabilities. In summary: the cheapest memory in which A can be stored is a 2-port memory, of which one port is a read port and the other port is a read/write port.

There is a hyper edge between basic groups A, B, and D. This hyper edge indicates that if all three of these basic groups are stored in the same memory, it should have at least three ports. In contrast, basic groups A, B, and C are also in conflict with each other, but are never accessed all three of them together (there is no corresponding hyper edge). Therefore, they can be stored together in a 2-port memory.

10.3.3.5 Conflict graph examples. Figure 10.7(b) shows an example of a conflict graph generated by the storage bandwidth optimization step, based on the ordering of figure 10.7(a).

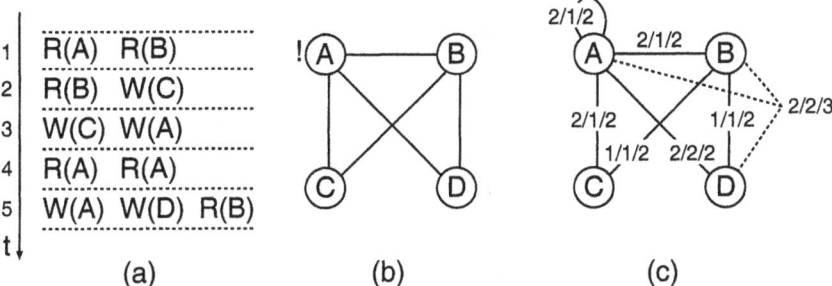

Figure 10.7. Example of a partial ordering and the corresponding (extended) conflict graph

Our extended conflict graph model makes it possible to cope with all these issues. The simple conflict graph is extended in three ways which will now be illustrated:

1. Self-conflicts are possible.

   Self-conflicts occur when a basic group is accessed several times in the same abstract storage cycle. Such a conflict forces a multi-port memory to be allocated for that basic group. Self-conflicts are represented by loops in the extended conflict graph.

   In figure 10.7(a) we see that in cycle 4, two accesses to basic group A take place simultaneously. In the extended conflict graph (10.7(c)) this is represented by a loop on node A.

2. Every edge is annotated with R/W/RW numbers.

   Edges are annotated with the type of conflicts that can occur. More specifically, one has to know for every conflict, the maximum number of reads (R), writes (W), and total number of data accesses

(RW) (i.e., read or write) that can occur simultaneously. In the figures this information is shown next to the conflict edges in the form $R/W/RW$. This information allows to decide which type of ports (Read, Write, or Read-Write) are required on a multi-port memories when certain basic groups are assigned to it.

For assignment, the R/W/RW numbers are interpreted in the following way. When two basic groups are assigned to the same memory, it must have at least RW ports in total, of which at least R must provide read capability and at least W must provide write capability.

For instance, we see that basic groups B and C are in conflict (cycle 2), but that there is at most one simultaneous read and one simultaneous write to these basic groups. Therefore, if they are assigned to the same memory, a memory with one read port and one write port is sufficient.

For the self-conflict of basic group A, there is a single read access in cycle 1, a single write access in cycles 3 and 5, and two parallel read accesses in cycle 4. So there are maximally two accesses in parallel: one write access or two read accesses. So a memory with one read/write port and one extra read port would be sufficient to store this basic group.

3. Hyper edges connecting three or more nodes are added.

When more than two data accesses are scheduled in the same cycle, this results in a conflict between more than two basic groups. This type of conflict can be represented in the conflict graph by hyper edges, i.e., edges between more than two nodes. Just like a normal edge, a hyper edge is annotated with R/W/RW numbers. These are used to check the validity of assigning the basic groups together into a single memory.

In the example of figure 10.7, there is a hyper edge between basic groups A, B, and D. This hyper edge indicates that if all three of these basic groups are stored in the same memory, it should have at least three ports. The hyper edge is annotated with the numbers 2/2/3, so 2 read/write ports and one additional read or write port are sufficient. In contrast, basic groups A, B, and C are also in conflict with each other, but are never accessed all three of them together (there is no corresponding hyper edge). Therefore, they can be stored together in a 2-port memory (1 read/write port and one extra read port are sufficient).

With an extended conflict graph, no additional information about the application algorithm is required to be able to decide whether an allocation or assignment is valid. Data dependencies and required parallelism are fully decided by the storage bandwidth optimization step, and access statistics are summarized in the basic group information.

Using the extended conflict graph as the intermediary model, the two major steps in our methodology, storage bandwidth optimization and memory allocation and assignment, can be fully decoupled. This leads to a significant reduction of overall design effort, both for systematic manual design as for automated tool support.

10.3.4 Resolving conflicts

The Extended Conflict Graph represents the constraints that have to be satisfied by the subsequent memory allocation and assignment tasks to be sure that the cycle budget can still be met later on during detailed scheduling. When two basic groups are in conflict, this conflict has to be resolved during memory allocation/assignment. This can be done in two ways: either the basic groups are assigned to two different memories, or they are assigned to a multi-port memory. In the latter case, the R/W/RW numbers associated with the conflict determine the number and type of ports that are minimally required on the multi-port memory to which these two basic groups are assigned: the memory must have at least RW ports, of which at least R must provide read capability and at least W must provide write capability. When more than two conflicting basic groups that are connected by a hyper edge in the ECG are assigned to a single memory, the R/W/RW number of the hyper edge determines the number and type of ports that are minimally required on the multi-port memory to which they are assigned.

10.3.5 Conflict graph optimization

The goal of SBO is to come up with an optimized extended conflict graph that puts the least constraints on the search space of the subsequent memory allocation and assignment tasks. To this end, a cost function for extended conflict graphs will be defined in Section 10.5, such that ECGs with a smaller cost are likely to lead to cheaper memory architectures after memory allocation and assignment. The task of SBO is then to order all data accesses within the cycle budget such that the resulting conflict graph is as cheap as possible.

10.3.6 Output: extended conflict graph

The main output of the SBO task is the extended conflict graph (ECG) as defined in subsection 10.3.3. It contains all *relevant* information from the data access ordering for the subsequent memory allocation and assignment tasks. Once the ECG is derived, the detailed data access ordering is no longer needed and can be thrown away.

The ECG represents an (optimized) set of basic group conflicts that have to be resolved during the memory allocation and assignment tasks. These tasks derive an optimal memory architecture within the constraints expressed by the ECG. Because the ECG is derived from a valid data access ordering, it is guaranteed that there is enough memory bandwidth available to schedule the application within the specified cycle budget afterwards. This will be illustrated on an example in the next section.

10.3.7 Scheduling freedom

The data access ordering obtained during the SBO process, is one possible schedule that meets the cycle budget requirements for the memory architecture that satisfies all its constraints. So it is guaranteed that a valid schedule exists. In practice, there are many more schedules compatible with the constraints expressed in the ECG. Moreover, the memory allocation and assignment tasks usually create even more freedom for the detailed scheduler, as they can assign basic groups that are not in conflict to different memories, thereby allowing that they are accessed in parallel. As the memory architecture, and not the ECG, defines the constraints for the detailed scheduling step, the final schedule can therefore be quite different from the partial ordering obtained during the SBO task. This has been illustrated in figure 10.8. Remark the large differences between the two examples of the final schedule and the ordering obtained during SBO.

10.4 THE EFFECT OF SBO: ILLUSTRATION

In this section, the effect of SBO on the outcome of the memory allocation/assignment tasks is illustrated with an example. Several partial orderings will be proposed to illustrate the effect of SBO on the area and power cost of the memory architecture (after the memory allocation/assignment tasks). Area figures are given for a 1.2 μ CMOS technology. The values for the single-port memories were obtained using an embedded SRAM generator from IMEC. The values for multi-port memories were obtained by adjusting the values for single port memories with the port dependent factor of Mulder's area model for on-chip memories [286]. Power figures are obtained with the TI memory power model discussed in subsection 4.4.2.2. Therefore only relative power figures will be given. Figure 10.14 at the end of this section collects the power results for each proposed partial ordering. The code and the basic group information of the example are shown in figure 10.9. The cycle budget used is 550 cycles.

Every array in the code corresponds to one basic group. It is assumed that all produced basic groups are still needed at the end of the code fragment, such that no in-place optimization [108, 106] can be performed. Only one level of hierarchy is assumed. The storage cycles are assumed to be long enough, such that the IO-profiles allow to read the operands of an operation, perform the operation, and write the results back to memory in a single storage cycle. The occupation times for reads and writes are assumed to be 1 storage cycle.

Notice that in the different stages we show structured code, whereas the technique really works on a (hierarchical) CDFG that corresponds to this code. We believe, however, that it is more clear to show the result of the SBO on the code rather than in CDFG format. One potential problem with this representation is that one could think that after SBO the code is fully ordered (or scheduled). This is

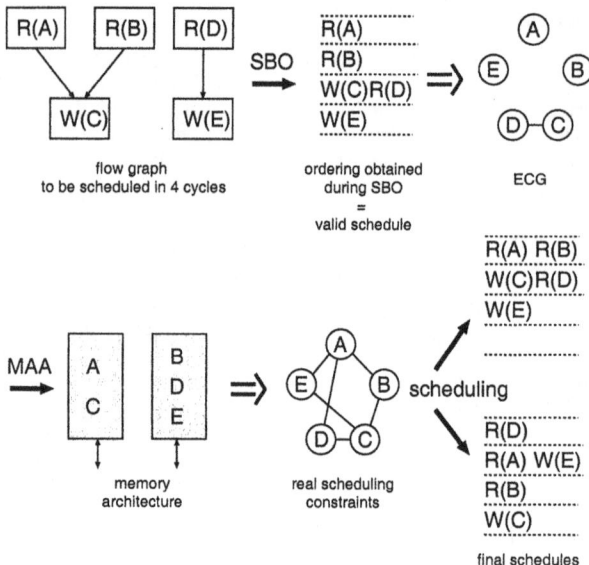

Figure 10.8.   Scheduling freedom after Memory Allocation and Assignment (MAA).

```
for (i = 1 to 50)
  tmp = input;
  A[2*i - 1] = tmp;
  A[2*i] = -tmp;
for (j = 1 to 50)
  B[j] = A[101 - j];
  G[j] = B[j] + A[2*j];
for (k = 1 to 50)
  C[2*k] = B[51 - k];
  C[2*k - 1] = B[k];
for (l = 1 to 50)
  D[l] = C[l] + A[l] + B[l];
for (m = 1 to 150)
  E[m] = m^2;
for (n = 1 to 50)
  F[n] = C[n] + C[2*n] + E[3*n];
```

```
A[]: 100 words of  8 bits
B[]:  50 words of 24 bits
C[]: 100 words of  8 bits
D[]: 100 words of 24 bits
E[]  150 words of 24 bits
F[]:  50 words of 24 bits
G[]:  50 words of  8 bits

Cycle budget = 550 cycles
```

Figure 10.9.   (left) source code of the example; (right) size of the BGs

not true. Only the constraints contained in the ECG are imposed. In the figures depicting the ECGs, the hyper edges are represented in dashed lines.

10.4.1 Partial ordering 1

As a first experiment, we have maintained the original procedural execution order as specified in the algorithm. As it is often believed that uniformly balancing the number of simultaneous read/write leads to good results [409], we have given more cycles to the expressions that have to read many operands such that these data accesses can be distributed over several cycles. In order to do this within the cycle budget, we have to reduce the number of cycles available to other expressions. For the m loop, this is done by unrolling it with a factor of two, such that two expressions can be scheduled in parallel. The resulting code and corresponding ECG is shown in figure 10.10. Data accesses to indexed arrays represent background memory accesses. The *tmp* variables are stored in registers and therefore not in the ECG. The ECG can be obtained from the code by assuming that each line in the code has to be executed in one clock cycle. Notice that the number of simultaneous data accesses is fairly well balanced.

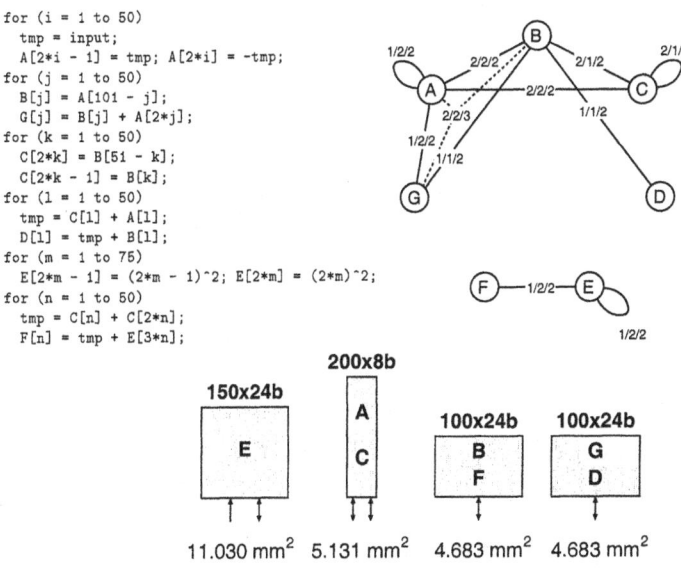

```
for (i = 1 to 50)
  tmp = input;
  A[2*i - 1] = tmp; A[2*i] = -tmp;
for (j = 1 to 50)
  B[j] = A[101 - j];
  G[j] = B[j] + A[2*j];
for (k = 1 to 50)
  C[2*k] = B[51 - k];
  C[2*k - 1] = B[k];
for (l = 1 to 50)
  tmp = C[l] + A[l];
  D[l] = tmp + B[l];
for (m = 1 to 75)
  E[2*m - 1] = (2*m - 1)^2; E[2*m] = (2*m)^2;
for (n = 1 to 50)
  tmp = C[n] + C[2*n];
  F[n] = tmp + E[3*n];
```

Figure 10.10. Partial ordering 1.

A good memory architecture that is compatible with this ECG is also given in figure 10.10. The area cost is 25.529 mm^2. It contains two dual-port memories, which consume a large part of the area. The dual-port memories are a direct consequence of the two self edges in the ECG. Note that these memories could be combined into one dual-port memory, as can be seen in the ECG because there is no hyper edge between basic groups A[], C[], and E[], which means that they can be stored together in a dual-port memory. However, this would waste a lot of bits because E[] has a much larger word-length than the other two basic groups.

As multi-port memories are very costly both in terms of power consumption and area, it is clear that self-edges in the ECG are costly and have to be avoided whenever possible (see cost function in section 10.5).

It is possible to arrive at a slightly better solution in terms of area when G[] is stored separately from D[], because then no bits are wasted by storing G[] in a too wide memory. The difference is small, however, and therefore we prefer a solution with less memories, because the number of memories is also a cost factor (more memories means more interconnect and increased design effort) that has to be taken into account.

10.4.2 Partial ordering 2

In this experiment, we have derived an ECG directly from the procedural execution order of the original code (figure 10.9), by assuming that every line has to be executed in one clock cycle. We did this to show that it is not necessarily bad to have a large unbalance in the number of simultaneous read/write operations. The resulting ECG and memory allocation/assignment scheme are shown in figure 10.11. Remark that this time there are less self-edges in the ECG. As a result the total area cost is now reduced to 17.153 mm², despite the unbalance.

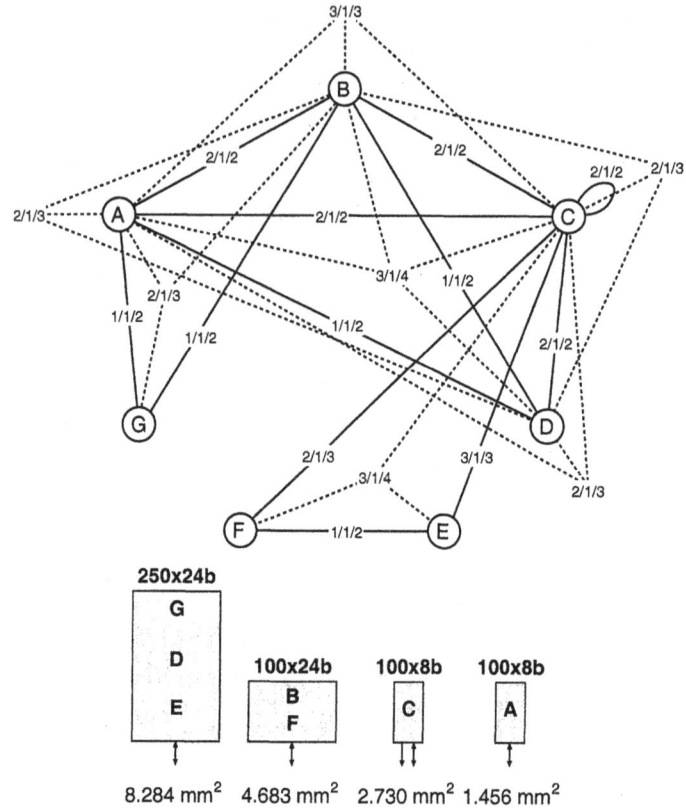

Figure 10.11.   Partial ordering 2.

The reason for this result is that it is also very important *which* data accesses are put in parallel, something that has not been incorporated in previous work.

10.4.3   Partial ordering 3

In the third experiment, we try to fully avoid the self-edges in the ECG and at the same time reduce the maximum number of simultaneous data accesses. We have done this by applying some cost optimizing loop transformations [110]. More specifically, we have applied loop splitting (loop m), body splitting (loop j), body merge (loop i and part of loop m), loop reordering (part of loop m and part of loop j), and loop body reordering (loop n) to obtain an execution order shown in figure 10.12.

The area cost is 16.292 mm², about 36% less than the first partial ordering. The architecture contains still many (single-port) memories. The reason for this is the relatively high chromatic number of the conflict graph (i.e., the graph obtained by removing the hyper and self-edges from the ECG). The chromatic number of this graph is 3, which means that we need at least three 1-port memories in a memory architecture that consists entirely out of 1-port memories. However, because of the difference in word-length it turns out to be more cost effective to introduce a fourth memory in this case.

```
for (i = 1 to 50)
  tmp = input;
  A[2*i - 1] = tmp; E[2*i - 1] = (2*i - 1)^2;
  A[2*i] = -tmp; E[2*i] = (2*i)^2;
for (j = 1 to 50)
  B[j] = A[101 - j];
for (k = 1 to 50)
  C[2*k] = B[51 - k];
  C[2*k - 1] = B[k];
for (l = 1 to 50)
  tmp = C[1] + A[1];
  D[1] = tmp + B[1];
for (m = 1 to 50)
  tmp = B[m] + A[2*m];
  G[m] = tmp; E[m + 100] = (m + 100)^2;
for (n = 1 to 50)
  tmp1 = C[n]; tmp2 = E[3*n];
  F[n] = tmp1 + C[2*n] + tmp2;
```

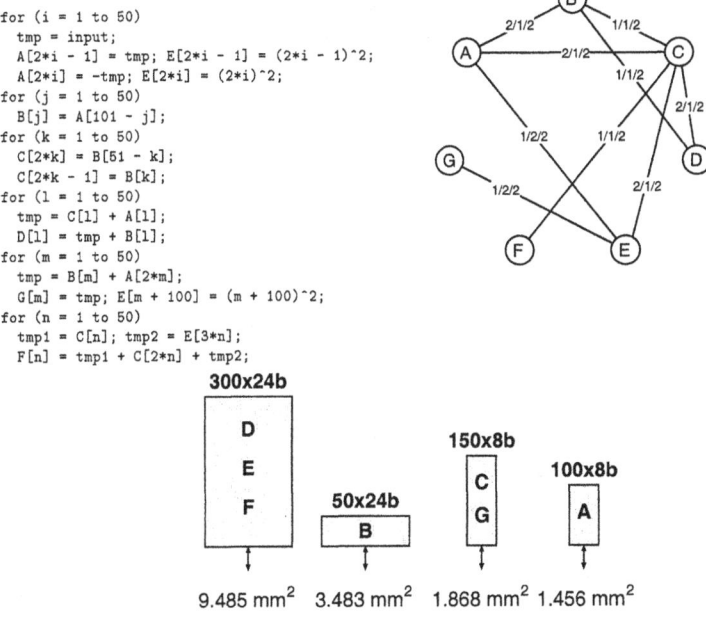

Figure 10.12.   Partial ordering 3.

10.4.4   Partial ordering 4

In this final experiment, we have taken the ordering of the previous experiment, with a little change in the ordering of the body in the loop producing D[]. The result is that the chromatic number of the graph is reduced to 2 instead of 3. This means that less memories are needed. The results are shown in figure 10.13. The area cost is 13.379 mm^2. This is a much better result than the one obtained in the first experiment: the area is about a factor 2 less; it contains only 2 memories instead of 4; and the total number of memory ports is 2 compared to 6, which means less interconnect.

This once more shows that it is very important to take into account *which* data accesses are done in parallel. Even small changes can have large consequences. This makes it particularly difficult to optimize this task by hand for large real-life applications.

10.4.5   Effect on power consumption

Figure 10.14 summarizes the effect of the different partial orderings with resulting ECGs on the power consumption of the memory architecture after allocation and assignment. The results are shown per amount of allocated memory ports. The reason is that, in general, the more memories are available, the less power is consumed because the memories can be made smaller and hence less power consuming. The designer decides how many memory ports he can afford in the design. The minimum number of ports that have to be allocated to meet the timing constraints is given by the maximal RW number of all (hyper) edges in the ECG. The figure clearly shows the positive effect of SBO on the memory power consumption.

10.5   COST FUNCTION

From the example in the previous section, we learn that the SBO task has to optimize the following three items of the extended conflict graph: the number of self-conflicts, the chromatic number of the

```
for (i = 1 to 50)
  tmp = input;
  A[2*i - 1] = tmp; E[2*i - 1] = (2*i - 1)^2;
  A[2*i] = -tmp; E[2*i] = (2*i)^2;
for (j = 1 to 50)
  B[j] = A[101 - j];
for (k = 1 to 50)
  C[2*k] = B[51 - k];
  C[2*k - 1] = B[k];
for (l = 1 to 50)
  tmp1 = A[l]; tmp2 = B[l];
  D[l] = C[l] + tmp1 + tmp2;
for (m = 1 to 50)
  tmp = B[m] + A[2*m];
  G[m] = tmp; E[m + 100] = (m + 100)^2;
for (n = 1 to 50)
  tmp1 = C[n]; tmp2 = E[3*n];
  F[n] = tmp1 + C[2*n] + tmp2;
```

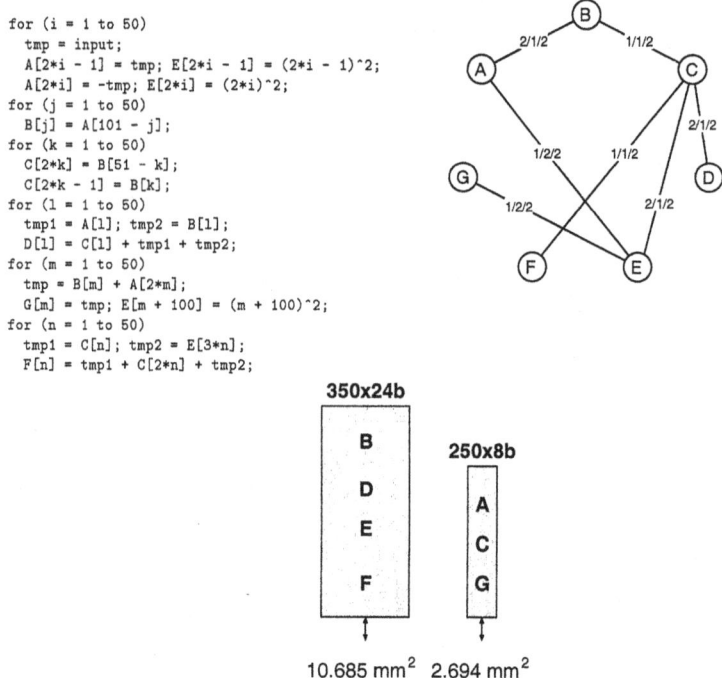

Figure 10.13.   Partial ordering 4.

conflict graph, and the number of conflicts in the conflict graph. Each of these are detailed in the following subsections. At the end of this section the complete cost function for extended conflict graphs is presented.

10.5.1 Self-conflicts

Clearly, self-conflicts have to be avoided as much as possible because they force the allocation of multi-port memories which are very costly both in terms of area and power. Especially, self-conflicts of large and frequently accessed basic groups should be avoided. Therefore, a weighting of the self-conflicts has to be introduced here to make the right trade-offs.

10.5.2 Chromatic number of conflict graph

The chromatic number of the conflict graph, i.e., the extended conflict graph without the hyper- and self-edges, corresponds to the minimal number of memories in the memory architecture.[2] For power reasons only, it is not that important to minimize the number of memories, as distributing the data over smaller memories usually results in lower power dissipation. However, having too much memories in the memory architecture is not good for several reasons (routing overhead, design complexity, test cost, number of I/O pins, ...). Therefore, it is always good to try to minimize the required number of

---

[2]This is only exact in case only single port memories are considered. However, as we try to avoid multi-port memories as much as possible, this is a good approximation.

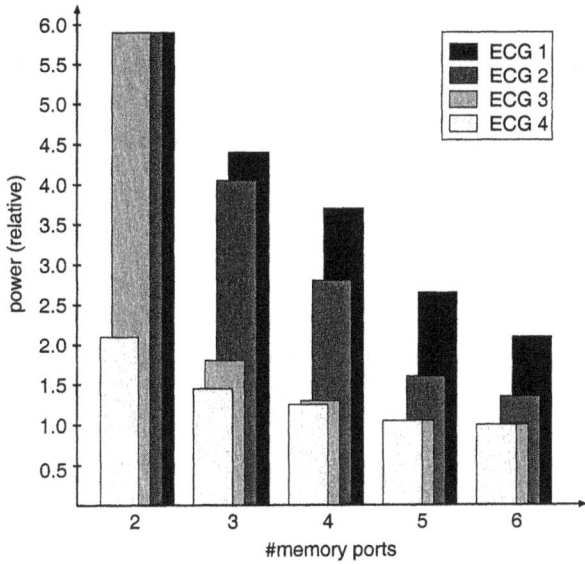

Figure 10.14.   Power results for the 4 partial orderings and corresponding ECGs.

memories. During the allocation and assignment tasks, more memories can always be added when this would lead to an important reduction in power and/or area.

10.5.3 Weighted number of conflicts in conflict graph

To leave as much freedom as possible for the memory allocation/assignment tasks, it is important to come up with an ECG with as few conflicts as possible. Not all conflicts are equally costly, though. For instance, when two basic groups have a large difference in word-length, it is not that bad that they have to be stored in different memories, because this saves bits that would otherwise be wasted. This means that such a conflict has to be preferred compared to a conflict between basic groups with equal word-length (especially when the basic group with the smaller word-length consists of a large number of words). Another example has to do with power optimization. It can be realistically assumed that a larger memory consumes more power than a smaller memory, and that the power consumption of a memory is proportional to the number of accesses to it (see section 4.4).[3] Therefore, it can be seen that it is not good practice to store a small, very frequently accessed basic group together with a very large, infrequently accessed basic group [435]. Hence, conflicts between these types of basic groups should be preferred over conflicts between basic groups for which there is no reason to store them in different memories. The previous two examples show that some conflicts have to be preferred compared to others because there is some gain when the two basic groups are stored in different memories. The contrary is also possible: some conflicts have to be avoided more than others. For instance, when two basic groups are part of the same logical structure (e.g., a user defined array is split in two basic groups), it is often better to store both basic groups in the same memory because this reduces the controller and addressing costs. Therefore, conflicts between this type of basic groups should be discouraged, such that the memory assignment task is allowed to store them in a common memory. This justifies the introduction of *pairwise basic group conflict costs* C_e corresponding to

---

[3]This is confirmed by several memory models that we have obtained from vendors. The exact dependence on size is between logarithmic and linear.

the binary edges $e \in E$ of an ECG $G(V, S, E, H)$. These pairwise conflict costs are calculated based on the properties of the two basic groups involved. This is an important difference with the scalar oriented techniques, where all scalars are considered to be more or less equal (although sometimes a distinction is made based on interconnect costs).

The *pairwise basic group conflict cost* can be obtained as a weighted sum of a number of cost terms that depend on the characteristics of the two BGs involved. Some of these cost terms are positive (marked with (+) in the following list), meaning that both BGs are stored preferentially together, others are negative (marked with (-)), meaning that both BG are preferentially stored in different memories. We have also included a conflict cost offset in order to make the cost of all conflicts positive.

The cost terms we consider are:

■ **Power cost (-)**
For power reasons it is sometimes better to split the data over different memories. For instance, assigning a small, frequently accessed, BG to the same memory as that of a large, infrequently accessed, BG can be bad for power. The following cost term reduces the cost of a conflict between two such BGs:

$$Size(BG_1) \times \#Accesses(BG_2) + Size(BG_2) \times \#Accesses(BG_1)$$

■ **Bits lost in too wide memories (-)**
The difference in word-length has to be multiplied by the number of words of the BG with the smaller word-length to take the memory loss into account. Because this is a pairwise conflict cost term, it is not easy to prevent overestimating the memory loss when more than two BG are assigned to the same memory.

$$\#Words(BG_{min\ bw}) \times [BitWidth(BG_{max\ bw}) - BitWidth(BG_{min\ bw}))]$$

■ **Possible memory sharing (+)**
Two basic groups can only be stored in-place when they are assigned to the same memory. Therefore BGs with a high possibility of in-place sharing should be allowed to be put in the same memory. This requires early knowledge about which basic groups have non-overlapping life times. This can be provided by an inter-array in-place estimate (see section 4.1.1).

$$\#Words(BG_{min\ wd})\ if\ life\ times\ are\ not\ overlapping,\ 0\ otherwise$$

■ **Separating an array / virtual memory segment (+)**
This term tries to prevent the splitting of an array / virtual memory segment over different memories. This is a fixed cost between every two BGs that are part of the same array / virtual memory segment. Because this is a pairwise conflict cost term, it is not easy to prevent overestimating the possible memory gain when more than two BG are assigned to the same memory.

$$1\ if\ BGs\ are\ part\ of\ the\ same\ array/VMS,\ 0\ otherwise$$

■ **Conflict cost offset (+)**
Used to make all conflict costs positive. Indeed, if conflict costs would be allowed to be negative, the corresponding conflicts would always be part of the optimized conflict graph because their inclusion would always reduce the total conflict cost. This contradicts our goal of minimizing the number of conflicts in the conflict graph. Hence the cost offset to make sure that conflicts costs are always positive.

It has to be noted that it is not easy to add these totally different costs together (adding apples and pears) and get still meaningful results. So many experiments have to be done to come up with good weighting factors for these cost terms.

10.5.4 Cost function

We propose the following cost function for optimizing the extended conflict graph:

$$Cost(G(V, S, E, H)) = \alpha \cdot \sum_{s \in S} RW_s + \beta \cdot ChromaticNumber(G(V, E)) + \gamma \cdot \sum_{e \in E} C_e \quad (10.1)$$

The first term penalizes self-edges in the ECG which reduces the number and size of multi-port memories in the final memory architecture, the second term reduces the number of required memories, and the last term minimizes the total weighted conflict cost of the extended conflict graph. The hyper edges are not included in this cost function because SBO it is not known at this stage whether a conflict will be resolved by assigning the conflicting basic groups to different memories or not. An optimistic scenario is assumed here. Only when they are actually assigned to a multi-port memory, the R_t, W_t, and RW_t values for $t \in E \cup H$ come into play. The hyper edges contain vital information for the memory allocation and assignment tasks, though.

10.6 SBO AND MEMORY HIERARCHY

In case of memory hierarchy, there is a separate conflict graph for each memory level. Every level is optimized independently from the others except for the ordering constraints imposed by the copying of data between layers. Figure 10.15 illustrates this for a copy of basic group S on level 2 to basic group S' on level 1.

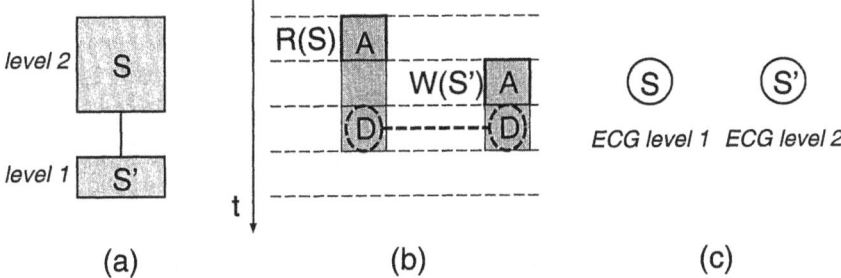

Figure 10.15. SBO and Memory Hierarchy: (a) hierarchical memory architecture, (b) ordering constraints for SBO, (d) one ECG per level in memory hierarchy.

Mode	Input			Output	Occupation
	Address	Data In	Write Enable	Data Out	Time
Read	0	×	×	2	3

Figure 10.16. IO-mode profile for reading basic group S (level 2)

Mode	Input			Output	Occupation
	Address	Data In	Write Enable	Data Out	Time
Write	0	1	1	×	2

Figure 10.17. IO-mode profile for writing basic group S' (level 1)

The IO-mode profiles for the read operation on level 2 and the write operation on level 1 are given in figure 10.16 and 10.17 respectively. Assuming that the copy from level 2 to level 1 does not pass through foreground memory (in the data paths), both operations must be perfectly synchronized: the data must be written in level 2 when it is available from level 1 (see figure 10.15b). For the given

IO-profiles this means that the write operation has to be scheduled exactly one storage cycle after the read operation. This constraint has to be taken into account during the SBO. Even though both operations are overlapping, there is no corresponding conflict in the conflict graph, because the two basic groups belong to different memory hierarchy levels with their own conflict graph (figure 10.15c).

10.7 STORAGE BANDWIDTH OPTIMIZATION FOR FLAT CONTROL DATA FLOW GRAPHS

The previous sections in this chapter have explained the importance of storage-bandwidth optimization for data-dominated applications and provides a formalized problem definition for it. It does not, however, present a solution for the problem. The subsequent sections discuss a storage-bandwidth optimization methodology for data-dominated applications without loops. Data-dependent conditions are allowed, though. An important class of applications that fits in this category are the network component applications targeted by the MATISSE project at IMEC [97]. These applications handle large amounts of data organized in dynamically allocated tables and records, and have to meet strict timing constraints. Storage-bandwidth optimization determines which groups of data should be made simultaneously accessible in the memory hierarchy such that the real-time constraints can be met with minimal memory cost. A custom basic group partitioning, compatible with the MATISSE design flow, is presented for the dynamically allocated data structures typically found in this type of applications. The core of the methodology, the conflict directed ordering algorithm, is explained in detail and has been implemented in a prototype tool. With some straightforward extensions, it can also be used as an important component in SBO for hierarchical flow graphs (see above). The results of applying the basic (flat) methodology on two applications from Alcatel, namely a Segment Protocol Processor and the "STORM" are presented at the end of the chapter. Using our methodology, only three 1-port memories are needed, whereas a conventional design methodology using Synopsys' Behavioral Compiler requires six memories for the same cycle budget. This clearly shows that our methodology allows to significantly reduce the bandwidth requirements.

10.8 DATA FLOW ANALYSIS REVISITED: BASIC GROUPS FOR NETWORK COMPONENTS

The data-flow analysis task defines the basic groups and derives a maximally parallel control/data-flow graph at the basic group level.

A virtual memory segment (VMS) is a chunk of memory allocated for storing data [376, 97]. In case of dynamically allocated data, a VMS is created for each dynamically allocated data *type*. Such a VMS contains memory space for storing all instances of the corresponding data type. In case of statically allocated data, a VMS is created for each statically allocated data *structure*.

For network components, described in the MATISSE methodology [97], we could use the Virtual Memory Segments (VMS) as basic groups.

Indeed, they obey the first three properties of the basic group partitioning (see section 10.3.1):

1. The virtual memory segments form a partitioning of all data that has to be stored in background memory.

2. The virtual memory segments are determined at compile time, and are therefore manifest.

3. Every read/write instruction is associated with a single virtual memory segment.

However, they do not necessarily obey the fourth property. Often the virtual memory segments can be partitioned further into basic groups while still satisfying the third basic group property. Examples of this are given in the next subsection.

Let us study some examples first. Consider the C-code fragment in figure 10.18. It defines a record type with four fields divided over three words: *Word1* contains *field1* and *field2*, *Word2* contains *field3*, and *Word3* contains *field4*. With this record type corresponds a virtual memory segment which is also shown in the figure. The VMS contains five frames. Due to the dynamic memory allocation we do not know at compile time to which frame the record pointers, *pRecord1* and *pRecord2*, point. However, if we look at the first data access instruction, *pRecord1* → *field1*, we see that it can only access the first words of the VMS frames. The second data access instruction, can only access the second words of the VMS frames. The last data access instruction, also can only access the first words of the VMS frames. Therefore, we can partition the VMS into basic groups as shown on the right side of figure 10.18. The first basic group (BG 0) collects all first words of the records. The second basic group collects all the second words, and so on. Note that we cannot split the basic groups further because this would violate the third basic group property. Indeed, a single data access instruction would then access more than one basic group.

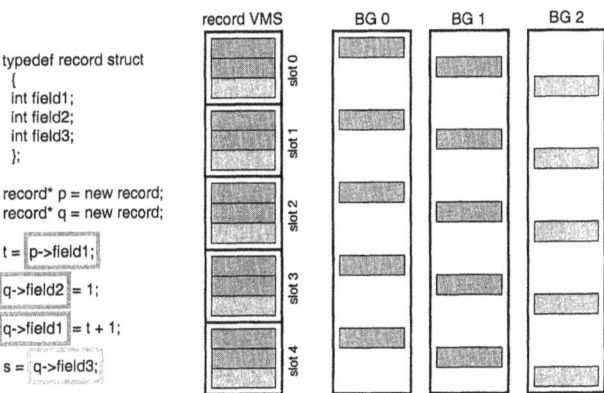

Figure 10.18. Basic Groups vs. Virtual Memory Segments.

Let us now consider the C-code fragment of figure 10.19. There is a VMS corresponding to the table data type. The first data access instruction of the code fragment accesses the first element of one of the tables as indicated by the hashed border around these data elements. However, we can not split the VMS according to this partition, because the second data access instruction is one with a data dependent index and can therefore access any data element in the VMS, as the frame is unknown (data dependent pointer value) and the index within the table is also unknown. Therefore, it can also access the first element of the tables. According to the third basic group property, all of these data elements must belong to the same basic group. In this case there is only one basic group which is accessed by both write instructions in the code fragment.

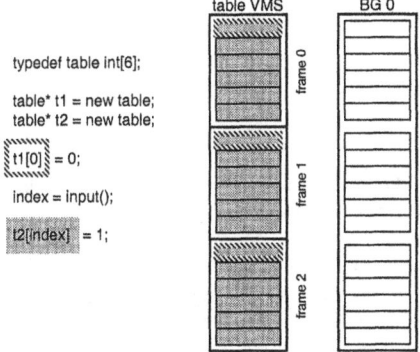

Figure 10.19. Virtual Memory Segments is Basic Group.

Finally, consider the C-code fragment of figure 10.20. There is again a VMS corresponding to the table data type. This time, however, the tables are statically allocated and therefore we know at compile time which table is being accessed by every data access instruction. Therefore, the VMS can be split according to the tables. The tables themselves can not be split here because they are accessed with a data dependent index which forces all data elements of a table to be in the same basic group.

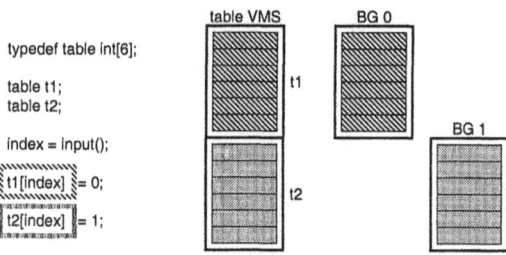

Figure 10.20.   Virtual Memory Segments split in Basic Groups.

Now we can propose an effective basic group partitioning methodology for dynamically allocated and/or data-dependent indexed data structures.

All compound data structures consist of a hierarchical composition of arrays and/or records:

- An *array* is a compound data structure consisting of an ordered set of objects. The objects all have the same type and are accessed via an index.

- A *record* is a compound data structure consisting of a number of named objects called fields. The fields do not have to be of the same type and are accessed by specifying their name.

The elements of an array and the fields of a record can themselves be arrays and records.

In the MATISSE design flow, all data that has to be stored in memory is contained in VMSes. These VMSes are compound data types as described above. For instance, the virtual memory management step assigns all list nodes (records) of a certain linked list type to a common VMS. The resulting VMS is an array of list nodes (records), of which the amount of slots is determined by means of analysis or simulation.

$$\text{var.field}_a \, [\text{index}_a].\text{field}_b \, [\text{index}_b].\text{field}_c \, [\text{index}_c] \dots .\text{field}_z$$
$$\quad \downarrow \quad \downarrow \quad\quad \downarrow \quad\quad \downarrow \quad\quad \downarrow \quad\quad \downarrow \quad\quad \downarrow \quad\quad\quad \downarrow$$
$$\quad cst \quad cst \quad var \quad cst \quad var \quad cst \quad var \quad\quad cst$$

$$p \to \text{field}_a \, [\text{index}_a].\text{field}_b \, [\text{index}_b].\text{field}_c \, [\text{index}_c] \dots .\text{field}_z$$
$$\quad \downarrow \quad\quad \downarrow \quad\quad \downarrow \quad\quad \downarrow \quad\quad \downarrow \quad\quad \downarrow \quad\quad \downarrow \quad\quad\quad \downarrow$$
$$\quad var \quad cst \quad var \quad cst \quad var \quad cst \quad var \quad\quad cst$$

Figure 10.21.   General addressing template. The top template is for statically allocated data. The bottom template is for dynamically allocated data.

Figure 10.21 shows a general addressing template for accessing a data element in such a compound data structure. Note that in generating specific symbolic addresses, parts in this template can be skipped if necessary: an index can follow an other index immediately without intervening field name, and a field name can follow an other field name immediately without intervening array index.

Parts of a symbolic address are fixed at compile time (indicated with 'cst' in 10.21), while others are variable (indicated with 'var' in 10.21) due to data dependent indexing, or iterator dependent indexing.

Assuming that all data elements with a symbolic address only differing in the variable parts, are mapped on the same basic group[4], we can derive the basic group being accessed from the symbolic address in the following way: take the data type of the compound data structure (corresponding to the VMS) and concatenate all fixed parameters in the symbolic address skipping the variable ones. The last field name has to be replaced by the word number to which this field belongs, because basic group partitioning stops at the word level (see third basic group property).

For the examples in the previous subsection this gives:

- figure 10.18: *record_word0, record_word1, record_word2*

- figure 10.19: *table*

- figure 10.20: *table_t1, table_t2*

10.9 CONFLICT DIRECTED ORDERING

Storage-bandwidth optimization is a very complex problem, even for flat CDFGs. It is very similar to scheduling for a given cycle budget, which is proven to be NP-complete. Hence, a heuristic is needed to obtain near optimal results in a reasonable amount of time for real-life applications. Therefore, we propose a heuristic algorithm called *conflict directed ordering* (CDO) for optimizing the storage-bandwidth of flat CDFGs. We have chosen for an iterative solution similar to Improved Force Directed Scheduling (IFDS) [409], which leads to very good schedules for a fixed cycle budget. The idea we have taken over from IFDS is to gradually refine the scheduling of operations (data accesses in our case), postponing the *definite* scheduling of operations as far as possible because then the scheduling of other operations can be estimated more accurately.

Figure 10.22 illustrates the CDO algorithm. The diagram in the top left corner of the figure shows for every data access instruction in the CDFG, the interval in which it can be scheduled. These intervals initially result from an ASAP-ALAP analysis of the CDFG and are gradually reduced during CDO. To steer the CDO algorithm towards an optimal solution, we use a cost function that predicts, from a given set of scheduling intervals, the final cost of the ECG in terms of cost function 10.1. Hence, our CDO cost function contains also three terms: one for estimating the total weighted conflict cost, one for estimating the chromatic number cost, and one for estimation the self conflict cost.

From the information in this diagram, the conflict probability between every pair of data access instructions is computed. These data access conflict probabilities are used to calculate the basic group conflict probabilities for every pair of basic groups ($P_{conflict}$ in figure 10.22). These basic group conflict probabilities are multiplied with their respective *conflict costs* and summed to get the *weighted conflict cost*, one part of the CDO cost function.

In order to estimate the chromatic number of the current state, real conflicts are needed instead of conflict probabilities. Therefore, a conflict graph is constructed that contains all conflicts corresponding to basic group conflict probabilities above a certain threshold value. The chromatic number of this conflict graph is then taken as an estimate for the bandwidth requirements, the second part of the CDO cost function.

To avoid self conflicts as much as possible, data access distribution graphs for every basic group are used. These are balanced using forces similar to those used in force directed scheduling [320, 409]. These forces form the third part of the CDO cost function.

The conflict directed ordering algorithm operates in an iterative way similar to IFDS [409]. Our cost function is completely different, however, from the one used in IFDS (because it takes into account *which* data is being accessed instead of only the number of simultaneous data accesses). Therefore, also the manipulations needed at each iteration are quite different. At every iteration, one of the data access intervals will be reduced by one cycle either at the beginning or the end of the interval. From all possible interval reduction candidates, the one that has the best effect on the overall cost is selected.

---

[4]This is a valid assumption for network component applications, where the indices are usually data dependent and span the whole index space. If this is not the case, a more elaborate analysis can partition the index space into a number of disjoint index sets. This has been explained in Section 6.2 for RMSP applications where this splitting is much more common.

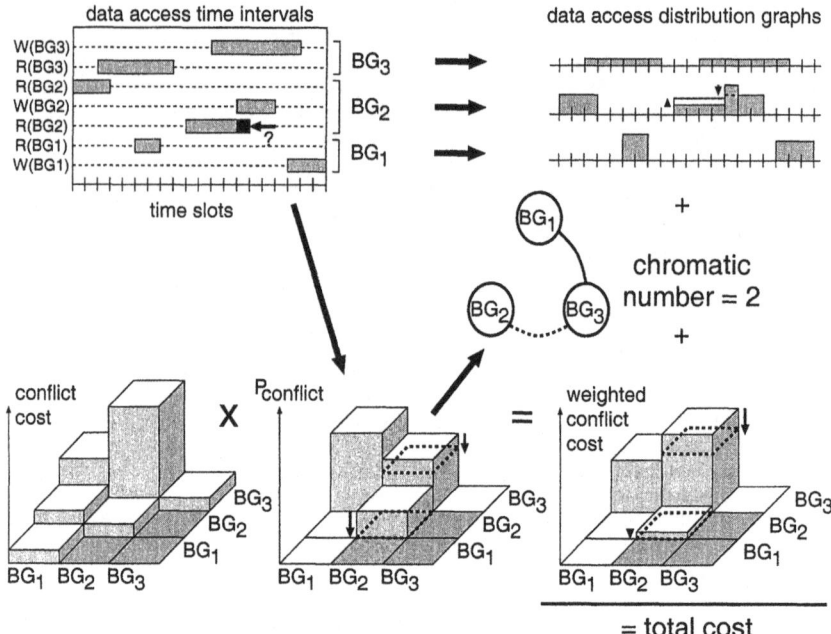

Figure 10.22.   Conflict Directed Ordering.

The iteration process stops when all basic group conflict probabilities are either 0.0 or 1.0, or in other words, when for every possible conflict it is known 100% sure whether or not it will occur, because we are only interested in the resulting (extended) conflict graph. This means that a complete ordering of all data access instructions is not necessary (as opposed to IFDS where the resulting schedule is the final goal).

10.9.1   CDO algorithm

This subsection lists and explains the different steps in the CDO algorithm.

10.9.1.1   Input.   The Conflict Directed Ordering algorithm takes the following inputs (see subsection 10.3.1):

- CDFG at basic group level;

- set of basic groups with their characteristics;

- timing constraints including the cycle budget.

10.9.1.2   Pre-processing steps.

1. *Calculate pairwise BG conflict costs*

   From the set of BGs and their characteristics, the pairwise basic group conflict costs are calculated as explained in Section 10.5.3.

2. *IO-profile assignment*

Table 10.1.   IO-profile in current prototype implementation.

Mode	Input			Output	Occupation
	Address	Data In	Write Enable	Data Out	Time
Read	0	×	×	1	1
Write	0	0	0	×	1

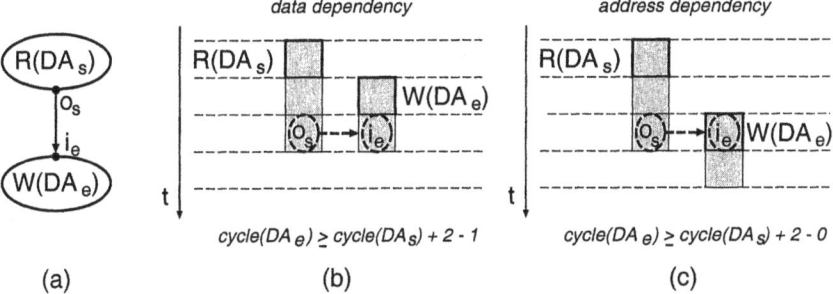

Figure 10.23.   Constraints for ASAP analysis: (a) dependency, (b) data dependency, (c) address dependency (e.g., pointer).

This step assigns an IO-profile to each BG based on the memory hierarchy level to which it has been assigned. The IO-profiles determine the minimum number of cycles that have to be scheduled between two data access instructions that have a data dependency between them (see subsection 10.3.2.1).

In the currently implemented prototype of SBO for flat graphs, the IO-profile shown in Table 10.1 is assumed. It corresponds to the IO-profile of an SRAM used in a pipelined way such that the occupation time is one storage cycle both for reads and writes. Data that is read becomes available one storage cycle after the read address has been supplied. Data that has to be written has to be supplied together with the address where it has to be stored.

3. *ASAP-ALAP analysis*

An ASAP-ALAP analysis is performed on the CDFG to find initial data access scheduling intervals.

Every data dependency starts at an output port, say o_s, of a data access instruction, say DA_s, and ends at an input port, say i_e, of another data access instruction, say DA_e.

The earliest cycle on which DA_e can be scheduled is then given by the following formula:

$$ASAP(DA_e) \geq ASAP(DA_s) + offset(o_s) - offset(i_e)$$

Of course, the formula for $ASAP(DA_e)$ must be evaluated for all dependencies that end at DA_e. $ASAP(DA_e)$ then has to be selected such that it satisfies all of them (this puts a minimum on $ASAP(DA_e)$).

Figure 10.23 illustrates this for a write operation that depends on a read operation, with IO-mode profiles shown in figure 10.16 and 10.17 for the read and write operations respectively. In case (b), the data value written depends on the data value read. The write operation has to be scheduled at least one cycle after the read operation. In case (c), the address to which something has to be written depends on the data value read (e.g., a pointer or data-dependent index). The write operation has to be scheduled at least two cycles after the read operation.

The latest cycle on which DA_s can be scheduled is given by the following formula:

$$ALAP(DA_s) \leq ALAP(DA_e) + offset(i_e) - offset(o_s)$$

Also here, the formula for $ALAP(DA_s)$ must be evaluated for all dependencies that start at DA_s. $ALAP(DA_s)$ then has to be selected again such that it satisfies all of them.

10.9.1.3   Iteration initialization.

1. *Calculate all data access conflict probabilities*

   During the conflict directed ordering, we have a scheduling interval for each data access instruction. From these scheduling intervals we can compute the probability of a conflict between two data access instructions.

   To keep the explanation simple, we only describe the calculation of the data access conflict probabilities for the IO-profile assumed in the currently implemented prototype tool (see table 10.1). With general IO-profiles, the computations become more complex. They are not necessary for understanding the proposed methodology, though. Therefore, they are not detailed here.

   From the scheduling intervals we can compute the probability of an overlap, which equals the conflict probability between the two corresponding data access instructions (for the assumed IO-profile). Not all data access instructions with overlapping scheduling intervals have a conflict probability larger than zero, though. Indeed, when there is a dependency between two data access instructions, one will always be scheduled before the other one, and therefore the conflict probability between these two data access instructions will always be zero. Also, when two data access instructions are mutually exclusive because they belong to two different branches of a condition, their conflict probability is always zero.

   The conflict probability between two data access instructions A_i and B_j, accessing basic groups A and B respectively, is given by (see figure 10.24a):

   - if A_i depends on B_j, or B_j depends on A_i:

     $$P_{conflict}(A_i, B_j) = 0.0$$

   - if A_i and B_j are mutually exclusive:

     $$P_{conflict}(A_i, B_j) = 0.0$$

   - otherwise:

     $$P_{conflict}(A_i, B_j) = \frac{c_{ij}}{a_i \cdot b_j}$$

   where a_i and b_j are the length (i.e., last cycle - first cycle + 1) of the scheduling intervals of data access instructions A_i and B_j respectively, and c_{ij} is the length of the overlap of these scheduling intervals (see figure 10.24a).

2. *Calculate all basic group conflict probabilities*

   To calculate the conflict probability between two BGs, all possible overlaps between the scheduling intervals of data access instructions to these BG have to be considered. To make the calculations tractable we assume that for every possible overlap, the probability for a conflict is independent from the conflict probabilities of the other overlaps (see figure 10.24b). Under this assumption it is possible to calculate the conflict probability between every two overlapping data access instructions. These conflict probabilities can then be combined to obtain the conflict probability between the two BGs.

   Under the assumption that all $P_{conflict}(A_i, B_j)$ are independent from each other, the conflict probability between two basic groups A and B is given by:

   $$P_{conflict}(A, B) = 1 - \prod_{i,j}(1 - P_{conflict}(A_i, B_j))$$

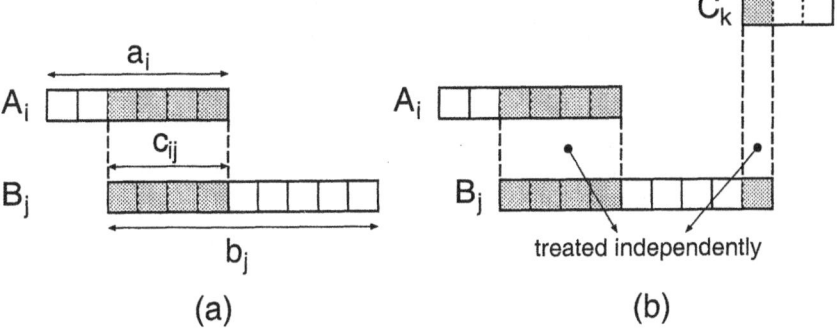

Figure 10.24. Calculation of the conflict probability: (a) definition of the constants, (b) approximation: different overlaps are assumed to be independent.

3. *Calculate chromatic number*

Because the chromatic number of a graph weighted with probabilities for the edges is not defined, we propose to introduce a threshold probability to obtain an estimate. A conflict graph containing all the conflicts between BGs with a probability above the threshold probability is constructed. The chromatic number is then calculated for this conflict graph.

4. *Calculate initial value of cost function of subsection 10.5*

5. *Initialize set of possible moves*

At each iteration of the CDO algorithm one data access scheduling interval is selected and reduced by one cycle, either at the beginning of the interval or at the end of the interval. Only scheduling intervals that can still be reduced (i.e., have a length larger than one cycle) and that can still have an effect on the cost function (i.e., overlap with other scheduling intervals) have to be considered. Each of these intervals leads to two possible moves: a reduction at the beginning or at the end of this interval. All of these are collected in a set of possible moves. During the CDO algorithm this set will shrink until no moves are possible anymore.

10.9.1.4   Iteration (until no more moves are possible).

1. For each possible move, calculate its effect on the cost function in the following way:

   (a) Determine which data access scheduling intervals are *indirectly* being reduced by the move. Due to dependencies between data access instructions, the reduction of one scheduling interval can force the reduction of other scheduling intervals as well. These indirect scheduling interval reductions affect the cost function in the same way as the direct schedule interval reductions. Therefore, it is very important to take them into account when determining the effect of a possible move.

   (b) For each data access scheduling interval that is being reduced:

      i. Calculate the change in conflict probability between the data access instruction of which the scheduling interval is being reduced, and all data access instructions that are possibly overlapping, i.e., those having scheduling intervals that overlap with the scheduling interval under consideration before it is reduced.

      ii. Calculate from the changes in conflict probability between data access instructions, the changes in conflict probability between BGs.

   (c) If at least one basic group conflict probability crosses the threshold value for inclusion in the conflict graph, the chromatic number of the conflict graph has to be recalculated.

(d) The effect on the cost function has to be calculated based on the changes in basic group conflict probabilities, and the change of chromatic number of the conflict graph.

2. Select the move that has the best effect on the cost function, and perform it.

3. Remove from the set of possible moves, all moves that directly reduce scheduling intervals that are not overlapping with other scheduling intervals anymore or that can not be reduced any further. In addition remove from the set of possible moves, all moves that directly reduce scheduling intervals that are only overlapping with intervals of basic groups that are known to be in conflict already. The latter moves have no effect on the cost function anymore but applying them would reduce the scheduling freedom.

10.9.1.5 Post-processing step: ECG construction.   Given the scheduling intervals of all data access instructions in the CDFG, the sets of mutually exclusive data access instructions, and the dependencies between the data access instructions, we can now calculate the maximal number of simultaneous read, simultaneous write, and simultaneous data accesses, between every set of basic groups that are in conflict. Not all overlapping scheduling intervals lead to a simultaneous data access because some of these data accesses are performed under mutually exclusive conditions and others are never performed simultaneously because there is a dependency between them, forcing one to be executed after the other. A careful analysis is needed here to obtain each of three numbers: one for the number of simultaneous reads, one for the number of simultaneous writes, and one for the number of simultaneous data accesses. This is illustrated in figure 10.25.

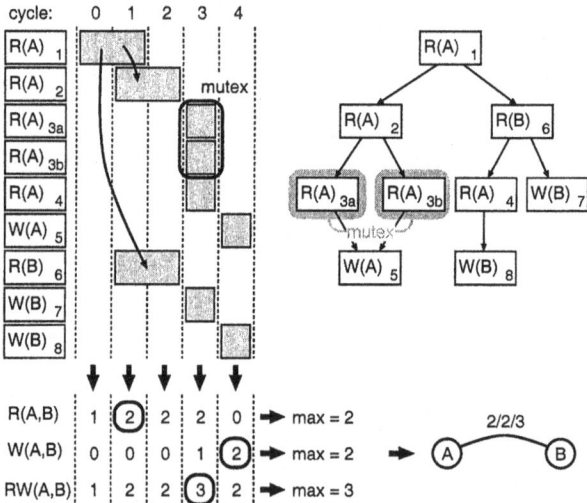

Figure 10.25.   Extended Conflict Graph construction.

If, after conflict directed ordering, there still exist overlapping intervals with length larger than one cycle, they can be reduced further to optimize the R/W/RW numbers on the (hyper)edges in the ECG. This optimization is, however, not of primary importance, as the best solutions tend to avoid multi-port memories anyway.

10.9.1.6 Output.   The main output of the Conflict Directed Ordering algorithm is of course the extended conflict graph. The partial ordering obtained during the CDO algorithm can also be written out if desired.

10.9.2 Differences with IFDS

Considering scheduling intervals that are gradually reduced until the desired result is obtained is an idea taken from IFDS (Improved Force-Directed Scheduling). The cost function used to determine *which* scheduling interval has to be reduced at each iteration is fundamentally different, though. This leads to a different optimization methodology.

The main difference is that our cost function takes into account *which* data (in terms of BGs) is being accessed in parallel, whereas IFDS only takes the *number* of parallel data accesses to reduce the required memory bandwidth. This allows for a much *more global* optimization compared to IFDS. For instance, when the decision is taken to schedule two data accesses in the same time slot this involves a certain cost (called the conflict cost), because the two corresponding BGs have to be stored either in two separate memories or in a multi-port memory. However, once this decision is taken, data accesses to these BGs can be scheduled in parallel many times without any additional cost. Therefore, this cost should be counted only once, which is done in CDO, but is impossible in (I)FDS because it does not take into account *which* data is being accessed. Optimizing the total conflict cost is a global optimization, whereas balancing the number of simultaneous data accesses is a local optimization which can be very bad globally. So, one very important advantage of CDO over IFDS is that CDO takes into account conflicts between BGs whereas IFDS does not.

Using the properties of the BGs, it is possible to weigh the conflict costs for every pair of BGs because some BGs are preferentially stored together while others are preferentially stored in separate memories. This is again an important advantage over IFDS where all data is treated equally.

Another important difference is that the chromatic number of the conflict graph is taken into account. Again, this is done in order to have a more global optimization of the memory cost. Indeed, the required number of memories cannot be estimated accurately by looking locally only, as is done in IFDS, because all conflicts of the whole algorithm have to be considered for this.

It should be clear by now that we try to optimize global measures (i.e., the total conflict cost and the size of the chromatic number of the conflict graph) of the algorithm, instead of locally balancing the CDFG by means of forces as is done in (I)FDS.

Only for minimizing the number of simultaneous accesses to *the same* basic group (self-conflicts), we use forces similar to IFDS. The difference here, however, is that we balance the number of simultaneous accesses *per basic group*, whereas IFDS balances the *total* number of simultaneous accesses. The resulting amount of self-conflicts can still be very bad in the case of IFDS, forcing the use of multi-port memories with an excessive amount of ports.

An other less important difference between IFDS and CDO is that the goal of the former is an optimal schedule, whereas the goal of the latter is an optimal conflict graph. This also leads to a different stop criterion for both algorithms: IFDS stops when everything is fully scheduled, CDO stops when for all possible BG conflicts it is known whether or not it is needed to meet the cycle budget constraint. Usually this means that CDO can stop *before* a full ordering is obtained.

10.9.3 Complexity

Let n be the number of data accesses, $b \leq n$ be the number of basic groups, F be the cycle budget, $l \leq F$ be the maximum length of a data access' schedule interval, and $m \leq n$ be the maximum number of simultaneously overlapping intervals.

The total number of moves is $\mathcal{O}(ln)$. For each move $\mathcal{O}(n)$ candidate moves are examined. For each candidate move $\mathcal{O}(n)$ intervals can be reduced. Each interval to be reduced can possibly overlap with $\mathcal{O}(m)$ intervals. Thus in the worst case:

- $\mathcal{O}(lmn^3)$ operations are needed to calculate the weighted basic group conflict costs,

- $\mathcal{O}(ln^2)$ chromatic number problems have to be solved to calculate the chromatic number costs,

- $\mathcal{O}(bFln^2)$ operations are needed to calculate the self conflict costs, and

- $\mathcal{O}(ln^2)$ operations are needed to decide on the move to be taken.

Clearly, the worst case complexity is quite high. Luckily, in practice, the number of operations is much lower than predicted by this number: the number of intervals that have to be reduced for each candidate move is usually much smaller than n. Moreover, the number of overlapping intervals is usually also much smaller than n because of dependencies, mutual exclusive operations, and the ASAP-ALAP analysis that determines that some intervals cannot be overlapping if the timing constraints have to be met. The chromatic number problems are themselves NP-complete. Luckily, in practice, solving them turns out to cause few problems.

10.9.4 Extensions

This subsection describes a number of extensions for the basic SBO described above. These extensions are easy to implement and allow to use SBO for flat CDFGs as a component for SBO for hierarchical CDFGs. They also allow for better user control.

10.9.4.1 Pre-defined conflicts. Allowing the user to specify a number of conflicts that have to be present in the conflict graph, is a simple but very useful extension. It gives the user more control, as he can decide which basic groups should be definitely in different memories. It is also required for using SBO for flat CDFGs as a component of SBO for hierarchical CDFGs (not handled in more depth in this book yet).

10.9.4.2 Pre-clustered basic groups. Allowing the user to specify which basic groups should be assigned to the same memory, is another useful extension. For SBO, this means that a certain number of basic groups are replaced by a clustered basic group. This clustered basic group has the following properties:

- its *word depth* is estimated by high-level in-place techniques taking into account the word depths of the BGs it replaces and the life-times of these BGs,

- its *word-length* is the maximum of the word-lengths of the BGs it replaces,

- the *average number of read accesses* is the sum of the average number of read accesses of the BGs it replaces, and

- the *average number of write accesses* is the sum of the average number of write accesses of the BGs it replaces.

The remaining characteristics should be the same for all BGs being clustered. Therefore, these characteristics of the clustered BG can be the same as those of the BGs it replaces.

10.9.4.3 (Partial) pre-assignment. The previous two extensions allow the user to do a (partial) pre-assignment, i.e., assign (part of) the basic groups to memories *before* performing the SBO. To do this, all basic groups assigned to the same memory should be clustered together, and a conflict should be defined between every two basic groups assigned to different memories. The SBO will take these conflicts into account by preferring the pre-defined conflicts over the others. It will add more conflicts and self conflicts if necessary to meet the cycle budget constraints.

10.9.4.4 (Partial) pre-allocation. Similar to specifying a pre-assignment, it can be useful to allow the user to specify a (partial) pre-allocation, i.e., specify a number of memories that should be present in the final memory architecture. This can lead to better results because SBO otherwise tries to minimize the bandwidth requirements, as shown in figure 10.26. This figure shows the effect of pre-allocation. Consider the conflict graph shown on the left hand side. Suppose one extra conflict is necessary to meet the cycle budget constraints. Two cases are considered: one without pre-allocation (rising arrow) and one with pre-allocation (descending arrow). In case no pre-allocation is done, SBO will select conflict A–D as the other conflicts would increase the chromatic number from 2 to 3. The resulting conflict graph allows for a memory architecture with only two 1-port memories. In this case, however, this leads to a large waste of bits because the conflicts force basic groups with different word-length to the same memory (top memory architecture). If three memories are allocated

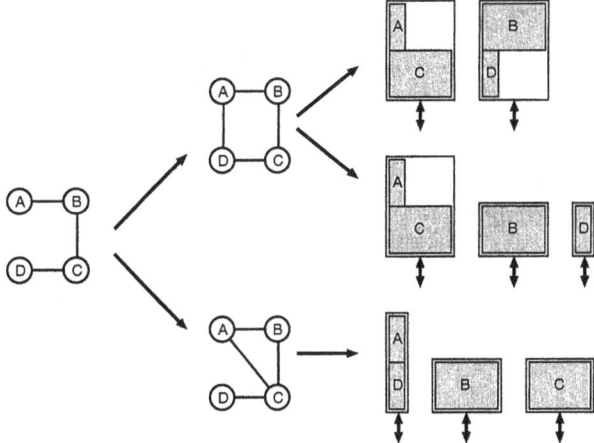

Figure 10.26.   Effect of pre-allocating three memories.

afterwards, the number of bits wasted is reduced, but the situation is still not ideal (middle memory architecture). If we compare this to the case where three memories are pre-allocated, we see that the latter leads to better results (bottom memory architecture), as SBO now selects the conflict $A-C$ (or $B-D$) because chromatic numbers up to 3 are not penalized and conflict $A-C$ (and $B-D$) are preferred over the conflict $A-D$ because of the difference in word-length (see pairwise BG conflict costs in section 10.5.3).

10.9.4.5 Multiple CDFGs. Allowing multiple CDFGs to be optimized simultaneously with a *common* conflict graph, is an extension that is especially useful for using SBO for Flat Graphs as a component of SBO for Hierarchical CDFGs. In this case, each CDFG represents part of algorithm (e.g., a loop body). If each of these CDFGs would be optimized separately, the conflict graph for the complete application would be the union of all the conflict graphs containing much more conflicts than necessary. If these CDFGs are optimized together, conflicts that have a large effect in many CDFGs can be preferred over conflicts that are only effective in one or a few CDFGs. In general this will lead to much cheaper conflict graphs for the application as a whole.

SBO for Flat Graphs can easily be extended for this, as shown in figure 10.27. The time line is divided into a number of sections, one for each FG to be optimized. Each section has a length equal to the cycle budget in which its corresponding CDFG has to be scheduled. An ASAP-ALAP analysis is done for each CDFG, taking the boundaries of its section into account. From then on, the basic SBO for Flat Graphs can be applied without modification. The scheduling intervals of data access instructions belonging to two different CDFGs are never overlapping, therefore there will never be conflicts between CDFGs.

10.10   RESULTS ON SEGMENT PROTOCOL PROCESSOR

The basic SBO methodology described in the previous section has been implemented in a prototype tool. Experiments with the prototype tool on Alcatel's Segment Protocol Processor (SPP) application [111] have shown some promising results.

The reduced SPP contains 14 basic groups and 38 read/write instructions. Figure 1.4 shows how the data of this application is organized in table structures with pointers and linked lists. For every data type, an amount of memory is reserved in what is called a virtual memory segment. The size of the virtual memory segments is determined either by static analysis or by simulation/profiling. The MID-tables, IPI-records, and RR-records are dynamically allocated and deallocated. This is done by the custom made dynamic memory managers constructed for each of the corresponding virtual

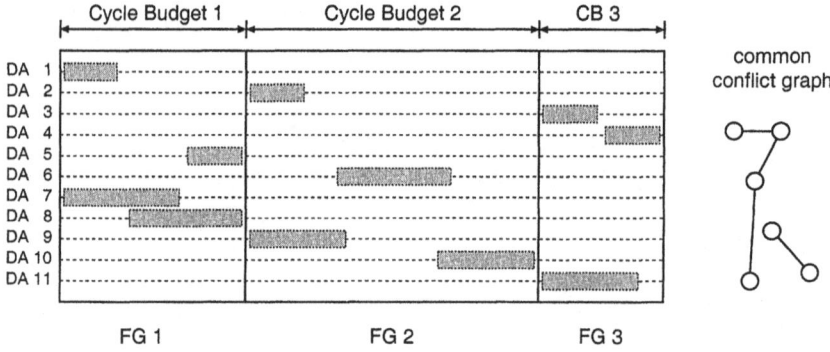

Figure 10.27.   Optimizing multiple CDFGs simultaneously.

memory segments [97]. The virtual memory segments are split up into basic groups, as far as possible, with the restriction that for every data access instruction it must be known at compile time which basic group it accesses. This means that tables that are accessed with data dependent indices belong as a whole to one basic group (e.g., the LID-table as a whole is one BG). This also means that data that is dynamically allocated cannot be separated from other data of the same type that is dynamically allocated, because it is accessed using pointers of which the value is not known at compile time (e.g., all data stored in MID-tables forms one BG). Remark that virtual memory segments containing records can be split into several BGs (one for each word of the corresponding record), because at compile time it is known for each data access instruction which field of the record is being accessed (even when it is not known which record is being accessed). This results in one BG for every word of a record of a given type.

An overview of the experiment is given in figure 10.28. The left-hand side shows the conventional design flow, which we use as a reference for this experiment. The right-hand side shows our proposed design flow. As a back-end behavioural synthesis tool we have used Behavioral Compiler (BC) of Synopsys, including its scheduler.

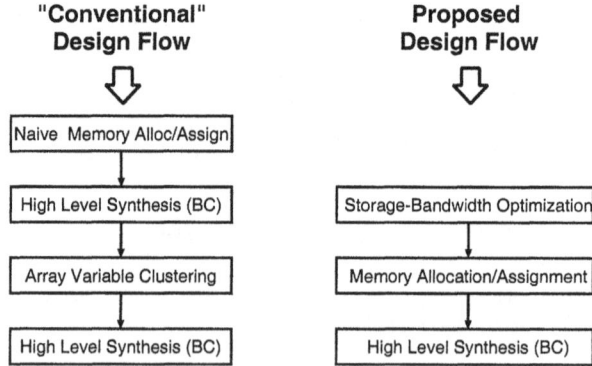

Figure 10.28.   Proposed design flow versus "conventional" design flow.

10.10.1 Conventional design flow

As current behavioural synthesis tools do not support memory management, memory allocation and assignment has to be done beforehand. Therefore, the first step in the conventional design flow is an allocation/assignment step that allocates exactly one memory for each basic group and assigns the basic group to it. [5] In the second step, BC schedules the application. A conflict graph is constructed from the resulting schedule, which is then used in the array variable clustering step [339] to combine memories that are never accessed simultaneously. If desired, BC can then be run again to take the new allocation and assignment into account.

For a cycle budget of 17, this design flow resulted in a conflict graph (figure 10.29) with 25 conflicts and a chromatic number equal to 6, requiring six 1-port memories (figure 10.30) after array variable clustering.

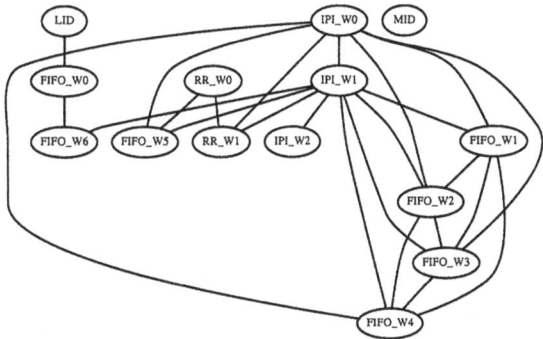

Figure 10.29.   Conflict graph obtained after scheduling with Synopsys' BC.

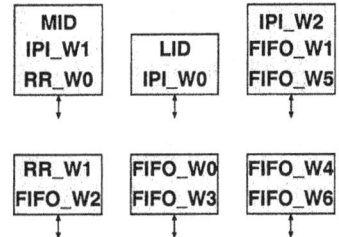

Figure 10.30.   Memory configuration with conventional design flow

10.10.2 Proposed design flow

In the proposed design flow, the first step is storage-bandwidth optimization, which minimizes the required memory bandwidth, followed by the memory allocation/assignment step [376], and finally the detailed scheduling by Synopsys' BC.

The conflict graph (figure 10.31) obtained with our SBO technique, for the same cycle budget of 17 cycles, contains only 13 conflicts and has a chromatic number equal to 3, which results in an optimal memory allocation of three 1-port memories (figure 10.32). Hence, the required memory bandwidth obtained with our methodology is only half that of the conventional one. All of the memories involved

---

[5]The conventional design flow normally does not use basic groups. But as we want to show here the benefit of applying SBO, and not the benefit of basic group splitting, we performed the basic group splitting for the conventional design flow also.

are too large to be put on-chip. Therefore, the reduced memory bandwidth obviously leads to a much smaller implementation cost because of the reduced amount of system components (off-chip memories) and I/O pins on the ASIC. Since the power consumption of off-chip memories is not very much dependent on the size but only on the amount of accesses, which the SBO task cannot change, the difference in power consumption will be negligible. The run-time of our CDO algorithm for this example on an HP9000-715/50 workstation was 130 sec. After expressing the memory architecture constraints in the VHDL code, Synopsys' BC found a valid schedule within the cycle budget of 17 cycles.

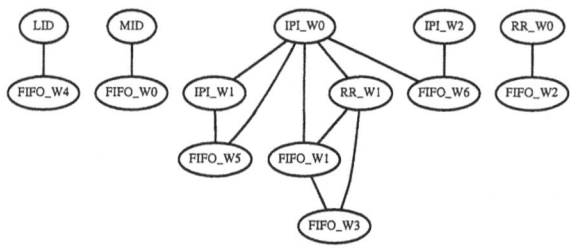

Figure 10.31.   Conflict graph obtained after storage-bandwidth optimization.

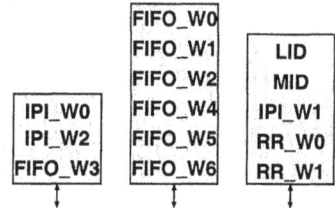

Figure 10.32.   Optimal memory configuration with proposed design flow

10.11   RESULTS ON STORM APPLICATION

We also applied our prototype SBO tool on a second test-vehicle, namely part of the STORM application specified by Alcatel.

10.11.1   The test vehicle

The STORM application, which stands for *STM1-level to Transputer-netwOrk Relayer with Multiple protocol handling*, deals with protocol conversion between an ATM transportation layer and a transputer network (see figure 10.33).

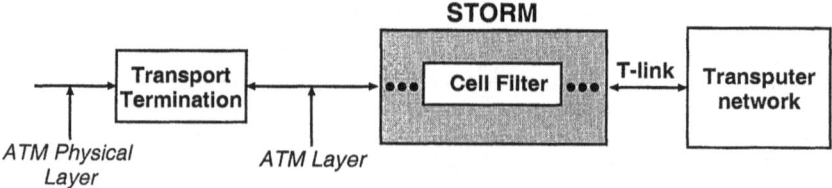

Figure 10.33.   Position of STORM in ATM context

We only focus on part of the STORM application, namely the cell filter. This part is representative, because, like the other subsystems, it is rather heavily IO-dominated: many background memory accesses have to be performed in real-time within a limited number of cycles. It relates incoming ATM cells to allocated internal streams, by means of two search algorithms (see figure 10.34), which can be combined in a pipelined way, i.e., they can be executed in parallel but acting on subsequent ATM cells.

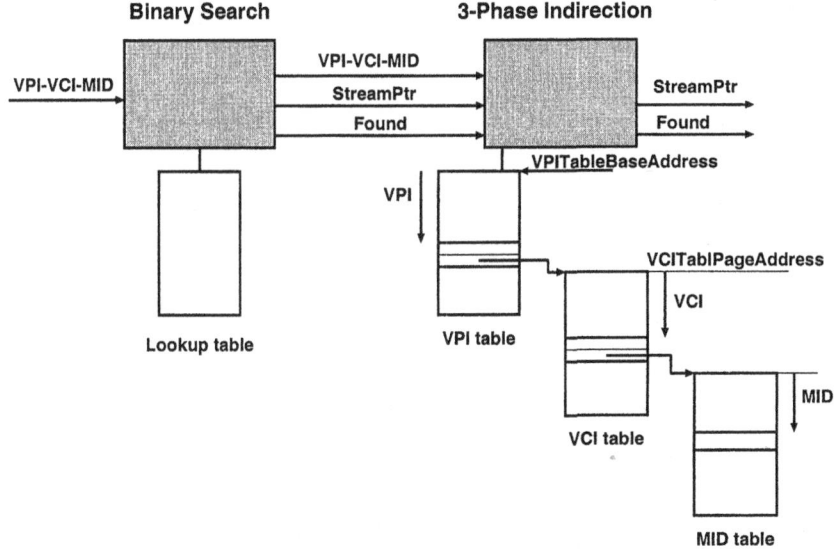

Figure 10.34.   Cell filter

Both search algorithms, binary search and 3-phase indirection, are looking for the same information but in a different way. The first algorithm uses the combination of some information fields of an ATM cell as one key into one big sorted table, stored in a *LookupRAM*. If the stream is not found in this table, the second algorithm looks for the stream in three steps. At each step, one of the subkeys (*VPI*, *VCI*, and *MID* extracted from the incoming ATM cell) is used as an index in the corresponding table, to find a pointer to the next table. The last table contains a pointer to the related stream.

10.11.2 Results

In this subsection, we show the results of our design space exploration experiments for the cell filter of the STORM application, by varying some of the design parameters.

To see the effect of the cycle budget on the resulting memory architecture, we have varied the cycle budget from the critical path length, 3 in this case, to the number of data access instructions in the CDFG of the application, 18 in this case. Assuming that the IO-profiles are such that every data access requires exactly one cycle, the minimal bandwidth is reached (i.e., one memory port) for this amount of cycles. Allocating more cycles will have no further effect on the memory architecture.

The graph at the top of figure 10.35 shows the extended conflict graph cost obtained with our SBO prototype tool in function of the cycle budget. Ideally the cost should be monotonically non-increasing, as is the case for this example. This is, however, not guaranteed with the proposed SBO methodology, as it is a heuristic that does not necessarily find the global optimum. The graph at the bottom of the same figure shows the power consumption of a number of power optimized memory architectures corresponding to the ECGs derived for each cycle budget. The memory configurations considered contain from one to six 1-port memories. The ECG obtained for $CB = 3$ contains a self-conflict. For larger cycle budgets the ECGs do not contain any self-conflicts anymore, and memory configurations consisting of only 1-port memories are feasible. The larger the CB the less memories are required in the memory configurations. The results illustrate the fact that the more memories are allocated, the lower the total power consumption. This is normal because more memories means smaller and hence less power consuming memories.[6] The graph also clearly shows the positive effect of SBO: when more freedom (i.e., a larger cycle budget) is given to the task, this is effectively used to

1. avoid self-conflicts which force the allocation of multi-port memories,

2. enable memory configurations with fewer ports and/or memories, and

3. obtain more power (and also area) efficient results for a given memory configuration.

This is only true for the general trend, though. Exceptions on this rule are very well possible. For instance, the bottom graph shows that the ECG obtained for $CB = 10$ leads to better results than the ECG obtained for $CB = 11$ when three memories are allocated. The exceptions are caused by two facts:

1. there is no one-to-one relation between the ECG cost and the memory architecture cost: an ECG with lower cost is likely to lead to a lower memory architecture cost, but this is not guaranteed.

2. the proposed SBO technique is a heuristic: it does not always find the optimal solution (with minimal ECG cost).

In the example of figure 10.35, two different ECGs are derived for cycle budgets of 10 and 11 cycles. These two ECGs have exactly the same cost, but one of them leads to a better solution after memory allocation and assignment. Nevertheless, the experiment shows that overall the proposed SBO methodology leads to very good results. For $CB = 18$ and larger, the ECGs do not contain any conflicts anymore. The resulting power cost shown at the right hand side of the graph is therefore the minimal reachable power cost for the given memory configurations. It is a lower bound for the power which is not necessarily reachable for the given timing constraints. We can see that the results after SBO optimization is never far above this lower bound and that they quickly converge towards the lower bound when the constraints are relaxed.

10.12 CONCLUSIONS

In this chapter we have shown that it is important to do a proper SCBD which as main substep SBO, to arrive at a good input for the memory allocation/assignment problem for large groups of data. We have also shown that SBO has to be done at the level of groups of scalars in such a way that the resulting conflict graph is optimized as opposed to most existing approaches. In addition, one has to take into account *which* data is being accessed in parallel, instead of only considering the *number* of parallel data accesses which is done in other approaches. This leads to the optimization of a conflict graph, for which an appropriate cost function has been derived. The result is a formalized definition of the problem in terms of inputs, outputs, constraints, and a cost function. In the next chapters, formalized solution methodologies will be proposed.

In addition, we have proposed an SBO methodology for data-dominated applications containing data-dependent conditions but *no loops*. The majority of the network component applications targeted by the MATISSE design flow fit in this category. A custom basic group partitioning methodology was presented for the complex dynamically allocated data types found in these applications. The conflict directed ordering algorithm was proposed and explained in detail. Also some straightforward extensions have been presented allowing to use CDO as a component for SBO for Hierarchical CDFGs as needed for RMSP applications. More details will not be provided here. The status of our research in this direction is partly discussed in [441]. The conflict directed ordering technique was implemented

---

[6]The influence of interconnect is not taken into account here, because it is very difficult to estimate before place and route. Moreover, adding more memories does not necessarily increase the power dissipated in interconnect, because it may allow to put the memories closer to the data paths. Anyhow, the contribution of on-chip interconnect to the power consumption is small compared to the power consumption of the on-chip memories themselves.

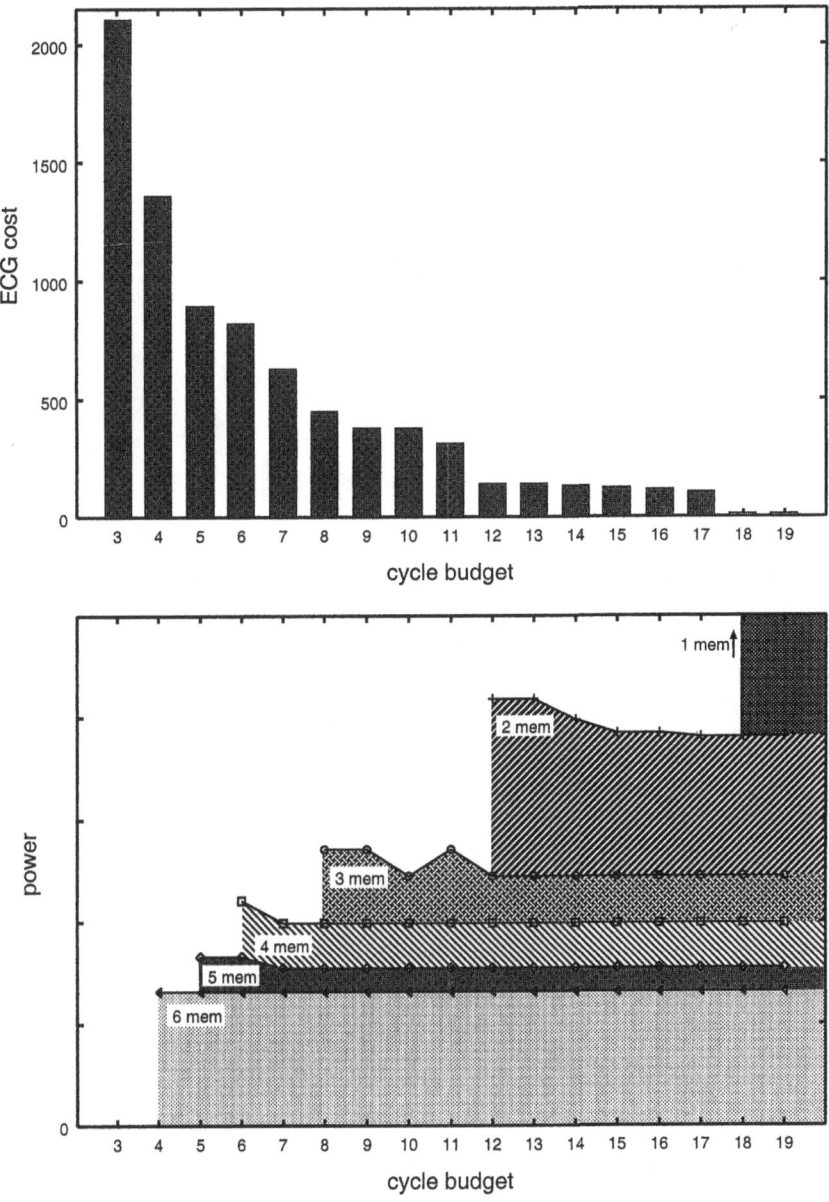

Figure 10.35.   ECG cost and power for different cycle budgets.

in a prototype tool. Experiments with the prototype tool on realistic ATM applications have shown very promising results. For instance, applying our methodology on the Segment Protocol Processor application, required only three 1-port memories to meet the cycle budget, whereas a conventional methodology using Synopsys' Behavioral Compiler and array variable clustering required 6 memories for the same cycle budget. This clearly shows that our methodology has a very positive effect on the required memory bandwidth.

11 MEMORY ALLOCATION AND ASSIGNMENT

In embedded systems the memory architecture can be more or less freely chosen. Different choices can lead to solutions with a very different cost, which means that it is important to make the right choice. Therefore, we have introduced a memory allocation and assignment step in our design flow, where trade-offs are made to arrive to an optimal memory architecture. In this step we take into account both the bandwidth constraints defined in chapter 10, and the target implementation technology.

This chapter defines the memory architecture problem in detail (section 11.1). Next, in section 11.2 it gives an overview of the complete memory allocation and assignment step, followed by four sections detailing each of the four substeps. Finally, section 11.8 illustrates the methodology with an example.

11.1 GOALS AND MOTIVATION

The memory allocation and assignment step exploits the freedom in the memory architecture to minimize its cost. This section explains which freedom the memory architecture offers, and how the freedom affects the cost of the memory architecture. Figure 10.2 has shown a target background memory architecture for a single processor unit.

The technique proposed here considers the *background* memory architecture separately for each processor unit. The foreground memory will be optimized together with the processor unit after our memory management is completed. The part of the background memory architecture shared between tasks and processors creates some additional issues which are currently under research. Chapter 14 considers some of these and gives directions for future work.

11.1.1 Memory architecture freedom

The assumption is that the memory architecture can be fully or partially chosen. Partial constraints occur for instance when a DSP processor with an on-chip cache is used, forcing the use of that particular memory. Nevertheless, there is always freedom in the further layers of hierarchy, and especially in the assignment of the signals to the available memories. The number of memories can be chosen, what type of memory will be used, and how large they should be.

In a distributed memory architecture, there is also freedom in the assignment of signals to the memories. This determines how large each of the memories needs to be: a memory can be chosen just large enough to accommodate all signals assigned to it. The signal-to-memory assignment freedom exists even if part of the memory architecture is fixed due to the target implementation.

The freedom of the memory architecture is constrained by the requirements of the application. These have been taken into account by the previous substeps and are refined into constraints for the memory allocation and assignment step. The most important example is the extended conflict graph resulting from the storage-bandwidth optimization step (see subsection 10.3.3).

11.1.2 Motivation

The goal of the memory allocation and assignment step is to make use of the memory architecture freedom to minimize the cost related to background memory storage and transfers. Possible cost measures include the power consumption (average or worst case), the active chip area, unused memory bits, number of board components, market price of external memory modules. These cost measures are estimated at a high level, but taking into account the target implementation technology.

The results in section 11.8 demonstrate the importance of making a good memory allocation and assignment decision. Consider for example the power consumption due to the memory architecture. Because the power consumption per access of SRAM memories increases with increasing memory size, it is important to keep memories small. One way of doing this is to use many small memories instead of one large one, but this will incur overhead in routing which also takes up power, and in active chip area. Another possibility is making sure that heavily accessed memories are small, while large memories are not accessed so often (as discussed in section 9.2). In conclusion, different alternatives will have to be explored to find one with an acceptable power consumption. Careful trade-offs with other cost measures have to be made, for instance with active chip area.

The memory allocation and assignment step is placed at this precise point in our DTSE methodology for two reasons. On the one hand, it is the final product of memory management, so it depends on all other steps. It must therefore be situated as late as possible in the overall methodology. On the other hand, the storage order optimization step (chapter 12) depends on the assignment of the arrays to the memories. Indeed, arrays can only be merged (see section 12.3) when they are stored in the same memory. Section 11.2.4 elaborates on the interaction between the memory allocation and assignment step and the storage order optimization step.

As discussed in chapter 2, related work does not consider this practical context of memory allocation and assignment, so we had to propose new solution techniques. Preliminary versions of our approach have been published in [27, 28, 32]. More recent results are disclosed in [376].

11.2 OVERALL METHODOLOGY OVERVIEW

To tackle the problem of finding a good memory architecture, we have subdivided the memory allocation and assignment step in four substeps. The two main substeps are called allocation resp. assignment. Two additional, optional substeps are introduced to exploit some extra freedom and especially to arrive at complete and valid solutions within the constraints of available memory library and the interconnection rules. This section will first outline the four substeps and how they interact. Next it details the inputs and the outputs of the allocation and assignment step. Finally, special attention is paid to the interaction with the storage order optimization step.

11.2.1 Substeps

Figure 11.1 summarizes the substeps in our memory allocation and assignment step. The order of the substeps is depicted on the left hand side. On the right hand side, the result of each of the substeps is represented. The following explains each of the substeps in more detail.

1. The **memory allocation substep**[1] (allocation for short) fixes the number of memories and the type of each of them. 'Type' includes the number of access ports to the memory, whether it will be an on-chip or off-chip memory. Also the vendor or the module generator (for on-chip memories) is chosen, and the model (e.g. a high-speed model or a low-power model).

---

[1]Note that in literature, the term 'allocation' is often used to indicate assignment as well.

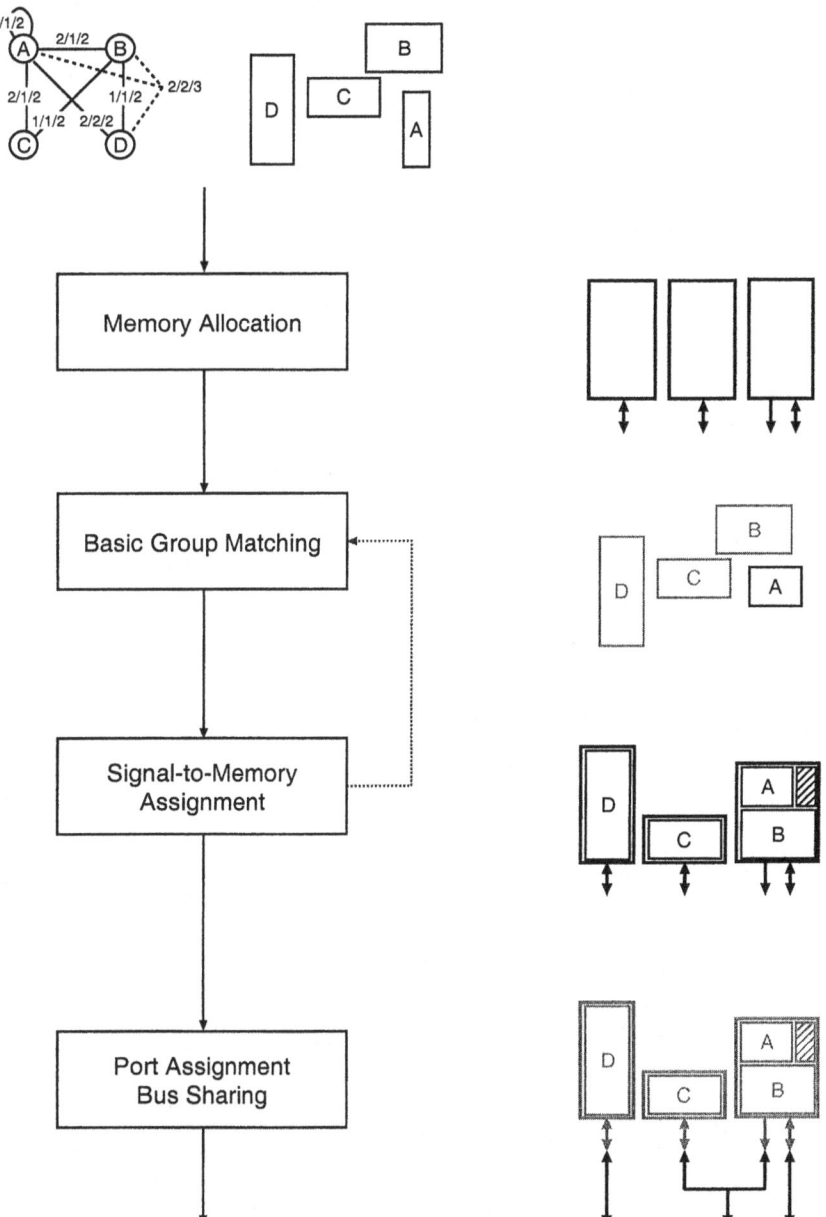

Figure 11.1.   Overview of the memory allocation and assignment substeps

The dimensioning of the memories is however not yet fixed in the allocation substep. A memory's word-length and size are chosen based on the data which will be stored in the memory, which is only decided in the assignment substep.

2. The **basic group matching substep** is a preprocessing of basic groups to make them fit better in the memory architecture. A basic group can have a word-length which is too large or too small, or it may simply be too large to fit in any memory. In this case, it must be restructured before it can be assigned to a memory.

   Note the feedback loop between the basic group matching substep and the assignment substep. The reason for this loop will be explained with basic group matching, in section 11.5.4, page 245.

3. The **signal-to-memory assignment substep** (assignment for short) decides for each of the application's signals in which of the allocated memories it will be stored. A memory's dimensions are determined by the characteristics of the signals assigned to it.

4. The **port assignment and bus sharing** substep optimizes the interconnections between memories and the processor unit. Connections are eliminated by multiplexing busses, making the interconnection architecture cheaper. This is called bus sharing. Access-to-port assignment (port assignment for short) is required for multi-port memories: before busses can be shared, a port has to be chosen for every access to the multi-port memory.

The first three substeps can be performed for every memory hierarchy layer independently. Indeed, the memories in one layer do not influence the memories in the other layers. This is because all basic groups have been unequivocally assigned to one memory hierarchy layer during the data reuse decision step.

The interconnect optimization substep, on the other hand, does have to take into account all layers at once, to make a global approach possible.

11.2.2 Inputs and constraints

11.2.2.1 Basic group information. The memory allocation and assignment step treats the application data at the granularity of basic groups. Basic groups have been explained in section 10.3.1 on page 199. Some information about the basic groups is required to make accurate cost estimations possible.

Memory cost can be estimated if the memory is completely specified: vendor, type, operating conditions, dimensions, access pattern, all of this has to be known for the estimates. Most of these are assumed to be specified in the memory allocation step, when the type of the memories is chosen. The last two parameters, however, must be derived from the basic groups assigned to the memory. To enable this, the following parameters have to be known for each basic group:

- word-length, i.e. the number of bits per word,

- size, i.e. number of memory words occupied by the basic group,

- access frequency, usually separately for read and write accesses.

A basic group's access frequency can be retrieved either by profiling the application, or by analysis. Statically allocated data with an access pattern independent of the data itself can be analyzed at compile time. Otherwise, when data-dependent execution paths exist or for dynamically allocated data, the access frequency can be estimated from profiling. Worst-case estimates are used if the worst-case power needs to be optimized; average cases are used to optimize average power.

In addition to the information necessary to estimate the memory cost, the assignment possibilities are limited by the basic group's IO-profile. For each basic group, an IO-profile is chosen already in an earlier step, as explained in section 10.3.2.1 on page 200. A memory also provides a certain IO-profile. A basic group can only be assigned to a certain memory if the memory is fast enough, i.e. if the two IO-profiles are compatible.

11.2.2.2 Data reuse decision. The data reuse step (chapter 9) simplifies the problem for memory allocation and assignment. Indeed, it assigns every basic group to one and only one memory hierarchy layer. Therefore, a basic group assigned to one layer can never be stored in the same memory as a basic group assigned to another layer. As a result, every layer of the memory hierarchy can be treated separately by allocation and assignment. The sequel of this chapter always assumes that a single memory hierarchy layer is being considered.

The port assignment and bus sharing substep makes an exception to the partitioning of memory hierarchy layers. It concerns precisely the connections between the different layers, so it is heavily suboptimal to treat the layers separately. The port assignment and bus sharing substep will bring all memory hierarchy layers together again, taking into account the memory allocation and assignment decisions for each of them.

11.2.2.3 Storage-bandwidth optimization. Our methodology relies on the storage-bandwidth optimization step (chapter 10) to make sure the real-time constraints will be met. That step summarizes its results in an extended conflict graph. This graph contains all information to decide which basic groups can be assigned together to a single memory. Both single- and multi-port memories are provided for. Section 10.3.3 explains all about extended conflict graphs.

11.2.2.4 Implementation constraints imposed by the designer. The target implementation technology is already taken into account in this step. Knowledge about the type of memories which will be used in the implementation allows better estimations of the cost factors. This knowledge should be collected in a memory library, with much flexibility in the information that can be expressed in it.

Other implementation constraints, e.g. an on-chip cache of the DSP processor that is going to be used, can also be taken into account by manually pre-allocating a memory which possesses those characteristics.

Finally, the designer can simply impose decisions before the memory allocation and assignment methodology is started. Often the designer can easily see that some assignment decision is unavoidable. For example, two large basic groups which are accessed often should be assigned to different memories. Another possibility is constraints arising from other considerations. An example is the use of burst mode access for a certain basic group: that basic group should be assigned to a memory which allows burst mode access, and no other basic groups should be assigned to the same memory if they are accessed in the middle of the burst. Our methodology makes it possible to take all these pre-assignment decisions into account.

11.2.3 Output of the allocation and assignment step

The result of the allocation and assignment step is a description of the memory architecture on the one hand, but also a transformed version of the original source code. The latter is required to propagate the basic group matching and the signal-to-memory and access-to-port assignment decisions.

A netlist of memories and interconnections is the result of the allocation and bus sharing decisions. Each of the memories is annotated with all necessary data to specify it: vendor, type, dimensions, port configuration, etc.

In the basic group matching step, the parameters related to the basic groups are modified. This can easily be reflected in the source code, as illustrated in the section (11.5) on basic group matching (page 242).

The signal-to-memory assignment decision is consumed in the port assignment and bus sharing substep. The latter step specifies for every access which of the ports of a multi-port memory will be used. This refines the signal-to-memory assignment decision: the port through which a basic group is accessed implies that the basic group is stored in that port's memory.

Just like the basic group matching decisions, the access-to-port assignment can be annotated in the source code. This is explained in section 11.7.3 on page 251.

11.2.4 Interaction with storage order optimization

Detailed storage order optimization and allocation/assignment suffer from a chicken-and-egg problem. Allocation and assignment needs information about the size of the basic groups, so an accurate estimation of memory cost can be made. Storage order optimization, however, reduces the size of the basic groups. But storage order optimization in turn needs to know which basic groups are stored in the same memory before it can perform inter-array placement (see section 12.3, page 287). In conclusion, there is a circular dependency between detailed storage order optimization and allocation/assignment.

Our solution for the interdependency with storage order optimization is to use high-level estimates for allocation and assignment, and perform the in-depth final storage order optimization afterwards. Estimating how much basic group size can be reduced can be done very fast. This estimate is made before the allocation and assignment is started. It is then used as the basic group size for the memory cost estimations. The inter-array reduction when placing two or more basic groups in one memory can similarly be estimated, e.g. with the techniques proposed in section 4.1. The latter estimate is used instead of a simple sum of the basic group sizes to dimension the memory.

Storage order optimization finds a placement for the arrays based on their original dimensions. But the memory allocation and assignment step linearizes the arrays when the decisions are back-annotated in the source code. Therefore, back-annotation should be postponed until after the storage order optimization step. The back-annotation is only done immediately if no storage order optimization step is performed after the allocation and assignment.

11.3 MEMORY LIBRARY

To determine the cost of an assignment, one could apply a fixed cost estimation formula, e.g. Mulder's formula for area [286], as discussed in section 4.3. The advantage of using a formula like this is that it is possible to formulate the problem as a integer linear (or quadratic) program.

We have, however, chosen for a more flexible approach. The cost function is stored in the memory library (section 11.3). This makes it possible to use a different cost function for different memory types. A formula is indeed only valid for a limited range of memory types — the Mulder formula, for example, is only valid for on-chip SRAM memories with a single memory bank.

In addition, the cost function is not limited to one specific parameter. One can use a weighing of area and power consumption for on-chip memories, and a weighing of power consumption and market price[2] for off-chip memories. Or it can be observed that solutions with minimal power will also have close to minimum area and eliminate area altogether, thereby eliminating the need to find good weighing factors. Or, finally, one might not care about power at all and instead concentrate on area, multiplying it with a fixed fabrication cost to make comparison with the off-chip memories possible.

To allow the memory allocation and assignment step to be performed as accurately as possible, a library containing information about the memories to be used is indispensable. For each memory in the memory allocation scheme, the memory library which will be used for it should be selected. Since allocation and assignment is done for each single level of memory hierarchy separately, the same library will usually be used for all memories in the allocation scheme. It is possible to mix though, for instance using on- and off-chip SRAMs in a single hierarchy layer.

A memory library contains memory types, which are an abstraction of a number of memories. This concept is introduced because a whole set of different memories can often be described by a single, parameterizable memory type. For example, an on-chip memory module generator can produce memories in a wide range of word-lengths and sizes. Also for off-chip memories, often a number of them are available which are essentially the same except for their size.

A memory library will contain a number of related memory types. For example, a library for an on-chip memory module generator vendor might contain a single-port memory type, a dual-port memory, a low-power single-port memory, a fast dual-port memory, a ROM, etc. All of these are

---

[2]Using the memory price involves a risk: by the time the chip is actually produced and the memories are bought, prices may have evolved drastically.

gathered in one memory library for that vendor and technology. The principle is that memories from the same library can be freely mixed in a single design.

In memory allocation and assignment, the memory library has two main uses:

1. Validating constraints: for an assignment scheme to be valid, actual memory modules have to be available to satisfy the constraints of the assignment scheme. These constraints include the following.

   - Number and type of ports: during assignment, for each memory in the allocation scheme a set of memory types with the right number of ports is chosen from the library.

   - The memory's dimensions: memory modules have limits on their dimensions. Module generators for on-chip memories usually cover a certain range of word-lengths and sizes. Off-chip memories are usually available only in a set of discrete dimensions, e.g. 1024×4, 1024×8, 2048×4, but not 1536×6.

   - The speed of the memory: this is represented by a memory IO-profile for every memory port. When a basic group is assigned to a memory, the memory's IO-profile needs to be faster in all respects than the basic group's IO-profile.

When evaluating an assignment scheme, a memory satisfying these constraints is selected from the library. If no such memory exists, the assignment scheme is invalid.

2. Evaluating cost functions: embedding the cost functions in the library makes it possible to use different cost models for different memory types, even within the same memory library. This makes it possible to cover a large number of memories with the same library, without compromising the accuracy of the models used.

   A problem still remains regarding the cost functions: the cost parameters to be used may depend on what kind of memory it is. For example, it does not make any sense to consider the area of an off-chip memory, but it does make sense to consider its financial cost. For on-chip memories the reverse holds. This means that the cost functions which are available may vary depending on the kind of memory concerned.

   An alternative would be to just define one single abstract cost function, which would be a combination of the relevant cost parameters of that specific memory type. The advantage of this approach is that the tools using the memory library do not need to be concerned with any details of the memory, including whether it is on- or off-chip. The problem, however, is that it is impossible to define an objective combination function for the different cost parameters. The chosen combination depends on the context of the optimization, and on what is important to the designer. A decision about this should not be taken in the shared memory library.

In addition to this information, the library must contain everything required to specify the memories for the tools that follow allocation and assignment. On the other hand, if any freedom is still left in the specification of the memory module to be used, this freedom should be made available to the subsequent tools.

The library makes it possible to take into account the target technology at a high level of abstraction. Details about the memories are hidden away, the only things the tools using the library have to take into account are the things that really matter to them: dimensions, cost parameters, etc.

11.4 MEMORY ALLOCATION

The first thing to do when designing a memory architecture is deciding how many memories will be used and what they will look like. The goal of the memory allocation substep is to make a choice which will result in a feasible and optimal memory architecture. This section discusses what precisely has to be decided at this stage, how the decision is constrained, and how the decision can be taken.

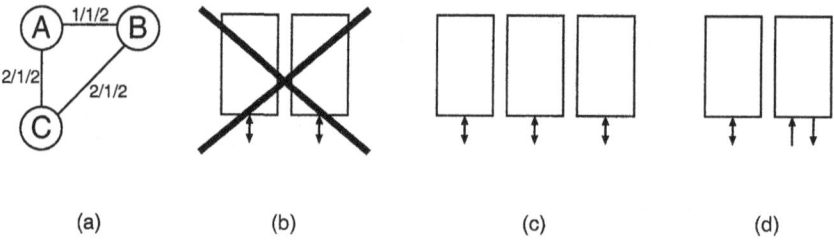

Figure 11.2.   Port configuration: invalid and valid example

11.4.1 Memory allocation schemes

11.4.1.1 Port configuration. First of all the number of memories and the number of ports accessing each memory have to be chosen. This is called a port configuration. Figure 11.2 shows some examples. A port configuration not only specifies how many ports are available. It also specifies which ports access the same memory. This is necessary to verify the validity of a port configuration.

A port configuration must provide enough bandwidth for the application. This can be checked with the extended conflict graph. The port configuration is valid only if an assignment of basic groups to memories of this port configuration is possible without violating the conflict graph.

Figure 11.2 shows an example of one invalid and two valid port configurations. Figure 11.2(b) is not a valid port configuration for the conflict graph given in figure 11.2(a). In this port configuration, two of the three basic groups will have to be assigned to the same memory, and the conflict graph does not allow this for one-port memories. In contrast, Figure 11.2(c) does show a valid allocation scheme: now all basic groups can be assigned to a different memory. Figure 11.2(d) is also valid: basic groups A and B can be stored in the dual-port memory, one read and one write port are all that is required to solve the 1/1/2 conflict.

A simple necessary condition for the validity can be derived from the extended conflict graph. Consider the annotations on the (hyper-)edges of the extended conflict graph. These indicate how many accesses occur in parallel, split into read, write and total number of accesses. Obviously, the total number of ports in the port configuration must be at least equal to the highest number of parallel accesses. Similarly, the total number of readable ports (i.e. read or read/write ports) must be at least equal to the highest number of parallel read accesses. The same goes for writable ports. As an example, the conflict graph of figure 11.2(a) requires at least two read ports, at least one write port, and at least two ports in total.

The easiest way to verify the validity of an allocation scheme (the port configuration as well as the memory types) is by performing a signal-to-memory assignment. Section 11.6.1 discusses how the validity of an assignment scheme is confirmed. This can be very fast in practice, once the signal-to-memory assignment substep has been automated (see section 11.6).

Given an extended conflict graph, many valid port configurations exist. Even if only one-port memories are considered, any number of memories at least equal to the chromatic number of the conflict graph may be allocated. Multiport memories introduce extra freedom in the type of the ports: read-only or write-only ports create additional possible port configurations. For example, for the conflict graph of figure 11.2(a), when the total number of ports is limited to three, there are still eight valid port configurations containing only a single memory, four with two memories and one with three memories. This shows that the number of alternative port configurations can quickly become very large.

11.4.1.2 Full memory allocation. A port configuration can be completed to a full memory allocation by deciding for each memory which type will be used for it. This can be done by selecting a memory library for each memory. It is also possible to select a number of memory types or libraries. The final choice is then delayed until the signal-to-memory assignment substep: the cheapest memory type is chosen for the given assignment decision.

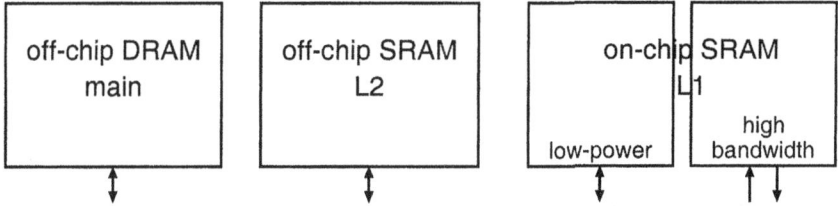

Figure 11.3.   Example memory allocation scheme

Note that the dimensions of the memories are *not yet* specified in the memory allocation scheme. The final dimensioning is only done after the signal-to-memory assignment step. Then the dimensions can be chosen such that the arrays closely fit into the memories. In practice this will mean that the smallest word-length and size of memory will be chosen which can still accommodate all basic groups assigned to it.

Choosing the right type of memory is also important for the validity of the memory allocation. The storage-bandwidth optimization step makes certain assumptions about the latency of the basic group accesses. This is modelled with an IO-profile for each basic group — see section 10.3.2.1. It must be possible to assign each basic group to a memory which satisfies this latency requirement. This also means that for all basic groups, at least one memory must exist with a compatible IO-profile.

Figure 11.3 shows an example of a full allocation scheme. The port configuration consists of four memories: three one-port memories and one dual-port one. Three layers of hierarchy have been selected in the data reuse step (chapter 9). The allocated memories are divided over these three levels, and depending on the level also the type of memory is chosen. So for the main memory an off-chip DRAM is chosen, while the second level of hierarchy will use an off-chip SRAM. For the first level of hierarchy two on-chip SRAMs are chosen: one fast dual-port RAM to cope with the bandwidth requirements, and a low-power memory for signals assigned to this level for power reasons.

11.4.2 Allocation methodology

We propose two possible approaches to tackling the memory allocation problem. One is to let the designer decide (with support to ensure a valid low-cost scheme), the other is to automatically search through the allocation possibilities.

11.4.2.1 Interactive methodology. In the first approach the designer simply decides on an allocation scheme. Taking into account the extended conflict graph, (s)he constructs a valid port configuration; next (s)he selects the types in a way that every basic group can be assigned to at least one memory. If (s)he makes a mistake and chooses an invalid allocation scheme, this is reported by the automated assignment tool (see further) and (s)he can try another scheme. Alternatively, the designer can try out a number of allocations, look at the cost figures resulting from the automated assignments and select the best one.

Making the memory allocation decision manually is easier than it may seem. First of all, the background memory is subdivided into a number of memory hierarchy layers in the data reuse step. Every basic group is already assigned to one of the layers in this step (see section 9.6, page 190). So the problem can be tackled separately for each of the hierarchy layers. In addition, usually only one- and two-port memories are available, and the two-port ones are much more expensive than the one-port ones. This may change in the future, however: memories with many ports are beginning to become available.

So the designer first selects a number of two-port memories to solve the self-conflicts (cfr. subsection 10.3.3, page 10.3.3). Then the allocation is completed with a number of one-port memories until the total number of memories is at least equal to the chromatic number of the conflict graph.

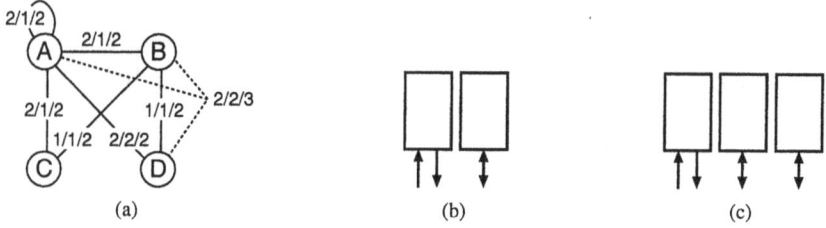

Figure 11.4.   Example of the interactive allocation methodology

The (automated) signal-to-memory assignment step can give useful feedback about the chosen allocation scheme. Many bits being wastedword-length waste in one of the memories, for example, indicates that it may be useful to allocate an extra memory to hold the smaller basic groups. Or when a large memory is accessed a lot and consumes much power, allocating an extra memory may allow one of the basic groups to be removed from this large memory, thus reducing the number of accesses and hence the power consumption. However, the extra interconnection cost of allocating an extra memory must always be traded off against the possible gain.

The methodology is clarified with an example, illustrated in figure 11.4. Looking at the conflict graph in figure 11.4(a), the first thing to note is that a self-conflict exists. A dual-port memory will therefore be required. One read/write port and one read port would be sufficient. Two read/write port are allocated, however, because only such dual-port memories are available in the library for the example.

The hyperedge between basic groups A, B and D indicates that a third port is required, so an additional one-port memory is allocated. It is not immediately clear whether this will be enough. Two solutions are retained: the first one with only one additional one-port memory, the second one with two one-port memories (figure 11.4(b) and (c)).

The signal-to-memory assignment substep will be performed for both allocation possibilities. This is a very fast procedure since the step can be automated (see section 11.6). With the feedback from the assignment substep the validity of the allocation can be verified and the two possibilities can be evaluated. In the end, one of the two allocation schemes is selected, together with the assignment scheme that was found, and this is passed to the port assignment and bus sharing substep.

11.4.2.2 Automated methodology. In the second, fully automated approach, port configurations are exhaustively enumerated and the cost that will result from each of these is estimated. Then a number of them are selected, and assignments to these memories are performed. The best of the resulting assignment schemes is retained, together with the port configuration on which it was based.

In most practical applications, when the timing constraints are sufficiently relaxed, the number of memories required is not too large and the enumeration is feasible. With very stringent timing constraints, however, many memories have to be allocated and the search space becomes too large. Manual interaction will be needed in this case. Alternatively, future research may be performed on pruning heuristics to avoid a fully exhaustive enumeration.

11.5 BASIC GROUP MATCHING

Because basic groups are defined based on the properties of the application, and not on the target technology, they may not be very well suited to fit in the memory architecture as such. In this substep, the elements of the basic groups can be packed in a different way to match them better to the available memories. The possibilities will be illustrated with some examples.

Introducing a basic group matching substep makes it possible for all other steps to make some assumptions about basic groups. It can be assumed that every basic group is always treated as a whole, and never split up. It can also be assumed that every element of the basic group takes up exactly one

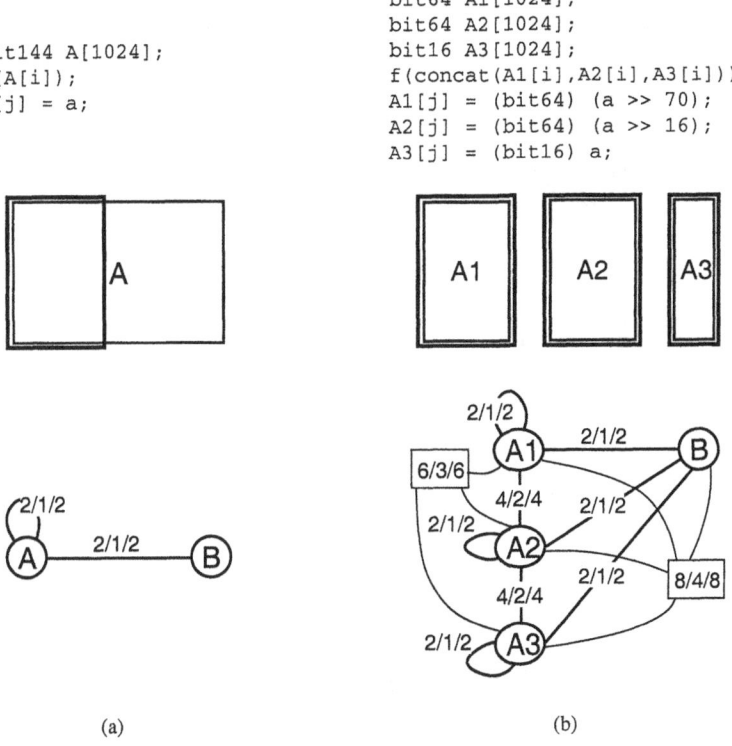

```
bit144 A[1024];
f(A[i]);
A[j] = a;
```

```
bit64 A1[1024];
bit64 A2[1024];
bit16 A3[1024];
f(concat(A1[i],A2[i],A3[i]));
A1[j] = (bit64) (a >> 70);
A2[j] = (bit64) (a >> 16);
A3[j] = (bit16) a;
```

(a)   (b)

Figure 11.5.   Matching basic groups when word-length is too large

memory word. If the memory words are larger than the word-length of the basic group, the superfluous bits are just wastedword-length waste.

11.5.1 Splitting wide words

The first case is when the word-length of a basic group is too large to fit in the available memories. The basic group should be split up in a number of smaller basic groups, so that they all fit in the available memories.

Figure 11.5 shows an example of a basic group with word-length 144, while the maximum word-length of the available memories is 64. This basic group should be split up into three smaller basic groups. Accesses to the original basic group are decomposed into parallel accesses to the three new basic groups. Some extra operations are required to recompose the original data, as shown in the example.

Note that also the extended conflict graph must be modified, because the accesses to all new basic groups occur in parallel. The original conflicts are propagated to all three new basic groups. The new basic groups are also in conflict with each other, because they have to be accessed in parallel. Also, hyperedges have to be added: every edge incident to the original basic group, becomes a hyperedge incident to the new ones.

The extra conflicts introduced by splitting a basic group can partially be avoided by splitting *before* the storage bandwidth optimization step. The accesses can then be sequentialized if time allows. This

```
bit2 A[1024];                 bit8 A[256];
f(A[i]);                      f(bit2(A[i/4] >> (i%4)));
A[j] = a;                     A[j/4] = update(A[j/4],j%4,a);
```

(a)   (b)

Figure 11.6.   Matching basic groups when word-length is very small

is the approach usually taken in software processors. For example, a 64-bit floating point number will be fetched from the 32-bit memory in two subsequent accesses.

11.5.2 Compacting small words

When the word-length of a basic group is very small compared with the word-lengths of all other basic groups, it may be useful to compact subsequent elements of the basic group into a single word. In this way word-length waste can be avoided.

Figure 11.6 illustrates this possibility with an example. The word-length of basic group A is much smaller than the word-lengths of the other basic groups. If A and B are assigned to the same memory, this would lead to an important word-length waste, as can be seen in the figure. The solution is to collect four two-bit words into one 8-bit word. This reduces the word-length waste to an acceptable level.

Some effort to choose such compaction has been proposed in [361], but it does not directly fit into this more practical context and should be extended.

Three changes are required to the accesses to this basic group. The two-bit data word needs to be extracted from the larger one. In figure 11.6(b) this is done by shifting over a number of bits and conversion to bit2. In addition, the addressing has to be changed, in this case with a division by four. Finally, for write accesses, only one of the four data words should be modified, so the entire memory word has to be read first and partially updated.

The extra read accesses for every write access imply that a trade-off has to be made between the their cost and the gain in word-length waste. Also, they invalidate the partial storage access ordering found in the storage bandwidth optimization step. That step has to be (partially) repeated before proceeding to signal-to-memory assignment. This can sometimes be avoided, when the data words are cached in a lower memory hierarchy layer. Then they can be assembled into a single word and copied to the next layer as a whole.

One final remark: this transformation has an important impact on the storage order optimization step. Every memory word is alive from the first time any of its constituent data words are written, until the last time any of them is read. This increases the lifetime of every memory word. Therefore, the detailed in-place optimisation step should use the original basic group instead of the transformed one. The transformation can then again be applied to the result of the in-place optimisation step.

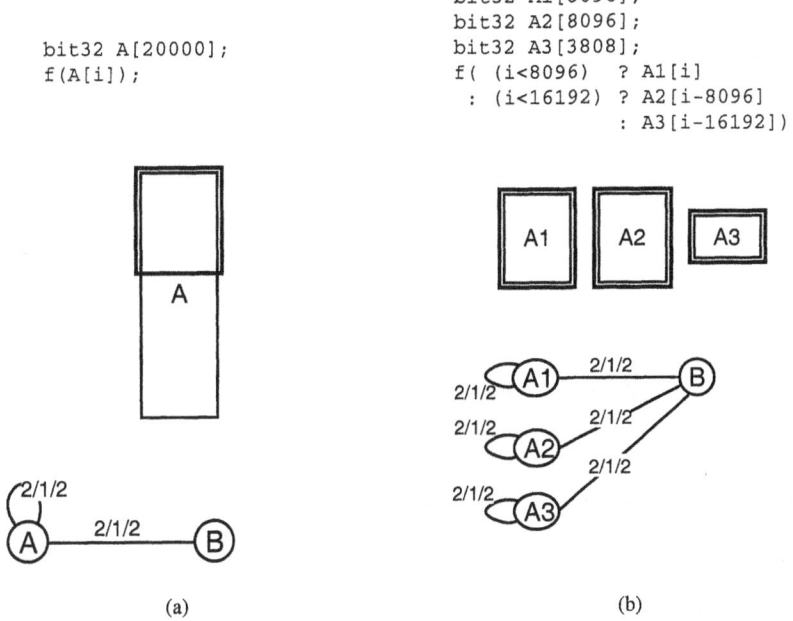

```
                                    bit32 A1[8096];
                                    bit32 A2[8096];
bit32 A[20000];                     bit32 A3[3808];
f(A[i]);                            f( (i<8096)   ? A1[i]
                                     : (i<16192)  ? A2[i-8096]
                                                  : A3[i-16192]);
```

(a)   (b)

Figure 11.7.   Matching a basic group when it is is too large

11.5.3 Splitting large basic groups

The last case is when a basic group contains more elements than can be stored in a single memory module. Figure 11.7 shows an example of a basic group containing 20,000 elements, while the maximum size of a 32-bit memory is only 8,096 words. In this case the basic group should be split into smaller basic groups which do fit in the memories.

There is a big difference with the word-length splitting of figure 11.5. In this case, every access is resolved to an access to only *one* of the constituent basic groups, and not split into three parallel accesses as for very wide basic groups. This can also be seen in the extended conflict graph in figure 11.7(b): there are no conflicts between the three new basic groups. Of course, this does not really matter because they are too large to be assigned to a single memory anyway.

Since every access is resolved to an access to only one of the new basic groups, the total number of accesses to the original basic group has to be divided over the new basic groups. This can either be done by re-profiling or re-analyzing the application with the new basic groups, or by heuristically dividing the number of accesses proportionally to the size of the new basic group. In the example of figure 11.7, basic groups A1 and A2 would each get 40% of the accesses, while A3 gets the remaining 20%.

11.5.4 Guiding basic group matching

Now that we have enumerated the possible basic group matching techniques, the designer still needs a decision strategy for when to apply them. The easiest way to guide the basic group matching decision is by using feedback from the signal-to-memory assignment substep, described in the next section.

Using feedback from a later substep introduces a loop in our design flow, which should be avoided as much as possible. In this case, however, the loop is not so much of a problem. First of all, it is only a local one, internal to the allocation and assignment step. In addition, the signal-to-memory assignment substep can be fully automated and is not overly CPU-intensive. Thus, the extra overhead of this loop will be relatively small.

For the problem of basic groups which are too wide or too large, the assignment tool will quickly indicate that it cannot find any memories which can accommodate this basic group. Then the designer can take a look at the limits of the memory library and split up this basic group accordingly.

Basic groups with small word-lengths will not result in invalid assignments, but in very costly ones. The assignment tool can indicate this in the form of *word-length waste*. Word-Length waste is the number of unused bits in a memory due to basic groups of different sizes. In figure 11.6 this is indicated by the hatched area. Basic groups with a combined word-length less than or equal to the memory's word-length are eligible for compaction.

With this feedback, the designer can easily decide which basic group matching techniques should be applied. Tool support for executing the necessary transformations on the source code and conflict graph can still be very useful.

11.6   SIGNAL-TO-MEMORY ASSIGNMENT

The memories are dimensioned by looking at the signals they will contain. The signal-to-memory assignment step decides for each basic group in which of the allocated memories it will be stored, allowing the dimensioning of the memories and thus an accurate estimate of the memory cost. Using this cost estimate, it is possible to explore different alternative assignment schemes and select the best one for implementation.

Subsection 11.6.1 considers the constraints for the assignment decision, mainly the extended conflict graph formalism. Subsection 11.6.2 explains how the cost of an assignment scheme is estimated in our methodology. With this, the search space and the target function are defined, and subsection 11.6.3 shows how the optimization is undertaken.

11.6.1   Validity of assignments

Basic groups are always assigned as a whole to a single memory: in this step it is assumed that all necessary analysis has been done to refine the granularity of the basic groups to make a good assignment possible. Chapter 6 explains how this analysis is done.

The freedom of assigning basic groups to memories is somewhat limited. First of all the assignment must respect the constraints generated by the storage-bandwidth optimization step, i.e. the extended conflict graph.

For one-port memories this constraint can be modeled as a graph coloring problem. Every memory is represented by a color, basic groups are 'colored' with the memories to which they are assigned. When two basic groups are in conflict, there is an edge in the conflict graph, which means they cannot have the same color. This is the way the assignment of scalars to registers is usually modeled, as summarized in section 2.1.

With multi-port memories, the conflict graph has to be extended with loops and hyperedges and ordinary coloring is not valid anymore. One could modify the coloring problem into an 'extended coloring' problem and try to model the signal-to-memory assignment like that. This approach does not look very promising, however, because in addition to the extended conflict graph, a cost function and other constraints have to be taken into account.

The second constraint is that the memory's IO-profile must match that of the basic group. Finally, memories which are large enough to accommodate all basic groups assigned to them must be available. The last two constraints imply that it is important to know the parameters of every memory in the allocation. These parameters can best be managed in a memory library (See subsection 11.3).

11.6.2   Assignment cost

We are not looking for just any signal-to-memory assignment, but for an optimal one. As mentioned before, the cost of a memory architecture depends on the word-length and the number of words of

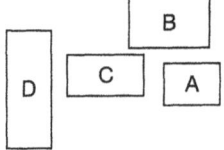

BG.	word-length	# words	# reads	# writes
A	8	256	3	2
B	10	448	3	0
C	12	282	0	2
D	6	1096	0	1

(a) Basic group information

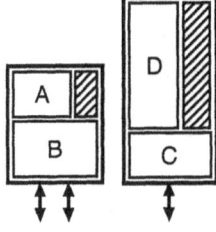

mem.	word-length	# words	# reads	# writes
1	10	702	6	2
2	12	1378	0	3

mem.	Area cost	Power Cost
1	14	16
2	16	8
total	30	24

(c) Memory cost

(b) Assignment

Figure 11.8.   Calculating the cost of memories from the basic groups assigned to them

each memory, and the number of times each of the memories is accessed. These parameters can easily be derived from the assignment.

Figure 11.8 shows an example of calculating the cost of an assignment scheme. First the memories' parameters are derived from the parameters of the basic groups assigned to them. For instance, memory 1 has word-length 10 because it must be able to accommodate basic group B. It needs at least 702 words to have enough space for both basic group A and B. A total of 6 read accesses (3 from A and three from B) and 2 write accesses (all to A) are done to memory one in every iteration of the algorithm.

From these numbers, an estimation can be made of the cost of the memories. In this example, chip area and power consumption are used as the cost figures.

11.6.3 Assignment methodology

Figure 11.9 illustrates signal-to-memory assignments. In (a) the inputs for this example are summarized: a memory assignment scheme, the basic group information and an extended conflict graph.

The basic groups are assigned to memories one by one. Beginning with basic group A, there is only one possibility because of the self-conflict. So in figure 11.9(b) basic group A is assigned to the dual-port memory. The area and power cost are noted below the figure.

For basic group B there are two possibilities. Figure 11.9 shows the two possibilities and the area and power cost they imply. Assigning basic group B to memory 2 is obviously the best choice at this point, but it is possible that globally this is not a good choice.

Because the constraints and the cost function of the assignment problem have been formally defined, the solution of the problem can be automated. The search space can be explored using either a greedy constructive heuristic or a full-search branch and bound approach. An illustration of the assignment methodology is given in figure 11.10.

In every step of the search procedure, a basic group is selected and assigned to a memory where it does not have conflicts. When no assignment is possible for the basic group, backtracking is started and other possibilities are tried for the basic groups previously assigned.

The cost of the assignment is evaluated at every step and the cheapest possibility is selected. This greedy approach quickly finds a suboptimal solution, as shown in figure 11.10(a). The constructive algorithm terminates as soon as a complete assignment scheme is found.

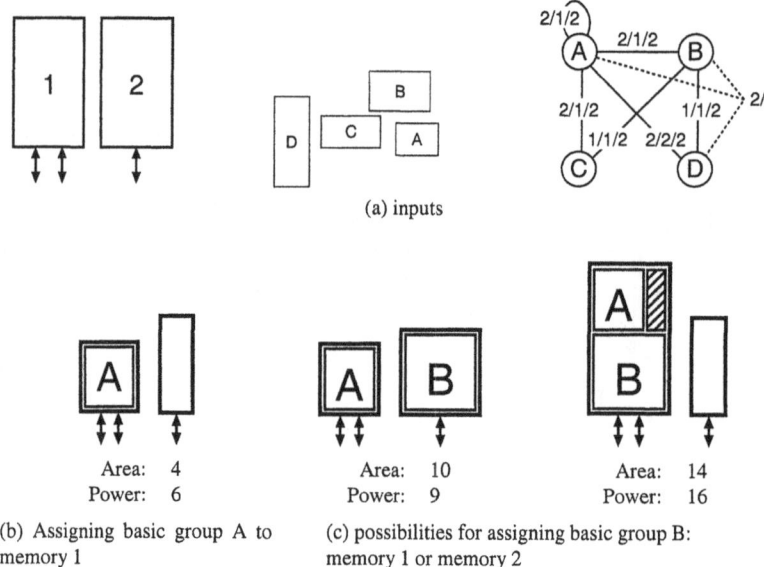

(a) inputs

Area: 4
Power: 6

Area: 10
Power: 9

Area: 14
Power: 16

(b) Assigning basic group A to memory 1

(c) possibilities for assigning basic group B: memory 1 or memory 2

Figure 11.9.   Example of the signal-to-memory assignments

The full-search algorithm, on the other hand, continues to backtrack and try finding other solutions. Paths which certainly lead to solutions with higher cost are immediately pruned away, so only solutions which are really better than the original one are fully explored. This is the branch-and-bound strategy. An example is given in figure 11.10(b): assigning basic group C to memory two results in a higher cost than the best solution found up till now, without even taken into account basic group D. This path can never result in a better solution so it is pruned away.

The full search algorithm can be sped up by using a heuristic for early pruning. By estimating the cost of a full assignment based on the one at a certain point in the search tree, the paths leading to higher-cost solutions can be pruned away earlier. If the estimation always underestimates the final cost, the solution is still guaranteed to be optimal: paths which are pruned away will have higher cost than the estimate, so certainly higher cost than the current best solution.

For real-life designs, pre-assignmentspre-assignment can often be used to speed up the full-search assignment. The designer can often see immediately that some basic groups should be kept separated. Another possibility is the following. First do the assignment for only the most important (costly) basic groups: the largest or most heavily accessed ones. Then use the resulting assignment scheme as a pre-assignment for the complete application. This cuts down the search space tremendously. The risk is, of course, that the solution is not optimal. Therefore the cost of the full assignment scheme should be compared to the partial one: if it is much larger (e.g. twice as much), the solution may be far from optimal.

11.7   PORT ASSIGNMENT AND BUS SHARING

11.7.1   Interconnection optimization

This substep concentrates on the interconnection of memories and the processing unit. As motivated in chapter 4, the area and power contribution of interconnection can still be ignored to a large extent during our optimisations, as a second order effect. So it does not (yet) need to be incorporated in the

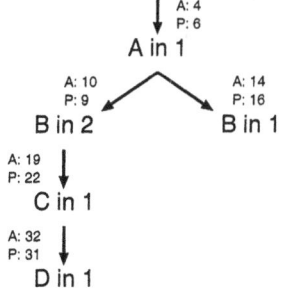

(a) constructive assignment

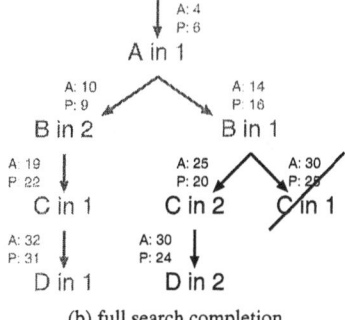

(b) full search completion

Figure 11.10.   Example of the assignment methodology

actual memory cost models. This may have to change for future deep submicron technology whenever power is a dominant factor in the design.

None of the other substeps consider the interconnection constraints. In fact, a full mesh between all memories and the processing unit is assumed. In this substep, the full mesh is refined to just the number of busses required to make all transfers possible, taking into account the constraints introduced by the signal-to-memory assignments.

The freedom available for interconnection optimization will be explained with a number of examples in this section. We do not yet formulate a methodology which is automatable, however. Probably, techniques from the high-level synthesis research on bus and register allocation and binding can be largely reused here. Appropriate tool support is needed for this tedious, error-prone substep, though.

11.7.2 Sharing a bus

The basic idea of bus sharing is illustrated in figure 11.11. Three memories have been allocated, even though only two accesses ever occur in parallel. This means that only two parallel busses are really required to meet the bandwidth requirements. Figure 11.11(c) shows the reduced interconnection architecture. Accesses to basic group C in memory m2 can go either via bus a or via bus b, depending on which of the two is not used at that time. In other words, every transfer will have to be assigned to a bus.

Sharing a bus trades off the added complexity of switches (the crosses in figure 11.11(c)) against the following advantages (ordered from high to low impact):

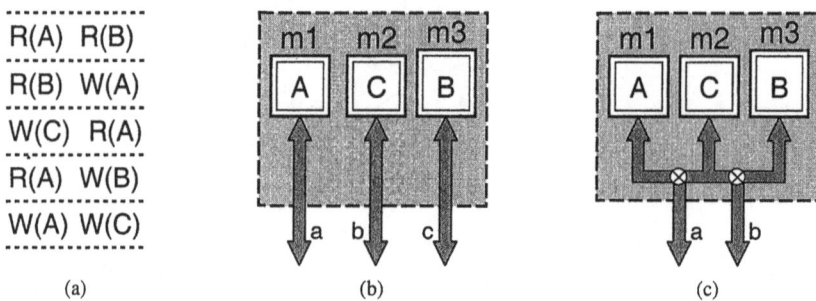

Figure 11.11.   Example of bus sharing possibilities

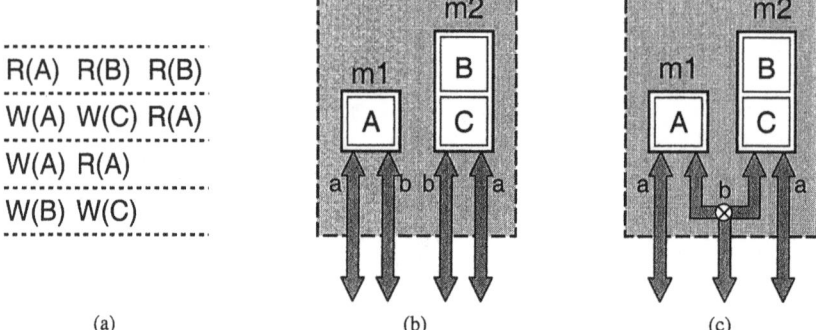

Figure 11.12.   Example of bus sharing for multi-port memories

- Pin count: the most obvious advantage lies in the fact that less pins are required for off-chip connections. This may make it possible to make a smaller chip, and certainly results in cheaper packaging.

- Power: as mentioned, the contribution of the interconnections in the total power budget is relatively small compared to that of the memories. This ceases to be the case, however, if there are many of them. In this case, reducing the number of busses can have a very positive effect on the overall power consumption.

- Layout complexity: without bus sharing, all memories should be placed as close as possible to the data-paths. Since this is impossible, finding a good layout will be a tough problem. Sharing can make the problem more tractable by creating clusters: the memories which share busses should be placed close together, but they can be some distance from the data-paths.

- Area: long busses take up much area at the package and board level for off-chip connections. When the memories sharing a bus can be placed close together, less long lines will be required to connect to the destinations.

11.7.3 Port assignment: sharing a bus for multi-port memories

When multi-port memories are sharing a bus, the assignment of the memory accesses to memory ports becomes important. For example, in figure 11.12, it is possible to reduce the number of busses from four to three, by letting memory m1 and m2 share the bus accessing their port b. This means that the

```
                                    bit8  m1[256];
                                    bit8* m1_a = &m1[0];
bit8   A[256];                      bit8* m1_b = &m1[0];
bit10  B[448];                      bit12 m2[730];
bit12  C[282];                      bit12* m2_a = &m2[0];
                                    bit12* m2_b = &m2[0];

tmp = A[i];                1        tmp = m1_a[i];
C[i] = f(tmp,B[i]);        2,1      m2_a[i+448] = f(tmp,m2_a[i]);
A[j] = f(tmp,B[j]);        2,1      m1_a[j] = f(tmp,m2_b[j]);
A[i] = A[i+j];             3,2      m1_a[i] = m1_b[i+j];
tmp = A[j+1];              3        tmp = m1_b[j+1];
B[j] = g(tmp);             4        m2_a[j] = g(tmp);
C[j] = g(tmp+1);           4        m2_b[j+448] = g(tmp+1);
```

Figure 11.13.   Source code annotation for the port assignment decision

port through which the memories will be accessed will have to be carefully chosen so the memories can be accessed in parallel as the ordering in figure 11.12(a) prescribes.

For memories with special access modes like burst mode, the port assignment decision becomes even more important. Burst mode can only be exploited if all the accesses in the burst go through the same port, and they are not interrupted by accesses to other basic groups. If special access modes are available, the designer should make a careful port assignment decision, to maximize the burst mode access possibilities. How to exploit burst mode has partly been researched in other work, e.g. [310].

Port assignment decisions can easily be annotated in the source code, together with the signal-to-memory assignment decision. Figure 11.13 shows an example, based on the ordering and bus sharing decisions of figure 11.12. The left-hand side shows the original code, the right-hand side shows the result of signal-to-memory and access-to-port assignments. The middle column indicates in which cycle the access is scheduled in figure 11.12(a).

Basic group declarations in the code are replaced with declarations of memories. For multi-port memories, additional pointers are declared, representing the ports. Every access in the original code is then translated to an access to one of these pointers. The addresses have also been modified to take into account the placement of the basic groups in the memories. Note that this array placement will be changed later in the storage order optimization step (chapter 12).

It is feasible to manually make good bus sharing decisions. The process of assigning every transfer to the right bus and memory port, however, is a tedious and error-prone process. We are currently investigating the formalization and automation of both the bus sharing and the bus and port assignment decisions. This will complete our methodology support for designing the background memory architecture.

11.8   DEMONSTRATOR RESULTS

As illustration, a realistic test-vehicle will be discussed now, namely a Binary Tree Predictive Coder application. This section shows the results of all memory allocation and assignment substeps.

11.8.1   Demonstrator application: Binary Tree Predictive Coder

Binary Tree Predictive Coding (BTPC) is a general-purpose image coding method for lossless or lossy compression of photographs/graphics [345]. The implementation of the encoder used in this experiment consists of about 3000 lines of C code. Before going to implementation, some of the application's parameters have been fixed. The input image, for example, is fixed at 144×176 pixels using 24-bit color. The application is constrained to encode 25 frames per second.

Figure 11.14 gives a sample of the application code. The important part of the application is a sequence of loop nests, which can roughly be divided in three blocks of more or less equal importance with respect to memory usage. In the bodies of these loop nests are a lot of nested conditionals, both manifest and non-manifest. In the bodies of these, the accesses to the signals can be found. These signals will form the basic groups which are treated by allocation and assignment.

```
/* Predictive coder */              /* Huffman coder */
for level                           for level
  for Y                               for Y
    for X                               for X
      if (X < Y)                          ...
        ...                             for j
      if (in[X,Y-level] == p && ...)      while (...)
        if (...)                            for i
          tmp = in[X,Y-level];                if (freq[i] == 1 && ...)
        else                                    ...
          ...                               freq[h1] += freq[h2];
      in[X,Y] = f(tmp);                     length[h1]++;
      pyr[X,Y] = g(tmp);                    while (descendents[h1] >= 0)
      ridge[X,Y] = h(tmp);                    h1 = descendents[h1];
      ...                                     length[h1]++;
                                          descendents[h1] = h2
/* Binary coder */                        while (...)
for Y                                       ...
  for X                                 ...
    if (pyr[X,Y] == 0)                  for i
      in[X,Y] = 1;                        for k
    ...                                     if (length[i] == k)
for level                                   ordered_symbols[l++] = k;
  for Y                                 for Y
    for X                                 for X
      if (in[X+1ch,Y+h1] == 1 &&           if (ridge[X,Y] == 1)
          pyr [X+1ch,Y+h1] == 0)             ...
        in[X,Y] = 1;                       else
      ...                                     ...
```

Figure 11.14. High-level structure of the BTPC algorithm. This code is executed three times for every frame: once for every colour plane.

The application uses 23 basic groups. Only 16 of these have been retained for our methodology, the others are very small (less than 32 words) and are treated as scalars. Table 11.1 gives the basic group information for the 16 basic groups. They have been grouped corresponding to the block of the application where they are used. The last two columns of the table show the number of read and write access to each array for processing one image. These numbers have been obtained through a combination of profiling and analysis.

Before starting allocation and assignment, the storage-bandwidth optimization step was applied to the BTPC algorithm. The conflict-directed ordering technique (section 10.9) was applied to each of the three main loop nests separately. Based on the ordering of these three loop nests, they were combined and allowed to partially overlap, introducing extra conflicts. The end result is the extended conflict graph shown in figure 11.15.

11.8.2 Memory allocation substep

The allocation for the BTPC application can be done fully manually. A few possible allocations will be retained. These can later be compared with each other, based on the results of assignment.

The extended conflict graph (figure 11.15) indicates that at least two dual-port memories are required. Indeed, the basic groups 'freq' and 'in' both have a self-conflict, and in addition there is a conflict between them for which four ports would be needed.

After the self-conflicts have been tackled, extra single-port memories can be considered. It is not immediately clear from the extended conflict graph whether two dual-port memories suffice for all basic groups. A quick check with the automated assignment tool learns that this is not the case.

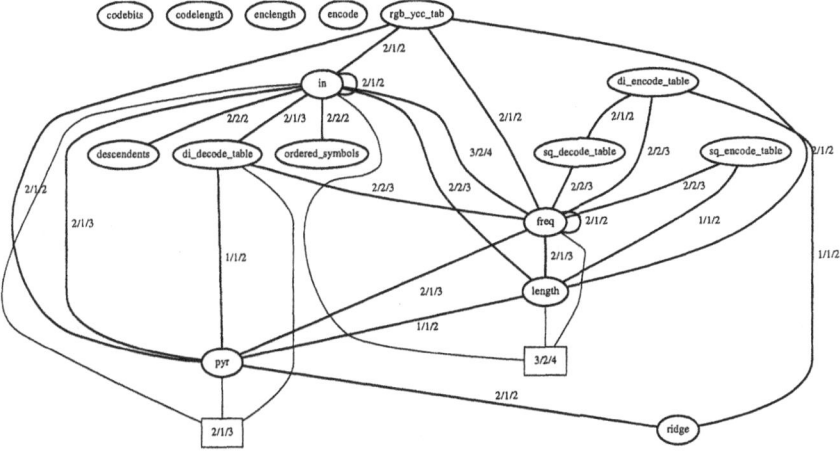

Figure 11.15.   Extended conflict graph of the BTPC application. Note the self-conflicts of 'in' and 'freq'.

Therefore, extra single-port memories must be allocated. Three alternatives are retained: one, two or three extra single-port memories.

For comparison, allocation schemes with an extra dual-port memory replacing two single-port memories are also tried. The figures will show that this is not advantageous. For some memory libraries it might be, though.

At this point, an implementation technology is chosen for the allocated memories. This makes it possible to verify if the available memories can accommodate the basic groups. The chosen library

Table 11.1.   Basic group characteristics of the BTPC application. 16 basic groups are retained, plus one containing a copy for data reuse.

basic group name	# words	bit width	#reads	#writes
codebits	256	16	467	467
codelength	257	8	796	530
enclength	256	8	6 844	467
encode	256	16	6 844	467
rgb_ycc_tab	2 048	32	228 096	2 304
in	25 344	8	50 688	50 688
pyr	25 344	8	123 852	50 832
ridge	25 344	2	37 719	37 863
di_decode_table	513	32	25 146	888
di_encode_table	513	32	25 146	5 100
sq_decode_table	513	32	12 573	980
sq_encode_table	513	32	12 573	5 100
descendents	1536	32	2 847	16 608
freq	1536	64	146 956	18 624
length	1536	32	412 030	18 495
ordered_symbols	1536	8	960	480

goes up to only 8 192 words of 32 bits, so basic groups 'in', 'pyr', 'ridge' and 'freq' will not fit in a single memory. Basic group matching is necessary.

11.8.3 Basic group matching

Four basic groups do not fit in the chosen memory modules. Basic groups 'in', 'pyr' and 'ridge' are too large: 25 344 words, while only 8 192 will fit in a memory. Basic group 'freq' is too wide: 64 bits, while only 32 will fit in a memory.

Basic groups 'in', 'pyr' and 'ridge' could be split to fit them in a single memory, but it is better to use compaction instead. Because they are only 8 bits or less wide, it is possible to assemble four data words into a single (32-bit) memory word. This reduces the required number of memory words to 6 336, small enough to fit into a 8192-word memory. For 'ridge' it is even possible to go one step further, collecting 8 words of 2 bits together into a single 16-bit memory word. This reduces the size to 3 168 memory words. Table 11.2 shows the basic group information for the new basic groups. In all cases, the extra read accesses could be avoided by collecting write accesses to subsequent words into one write.

For basic group 'freq', splitting has to be applied. Two new basic groups are created: 'freq_1' and 'freq_2'. Both are 32 bits wide and have 256 words. Two memory accesses have to be made now whenever data from 'freq' is required. Those two have to take place in parallel, so all conflicts for 'freq' are propagated to 'freq_1' and 'freq_2'. The same applies for the self-conflicts and the hyperedges: both 'freq_1' and 'freq_2' have the self-conflict, and a copy of the hyperedge is created. Finally, there is also a conflict between the two new basic groups. The annotation for this conflict is simply the double of the self-conflict: indeed, in the cycle where 'freq' was accessed in parallel, now both 'freq_1' and 'freq_2' have to be accessed *twice*. Figure 11.16 shows the resulting conflict graph, Table 11.2 shows the basic group information for the new basic groups.

The new basic group also introduces a new self-conflict. Therefore, two dual-port memories are not sufficient anymore, a third one must be allocated.

11.8.4 Signal-to-memory assignment

Five different memory allocation schemes have been evaluated in the signal-to-memory assignment step. Table 11.3 summarizes the results. The cost figures have been obtained using a polynomial approximation function provided by the memory vendor. The first allocation scheme, using three dual-port memories and one single-port memory, was not even valid. For the others, the entire assignment search space has been covered to find the optimal solution given in the table. The full search has taken several hours, but each time a solution within 5% of the optimum has already been found after a couple of minutes.

Looking at the table, the solution with six memories is obviously the best choice. Allocating even more memories may decrease the power a little, but probably not much. It can even be worse, due to the extra overhead in interconnection. Hence, the solution with six memories is retained. Figure 11.17 shows the optimal assignment scheme for six memories.

Table 11.2.   Basic group information for the reorganized basic groups of the BTPC application.

basic group name	# words	bit width	#reads	#writes
in	6 336	32	50 688	50 688
pyr	6 336	32	123 852	50 832
ridge	3 168	16	37 719	37 863
freq_1	1536	32	146 956	18 624
freq_2	1536	32	146 956	18 624

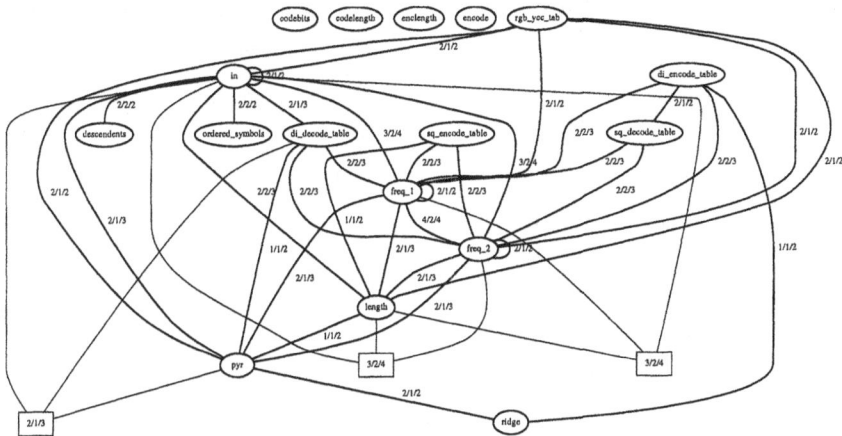

Figure 11.16. Extended conflict graph for BTPC, after splitting 'freq' into 'freq_1' and 'freq_2'. All conflicts, including the self-conflicts and hyperedges, are propagated to the new basic groups.

Table 11.3. Results of signal-to-memory assignment. These were obtained after a full search of the possible assignments.

total # memories	dual-port memories	single-port memories	total power (mW)	total area (mm²)
4	3	1	—	—
5	4	1	1235	636
6	3	3	885	573
4	4	0	1615	706
5	3	2	1032	590

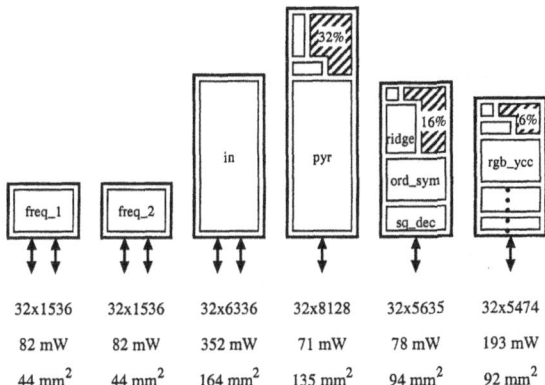

Figure 11.17. Final signal-to-memory assignment for the BTPC. The sixth memory contains all basic groups not mentioned in the others.

11.8.5 Interpretation of the results

Figure 11.17 suggests two conclusions. First, the 'in'-array is the most important power consumer. It may be useful to apply additional higher-level optimizations on this array, for example data reuse. Second, the fourth and fifth memory waste quite a lot of bits: 32% resp. 16%. Adding still another memory might alleviate this problem. However, the share of power used by these memories is very limited, and the area reduction would also probably be offset by the interconnection overhead. A more promising solution is to do more basic group matching, for example compacting 'ridge' and 'ordered_symbols' to 32 bits.

The chosen memory allocation scheme offers nine memory ports, while only six are required according to the storage cycle budget optimization (cfr. the extended conflict graph). This indicates that bus sharing can be useful. However, because of the complexity of the problem and lack of automated support, we have not worked out the bus sharing substep for this application.

11.9 CONCLUSIONS

The ultimate goal of our DTSE methodology is to arrive at an optimized background memory architecture organisation. The final organisation in terms of memory units and signal assignments is reached after the memory allocation and assignment step.

The allocation of memories is heavily based on the bandwidth requirements derived in the storage-bandwidth optimization step to meet the real-time constraints. The memory dimensions (word-length, size) are derived from the assignment of the application data to the memory. In addition, the application data may be reorganized somewhat to fit better into the allocated memories. Finally, the memory architecture is completed with the design of the interconnections between memories and the processing unit. All substeps use shared information, stored in a parameterizable, easily extensible memory library.

Our systematic 4-step methodology allows to find a near-optimal memory architecture based on cost estimators specific for the target technology. The cost estimators used can be very accurate because the memory architecture is fully specified after the memory allocation and assignment step.

We have also proposed a fully automated signal-to-memory assignment substep which is implemented in a CAD tool. The other aspects of the memory architecture can easily be explored manually by the system designer with the support of this assignment tool.

12 STORAGE ORDER OPTIMIZATION FOR REDUCING STORAGE SIZE REQUIREMENTS

Earlier, we have identified a crucial problem in an embedded multimedia application context: finding a good storage order for the data in a program such that the storage size requirements are minimized for a given execution ordering (largely determined in the previous steps). Before developing a practical optimization strategy, we have derived a powerful model of the problem in chapter 5 and gained some vital insights. Now we are ready to proceed with the development of a practical optimization strategy. In this chapter we present our solutions for the storage order optimization step. It mainly aims at reducing storage size requirements for data-dominated applications which can be statically analyzed. A two-step approach is selected from many possible alternatives. In the first substep we try to optimize the intra-array storage order (through the equivalent of data-reverse and data-interchange transformations) in order to obtain an as small as possible address reference window for each array. In the second substep, we approximate the shapes of the occupied address/time domains of the different arrays by rectangles. This allows us to reduce the second substep to a relatively simple placement.

12.1 GENERAL STORAGE ORDER OPTIMIZATION FORMULATION

In this section, a few pragmatic solution strategies are presented for the problem of storage order optimization for reducing storage size requirements. The goal of this strategy is to *reuse* memory locations as much as possible and hence reduce the storage size requirements. This means that several data entities can be stored at the same locations (at different times). Therefore the problem is also referred to as *in-place optimization* [404]. We start by motivating that it is practically infeasible to come up with an ideal global optimization strategy that can find the overall optimal execution and storage order of a program for a given cost function (such as area and/or power cost). Even if we separate the execution order optimization and the storage order optimization, as it is done in our DTSE approach, the general storage order optimization problem remains intractable for practical problem sizes. Therefore we have to come up with a pragmatic solution strategy. We present a few possible alternatives, and select the most promising one. After this introduction we clearly define the problem and discuss the assumptions that we make in subsection 12.1.1. Next, in subsection 12.1.2, we motivate the adoption of a pragmatic solution strategy and discuss several alternatives of which the most promising one is selected. In subsection 12.1.3 we discuss the relation between that strategy and the concept of data transformations.

As motivated in chapter 2, scalar oriented techniques for our problem quickly become infeasible when the number of scalars becomes too large, which is the case in data-dominated multimedia applications. Also, straightforward extensions of these techniques that would treat arrays as scalars, would lead to suboptimal results due to the very different nature of arrays and scalars (e.g. different arrays generally have non-equal sizes, elements of the same array in general have different and possibly even non-overlapping life-times, ...). Also other related techniques do not provide a solution for the full storage order optimization problem as we have formulated it.

12.1.1 Problem definition and assumptions

In the ATOMIUM methodology [53], two of the main optimization parameters are the execution order of the operations and the storage order of the data variables. However, for practical applications it is infeasible to find the optimal execution date for every single operation and the optimal storage location for every single variable, because there are typically millions of operations and variables present in a multimedia application. Therefore we try to find the optimal execution and storage order for *groups* of operations and variables respectively. The modeling techniques for dealing with groups of operations and variables are described in chapter 5. These models allow us to assign an execution order function to each statement (which corresponds to a group of operations) and a storage order function to each array variable (which corresponds to a group of variables).

Unfortunately, even when we group operations and variables, finding *the* optimal solution for a given cost function, satisfying all boundary constraints, is still infeasible for real-life applications. One of the main reasons is that the validity constraints expressed by equation 5.14 on page 113 and equation 5.16 on page 115 are very difficult to take into account during an optimization process.

These problems motivate a more pragmatic approach, in which we introduce several simplifications, and which allows us to come up with a good solution, but not necessarily the best one. This is also the reason why the optimization of the execution order and the optimization of the storage order are separated as much as possible in the ATOMIUM methodology: first the execution order is optimized, mainly through loop transformations, memory hierarchy exploitation, and SCBD; afterwards, among other memory organisation issues, the storage order optimization problem is tackled (through the equivalent of data transformations).

This execution and storage order have a direct impact on the main two cost factors in the ATOMIUM context, namely the area occupation by memories and power consumption by data transfers. However, as indicated in chapter 1 and also in chapter 3, the storage order has a much larger effect on the area requirements than on the power requirements. By choosing a good storage order, we can increase the reuse of storage locations for different variables, and hence reduce the storage size requirements, which results in a smaller area occupation by the memories. Indirectly, the power budget due to memory transfers may also be decreased because of the lower capacitive load of smaller memories. However this power budget is much more sensitive to the introduction of memory hierarchy, as indicated also in chapter 9. This is also why the decisions on memory hierarchy, memory allocation and array-to-memory assignment are taken during separate preceding steps in the ATOMIUM methodology. For these reasons the storage order optimization phase focuses on storage size reduction.

The storage order optimization techniques presented in this and the following sections can be either used on a stand-alone base, or in combination with other optimization tasks in the ATOMIUM methodology, such as execution order optimization. In the latter case, it is important that the storage order optimization step is executed as the last step, for several reasons:

1. If the execution order has not been fixed, one has to make some worst case assumptions during the storage order optimization, which may lead to (highly) suboptimal results.

2. In contrast, as long as the storage order has not been fixed, there is a lot of freedom that can be exploited during the other optimization steps because the storage order can be virtually ignored. When the storage order has been fixed on the other hand, the constraints on the other tasks would become far too complex. For instance, checking whether a loop transformation is valid would be very difficult in that case (it would require taking into account equation 5.16 on page 115).

3. The effect on loop and memory hierarchy related transformations on the optimal storage order (and hence the required storage sizes) is quite large. For instance, loop transformations can enable more opportunities for storage location reuse by influencing the life-times of the data. The effect of the storage order on other cost factors, such as parallelism, performance, and/or power consumption, is much smaller. Therefore it is more natural to tackle the other problems first, before the storage order optimization. Still, it should be stressed that after SCBD, quite a bit of execution order freedom is left, even for a given conflict graph (see section 10.1.2). We can potentially use part of this freedom to further optimize the starting point of the in-depth storage order optimization done here. This can lead to a local iteration with the actual storage order optimization substeps.

Still it is important to be able to make an estimate of the storage requirements during the early stages of the ATOMIUM methodology. For this purpose, some of the techniques presented in this and the following sections can be used as a basis for quick estimators, next to the very high-level estimators proposed in section 4.1. More research on this topic is needed though (see also section 14.3).

12.1.1.1 Assumptions. Based on the above reasoning we can now present the assumptions that we start from:

1. The data-flow of the program has been completely fixed and cannot be changed anymore (e.g. data-flow transformations cannot be applied anymore).

2. The execution order of the program has been mostly fixed. For instance, the code may have been parallelized in advance. In case there is still some freedom left, we do not try to exploit it. Future techniques may try to also exploit this remaining freedom.

   In case the program is to be implemented on a parallel architecture with several processors, we assume that there is sufficient synchronization between the different processors. In that case we can accurately model the relative execution order of the memory accesses by different processors.

3. The memory architecture has been fixed already, *except for the sizes of the memories*. For instance, memory hierarchy decisions have been taken already.

4. The order of the memory accesses is deterministic for a given application code and a given set of input data. This rules out the presence of hardware-controlled caches, because we usually do not know the exact caching algorithm implemented in hardware and therefore we cannot accurately predict the order of the memory accesses in that case.

5. Each (part of a) data array in the program has already been assigned to one of those memories, i.e. we assume that data-distribution decisions have already been taken (array-to-memory assignment). In other words, we assume that we know what data are stored in what memory (but not how they are stored).

6. The storage order of the arrays in the memories has not been fixed yet and can be freely chosen.

These assumptions are compatible with the context of the in-place optimization task in the global ATOMIUM approach.

Finally, an assumption of an entirely different nature, is that the program to be optimized is given in single-assignment form. In this way the freedom present in the program is maximally exposed and our analysis is simplified. This assumption should not be seen as a restriction because, as mentioned in chapter 5, data-flow analysis allows us to convert non-single-assignment programs to single-assignment ones. It may be time-consuming however for complex applications.

12.1.1.2 Problem definition. We can now define our optimization problem as follows:

Given the assumptions mentioned above, find a storage order for each (part of an) array such that the overall required size (number of locations) of the memories is minimal.

In other words, we try to find an optimal layout of the arrays in the memories such that the reuse of memory locations is maximal. The situation before and after storage order optimization is symbolically depicted in Fig. 12.1a and Fig. 12.1b respectively for an architecture with 3 memories and 2 processing elements (PEs).

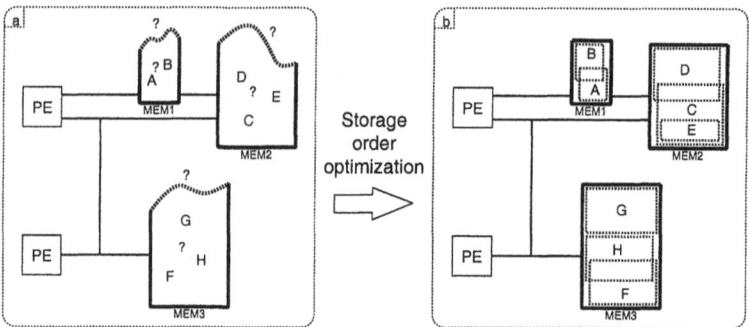

Figure 12.1.   The storage size reduction problem.

An important consequence of the assumption that the execution order has been fixed is the fact that we can treat each memory separately. Because the required size of a memory depends only on the storage order of the arrays or parts of arrays stored in it on the one hand, and the storage orders of arrays stored in different memories are independent on the other hand, there is no interference between memories. This allows a significant complexity reduction for complex memory organizations such as employed in large multimedia applications.

The techniques presented in this and following sections can be easily extended to *take into account* ($\neq$ optimize) a partially unfixed execution order, although the opportunities for storage location reuse will decrease when the uncertainty on the execution order increases. In case the remaining freedom on the execution order would have to be exploited also during the storage optimization step, the problem would become much more complex because the execution order may affect the life-times of different arrays in different memories, and therefore the memories would no longer be independent.

That the assumptions presented above are realistic, is illustrated by the fact that we have applied this approach successfully in the past. For instance, the H263 video conferencing application in [291] has been effectively optimized in this way, resulting in almost a factor of two savings in memory size compared to traditional approaches.

Also the memory-efficient SIMD-type architecture template that we presented in [102, 104] for implementing block-oriented video algorithms such as motion estimation, which we briefly describe in section 13.5, has been designed in this way. After deciding on a detailed data-distribution and a detailed data-transfer schedule, both for the transfers between the shared memory and the local memories and for the transfers between the local memories and the PEs, we exactly knew what data were stored in what memory, and when these data where accessed. Given this information, we have then successfully (manually) applied storage order optimization techniques to obtain minimal sizes for each of the memories (both shared and local, see figure 13.20), resulting in a size reduction by a factor of three for the shared memories. This was done for each memory separately as, for a given schedule, there is no interference between different memories. Of course in this case, due to symmetry reasons, it was sufficient to perform this step only for a few memories, and apply the results to the similarly accessed memories.

For MIMD-type architectures with shared memories similar observations can be made, provided that the memory accesses can be synchronized and that their relative order is (mostly) known at compile time.

Several realistic applications substantiating these claims are also discussed in the experimental result sections of section 12.2 and 12.3.

Our techniques are currently limited to synchronous architectures, i.e. architectures for which the order of memory accesses is only determined by the application and the input data. This even includes architectures in which software-controlled data-caches are present. In the presence of hardware-controlled caches accessed by multiple processors however, the relative order of memory accesses may be unpredictable. Moreover, the organization of the data in the memories may then affect the caching effectiveness, so in that case we cannot claim any longer that our storage order optimization techniques do not interfere with other steps (such as cache performance optimization). Recent research has lead to extensions of our methodology to overcome this problem but this is not discussed further in this book (see also section 14.3).

12.1.2 Pragmatic solution strategy

As explained in the previous section, our intention is to optimize the storage order for each array, such that the required size of each memory is minimal. In section 5.6 we have derived the necessary and sufficient constraints that have to be satisfied by the execution and storage order in order to be valid. Unfortunately, as indicated on page 115, these constraints can be very hard to evaluate in practice. Therefore, an optimization strategy that takes them explicitly into account would probably not be feasible for realistic problems. Hence we have to use a more pragmatic approach that avoids the evaluation of these constraints as much as possible, but that may lead to suboptimal results.

Without loss of generality, we concentrate on the size reduction of only one (shared) memory. In general, multiple memories can be present, but as motivated in subsection 12.1.1, our techniques can be applied to each memory separately, as there is no interference between the optimizations for different memories for a given data distribution and execution order.

12.1.2.1 Observations.   Our pragmatic solution strategy is based on the fact that we can identify two components in the storage order of arrays:

- the *intra*-array storage order, which refers to the internal organization of an array in memory (e.g. row-major or column-major layout);

- the *inter*-array storage orderinter-array storage order, which refers to the relative position of different arrays in memory (e.g. the offsets, possible interleaving, ...).

In that sense we can see the mapping of array elements on memory locations as a two-phase mapping: first there is a mapping from the (multi-dimensional) variable domains to a one-dimensional **abstract address space** for each array, and then there is a mapping from the abstract address spaces onto a joint **real address space**[1] for all arrays assigned to the same memory.[2] This is symbolically depicted in Fig. 12.2.

This observation has stimulated us to come up with a two-phase optimization strategy. In a first phase, we try to find an optimal intra-array storage order for each array separately. This storage order then translates into a partially fixed address equation, which we refer to as the **abstract address equation**, denoted by the function $\mathbf{A^{abstr}}()$. In the second phase, we look for an optimal inter-array storage order, resulting in a fully fixed address equation for each array. This equation is referred to as the **real address equation**, denoted by the function $\mathbf{A^{real}}()$.

In the following section we outline a few strategies that can be used during the *second* mapping phase, i.e. the phase that transforms the $\mathbf{A^{abstr}}()$ of each array into an $\mathbf{A^{real}}()$. Then we select the most promising strategy, and derive from this how we can steer the *first* mapping phase.

12.1.2.2 Inter-array storage order.   Given the abstract address equations of each array, which represent a partial storage order, and the execution order of the program, we can already describe the occupation of the abstract addresses, i.e. we can describe (binary) occupied address/time domains (for the abstract addresses) by means of equations 5.11 and 5.12 in chapter 5, in which we substitute the storage order function $O_x^{addr}()$ by $\mathbf{A}_x^{abstr}()$.

---

[1] This "real" address space can still be virtual. For instance, an extra offset may be added during a later mapping stage.
[2] Note that in chapter 5 we make a distinction between storage addresses and memory addresses. In practice, the relation between these two is always linear, so we do not continue to make this distinction anymore.

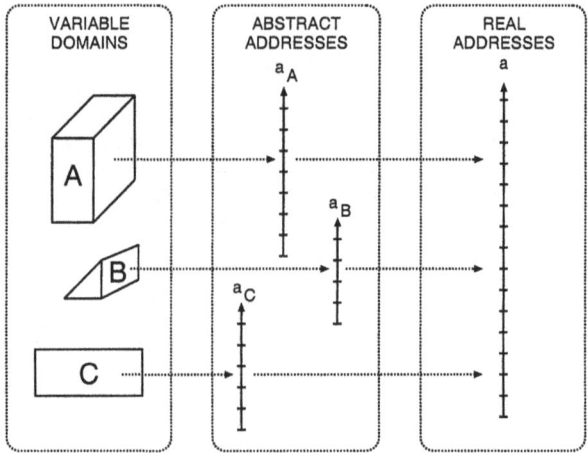

Figure 12.2.   The two-phase mapping of array elements onto (real) memory addresses.

In general, the "shape" of these domains is not known, as we have only implicit descriptions. It is however possible to analyze these descriptions and to extract certain properties (such as the width or the height), which allow us to approximate the shapes. These approximations can then be used in several ways, depending on their accuracy, as indicated next.

We will now illustrate five alternative strategies for the second mapping phase by means of an example. In Fig. 12.3 the occupied address/time domains are shown for five arrays of which the intra-array storage order (i.e. the $\mathbf{A}_x^{\mathtt{abstr}}()$ function) is assumed to be known.

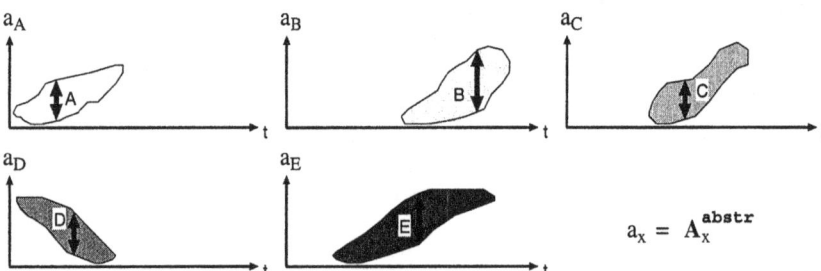

Figure 12.3.   Abstract OAT-domains of 5 arrays.

We can then obtain an $\mathbf{A}_x^{\mathtt{real}}()$ in the following ways:

1. The simplest way to allocate memory for these arrays is to assign a certain address range to each of them in such a way that the different address ranges do not overlap. The size of the address range allocated for an array then corresponds to the "height" of its occupied address/time domain and this typically equals the size of the array. This is depicted in Fig. 12.4a. The difference between the abstract and the real address equation is then simply a constant offset C_x as indicated at the bottom of the figure. The choice of the C_x values is straightforward, i.e. the arrays are simply stacked in the memory and the address ranges are permanently assigned to the different arrays. This strategy is commonly referred to as **static allocation**. This is the standard approach taken by compilers (for global variables). Note that this approach does not result in any memory reuse at all, but that it can be implemented very easily (one only needs to know the size of the arrays). Moreover, the

intra-array storage order (which influences the shape of the occupied address/time domains) has no effect on the required memory size (assuming that for each array a range equal to its size is allocated).

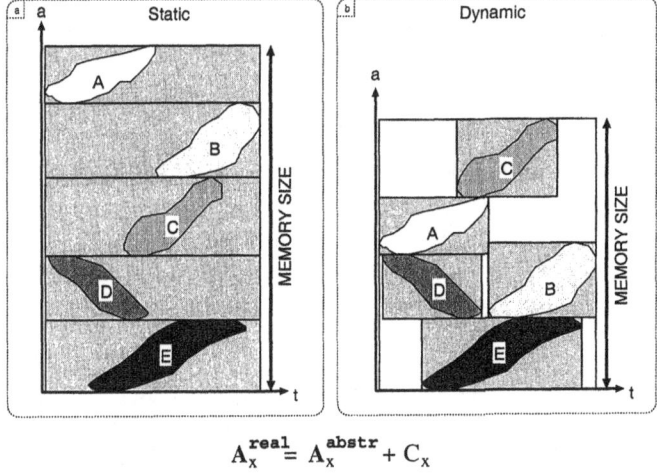

$$A_x^{real} = A_x^{abstr} + C_x$$

Figure 12.4.   Static and dynamic allocation strategies.

2. A potentially better strategy is illustrated in Fig. 12.4b. Here the address range assigned to an array is allocated only during the time that the array is in use. The shapes of the occupied address/time domains are approximated by rectangular bounding boxes.[3] This allows sharing of certain address ranges by more than one array. We refer to this strategy as **dynamic allocation**, but it is important to note that this allocation can be performed at *compile* time, in contrast to, for instance, heap-allocation in a language such as C. In general, a dynamic strategy requires less memory than a static one. Unfortunately, it requires life-time analysis of the arrays, and the placement of the arrays in memory is no longer straightforward, as the required memory size depends on it. The relation between the abstract and real address equations is similar to that of the first alternative, and also in this case the intra-array storage order has no effect on the required memory size (under the same assumptions as mentioned above).

3. The previous strategy enables memory reuse between different arrays, but we can also exploit memory reuse between elements of the same array. Many times this is possible because an array may not need its complete address range all the time. This has lead to the definition of an **address reference window** [413] as the maximum distance between two addresses being occupied by the array during its life-cycle. This address reference window W_x is indicated for each of the arrays in Fig. 12.3 by means of a vertical double arrow. If we know the size of the window, we can "fold" the occupied address/time domains of the arrays by means of a modulo operation in the address equations. The result for a static windowed allocation strategy is depicted in Fig. 12.5a. The relation between the $A^{abstr}()$ and the $A^{real}()$ of an array is again relatively simple, as indicated at the bottom of the figure. Note that a folded occupied address/time domain can never overlap with itself, provided that the window equals at least the maximum "thickness" of the domain, so the folding can never introduce memory occupation conflicts.

4. Of course, we can also combine the windowing approach with dynamic allocation, as depicted in Fig. 12.5b. This results in memory reuse between elements of the same array *and* elements of

---

[3] In section 12.3 we discuss some extensions.

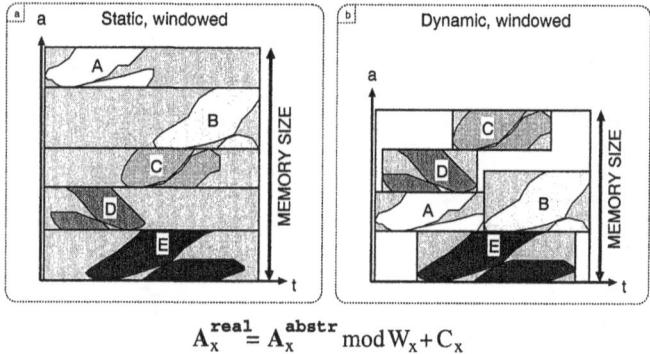

$$A_x^{real} = A_x^{abstr} \bmod W_x + C_x$$

Figure 12.5.   Windowed allocation strategies.

different arrays, and in general leads to the smallest memory size requirements of these first four alternatives. An important difference between the last two alternatives and the first two is the fact that for the windowed approaches, the intra-array storage order is important, as it directly influences the shape of the occupied address/time domains and hence the size of the window. Moreover, the exact evaluation of the window size and especially the choice of the best intra-array storage order is far from trivial. Techniques for this are presented in section 12.2.

5. A possibly even better strategy is depicted in Fig. 12.6. In a first step, the occupied address/time domains are shifted (and possibly even vertically scaled or flipped by a factor S_x) such that their common address reference window is minimal. After that, the complete address range is folded by the common window W. Note that for the example this last strategy is the best one, but this is *not* true in general. The previous strategy with separate windows can sometimes yield better results (e.g. when the domains don't "fit" together very well). Moreover, since we do not have an explicit description of the exact shapes of the domains, it would be very difficult to find an optimal placement.

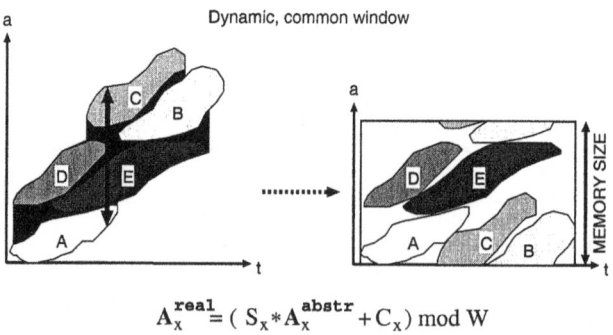

$$A_x^{real} = (S_x * A_x^{abstr} + C_x) \bmod W$$

Figure 12.6.   Dynamic allocation strategy with common window.

Note that for the first four alternatives we can completely avoid the evaluation of the complex correctness constraints represented by equation 5.16 on page 115. In each case, the shape of the occupied address/time domains is approximated by a rectangle. We only have to make sure that the rectangles do not overlap, which results in much simpler (sufficient) constraints.[4]

An efficient implementation of the fifth strategy on the other hand would require that we can calculate the distances between the different domains. In section 12.3 we describe how this can be done. Unfortunately these distance calculations are very costly and also turn out to be very difficult from a numerical point of view (see subsection 12.3.4).

Therefore we have selected the fourth alternative (dynamic, with separate windows, as depicted in Fig. 12.5b) as the base for our placement strategy, as it offers the best compromise between optimality and complexity. Our full placement strategy, which is outlined in section 12.3, is in a sense a hybrid approach, because it uses elements of the fifth alternative too (i.e. OAT-domains are not always approximated by rectangles).

12.1.2.3 Intra-array storage order. Given the fact that we have chosen the fourth alternative as the basic strategy for the second mapping phase, the strategy for the first mapping phase becomes rather straightforward: we have to find the storage order for each array (or part of an array) that results in the smallest window size. Doing this results in the smallest rectangular bounding boxes for the OAT-domains, and most likely also in the smallest required memory size. Our techniques for optimizing the intra-array storage order (and calculating the window sizes) are presented in section 12.2.

The number of possible storage orders is huge, even if we restrict ourselves to the affine ones. Moreover, checking whether a storage order is valid generally requires the evaluation of the complex constraints represented by equation 5.16 on page 115 and to our knowledge no practically feasible strategy exist for choosing the best order while taking into account these constraints. Therefore, we restrict the search space drastically.

First of all, we require that each element of an array is mapped onto an abstract address that is unique w.r.t. the address of the other elements of the same array.[5] This property is also called **unambiguity** in [74]. By requiring this we get an abstract address equation that is correct by construction, i.e. we can avoid expensive checks for intra-array memory occupation conflicts. Later on, during the second mapping stage, several array elements can be mapped on the same real addresses, provided that they have non-overlapping life-times. In case this unambiguity requirement for the abstract addresses is too restrictive, one can always split arrays at the specification level to provide more freedom (or merge/interleave them to limit the freedom).

A second constraint that we impose (and that is also imposed in [74]), is that the storage order should be dense, i.e. the range of abstract addresses occupied by an array should be as small as possible (taking into account the limitations of an affine storage order). The reason for this is that it would in general be very difficult, if not impossible, to reuse the "holes" in a non-dense address range during the following mapping phase (unless they are periodical). A standard row-major order, as used in C, usually satisfies this requirement, but not always. For instance when only the even elements of an array are accessed, one can compress the address range by using the appropriate scaling factors in the address equation. In section 12.3 we elaborate on this, but for now we assume that no such compression is possible or that his has been done in advance, so a row-major or column-major order is a valid solution.

However, for multi-dimensional arrays, we consider *all* possible orders of the dimensions, and also both directions (i.e. forward and backward) for each dimension. Consequently the $A_x^{\mathtt{abstr}}()$ functions

---

[4]At least if we assume that no "orphan" writes occur, i.e. writes of values that are never read. These do not show up in the (B)OAT-domains and consequently also not in the approximations (in contrast to the validity constraints), and may result in the overwriting of other data if we do not take them into account. Checking for the occurrence of these writes is relatively simple and we assume that they have been removed from the code as they have no purpose. Another possibility is to add a "dummy" read after each orphan write *in the geometrical model*. In that way the orphan writes are also taken into account in the (B)OAT-domain descriptions.

[5]Although we violate this requirement in certain cases, as indicated on page 281, but for now we assume that this is not the case.

of an array have the following format:

$$\mathbf{A}_x^{\text{abstr}}(a_{1x}, a_{2x}, \ldots, a_{D_x x}) = \sum_{j=1}^{D_x} N_{jx}|a_{n_j x} - B_{n_j x}^{\text{up/lo}}| \tag{12.1}$$

$$\text{where} \quad N_{jx} = \prod_{k=j+1}^{D_x} S_{n_k x}, \quad N_{D_x x} = 1 \tag{12.2}$$

$$\text{and} \quad S_{n_k x} = B_{n_k x}^{\text{up}} - B_{n_k x}^{\text{lo}} + 1 \tag{12.3}$$

In these equations, the a_{ix} correspond to the dimensions of an array x and D_x is the number of dimensions. The order of the dimensions is determined by the values of the n_{jx}: $(n_{1x}, n_{2x}, \ldots, n_{D_x x})$ is a permutation of $(1, 2, \ldots, D_x)$. The constant $B_{n_j x}^{\text{up/lo}}$ coefficients are either the upper or the lower bounds of the corresponding dimensions, depending on the direction that is chosen for a dimension. By using an absolute value in equation 12.1, we guarantee that each of the terms in the sum is positive.[6]

The constant N_{jx} coefficients represent scaling factors obtained as a product of the sizes of the dimensions $S_{n_j x}$. For a given dimension order we can easily calculate these coefficients, starting with the last one.

An example of the possible orders we consider for a 2-dimensional array of 2×3 is given in Fig. 12.7.

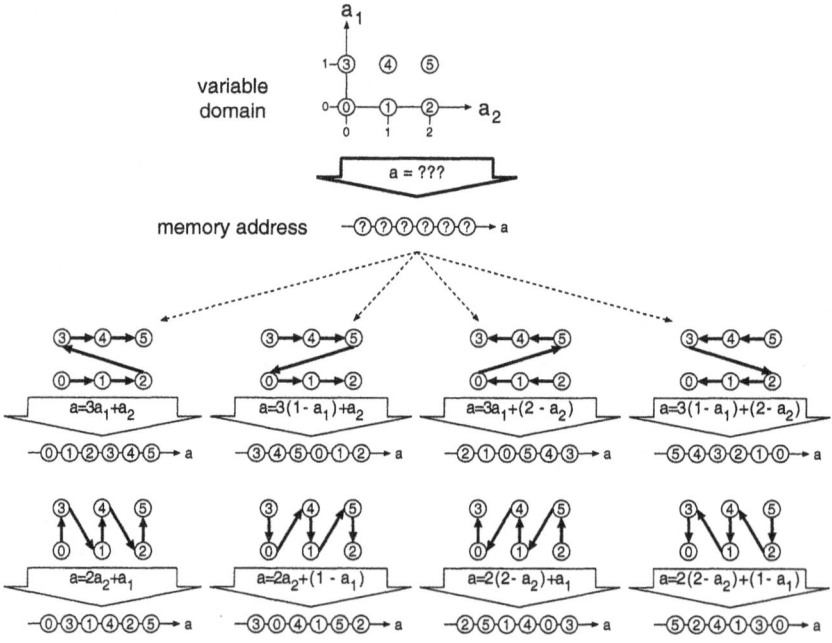

Figure 12.7.   The possible storage orders for a 2×3 2-dimensional array.

In general, for a D-dimensional array, we consider $2^D D!$ possibilities. For a 6-dimensional array for instance, there are no less than 46080 possibilities! In [74] it is claimed (in another context) that a

---

[6]In practice we do not need an absolute value once we know the direction of a dimension, because we can add appropriate signs in that case (see also Fig. 12.7).

brute-force search strategy over all these possibilities is practically feasible because the vast majority of arrays do not have more than 3 dimensions. However, as motivated in subsection 12.1.1, we assume that we start from single-assignment code (which maximally exposes the available freedom), and in that case it is not unusual to encounter a 6-or-more-dimensional array. It must be clear that even though this is only a very limited subset of all possible storage orders, evaluating each order in this set is infeasible for arrays with many dimensions. So we have to come up with a more intelligent search strategy. This search strategy is presented in section 12.2.

In subsection 12.2.5 we extend the search space even further by also allowing orthogonal projections of dimensions of the array. The effect of these additional possibilities on the complexity of the optimization process is however limited. Therefore, we ignore these possibilities for now. Taking them into account may even help us to reduce the effective size of the search space, as shown in subsection 12.2.5.

12.1.2.4   Addressing complexity.   Finally, we can remark that the complexity of the address equations obtained by each of the strategies presented in subsection 12.1.2.2 is relatively low, provided that the abstract address equation is not too complex (and this is usually the case, because we limit the search space). The only possible difficulty is the presence of the modulo operation in the equations (at least for the windowed approaches). In practice this poses no problems for the following reasons:

■ In many cases the modulo operation can be avoided completely, e.g. by substituting it by a projection. This is explained in more detail in section 12.2 on page 280.

■ In case the modulo operation cannot be avoided in this way, it is usually possible to generate the address sequences by means of incremental pointer updating, with a simple boundary check. This requires induction analysis of the instantiated address expressions, but as most address expressions in multimedia applications are linear or largely linear, this is usually relatively simple. Moreover, this type of pointer updating matches the addressing capabilities of DSP processors very well. DSP processors usually have one or more address generators which can implement this bounded incremental pointer updating in a very efficient way (e.g. one address per clock cycle).

■ In case the modulo operation cannot be avoided by one of the previous techniques, it may be worthwhile to increase the window size somewhat until the next power of two. By doing this, the modulo operation can be reduced to a simple bit masking operation (assuming a two's-complement number system). Of course this may result in a relatively large memory waste. For instance, if the window size would be 1025, we would (at least temporarily) waste 1023 locations by rounding it up to 2048. However, if the program is to be implemented on a general purpose processor, executing modulo operations may be very costly in terms of cycles. In that case a trade-off between performance and memory cost must be made. In case the program is mapped on a custom processor, the cost of a modulo operation is much lower. In that case it could even be better to use a plain modulo operation instead of incremental pointer updating.

So, the optimal address generation strategy is highly dependent on the target architecture, but in practice the address model that we use is well suited for most architectures. Therefore, we do not come back to this issue again.

12.1.3   Relation with data transformations

There exists a strong relation between the intra-array storage orders we consider and data transformations. In fact, the storage orders described in subsection 12.1.2.3 can be seen as the result of a data transformation composed of a sequence of unimodular transformations of the variable domains, followed by a standard (singular) storage mapping such as row-major or column-major:

$$\mathbf{A}_x^{\mathtt{abstr}}(a_{1x}, a_{2x}, \ldots, a_{D_x x}) = \sum_{j=1}^{D_x} N_{jx}^* |a_{1x}^*| + C_x \qquad (12.4)$$

$$\text{where} \qquad N^*_{jx} = \prod_{k=j+1}^{D_x} S^*_{kx}, \quad N^*_{D_x x} = 1 \qquad\qquad (12.5)$$

$$\text{and} \qquad S^*_{kx} = B^{*\text{up}}_{kx} - B^{*\text{lo}}_{kx} + 1 \qquad\qquad (12.6)$$

$$\text{and} \qquad \begin{bmatrix} a^*_1 \\ a^*_2 \\ \dots \\ a^*_{D_x} \end{bmatrix} = \underbrace{U_{m_x x} \times U_{(m_x-1)x} \times \dots \times U_{1x}}_{\text{unimodular } (D_x \times D_x)} \times \begin{bmatrix} a_1 \\ a_2 \\ \dots \\ a_{D_x} \end{bmatrix} \qquad (12.7)$$

In these equations the U_{ix} matrices represent unimodular transformations (m_x in total) and the a_{jx} and a^*_{jx} coefficients represent the original and transformed dimensions respectively. The $B^{*\text{up/lo}}_{kx}$ and S^*_{kx} coefficients represent the upper and lower bounds and sizes of the transformed dimensions respectively. A constant value C_x, which is a combination of the N^*_{kx} and $B^{*\text{up/lo}}_{kx}$ coefficients, is added to equation 12.4 to make the address range start at zero.

The unimodular transformations that we currently consider are data-reverse and data-interchange transformations (which are similar to loop-reverse and loop-interchange transformations). Because these transformations are unimodular, we know that we satisfy the requirements that we have imposed on the storage order: the mapping onto abstract addresses should be unambiguous and dense.

In [74] it is stated that data transformations are always legal, provided that they satisfy the unambiguity requirement. However, this is only a sufficient condition, not a necessary one (in contrast to what is claimed in [74]). If we would impose this condition on the final storage order, *we would not be able to reuse memory locations!* Therefore, we only impose it during the first mapping stage. Later on, during the second mapping stage, we even explicitly rely on the fact that array elements can share memory locations, either through windowing (intra-array reuse), or through dynamic allocation (inter-array reuse). Unfortunately, if we drop the unambiguity requirement, the legality of the data transformations is no longer guaranteed as it also depends on the execution order. This leads to the (complex) constraints represented by equation 5.16 on page 115, but as indicated in subsection 12.1.2.2, we can also avoid having to evaluate them by relaxing them to simpler (sufficient) ones.

In theory we could allow any type of unimodular transformation, so also data-skew transformations are valid. However, if we would also allow those, the number of possible storage orders explodes (and even becomes infinitely large). Moreover, for practical applications the number of cases where a skewing transformation would result in lower storage size requirements is probably relatively low. This has been experimentally verified on several realistic applications, where the obtained solutions came very close to the theoretical optimum without using data-skewing (see subsection 12.2.8).

It is even possible to allow other (unambiguous) data transformations such as data-tiling (or data-blocking). These transformations alter the number of dimensions of the variable domain before the mapping onto an abstract address space takes place. The number of possibilities is then exploding again, while the practical relevance is probably also limited in a context without memory hierarchy. However, when memory hierarchy is present we have another task in our ATOMIUM methodology which performs this step prior to the storage order optimisation mapping task. In this way the global optimum is of course not ensured, but it provides a very good practical approach as demonstrated by the manual results in [439].

In a practical optimisation framework, it is perfectly possible and even desirable to allow data-skewing and non-unimodular transformations to be imposed interactively by a designer.

12.1.4 Illustration of code generation

The memory assignment and in-place optimisation decisions are incorporated by transforming the application specification. All accesses to arrays are replaced with accesses to the memory to which they are assigned. Addresses need to be changed to reflect the offsets into the memory. These addresses can be updated again, after iterations of the in-place optimisation substeps. After the transformation, the only arrays left in the application code will be one-dimensional ones representing the memories to be used. An example of this transformation is given in figure 12.8 for a C-like specification.

```
bit8 A[100][100];
bit6 B[20][20];
for (int i, j, k, l; ...) {
  B[i][j] = f(B[j][i],
              A[i+k][j+1]);
}
```

(a) Original code

```
bit8 mem1[10400];
for (int i, j, k, l; ...) {
  mem1[10000 + i + 20*j] =
    f(mem1[10000 + j + 20*i],
      bit6(mem1[i+k + 100*(j+1)]));
}
```

(b) Transformed code

Figure 12.8. Signal-to-memory assignment and in-place optimisation transformation expressed in C code

In the example, all references to the arrays A and B are transformed into references to the memory array mem1. The array takes the largest word-length of all arrays which are assembled in it. In this case that is 8, the word-length of array A. The mem1 array is fully linearized: the address A[i+k][j+1] becomes mem1[i+k + 100*(j+1)] because the row-length of array A is 100. Finally, an offset is added to the addresses to separate the different arrays from each other: B[i][j] becomes mem1[10000 + i + 20*j] to skip over the 10, 000 words of array A. If both arrays can (partially) overlap then this is expressed in an appropriate way in the one-dimensional mem1 array.

The quality and feasibility of the proposed two-phase mapping strategy still have to be proven. In section 12.2 we describe the first mapping stage and the accompanying techniques in more detail. The second mapping stage is described in section 12.3. In these sections we also present experimental results to assess the quality and feasibility of our approach.

12.2   INTRA-ARRAY WINDOWING

In this section we describe the first substep of our optimization strategy, namely the intra-array storage order optimization, in more detail. During this step we try to maximize the reuse of memory locations between elements of the same array. A key concept is the calculation of an address reference window for a given storage order of a multi-dimensional array. This concept has been introduced by other researchers, but we present a technique to calculate an exact window, based on the models presented in earlier chapters. Up to now, only upper bound estimation techniques based on simulation were available, and these techniques break down when the sizes of the arrays become too large. The technique that we present here is analytical and is far less sensitive to the array sizes. Moreover, we indicate how one can calculate upper and lower bounds on the window size of an array for which only a partial storage order is known. This allows us to traverse the (sometimes huge) search space in a very efficient way, especially when combined with several speedup-heuristics (presented in [109]. After this introduction we review some related work in subsection 12.2.1. Next, we present an exact address reference window evaluation technique in subsection 12.2.2. After that, we show how we can calculate upper and lower bounds on the window size for a *partial* storage order in subsection 12.2.3. This allows us to traverse the search space more efficiently, as indicated in subsection 12.2.4. Next, we demonstrate how we can prune the search space by making use of projections in subsection 12.2.5 and hence further improve the efficiency. Then, we present some experimental results in subsection 12.2.8 that show that applying only the intra-array storage order optimization techniques can already result in a considerable memory size reduction compared to classical storage schemes.

12.2.1   Detailed related work

In [288, 413] the concept of an address reference window was introduced. Given the size of this window, we can "fold" array elements onto the same locations and hence increase the memory reuse. However, the presented method for calculating the window based on life-time analysis of the array elements is not exact and provides only an upper bound. Moreover, the technique is based on symbolic simulation of the application, which is infeasible for e.g. large video applications. Also, no method is provided to *optimize* the storage order in order to minimize the window sizes. In contrast, in this section we present techniques to evaluate the *exact* window sizes based on the formal models presented in chapter 5, *without* symbolic simulation, and efficient techniques to find the storage order that results in the smallest window size. The storage size estimation techniques for multi-dimensional arrays that are discussed in [404] and [27] are also not suited for our purpose. Even though these techniques are based on an accurate life-time analysis of the array elements, they cannot provide exact storage size requirement figures, but rather estimates on the maximal *number* of elements alive. The main reason for this is the fact that they do *not* take into account the storage order. Consequently, they neglect the presence of unoccupied memory locations and obtain only (estimates on) lower bounds. The main purpose of these techniques is to provide other tasks with quick estimations on storage size requirements though.

Finally, the projection approach described in [430] is partly overlapping with and partly complementary to our approach. They provide no strategy for choosing good projections, but in subsection 12.2.5 we indicate how we can use orthogonal projections to drastically decrease the effective size of the search space that we consider. Nevertheless, the address reference window approach that we use is usually able to perform more aggressive optimizations than a pure projection-based approach. Only in exceptional cases a projection approach can obtain better results (see subsection 12.2.7).

12.2.2 Address reference window evaluation

Before we can start thinking about optimizing the intra-array storage order of an array to obtain a minimal address reference window size, we should first be able to evaluate the window size for a given storage order.

Conceptually, the evaluation is relatively simple. We have to find the maximal distance between two addresses being occupied by the array at the same time. We cannot calculate this distance in one shot, but the binary occupied address/time domain descriptions of equation 5.11 on page 104 (where we substitute the $O_m^{\tt addr}()$ functions by $\mathbf{A}_m^{\tt abstr}()$) allow us to calculate the window as follows:

$$W_m = \max_{i_1 j_1 k_1 l_1 i_2 j_2 k_2 l_2} |a_1 - a_2| + 1 \quad \text{for which}$$

$$\exists\, t \text{ s.t. } (a_1, t) \in \mathbf{D}_{i_1 j_1 k_1 l_1 m}^{\tt BOAT} \ \wedge\ (a_2, t) \in \mathbf{D}_{i_2 j_2 k_2 l_2 m}^{\tt BOAT} \tag{12.8}$$

In other words, if we evaluate for each pair of binary occupied address/time domains of an array m the maximum absolute distance between two occupied address/time tuples (with a common time t, one tuple in each domain), and then take the overall maximum incremented by one, we get the window size of the array.

In general, the calculation of these maximum distances is a very difficult problem on its own, especially when the program is non-manifest. In practice however, in many multimedia applications most array accesses use affine indices, resulting in BOAT-domain descriptions consisting only of affine expressions. As a result, the BOAT-domains are linearly bounded lattices (LBLs) [391], in case only conjunctions are present, or unions of LBLs, in case also disjunctions are present.

In case these BOAT-domains are LBLs, the distance for each pair of domains can be found by solving two[7] integer linear programming (ILP) problems. The number of variables and constraints in the ILP problems depends on the number of dimensions of the array and the depth of the loop nests around the array accesses. In practice, there are usually at most a few dozen variables and constraints. In case a BOAT-domain consists of a union of LBLs (which is usually *not* the case), we can decompose it into its composing LBLs, and proceed in the same way for each component, so we do not come back to this case.

The following example demonstrates the window calculation process (for BOAT-domains consisting of one LBL).

Example 12.2.1

```
int B[10];
```

---
[7]Two because of the absolute value.

```
    for ( i = 0; i < 10; ++i )
    {
S1:   B[i] = ...;
      if ( i < 7 )
      {
S2:       ... = B[i];
      }
    }
    for ( j = 5 ; j < 10; ++j )
    {
S3:     ... = B[j];
    }
```

The iteration, definition, operand and variable domains for this example are the following:

$$\mathbf{D}_B^{\text{var}} = \{ b \mid 0 \leq b \leq 9 \ \wedge \ b \in \mathbb{Z} \}$$
$$\mathbf{D}_1^{\text{iter}} = \{ i \mid 0 \leq i \leq 9 \ \wedge \ i \in \mathbb{Z} \}$$
$$\mathbf{D}_{11B}^{\text{def}} = \{ b \mid \exists\, i \in \mathbf{D}_1^{\text{iter}} \ \wedge \ b = i \ \wedge \ b \in \mathbb{Z} \}$$
$$\mathbf{D}_2^{\text{iter}} = \{ i \mid 0 \leq i \leq 6 \ \wedge \ i \in \mathbb{Z} \}$$
$$\mathbf{D}_{21B}^{\text{oper}} = \{ b \mid \exists\, i \in \mathbf{D}_2^{\text{iter}} \ \wedge \ b = i \ \wedge \ b \in \mathbb{Z} \}$$
$$\mathbf{D}_3^{\text{iter}} = \{ j \mid 5 \leq j \leq 9 \ \wedge \ j \in \mathbb{Z} \}$$
$$\mathbf{D}_{31B}^{\text{oper}} = \{ b \mid \exists\, i \in \mathbf{D}_3^{\text{iter}} \ \wedge \ b = i \ \wedge \ b \in \mathbb{Z} \}$$

Let us assume a purely sequential execution order and a linear storage order:

$$O_1^{\text{time}}(i) = 2i \qquad O_1^{\text{vtime}}(x) = x$$
$$O_2^{\text{time}}(i) = 2i + 1 \qquad O_2^{\text{vtime}}(x) = x \qquad O_B^{\text{addr}}(b) = b$$
$$O_3^{\text{time}}(j) = j + 20 \qquad O_3^{\text{vtime}}(x) = x$$

Since there are 2 flow dependencies present, we also have two BOAT-domains:

$$\mathbf{D}_{1211B}^{\text{BOAT}} = \{ a,t \mid \exists\, i \in \mathbf{D}_1^{\text{iter}}, i' \in \mathbf{D}_2^{\text{iter}}, b \in \mathbf{D}_B^{\text{var}} \text{ s.t.}$$
$$b = i = i' \ \wedge \ a = b \ \wedge \ t \geq 2i \ \wedge \ t \leq 2i' + 1 \}$$
$$= \{ a,t \mid 0 \leq a \leq 6 \ \wedge \ 2a \leq t \leq 2a+1 \ \wedge \ a \in \mathbb{Z} \}$$
$$\mathbf{D}_{1311B}^{\text{BOAT}} = \{ a,t \mid \exists\, i \in \mathbf{D}_1^{\text{iter}}, j \in \mathbf{D}_3^{\text{iter}}, b \in \mathbf{D}_B^{\text{var}} \text{ s.t.}$$
$$b = i = j \ \wedge \ a = b \ \wedge \ t \geq 2i \ \wedge \ t \leq j + 20 \}$$
$$= \{ a,t \mid 5 \leq a \leq 9 \ \wedge \ 2a \leq t \leq a + 20 \ \wedge \ a \in \mathbb{Z} \}$$

Now we have to calculate the maximal distance between each pair of BOAT-domains[8]:

$$\max |a_1 - a_2| + 1 \quad \text{for which} \quad [a_1, t] \in \mathbf{D}_{1211B}^{\text{BOAT}} \ \wedge \ [a_2, t] \in \mathbf{D}_{1211B}^{\text{BOAT}}$$
$$\max |a_1 - a_2| + 1 \quad \text{for which} \quad [a_1, t] \in \mathbf{D}_{1311B}^{\text{BOAT}} \ \wedge \ [a_2, t] \in \mathbf{D}_{1311B}^{\text{BOAT}}$$
$$\max |a_1 - a_2| + 1 \quad \text{for which} \quad [a_1, t] \in \mathbf{D}_{1211B}^{\text{BOAT}} \ \wedge \ [a_2, t] \in \mathbf{D}_{1311B}^{\text{BOAT}}$$

We could find these distances by solving the corresponding ILP problems, but for this simple example we can read the solution from the graphical representation of the BOAT-domains in Fig. 12.9.

We can see that the distances are 1, 5, and 2 respectively. The overall maximum is 5, so the window size of B is 5, and we can fold the OAT-domain by means of a modulo operation, as depicted in Fig. 12.10.

So equation 12.8 offers us a way to calculate the exact window size of an array with F flow dependencies by calculating *at most* $F \times (F + 1)/2$ distances, or solving *at most* $F \times (F + 1)$ ILP

---

[8]Note that we also have to consider self-pairs, i.e. pairs consisting of twice the same BOAT-domain.

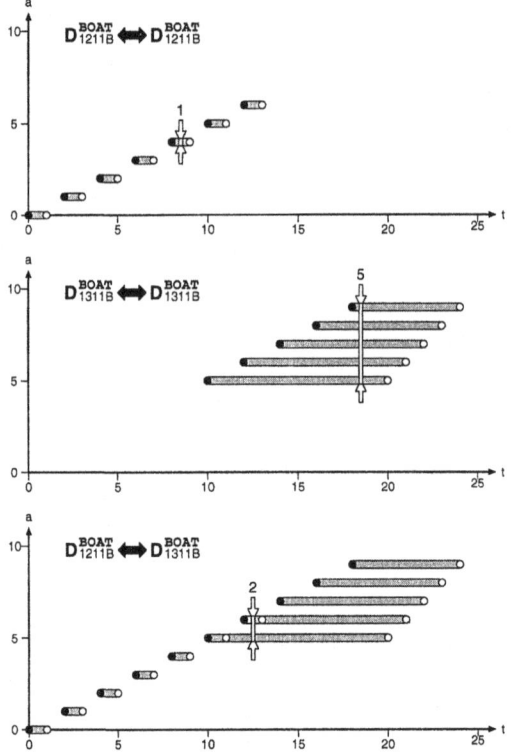

Figure 12.9.   The address reference window calculation.

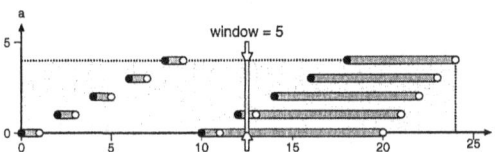

Figure 12.10.   Folding of the OAT-domain by means of a modulo operation.

problems (assuming LBLs). In practice, a (large) number of ILP problems can be avoided or simplified by applying several heuristics that are commonly used during dependency analysis. These techniques are discussed in more detail in [109]. In subsection 12.2.8 we already present some experimental results which indicate that the window calculation can be performed with reasonable run-times.

In case the BOAT-domain descriptions are not LBLs (or unions of LBLs), e.g. because of non-manifest or non-linear expressions, specific solution techniques must be used. An advantage that we can exploit in these cases, is the fact that we do not need the exact value of the window to obtain a valid solution. If we can find an upper bound on the window size, we are sure that the folding of an OAT-domain does not result in occupation conflicts. Of course, working with an upper bound can potentially result in a waste of memory locations.

The following example illustrates how we can deal with non-manifest behaviour.

Example 12.2.2

```
int A[10], B[10];
for ( i = 0; i < 10; ++i )
{
     A[i] = ...;
S1:  B[i] = ...;
     if ( i > 0 && A[i] > 0 )
     {
S2:       ... = f(B[i-1]);
     }
}
```

In this program the read accesses to B are non-manifest. We will therefore only look at domains related to B:

$$\mathbf{D}_B^{\mathtt{var}} = \{ b \mid 0 \le b \le 9 \land b \in \mathbb{Z} \}$$
$$\mathbf{D}_1^{\mathtt{iter}} = \{ i \mid 0 \le i \le 9 \land i \in \mathbb{Z} \}$$
$$\mathbf{D}_{11B}^{\mathtt{def}} = \{ d \mid \exists\, i \in \mathbf{D}_1^{\mathtt{iter}} \text{ s.t. } d = i \land d \in \mathbb{Z} \}$$
$$\mathbf{D}_2^{\mathtt{iter}} = \{ i \mid \exists\, p \text{ s.t. } 0 \le i \le 9 \land i \ge 1 \land p = F(i) \land p > 0 \land i \in \mathbb{Z} \}$$
$$\mathbf{D}_{21B}^{\mathtt{oper}} = \{ o \mid \exists\, i \in \mathbf{D}_2^{\mathtt{iter}} \text{ s.t. } o = i - 1 \land o \in \mathbb{Z} \}$$

For the following order functions:

$$O_1^{\mathtt{time}}(i) = 2i \qquad\qquad O_1^{\mathtt{vtime}}(x) = x \qquad\qquad O_B^{\mathtt{addr}}(b) = b$$
$$O_2^{\mathtt{time}}(i) = 2i + 1 \qquad\qquad O_2^{\mathtt{rtime}}(x) = x$$

we obtain the following BOAT-domain description:

$$\mathbf{D}_{1211B}^{\mathtt{BOAT}} = \{\, a, t \mid \exists\, i \in \mathbf{D}_1^{\mathtt{iter}}, i' \in \mathbf{D}_2^{\mathtt{iter}}, b \in \mathbf{D}_B^{\mathtt{var}},$$
$$b = i = i' - 1 \land a = b \land t \ge 2i \land t \le 2i' + 1 \,\}$$
$$= \{\, a, t \mid 0 \le a \le 8 \land 2a \le t \le 2a + 3 \land F(a + 1) > 0 \land a \in \mathbb{Z} \,\}$$

The window size is given by the following formula in this case:

$$\max |a_1 - a_2| + 1 \quad \text{for which} \quad [a_1, t] \in \mathbf{D}_{1211B}^{\mathtt{BOAT}} \land [a_2, t] \in \mathbf{D}_{1211B}^{\mathtt{BOAT}}$$

Due to the non-manifest read accesses, the BOAT-domain is also non-manifest, and consequently we cannot evaluate this formula until run-time. However, we can take worst-case assumptions. The condition $F(a + 1) > 0$ in the description of $\mathbf{D}_{1211B}^{\mathtt{BOAT}}$ indicates that each of the addresses may or may not be occupied. In the worst case they are all occupied. Hence we can calculate the window with the assumption that $F(a + 1)$ is always larger than zero. This results in a worst-case window size of 2, as indicated in Fig. 12.11

12.2.3 Partial window evaluation

During our search for the optimal intra-array storage order for an array, it would be useful if we could already estimate the size of the window when we have not yet completely fixed the storage order. This would allow us to traverse the search space in a more efficient way.

It turns out that because of the special properties of the abstract address equation 12.1 on page 266an calculate an upper and a lower bound on the window size when we have fixed the order and direction of only some of the dimensions, no matter what the order and direction of the remaining dimensions is. This can be understood as follows.

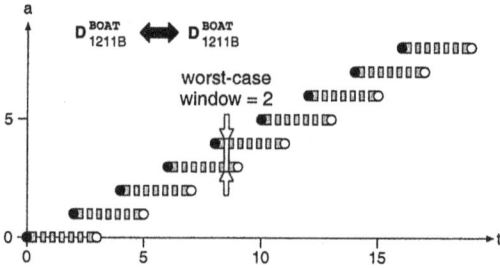

Figure 12.11.   The address reference window calculation for a non-manifest BOAT-domain.

First, let us define the following set of *partial* abstract address equations for an array[9]:

$$\tilde{A}_i^{\text{abstr}}(a_1, a_2, \ldots, a_D) = \sum_{j=1}^{i} \tilde{N}_j^i |a_{n_j} - B_{n_j}^{\text{up/lo}}| \tag{12.9}$$

$$\text{where} \quad \tilde{N}_j^i = \prod_{k=j+1}^{i} S_{n_k} \quad \text{and} \quad S_{n_k} = B_{n_j}^{\text{up}} - B_{n_j}^{\text{lo}} + 1 \tag{12.10}$$

As in equation 12.1 on page 266 n_j coefficients determine the order of the dimensions chosen, and the choices between the upper and lower boundaries $B_{n_j}^{\text{up/lo}}$ determine the directions of the dimensions. Note that the values of the n_j coefficients in equation 12.9 are not depending on i.

In other words, these are the abstract address equations that would be obtained when only the dimensions $a_{n_1}, a_{n_2}, \ldots, a_{n_i}$ of the array would be present, and when we would take the same dimension order and directions as for the full-dimensional array. Obviously, when we take into account all the dimensions, we end up with the full abstract address equation for a given dimension order and given dimension directions:

$$\tilde{A}_D^{\text{abstr}}(a_1, a_2, \ldots, a_D) = A^{\text{abstr}}(a_1, a_2, \ldots, a_D) \tag{12.11}$$

We can then define the following partial windows $\tilde{W}^i$:

$$\tilde{W}^i = \text{window}(\tilde{A}_i^{\text{abstr}}(a_1, a_2, \ldots, a_D)) - 1 \quad \text{for } i \in [1..D] \tag{12.12}$$

where the exact definition of this partial window can be obtained from equation 12.8 through substitution of the full abstract address equation in the binary occupied address/time domains by equation 12.9.[10]

Note that the different partial windows are strongly related because they share the same dimension order and dimension directions (at least for those dimensions that are taken into account), i.e. if $\tilde{W}^i$ is the partial window for the case where i dimensions are taken into account, then $\tilde{W}^{i+k}$ is the partial window for the case where k *additional* dimensions are taken into account.

These (partial) window sizes can also be seen as the maximal absolute value of the weighted sum of distances between the indices of two elements of the array that are simultaneously alive (the weights correspond to the $\tilde{N}_j^i$ coefficients in the partial abstract address equations). These distances are indicated in Fig. 12.12 for an example with a 3-dimensional array.

---

[9]Since we are concentrating on only one array, we drop the indices referring to the array in the sequel to simplify the notation.
[10]Note that we have added a term equal to "-1" to compensate for the "$+1$" term in the definition of the window. This allows us to simplify our calculations in the sequel.

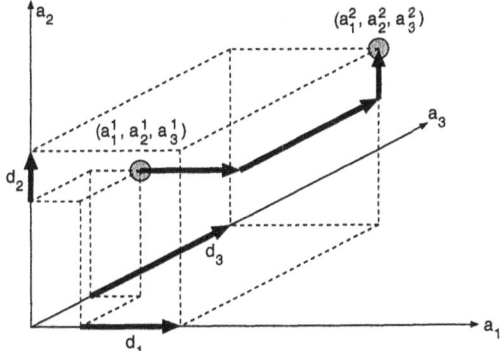

Figure 12.12.   The distance components for two array elements.

So let us assume that the distance vector between these two elements that maximizes the absolute weighted distance $\tilde{W}^i$ is represented by $(d^i_{n_1}, d^i_{n_2}, \ldots, d^i_{n_D})$. Consequently, we can express the partial windows as follows:

$$\tilde{W}^i = |\sum_{j=1}^{i} \tilde{N}^i_j d^i_{n_j}| \tag{12.13}$$

Note that only the first i components contribute to $\tilde{W}^i$. Since the components $d^i_{n_j}$ of this vector represent distances in the different dimensions of the array, we know that $|d^i_{n_j}| \leq S_{n_j} - 1$, where S_{n_j} is the size of the n_j'th dimension.

Of course, $\tilde{W}^D$ must equal the window size W for the full abstract address equation, minus 1 (due the "-1" in equation 12.12). Our intention is now to find a relation between $\tilde{W}^i$ and $\tilde{W}^D = W - 1$. We can proceed as follows:

$$\tilde{W}^D = |\sum_{j=1}^{D} \tilde{N}^D_j d^D_{n_j}|$$

$$\Downarrow \quad (d^D_{n_1}, d^D_{n_2}, \ldots, d^D_{n_D}) \text{ maximizes } \tilde{W}^D$$

$$\tilde{W}^D \geq |\sum_{j=1}^{D} \tilde{N}^D_j d^i_{n_j}|$$

$$\Downarrow \quad |a + b| \geq |a| - |b|$$

$$\tilde{W}^D \geq |\sum_{j=1}^{i} \tilde{N}^D_j d^i_{n_j}| - |\sum_{j=i+1}^{D} \tilde{N}^D_j d^i_{n_j}|$$

$$\Downarrow \quad \tilde{N}^D_j = \prod_{k=j+1}^{D} S_{n_k} = \tilde{N}^i_j \tilde{N}^D_i \text{ for } i \in [j, D]$$

$$\tilde{W}^D \geq \tilde{N}^D_i |\sum_{j=1}^{i} \tilde{N}^i_j d^i_{n_j}| - |\sum_{j=i+1}^{D} \tilde{N}^D_j d^i_{n_j}|$$

$$\Downarrow \quad |\sum_{j=1}^{i} \tilde{N}^i_j d^i_{n_j}| = \tilde{W}^i \text{ and } |d^i_{n_j}| \leq S_{n_j} - 1$$

$$\tilde{W}^D \geq \tilde{N}_i^P \tilde{W}^i - \sum_{j=i+1}^{D} \tilde{N}_j^P (S_{n_j} - 1)$$

$$\Downarrow \quad \sum_{j=i+1}^{D} \tilde{N}_j^P (S_{n_j} - 1) = \sum_{j=i+1}^{D} (\tilde{N}_{j-1}^P - \tilde{N}_j^P) = \tilde{N}_i^P - 1$$

$$\tilde{W}^D \geq \tilde{N}_i^P \tilde{W}^i - \tilde{N}_i^P + 1 \tag{12.14}$$

So, now we can calculate a lower bound for $\tilde{W}^D$, provided that we know the value of $\tilde{W}^i$. In a similar way, we can find an upper bound:

$$\tilde{W}^D = |\sum_{j=1}^{D} \tilde{N}_j^P d_{n_j}^D|$$

$$\Downarrow \quad |a + b| \leq |a| + |b|$$

$$\tilde{W}^D \leq |\sum_{j=1}^{i} \tilde{N}_j^P d_{n_j}^D| + |\sum_{j=i+1}^{D} \tilde{N}_j^P d_{n_j}^D|$$

$$\Downarrow \quad \tilde{N}_j^P = \prod_{k=j+1}^{D} S_{n_k} = \tilde{N}_j^i \tilde{N}_i^P \text{ for } i \in [j, D]$$

$$\tilde{W}^D \leq \tilde{N}_i^P |\sum_{j=1}^{i} \tilde{N}_j^i d_{n_j}^D| + |\sum_{j=i+1}^{D} \tilde{N}_j^P d_{n_j}^D|$$

$$\Downarrow \quad (d_{n_1}^i, d_{n_2}^i, \ldots, d_{n_D}^i) \text{ maximizes } \tilde{W}^i = |\sum_{j=1}^{i} \tilde{N}_j^i d_{n_j}^i|$$

$$\tilde{W}^D \leq \tilde{N}_i^P \tilde{W}^i + |\sum_{j=i+1}^{D} \tilde{N}_j^P d_{n_j}^D|$$

$$\Downarrow \quad |\sum_{j=i+1}^{D} \tilde{N}_j^P d_{n_j}^D| \leq \tilde{N}_i^P - 1 \text{ (see derivation of equation 12.14)}$$

$$\tilde{W}^D \leq \tilde{N}_i^P \tilde{W}^i + \tilde{N}_i^P - 1 \tag{12.15}$$

If we combine equation 12.14 and equation 12.15, we get the following:

$$|\tilde{W}^D - \tilde{N}_i^P \tilde{W}^i| \leq \tilde{N}_i^P - 1 \tag{12.16}$$

or equivalently:

$$|\tilde{W}^D - (\prod_{j=i+1}^{D} S_{n_j}) \tilde{W}^i| \leq \prod_{j=i+1}^{D} S_{n_j} - 1 \tag{12.17}$$

So these equations provide us with a relation between the partial window $\tilde{W}^i$ and the full window $W = \tilde{W}^D + 1$. Now let us suppose that we have chosen the order and directions of the first i dimensions of an array. If we then calculate the partial window for the corresponding partial abstract address equation, we can also calculate an upper and lower bound on the full window, even though we do not know the order and directions of the remaining dimensions yet. The only thing we need are the sizes of the remaining dimensions, and those are known. The more dimensions we fix, the tighter the upper and lower bounds become, as the product $\prod_{j=i+1}^{D} S_{n_j}$ of the sizes of the remaining dimensions decreases. The reader can verify that when all dimensions are fixed (i.e. $i = D$), the right-hand side of equation 12.17 evaluates to zero (assuming that $\prod_{j=D+1}^{D} S_{n_j} = 1$).

In the following subsection we indicate how we can exploit this possibility to calculate upper and lower bounds on the window size, given a partial storage order.

12.2.4 Search space traversal

In subsection 12.2.2 we describe how one can calculate the exact window size for a given intra-array storage order of an array. A straightforward way of selecting the best storage order is to evaluate the window size for each of them and to select the one with the smallest window size, i.e. we could perform a **full search**.

However, as indicated in subsection 12.1.2, the number of possible storage orders that we consider even after our earlier simplifications, can be huge, especially for arrays with many dimensions. For an array with D dimensions, there are no less than $2^D D!$ possibilities. Moreover, as stated in subsection 12.2.2, the number of distances that we have to calculate for *one* window evaluation is proportional to the square of the number of dependencies of an array. More in particular, in case the BOAT-domains are LBLs, the number of ILP problems to solve to evaluate the window of an array with F dependencies equals $F(F + 1)$. Therefore, a full search is out of the question for arrays with many (e.g. 6 or more) dimensions.[11] For instance, for an array with 6 dimensions and (only) 10 dependencies, this would result in about 5 million ILP problems to be solved (for only one array!), which is clearly unacceptable.

So it must be clear that we need a more intelligent way to traverse the search space in case there are many possibilities. We can do this by making use of the property that we have derived in the previous subsection for the partial windows: given that we have fixed the order and direction of some of the dimensions, we can already calculate an upper and a lower bound on the window, no matter what the order and direction of the remaining dimensions is.

This property is ideally suited for the implementation of a **branch-and-bound** (B&B) search strategy. The search space can be seen as a tree in which we have to find the optimal leaf. An example of the search tree for a 3-dimensional array is shown in Fig. 12.13.

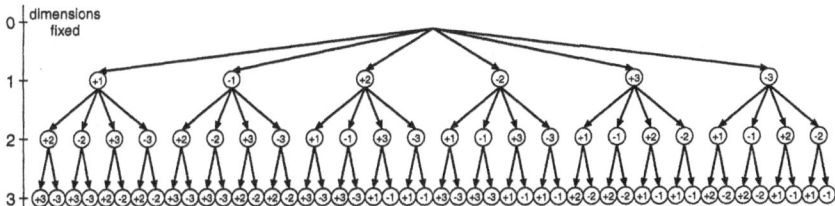

Figure 12.13. The intra-array storage order search tree for a 3-dimensional array. The numbers in the circles correspond to the dimension that is fixed, and the signs correspond to the direction that was chosen for that dimension.

We can start by evaluating the partial windows for the cases where we fix only one dimension and its direction. We can then calculate the corresponding upper and lower bounds and prune the cases that can not lead to an optimal solution. Next, for the remaining cases, we fix an additional dimension, calculate new bounds, and prune again. This process is then repeated until all dimensions have been fixed, and we end up with the optimal solution (within the boundaries of our search space). Of course the worst-case complexity of a full B&B strategy ($D! \sum_{i=1}^{D} \frac{2^i}{(D-i)!}$ partial window calculations) is worse than that of a full search in which we only evaluate the leaf-nodes of the tree ($2^D D!$ window calculations), but in practice it is highly unlikely that the complete B&B tree has to be searched as there are usually some dominant dimensions. Moreover, in the next subsection we present some additional

---

[11] As already mentioned before, it is not unusual to encounter arrays with many dimensions in single-assignment code.

techniques to prune the search space even further. In our experiments (see subsection 12.2.8), the number of evaluated nodes in the search tree always was relatively small. The reason for this is that the (partial) window difference between a good and a bad choice is usually relatively large, especially at the higher levels of the search tree.

A third possible search strategy could be called **greedy tree descent**. Just like the B&B strategy it is based on a gradual fixing of dimensions and partial window evaluations. However, unlike for the B&B strategy, at each level of the search tree only the most promising branch is explored further (even though other branches might lead to better solutions). This strategy requires only $D(D + 1)$ window evaluations. Of course, it is not guaranteed that this strategy leads to the optimal leaf node. Nevertheless, during our experiments, which are described in subsection 12.2.8, this strategy always turned out to lead to very good and most of the time even optimal results.

In the following subsection we describe some additional techniques for pruning the search space, which can speed up the search process for each of the three strategies (full search, B&B, and greedy tree descent). Next to these pruning techniques, there are also several other ways to reduce the complexity of the search problem by simplifying or even avoiding the ILP problems that have to be solved. These techniques are presented in [109].

12.2.5 Search space pruning

Although the search techniques presented in the previous subsection allow us to traverse the sometimes huge search space in a reasonably efficient way, it would still be worthwhile if we could prune the search space in advance, i.e. if we could rule out certain possibilities without having to evaluate the corresponding window sizes. In this way several (expensive) window evaluations could be avoided. This is explained next.

12.2.6 Projection of invariant dimensions

We will now illustrate how we can prune the search space by means of the following example:

Example 12.2.3

```
    int A[5][5];
    for ( i = 0; i < 5; ++i )
    {
        for ( j = 0; j < 8; ++j )
        {
            if ( j < 5 )
S1:             A[i][j] = ...
            ...
            if ( j >= 3 )
S2:             ... = A[i][j-3];
        }
    }
```

If we assume a sequential execution order and a storage order $O_A^{\text{addr}}(a_1, a_2) = 5a_1 + a_2$ for A, we obtain the OAT-domain (which equals the only BOAT-domain in this case) depicted in Fig. 12.14a. The window size equals 4, so we can fold the domain as shown in Fig. 12.14b.

However, if we look at the OAT-domain in more detail, we can recognize several parts that are not overlapping in time (indicated by the vertical striped lines). These parts correspond to the different executions of the i-loop. The reason why these parts do not overlap is that the dependency between S1 and S2 is not carried[12] by any loop, i.e. the elements of A that are produced during a certain execution of the i-loop are consumed only during the same execution of this loop, and not during any other execution. Therefore we call this loop an invariant loop.

This also means that the memory locations being occupied by elements of A during one execution of the i-loop can be reused during another execution of this loop. If we then also look at the index expressions used in the

---

[12]The concepts of carrying and invariant loops are very important in the parallel compiler world: loops that are not carrying any dependencies can be executed in parallel.

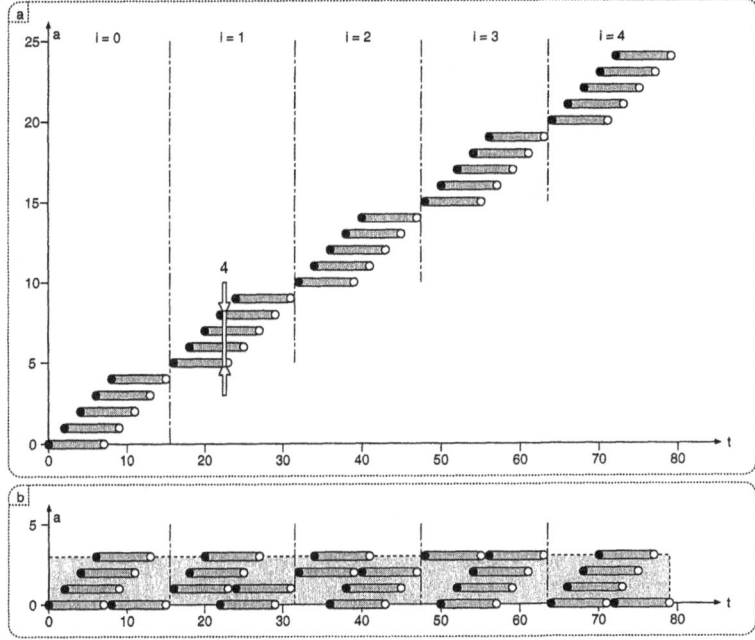

Figure 12.14.   OAT-domain and folded OAT-domain for example 12.2.3.

accesses of A, we can see that one of the indices, namely the first one, is only depending on the i-loop. Since we never simultaneously need elements of A produced during different executions of this loop, we can simply "ignore" the first dimension of A during our optimizations. This is equivalent with an orthogonal projection of the variable domain of A along the axis of its first dimension. In this way the number of dimensions of A is effectively reduced by one. We refer to this type of dimension as an **invariant dimension**. The effect of this projection on the OAT-domain is depicted in Fig. 12.15a.

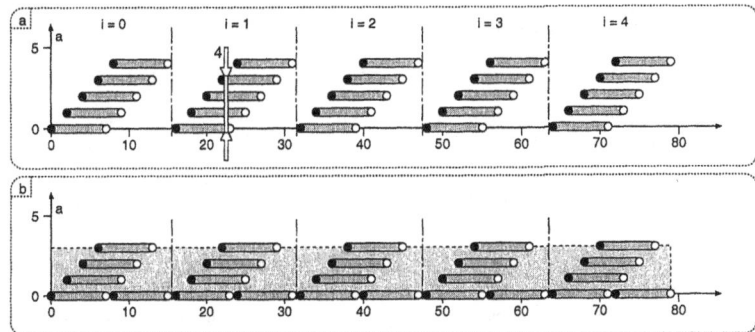

Figure 12.15.   OAT-domain and folded OAT-domain for example 12.2.3 obtain after a projection of the first dimension.

Now we can see that the window size still equals 4, and consequently we can fold the OAT-domain as depicted in Fig. 12.15b. If we compare this solution with the one shown in Fig. 12.14b, we can see that the organization of the array elements in the memory is somewhat different, but the window size is the same. So if we apply *the optimization strategy that we selected in section 12.1, in which we approximate the folded OAT-domains by rectangles, the resulting memory size is the same in both cases.*

On the one hand one can easily verify that in case of invariant dimensions, no matter what the storage order is, the window size of the projected and non-projected case are always the same because the parts of the OAT-domain that are not overlapping in time can never overlap as long as the execution order remains fixed. The maximum distance of equation 12.8 therefore cannot occur at a transition between different executions of invariant loops. So the relative position of the different parts of the OAT-domain (which is the only difference between the projected and non-projected cases) cannot influence the size of the window.

On the other hand, one can verify that an exact search strategy, such as a full search or a B&B search, always puts the invariant dimensions "at the outside", i.e. assigns them the largest coefficients in the storage order function (if the storage order influences the window size). In that way the non-invariant dimensions, which are the only ones that matter for the window size, are packed more densely, resulting in the smallest window size. Failing to do so would result in a "spreading" of these dimensions and therefore a suboptimal solution. This is depicted for example 12.2.3 in Fig. 12.16 for a storage order $O_A^{\text{addr}}(a_1, a_2) = a_1 + 5a_2$.

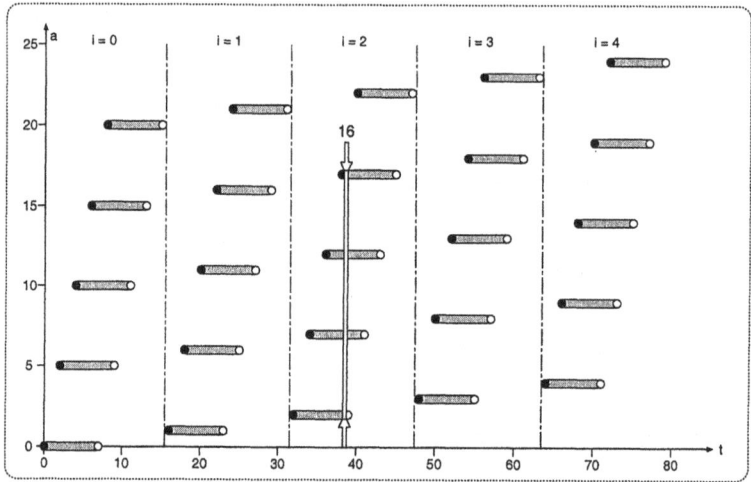

Figure 12.16.   OAT-domain of example 12.2.3 for a suboptimal storage order.

So, given these facts, if we can detect an invariant dimension, we can safely ignore it during our search for the optimal intra-array storage order, and effectively reduce the height of the search tree, because the resulting window size is the same as when we would take it into account.

12.2.7   Projection versus folding

An additional advantage of using projections is the fact the invariant dimensions no longer appear in the address expression (for example 12.2.3 we would obtain $O_A^{\text{addr}}(a_1, a_2) = a_2$), and therefore the generation of the address sequences is cheaper (either in terms of hardware cost, or cycle count, or both). Moreover, as already mentioned on page 267, the modulo operation may even become superfluous if the window size equals the product of the sizes of the remaining dimensions, which is often the case. In that case, the values generated by the abstract address equation can never exceed the window size, so there is no need for folding.

The additional effort needed for the detection of invariant loops and invariant dimensions is virtually non-existent. The reason for this is that the speed-up heuristics, which are presented in [109], require an even more detailed analysis of the dependencies. We can therefore use these results to prune the search space with almost no additional effort, while the effect on the size of the search space can be very large.

It is important to note that in case we apply a projection, we violate the unambiguity constraint that we imposed on page 265. After the projection, several array elements may be mapped onto the same abstract addresses, but as indicated above, in this context this poses no problems as it cannot lead to memory occupation conflicts.

Finally, we should also mention that a pure address reference window approach has an important shortcoming. In some cases one can obtain a better solution by applying a projection or a projection combined with a window instead of a window only. Moreover, a simple analysis of the dependencies as described above may not be sufficient to detect the possibility of this projection. An example of this is presented next.

Example 12.2.4

```
    int A[3][4];
    for ( i = 0; i < 3; ++i )
    {
S1:    A[i][0] = ...;
    }
    for ( j = 0; j < 3; ++j )
    {
        for ( k = 1; k < 4; ++k )
        {
S2:        A[j][k] = A[j][k-1];
        }
    }
    for ( l = 0; l < 3; ++l )
    {
S3:    ... = A[l][3];
    }
```

If we consider the two possible storage orders[13] for A, namely $O_A^{addr}(a_1, a_2) = a_1 + 3a_2$ and $O_A^{addr}(a_1, a_2) = 4a_1 + a_2$, we would end up with the OAT-domains presented in Fig. 12.17a and Fig. 12.17b respectively (as usual, we assume a sequential execution order). The latter possibility would be selected as it results in the smallest window size, namely 6.

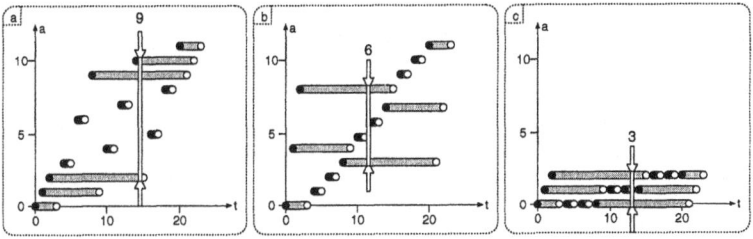

Figure 12.17.   OAT-domain of example 12.2.4 for 3 different storage orders.

However, if we look at the access pattern in the variable domain of A, which is shown in Fig. 12.18, we can see that at any moment in time there is at most one element alive in every row. Therefore all the elements in a row could be stored in the same location. In other words, a projection along the a_2-axis would not result in memory conflicts. If we apply this projection, i.e. use a storage order $O_A^{addr}(a_1, a_2) = a_1$ we end up with the OAT-domain shown in Fig. 12.17c, which has a window size of only 3. Moreover, we do not need to apply the modulo operation because the size of the projected array equals the size of the window, so folding is not necessary.

---

[13]We ignore the possibility of choosing a negative direction for a dimension as the results are similar.

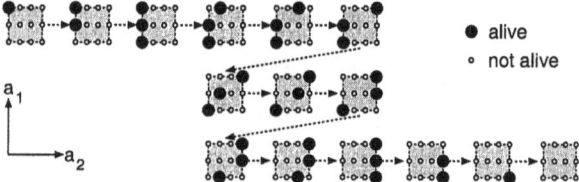

● alive

○ not alive

Figure 12.18.   Access pattern of A of example 12.2.4.

This solution is definitely better than the solution obtained for the other two storage orders. Unfortunately, a simple analysis of the dependencies as described above, cannot reveal that this projection is valid because there is no common invariant loop for all the dependencies (so we cannot make a clean vertical cut between different parts of the OAT-domains in Fig. 12.17).

So this example illustrates that a pure window-based approach may result in (highly) suboptimal results. A combination with projections (or sometimes even a pure projection approach, as in the example) can be a worthwhile alternative. The detection of the projection opportunities for which no equivalent window approach exists, requires some extra analysis effort. In [430] the constraints to be satisfied by a valid projection are described for the general case. However, checking these constraints for every dimension of every array may slow down the optimization process considerably, while it is to be questioned whether these additional opportunities occur frequently in practice. In case we could not use the dependency analysis results of other steps, which allow us to perform the simple test virtually for free, the situation would be different. In that case the difference in complexity between the "simple" test and the extended test might almost vanish.

During our experiments with several multimedia applications, which are described in subsection 12.2.8, we have observed that the vast majority of valid projections can be detected with the simple dependency test based on invariant loops (with almost no effort). For only two arrays in one application it was observed that the simple test could not detect the possibility of a projection. Therefore we decided not to implement the extended test in our prototype tool.

Our experiments also revealed that in practice many dimensions can be pruned in single-assignment code, such that effective number of remaining dimensions is usually quite small (e.g. 3 or less). Therefore, the difference between the different search strategies is not that large in practice. Full search might still explode in rare cases, whereas this is highly unlikely for B&B and even impossible for greedy tree descent. These last two are therefore preferable.

12.2.8 *Experimental results*

In this subsection we present some of the results that we obtained with our storage requirements reduction strategy for a few (multimedia) applications by means of a prototype tool.

Our strategy is intended to be applied to each memory separately as one of the last steps in the ATOMIUM methodology, i.e. when the execution order, the memory architecture and the array-to-memory assignment are known. The actual *nature* of the execution order (i.e. parallel or sequential) is not relevant for our techniques, as we are only interested in the relative memory access order. We therefore have assumed in each of our experiments that all of the arrays had to be stored in one (shared) memory, and that the applications had to be executed sequentially, but the techniques would work equally well in case of a parallel execution order and multiple memories.

In Table 12.1 we present the most relevant properties of the (multimedia) applications and application kernels that we used as test vehicles. The first one is an updating singular value decomposition algorithm that can for instance be used in radar applications. The second one is the kernel of a 2-dimensional wavelet compression algorithm that can be used for image compression, just like the third one, which is an advanced edge detection algorithm. The next one is a 3-dimensional volume rendering algorithm. For this application we also looked at a version with (heavily) reduced size parameters. Finally we have looked at a voice coder algorithm and a public domain GSM autocorrelation algorithm. Several of these applications consist of many pages of code. The applications are mostly manifest and even linear. In cases where this was not true, we have manually "linearized" them using worst-case assumptions, such that a valid solution could be obtained.

Table 12.1.   Relevant application properties.

Application	# Arrays	Max. #Dep. per array	Max. #Dim. per array	Multiple Assign. size [words]	Single Assign. size [words]	Scalar Minim. size [words]
Updating SVD	6	27	4	6013	6038	211
2D wavelet	11	18	5	1186	8704	514
Edge detection	18	17	4	724	5647	116
3D volume rendering	22	11	8	26581	216M	infeas.
Reduced 3D vol. rend.	22	11	8	166	6976	134
Voice coder	201	41	6	2963	314K	905
GSM autocorrelation	17	35	3	532	1279	209

The table contains the array memory sizes required for the original multiple-assignment versions and the single-assignment versions of the algorithms (with the same assumptions: one memory and a sequential execution order). The last column also indicates the maximal number of scalars that is simultaneously alive during the execution of the algorithms. These numbers are obtained by symbolic simulation and represent lower bounds on the required memory sizes, since these are the sizes that would be required if all the arrays would be split into scalars. Doing this is however unacceptable for realistic multimedia applications, as the control and address generation overhead would be prohibitively large, but we can use these numbers to assess the quality of our optimization strategy. The closer we come to them, the better our solutions are. Unfortunately, for larger examples, such as the full 3D volume rendering, symbolic simulation is infeasible.

12.2.9   Required memory sizes and optimization run-times.

During our experiments we let our prototype tool decide on an optimal intra-array storage order for each array, using different search strategies (full search, B&B, and greedy tree descent). We also looked at the cases were we just selected a row-major or column-major storage order and calculated the windows. In each case we assumed a static windowed allocation approach (as depicted in Fig. 12.5a on page 264). The additional techniques and results for a dynamic allocation strategy are presented in section 12.3.

The resulting memory sizes (in terms of the number of words) are indicated in Table 12.2 (the sizes correspond to the sum of the windows of all the arrays).

From this table we can see that the savings in memory size for the static windowed strategies can be considerable. The results also indicate that an arbitrary choice of the storage order (i.e. row-major or column-major) can easily lead to suboptimal results.

The row-major storage order performs quite well for most of the experiments, but this can be explained by the fact that most of these examples have been written manually, and humans seem to tend to use the dimensions of an array in a left-to-right manner, for which a row-major order is ideally suited. However, in our ATOMIUM memory and power optimization context, the storage size reduction step follows several other steps that make extensive use of loop transformations, which heavily increase the temporal locality of the data accesses compared to the original specification to reduce the storage size requirements after in-place optimization. After these crucial loop transformations, the array access order is usually changed drastically and consequently the row-major storage order is in general no longer likely to be near-optimal. The updating SVD algorithm is such an example that has been subject to loop transformations before our optimizations, and in that case the difference between the optimal storage order and a row-major or column-major one is very large.

Table 12.2.   Experimental results: required memory sizes.

Application	Row-major size [words]	Column-major size [words]	Full Search size [words]	B&B size [words]	Greedy tree descent size [words]
Updating SVD	3067	484	314	314	314
2D wavelet	3500	8348	3038	3038	3038
			(1024)	(1024)	(1024)
Edge detection	580	1021	576	576	576
3D volume rendering	26576	216M	26576	26576	26576
Reduced 3D vol. rend.	166	4756	166	166	168
Voice coder	2417	38537	2403	2403	2403
GSM autocorrelation	667	1096	529	529	529

Note that in most cases we were able to effectively remove the overhead from single-assignment, and that for several cases we already obtain results that are substantially better than what would be obtained by a standard compiler (i.e. the multiple assignment column in Table 12.1), even though we only applied a static allocation[14]. Only for one example, namely the 2-D wavelet algorithm, we could not completely remove the overhead. Detailed inspection of the example revealed this was a case where the simple projection test of subsection 12.2.5 failed, and a window-only approach leads to a suboptimal result. As mentioned in the previous subsection, by extending our techniques with a more extended test as described in [430] this overhead can be removed too. A similar result could be obtained by applying a loop transformation instead. The result in both cases would be in a memory size of 1024 words (indicated between brackets in Table 12.2).

Apart from the single-assignment overhead removal, the additional size reductions are definitely worthwhile. For the updating SVD example for instance, no orthogonal projection could be performed, while the reduction by means of a window is quite large. The IMEC tool "s2p", which implements the strategy described in [413], and to our knowledge the only other automated tool that tries to perform our type of optimizations,[15] could not reduce the memory usage to less than 3285 words, since the tool actually calculates an upper bound for the windowed row-major storage order, while we were able to reduce it to 312 words.

The optimization run-times of our prototype tool are shown in Fig. 12.19. All of our experiments were run on a Hewlett-Packard K260/4 workstation with 1.6 GB of physical RAM and 4 PA-8000 CPUs[16] with a clock speed of 180 MHz. As an ILP solver we used the public domain package LP_SOLVE [41]. The run-times were measured for the five different (search) strategies, using four different configurations:

1. no search space pruning and no speed-up heuristics [109] applied;

2. search space pruning applied but no speed-up heuristics;

3. no search space pruning but speed-up heuristics applied;

4. search space pruning and speed-up heuristics applied.

[14]The results for a dynamic approach, which are presented in the next subsection, are even better.
[15]Of course, several scalar memory/register reuse approaches exist, but it must be clear that these approaches are infeasible for data-intensive multimedia applications, where the number of scalars is huge.
[16]The availability of multiple CPUs was not exploited.

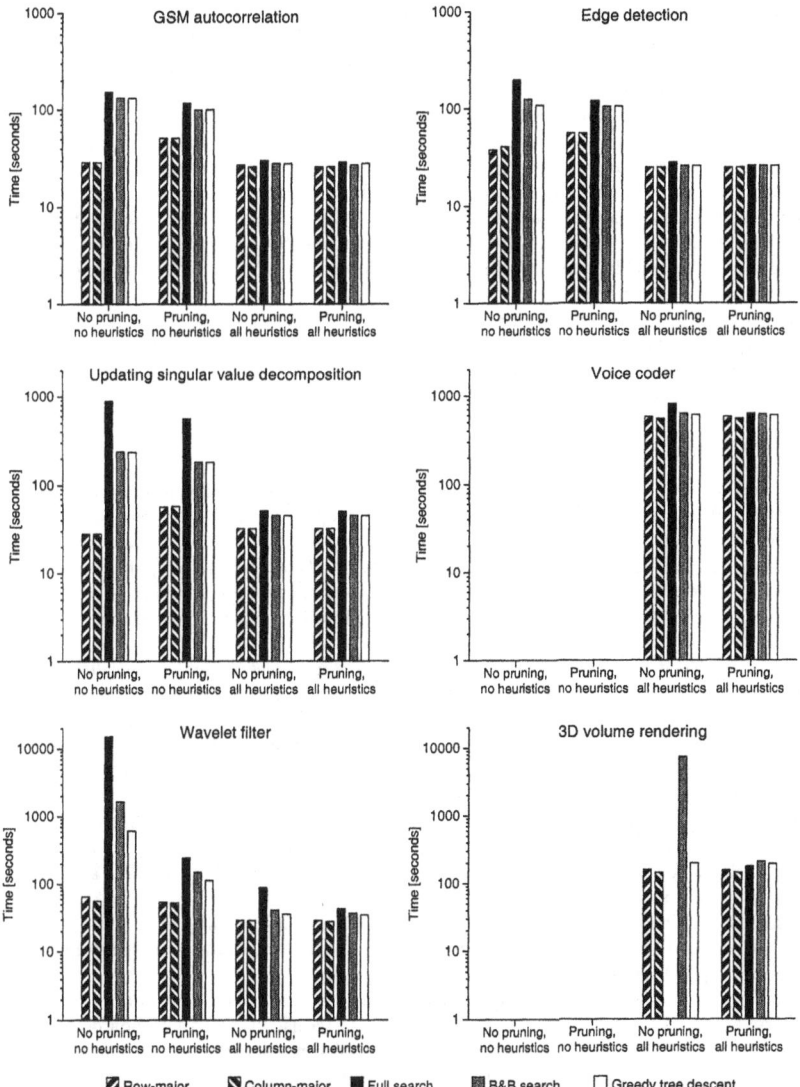

Figure 12.19.   Optimization run-times for various applications and different (search) strategies. Missing bars correspond to experiments that were aborted due to numerical instability problems in the ILP solver, or due to unacceptably high run-times (≫ 10000 seconds).

The run-times are certainly acceptable in an embedded application design context, especially when both the search space pruning and the speed-up heuristics are applied.

From these figures we can also conclude that the effect of the speed-up heuristics is usually larger than the effect of the search space pruning (typically a factor 5 for the heuristics, and about 20 % for the pruning). Moreover, for two test vehicles, namely the voice coder and the 3D volume rendering, no solution could even be found without applying the speed-up heuristics. Nevertheless, both techniques are necessary to reduce the chance of getting stuck to a negligible level. When we applied them both, we have always been able to obtain a solution in an acceptable time for each of the strategies.

We can also see that, in contrast to what might be expected, the full search, B&B, and greedy tree descent techniques have comparable run-times, even for applications with arrays with a large number of dimensions, especially when both the pruning and the speed-up heuristics are applied. The reason why the full search strategy run-times do not explode even in these cases, is that both the pruning technique presented in subsection 12.2.5 and the speed-up heuristics (especially the ones presented in [109]) result in a considerable reduction of the number of ILP problems to be solved. Nevertheless, the B&B and greedy tree descent techniques tend to be faster and they are probably the best choice. The B&B version is probably preferable as it guarantees the optimal solution (within the search space), in contrast to the greedy tree descent.

Finally, the figures also illustrate that the time penalty for searching the complete search space is relatively small. When both the pruning and the speed-up heuristics are applied, the optimization run-times for the full search, B&B and greedy tree descent strategies are only slightly larger than for the row-major or column-major cases, where only one storage order is evaluated.

12.2.10 Sensitivity to size parameter values

Our optimization techniques are heavily based on the geometrical modeling techniques presented in chapter 5. These models allow us to describe groups of operations and variables in a compact way that is independent of the program size parameters such as the values of loop boundaries. Also the techniques themselves are independent of these parameters. Nevertheless, the optimization run-times are likely to be affected by the size parameters, as these parameters also appear in (some of) the ILP problems that we have to solve, and the run-times of ILP solvers tend to be dependent on the size of the coefficients in the ILP problems (see also [109]).

Nevertheless, it can be expected that our optimization run-times are far less sensitive to the size parameters than the ones obtained by simulation-based techniques and certainly scalar optimization techniques. To verify this, we have taken the updating SVD test vehicle and performed several experiments with a varying size parameter N. The sizes of the most important arrays in this test vehicle and the number operations are proportional to N^3. Therefore it can be expected that the run-time of a simulation-based tool, such as s2p, is also proportional to N^3. In Fig. 12.20 we compare the run-times of s2p with the run-times of our prototype tool using different ILP solvers: LP_SOLVE [41] and an arbitrary precision version of PIP [132] called MPPIP. LP_SOLVE is a *mixed* integer linear problem solver, using a floating point number representation internally. In contrast, MPPIP is a parametric[17] ILP solver using an integer number representation internally.

Because s2p only evaluates a row-major storage order, we also configured our prototype tool to evaluate only that order, i.e. we did not search for the optimal storage order. The speed-up heuristics presented in [109] and search space pruning were activated.

At the left side of Fig. 12.20 the run-times are shown for s2p and our prototype tool using both ILP solvers. At the right side we show the number of times that we had to rely on a window decomposition procedure described in [109] to find the solution to an ILP problem that we encountered with our prototype tool. This procedure is used as a backup whenever the ILP solver gets into trouble, e.g. due to numerical instabilities (LP_SOLVE) or due to an explosion of the memory usage (MPPIP). The backup procedure was called each time an ILP solver could not find a solution in less than 1 second.[18]

The figure illustrates that for small parameter sizes, a simulation-based approach can be orders of magnitude faster than ILP-based approach. In contrast, the opposite is true for larger size parameters. As expected, the simulation run-time of s2p is proportional to N^3 for this example. In contrast, the run-times of our prototype tool are far less sensitive to N. When we use MPPIP the run-times are even nearly constant. LP_SOLVE is also performing very well for values of $N \leq 50$, but for larger

---

[17]We did not use the parametric capabilities.

[18]The average solving time for 1 ILP problem is in the order of 10 milliseconds (see also [109]). Therefore a solving time of more than 1 second is considered to be a practical indication that the solver is having some serious problems.

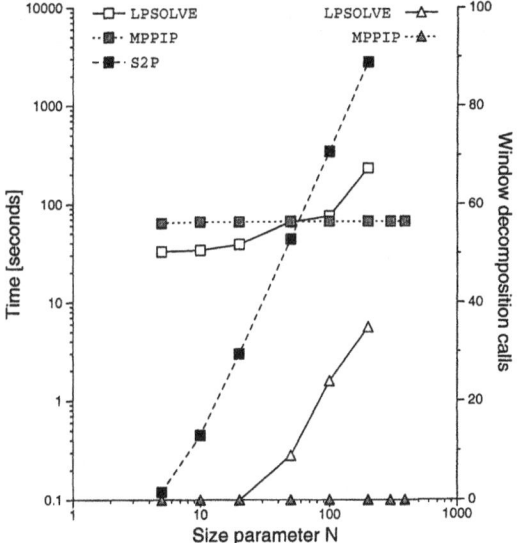

Figure 12.20. Optimization run-times for the updating singular value decomposition application with a varying size parameter.

values it experiences serious numerical problems (as indicated by the number of calls to the window decomposition backup procedure).

MPPIP is inherently slower than LP_SOLVE for smaller parameter sizes, especially due to its unlimited precision number representation and its integer-based solution method. However, these properties also make MPPIP far less sensitive to parameter sizes, such that the performance remains nearly constant (within the measured range). This is also illustrated by the fact that we did not have to fall back on the window decomposition procedure when using this solver. LP_SOLVE on the other hand uses a floating-point number representation and is therefore subject to numerical instability problems that come into play for values of N $\geq$ 50 and that result in a slow-down. For values of N > 200, LP_SOLVE even returned wrong results!

With MPPIP we experienced no problems for values of N up to 390. For larger values, we encountered integer overflow problems[19] in our prototype tool, *but not in the solver*. Therefore this limitation is due to the prototype implementation, and not due to MPPIP. There is no evidence that MPPIP would run into problems soon for values of N > 390.

Finally, something that is not indicated in the figures, is the fact that the virtual memory consumption on the workstation used also increased proportionally to N^3 when we used s2p. As a result, s2p ran out of memory for sizes of N larger than 200, *even though the workstation has 1.6 GB of physical memory!* In contrast, our prototype tool and the ILP solver never required more than 5 MB *together*.[20]

12.3 INTER-ARRAY PLACEMENT

In this section we describe the second substep of our storage size reduction strategy, namely the array placement step. During this substep we try to determine an optimal location of the arrays in the memories such that the reuse of memory locations between different arrays is maximized. After this introduction, in subsection 12.3.1, we first investigate the constraints that have to be satisfied and

[19]The coefficients in the execution order functions became larger than 2^{31}.
[20]The backup procedure prevented a memory explosion of MPPIP.

derive a few criteria that allow us to distinguish between valid and invalid placements. These criteria are then used in subsection 12.3.6, where we describe our placement algorithm. After that, we present some experimental results in subsection 12.3.8.

12.3.1 Placement constraints

Given that we have found the optimal intra-array storage order for each array, for most of the placement strategies presented in subsection 12.1.2 there is only one remaining parameter to fix in each of the address equations, namely the base address or offset of the array in memory. However, especially for the dynamic approaches, making a good choice for these offsets is not trivial.

In the scalar context, (binary) ILP formulations [7, 26], (iterative) line packing [224, 168], graph colouring [381], or clique partitioning [398] techniques have proven to be valuable in a register and register-file allocation and signal-to-register assignment context. Unfortunately, these techniques are not feasible when the number of scalars becomes too large, which is the case for data-dominated multimedia applications. Extensions of these scalar techniques towards periodical stream-based applications, such as in the left-edge algorithm based P HIDEO approach [244] also have only a limited scope.

Due to the very different nature of arrays and scalars, there is also little hope that any of the scalar techniques would obtain near-optimal results when extended in a straightforward way to treat arrays as scalars. Array storage management requires taking into account several aspects that are not even present in a scalar context. For instance, in contrast to scalars, different arrays may have different sizes and internal structures and exhibit even "holes". Moreover, life-times of arrays can *partially* overlap with those of other arrays. We therefore had to come up with a custom placement algorithm.

Before we can start placing the arrays, we have to be able to decide what (relative) placements are allowed. We therefore define two properties that can be defined for each pair of arrays: *compatibility* and *mergability*, which are explained next. Moreover, we can also calculate the minimal *distance* between two arrays in memory.

12.3.2 Compatibility

If two arrays have totally disjoint global life-times, i.e. if there is no moment in time at which elements of both arrays are alive, we say that they are **compatible**. This means that there is no chance of a memory occupation conflict between them and consequently, we can freely choose their relative offsets.

Checking for compatibility between two arrays comes down to checking whether the orthogonal projections of their OAT-domains on the (relative) time axis overlap or not. Mathematically we can express this condition as follows for a pair of arrays m_1 and m_2:

$$\{ \ t \mid \exists \ a_1, a_2 \text{ s.t. } [a_1, t] \in \mathbf{D}_{m_1}^{\mathtt{OAT}} \ \wedge \ [a_2, t] \in \mathbf{D}_{m_2}^{\mathtt{OAT}} \ \} = \phi \tag{12.18}$$

Since the OAT-domains in general consist of several BOAT-domains, one for each flow dependency of the corresponding arrays, the problem translates into a check of the projections of each possible pair of BOAT-domains, consisting of one BOAT-domain of both arrays:

$$\forall \ i_1, j_1, k_1, l_1, i_2, j_2, k_2, l_2 :$$
$$\{ \ t \mid \exists \ a_1, a_2 \text{ s.t. } [a_1, t] \in \mathbf{D}_{i_1 j_1 k_1 l_1 m_1}^{\mathtt{BOAT}} \ \wedge \ [a_2, t] \in \mathbf{D}_{i_2 j_2 k_2 l_2 m_2}^{\mathtt{BOAT}} \ \} = \phi \tag{12.19}$$

In case the BOAT-domains are linearly bounded lattices (LBLs), each of these checks can be easily translated into a relatively simple ILP feasibility problem. Unfortunately, the worst case number of ILP problems to be solved can become relatively large, as it equals the product of the number of flow dependencies of both arrays. Moreover, if this check has to be performed for each possible combination of arrays, the number of ILP problems to solve could become very large for real life applications.

In practice however, there are a few scenarios for which we can decide more easily whether two arrays are compatible or not:

■ During the execution of the algorithm, the last read access to one of the arrays occurs before the first write access to the other array. An example of this scenario is depicted in Fig. 12.21a. This case is very easy to detect and requires only a few relatively simple calculations.

■ During the execution of the algorithm, the arrays are periodically and alternately alive. This is for instance the case in the example of Fig. 12.21b. This case is also relatively easy to detect: first, we can use common dependency analysis techniques to find out which common loops are invariant and do not carry any of the dependencies of both arrays. After "removing" these loops, we can proceed with the method described in the first item.

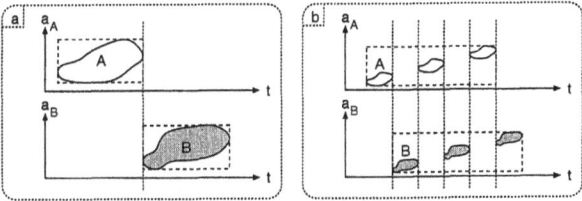

Figure 12.21.   Two simple compatibility scenarios.

Note that checking for compatibility allows us to go further than simply approximating an OAT-domain by a rectangle. In fact, if two arrays are compatible, we know that their OAT-domains can never overlap, even if their approximating rectangles do, as in Fig. 12.21b.

In our prototype tool, we first assume the simplest scenario. If that compatibility test fails, we try the second one. If that one fails too, we fall back on the exhaustive pairwise BOAT-domain projection overlap test. In many cases we can obtain a definite answer with only the simpler tests. Moreover, even during the exhaustive test we can often use similar simplifications. These are described in more detail in [109]. As a result of all of this, the run-times turn out to be acceptable for realistic applications (see subsection 12.3.8).

12.3.3 Mergability

If the tests indicate that two arrays are not compatible, in general we have to make sure that their address ranges do not overlap. Otherwise we cannot guarantee correct behavior, without checking the complex constraints described by equation 5.16 on page 115.

However, there is a special case that is relatively easy to check and that has some practical relevance. A commonly occurring pattern in practice is the calculation of elements of a new array based on the corresponding elements of another array. If the elements of the original array are no longer needed after these calculations, it may be possible to reuse their memory locations immediately for storing elements of the new array. If this is possible, we consider these arrays to be **mergable**. An example of this is shown in Fig. 12.22.

Note that if we would only check for compatibility between the arrays, we would not be able to overlap their address ranges, because then we effectively approximate the OAT-domains by rectangles. If the mergability check succeeds, we know that even though the rectangles overlap when aligned at the bottom, there is no overlap of the actual OAT-domains (of which we don't know the exact shape in general). So during the placement, we have to make sure that the rectangles are either perfectly aligned at the bottom, or totally non-overlapping.

Mathematically we can express this check as follows for a pair of arrays m_1 and m_2:

$$\mathbf{D}_{m_1}^{\mathtt{OAT}} \bigcap \mathbf{D}_{m_2}^{\mathtt{OAT}} = \phi \qquad (12.20)$$

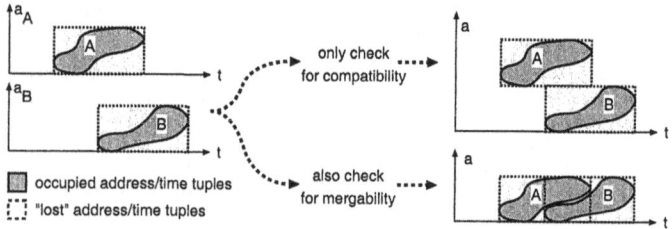

Figure 12.22.   Additional opportunities by checking for mergability.

or in terms of the composing BOAT-domains:

$$\forall i_1, j_1, k_1, l_1, i_2, j_2, k_2, l_2 : \quad D^{BOAT}_{i_1 j_1 k_1 l_1 m_1} \bigcap D^{BOAT}_{i_2 j_2 k_2 l_2 m_2} = \phi \qquad (12.21)$$

The check for mergability is slightly more complex than the worst-case check for compatibility, because we also have to take the address dimension into account, but we currently only perform this check for a pair of arrays when the following preconditions hold:

- the arrays are *not* compatible;

- the arrays are *not* folded, i.e. their window sizes equal the product of the sizes of the non-projected dimensions.

The last precondition is necessary because the test of equation 12.20 only guarantees that the non-folded OAT-domains don't overlap. If one or both of the OAT-domains are folded by means of a modulo operation, this check is not sufficient to make sure that the folded OAT-domains don't overlap. This is illustrated in Fig. 12.23.

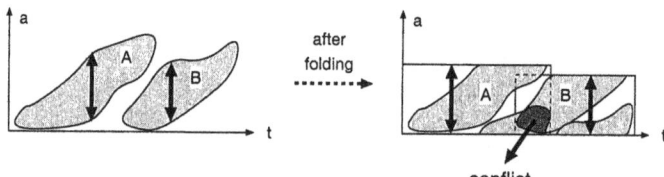

Figure 12.23.   Conflict after folding of seemingly mergable OAT-domains.

In practice, this means that the full mergability check is usually triggered only a few times. We could extend the test of equation 12.20 to take into account folding. For a pair of arrays m_1 and m_2 with window sizes W_1 and W_2 respectively, the extended test would be the following:

$$\{ [a_1 \bmod W_1, t] \mid [a_1, t] \in D^{OAT}_{m_1} \} \quad \bigcap \quad \{ [a_2 \bmod W_2, t] \mid [a_2, t] \in D^{OAT}_{m_2} \} = \phi$$

We can get rid of the modulo operations by applying the technique presented in subsection 5.2.1.

$$\{ [f_1, t] \mid \exists k_1 \in \mathbb{Z} \text{ s.t. } a_1 = k_1 W_1 + f_1 \wedge [a_1, t] \in D^{OAT}_{m_1} \} \quad \bigcap$$
$$\{ [f_2, t] \mid \exists k_2 \in \mathbb{Z} \text{ s.t. } a_2 = k_2 W_2 + f_2 \wedge [a_2, t] \in D^{OAT}_{m_2} \} = \phi \qquad (12.22)$$

So in case the BOAT-domains are LBLs, this extended test can also be performed by solving a set of ILP feasibility problems. It is questionable whether this extended test succeeds in many practical cases. The probability that two folded OAT-domains overlap is probably much higher than for non-folded domains, because they are more dense and irregular when folded.

12.3.4 Distance calculation

The (extended) mergability test described above is able to determine whether the (folded) OAT-domains of two arrays overlap or not. We could easily extend this test to check whether two OAT-domains, each *shifted by an arbitrary vertical offset*, overlap. However, such a test only provides a yes-or-no answer, and it gives no indication about the amount of overlap or the amount of space between the domains. A placement strategy that uses this test is therefore likely to call it many times (i.e. for many different offsets) to find an optimal solution. Since this test is fairly expensive it could have a disastrous effect on the time needed for the placement.

Therefore it would be better if we could have an indication of how far two OAT-domains have to be spaced from each other to avoid memory occupation conflicts. In fact we need to know two minimal **distances**: one distance for the case where one OAT-domain is placed "on top of" the other one, and one distance for the opposite case. This is depicted in Fig. 12.24.

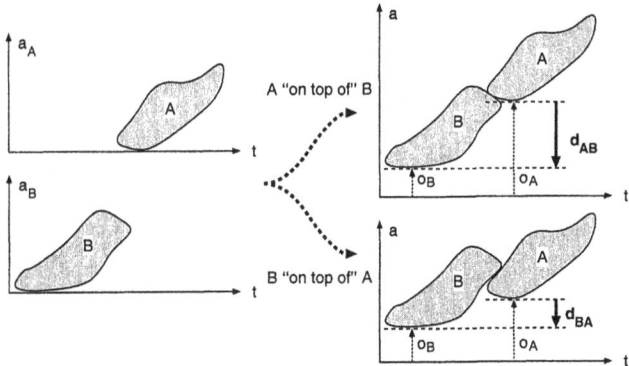

Figure 12.24.   Distances between OAT-domains.

If we know the distances d_{AB} and d_{BA} in this figure, we know that there can be no conflict as long as the difference in offsets $o_A - o_B$ between A and B is either larger than or equal to d_{AB} or smaller than or equal to d_{BA}. For differences between these two values there are potential memory occupation conflicts.

We can calculate these distances for a pair of arrays m_1 and m_2 as follows, given their OAT-domain descriptions:

$$d_{m_1 m_2} = \max\{ \ d \ | \ \exists \ a, t \ \text{s.t.} \ [a+d, t] \in \mathbf{D}^{\text{OAT}}_{m_1} \ \wedge \ [a, t] \in \mathbf{D}^{\text{OAT}}_{m_2} \ \} + 1 \qquad (12.23)$$

$$d_{m_2 m_1} = \min\{ \ d \ | \ \exists \ a, t \ \text{s.t.} \ [a+d, t] \in \mathbf{D}^{\text{OAT}}_{m_1} \ \wedge \ [a, t] \in \mathbf{D}^{\text{OAT}}_{m_2} \ \} - 1 \qquad (12.24)$$

In other words, we can find these distances by shifting the OAT-domains as far from each other as possible (in both directions) while there is still an overlap between them. These distances, incremented and decremented by one respectively, then are the maximal and minimal safe distances. Note that the offsets of both domains are not important; only the difference in offset is.

So the distance criterion that we can impose is that the difference in offset between arrays m_1 and m_2 must be either larger than or equal to $d_{m_1 m_2}$ or smaller than or equal to $d_{m_2 m_1}$. This guarantees that there cannot be a memory occupation conflict between both arrays.

Again, in case the BOAT-domains are LBLs (or unions of LBLs), these distances can be found by solving a finite number of ILP problems. The following example illustrates this.

Example 12.3.1

```
    int B[5], C[10];
    for ( i = 0; i < 5; ++i )
S1:   B[i] = ...;
```

```
     for ( j = 0; j < 10; ++j )
S2:    C[j] = f(B[j div 2]);
     for ( k = 0; k < 10; ++k )
S3:    ... = g(C[9-k]);
```

The domain descriptions for this program are the following:

$$\mathbf{D}_B^{\mathtt{var}} = \{\ b \mid 0 \leq b \leq 4 \ \wedge\ b \in \mathbb{Z}\ \}$$
$$\mathbf{D}_C^{\mathtt{var}} = \{\ c \mid 0 \leq c \leq 9 \ \wedge\ c \in \mathbb{Z}\ \}$$
$$\mathbf{D}_1^{\mathtt{iter}} = \{\ i \mid 0 \leq i \leq 4 \ \wedge\ i \in \mathbb{Z}\ \}$$
$$\mathbf{D}_{11B}^{\mathtt{def}} = \{\ b \mid \exists\, i \in \mathbf{D}_1^{\mathtt{iter}} \ \text{s.t.}\ b = i\ \}$$
$$\mathbf{D}_2^{\mathtt{iter}} = \{\ j \mid 0 \leq j \leq 9 \ \wedge\ j \in \mathbb{Z}\ \}$$
$$\mathbf{D}_{21C}^{\mathtt{def}} = \{\ c \mid \exists\, j \in \mathbf{D}_2^{\mathtt{iter}} \ \text{s.t.}\ c = j\ \}$$
$$\mathbf{D}_{21B}^{\mathtt{oper}} = \{\ b \mid \exists\, j \in \mathbf{D}_2^{\mathtt{iter}}, n \in \mathbb{Z} \ \text{s.t.}\ 2*b+n = j \ \wedge\ 0 \leq n \leq 1\ \}$$
$$\mathbf{D}_3^{\mathtt{iter}} = \{\ k \mid 0 \leq k \leq 9 \ \wedge\ k \in \mathbb{Z}\ \}$$
$$\mathbf{D}_{31C}^{\mathtt{oper}} = \{\ c \mid \exists\, k \in \mathbf{D}_3^{\mathtt{iter}} \ \text{s.t.}\ c = 9 - k\ \}$$

Note that we got rid of the integer division operation $\mathtt{div}$ in the description of $\mathbf{D}_{21B}^{\mathtt{oper}}$ using a similar technique as for removing modulo operations. Again we assume a sequential execution order and a simple linear storage order:

$$O_1^{\mathtt{time}}(i) = i \qquad\qquad O_{11B}^{\mathtt{rtime}}(x) = x$$
$$O_2^{\mathtt{time}}(j) = 2j + 5 \qquad\qquad O_{21C}^{\mathtt{rtime}}(x) = x + 1 \qquad\qquad O_{21B}^{\mathtt{rtime}}(x) = x$$
$$O_3^{\mathtt{time}}(k) = k + 15 \qquad\qquad\qquad\qquad\qquad\qquad\qquad O_{21C}^{\mathtt{rtime}}(x) = x$$

$$O_B^{\mathtt{addr}}(b) = b \quad and \quad O_C^{\mathtt{addr}}(c) = c$$

The resulting BOAT-domain descriptions are the following (after simplification):

$$\mathbf{D}_{1211B}^{\mathtt{BOAT}} = \{\ [a,t] \mid \exists\, n \in \mathbb{Z} \ \text{s.t.}\ 0 \leq a \leq 4 \ \wedge\ 0 \leq n \leq 1 \ \wedge$$
$$t \geq a \ \wedge\ t \leq 4a + 2n + 5 \ \wedge\ a \in \mathbb{Z}\ \}$$
$$\mathbf{D}_{2311C}^{\mathtt{BOAT}} = \{\ [a,t] \mid 0 \leq a \leq 9 \ \wedge\ t \geq 2a + 6 \ \wedge\ t \leq 24 - a \ \wedge\ a \in \mathbb{Z}\ \}$$

Since we have only one dependency per array, the OAT-domain descriptions coincide with the BOAT-domain descriptions.
We can now calculate the minimum and maximum distances between both domains:

$$d_{BC} = \max\{\ d \mid \exists\, a,t \ \text{s.t.}\ [a+d,t] \in \mathbf{D}_{1211B}^{\mathtt{BOAT}} \ \wedge\ [a,t] \in \mathbf{D}_{2311C}^{\mathtt{BOAT}}\ \} + 1$$
$$= \max\{\ d \mid \exists\, a \in \mathbb{Z}, t, n \in \mathbb{Z} \ \text{s.t.}\ 0 \leq a+d \leq 4 \ \wedge\ 0 \leq n \leq 1 \ \wedge$$
$$t \geq a+d \ \wedge\ t \leq 4a + 4d + 2n + 5 \ \wedge\ 0 \leq a \leq 9 \ \wedge$$
$$t \geq 2a + 6 \ \wedge\ t \leq 24 - a\ \} + 1$$
$$d_{CB} = \min\{\ d \mid \exists\, a,t \ \text{s.t.}\ [a+d,t] \in \mathbf{D}_{1211B}^{\mathtt{BOAT}} \ \wedge\ [a,t] \in \mathbf{D}_{2311C}^{\mathtt{BOAT}}\ \} - 1$$
$$= \min\{\ d \mid \exists\, a \in \mathbb{Z}, t, n \in \mathbb{Z} \ \text{s.t.}\ 0 \leq a+d \leq 4 \ \wedge\ 0 \leq n \leq 1 \ \wedge$$
$$t \geq a+d \ \wedge\ t \leq 4a + 4d + 2n + 5 \ \wedge\ 0 \leq a \leq 9 \ \wedge$$
$$t \geq 2a + 6 \ \wedge\ t \leq 24 - a\ \} - 1$$

The reader certainly has found the solutions to these ILP problems by now: 5 and -5. We can verify this graphically in Fig. 12.25a and Fig. 12.25b respectively.

Unfortunately, these distance calculations turn out to be quite costly in practice for several reasons:

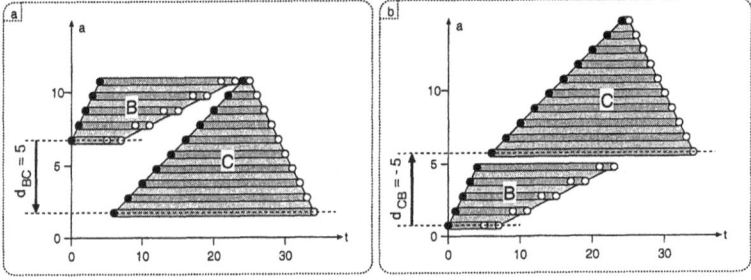

Figure 12.25.   The distance calculation for example 12.3.1.

1. If we want to know the distances between all pairs of arrays (which are not compatible), the number of distances to be calculated is proportional to the square of the number of arrays.

2. In the simplest case, where all BOAT-domains are LBLs, each distance calculation for a pair of arrays requires a number of ILP problems to be solved that is proportional to the product of the number of dependencies of both arrays.

3. The ILP problems of this kind turn out to be the most difficult ones to solve for an ILP solver (in contrast to all the other kinds of ILP problems that we encounter with our techniques). Moreover, the simplification heuristics presented in [109], that help us to reduce the complexity of the other kinds of ILP problems, turn out to be less effective in this case.

4. Taking these distances into account during the actual array placement step (presented in the next subsection) turns out to have a disastrous effect on the performance in case many arrays have to be placed.

Just as for the mergability check, the distance calculation described by equations 12.23 and 12.24 only makes sense in case the OAT-domains are not folded. In case one or both of the domains are folded, we can extend the test in a similar way as for the extended mergability test.

Note that a violation of the distance criterion not necessarily means that there is a memory occupation conflict. In other words, the distance criterion is a sufficient but not a necessary condition. An example of this is depicted in Fig. 12.26. In practice this case is probably unlikely to occur, and if it would occur, it would be very difficult to detect/exploit. Therefore we do not consider this case any further.

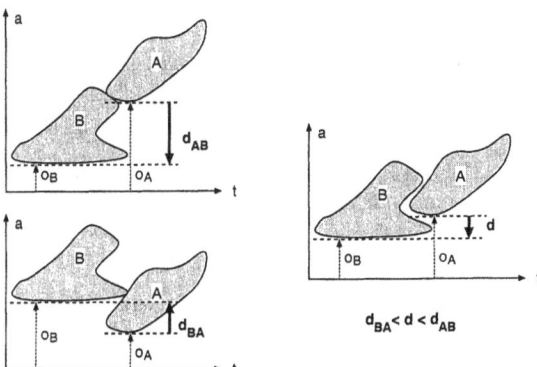

Figure 12.26.   A valid intermediate solution, not satisfying the distance criterion.

Finally, it is important to note that, at least in theory, if we decide to calculate the minimal and maximal distances, the mergability criterion does *not* become superfluous. It could be that a relative offset of 0 (which corresponds to the mergability criterion), violates the distance criterion, but still represents a valid solution.

12.3.5 Array placement

In the previous subsection we derived several conditions that allow us to check whether a certain relative placement of arrays is valid or not. In this subsection we present a placement algorithm and some possible extensions. For complexity reasons our current placement algorithm does *not* take into account minimal and maximal distances.

12.3.6 Placement algorithm

Our placement algorithm, which tries to find an optimal offset for each array in a memory, proceeds as follows:

1. We start by setting up two sets of arrays: those that have been placed (i.e. the ones whose offset has been fixed) and those that still have to be placed. Originally, all arrays belong to the second set, and we assume that all their offsets are zero.

2. We select from the arrays-to-place set the array that currently has the lowest offset. If two arrays have the same offset, we select the one with the largest window size. By doing so we decrease the probability that the array "sticks out" later on and increases the peak storage size requirements. If the arrays also have an equal window size, we select the one which has the smallest number of compatible and mergable arrays, as it has to largest number of constraints and is therefore likely to be difficult to place later on. Finally, if this selection procedure does not yield a new candidate, we pick a random one from the set.

3. We then compare the selected array with those that have already been placed, and check for conflicts. Two arrays have *no* conflicting positions if:

   ■ they are compatible, or

   ■ they have non-overlapping address ranges, or

   ■ they are mergable and aligned at the same offset.

If there is no conflict with any of the arrays placed, we place the selected array at its current position, and transfer it to the placed-array set.

If there is a conflict, we move the array upwards until its base crosses the top of an already placed array, and *leave it in the arrays-to-place set.*

In both cases, we return to step 2, unless there are no more arrays to place.

In Fig. 12.27 the different moves of the placement algorithm are shown for a small example.

The algorithm is greedy in the sense that when placing an array it selects the one that firstly can be placed as low as possible, and secondly has the largest window (step 2).

Implementing this algorithm in a naive way would result in a relatively high run-time complexity due to a large number of array comparisons, but by sorting the array sets in an intelligent way for instance, we have been able to keep the placement run-times negligibly small, compared to the intra-array optimization substep (see subsection 12.3.8).

12.3.7 Further enhancements

We can further enhance our overall storage size reduction strategy, by allowing an additional degree of freedom. Namely, in many practical cases it is possible to identify subsets of an array that are completely "independent" of each other. For instance, it is possible that the odd and even elements of an array are treated separately. If we then would separate these subsets *as the first step in our*

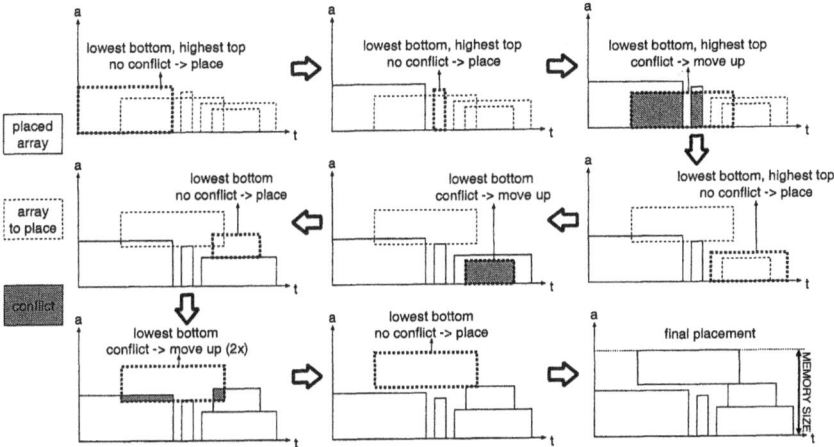

Figure 12.27.   The placement algorithm at work. Note that the steps are presented in zigzag order.

global optimization strategy, i.e. before the intra-array substep, we could optimize and also place them separately, which could possibly lead to a better overall solution.

The identification of good subsets is not trivial, especially because we want to keep the effect on the address generation overhead minimal. In a worst case scenario, an array would be split into nothing but scalars, each being addressed separately. This is obviously unacceptable for data-dominated multimedia applications. Therefore we impose the following constraint on the identification of subsets: if an array is split, it should not introduce extra conditions in the program to select the correct subset being accessed by a statement, i.e. the split should coincide with the "statement boundaries". For instance, we do not allow a split as indicated in Fig. 12.28a. In contrast, splits as indicated in Fig. 12.28b are allowed, as they do not affect the addressing and control overhead.

We can easily identify the candidate subsets from the results obtained by our original data-flow analysis. It is also possible to allow user interaction with respect to these decisions (e.g. a designer may indicate where an array should be split, even if it would introduce extra conditions).

After the identification of the potential sub-arrays, we perform a contraction step, which detects periodical "holes" in the different dimensions of the sub-arrays (e.g. due to an odd/even split) by means of a simple GCD-based algorithm and which contracts these dimensions to make them dense again if possible.

However, it is not guaranteed that a split of an array leads to better results. In some cases it might even lead to much worse results (e.g. when the resulting subsets are no longer dense and cannot be contracted). Therefore we perform the intra-array storage order optimization step for each of the sub-arrays, but also for the original non-split array, followed by the compatibility and mergability calculations described above. Then, before we start the actual placement, we decide for each array that can be split whether it should be split or not. We make this decision as follows:

- If the average window height over time of the sub-arrays is smaller than the average window height of the original array and the difference is relevant (a difference of more than 10% seems to be a good practical criterion), we decide to split the array, as it is likely to lead to a better overall result.

- If, on the other hand, it is larger and the difference is relevant, we decide to keep the array together, as a split is likely to give worse results.

- If it is about the same size, we perform the following steps:

    - If the different sub-arrays of the array are all mergable, we keep them together. Splitting them would increase the number of "lost" address/time tuples (see Fig. 12.22).

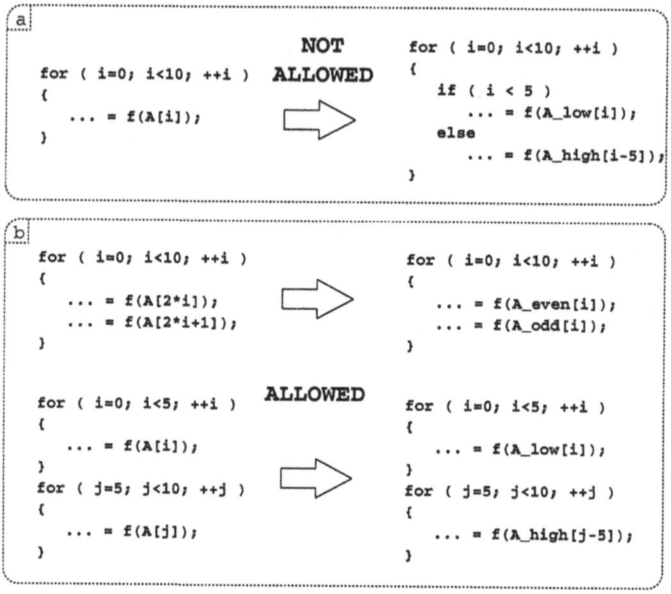

```
 a
                            NOT          for ( i=0; i<10; ++i )
   for ( i=0; i<10; ++i )  ALLOWED      {
   {                                      if ( i < 5 )
     ... = f(A[i]);          ===>           ... = f(A_low[i]);
   }                                      else
                                            ... = f(A_high[i-5]);
                                        }
```

```
 b
   for ( i=0; i<10; ++i )               for ( i=0; i<10; ++i )
   {                                    {
     ... = f(A[2*i]);        ===>         ... = f(A_even[i]);
     ... = f(A[2*i+1]);                   ... = f(A_odd[i]);
   }                                    }

                            ALLOWED
   for ( i=0; i<5; ++i )                for ( i=0; i<5; ++i )
   {                                    {
     ... = f(A[i]);                       ... = f(A_low[i]);
   }                                    }
   for ( j=5; j<10; ++j )   ===>        for ( j=5; j<10; ++j )
   {                                    {
     ... = f(A[j]);                       ... = f(A_high[j-5]);
   }                                    }
```

Figure 12.28.   Examples of allowed and non-allowed array splits.

–   Otherwise, if the split allows additional merging with *other* arrays or their sub-arrays, we split the array, as it increases the memory reuse opportunities.

–   Otherwise, if at least some of the sub-arrays are compatible with each other, we decide to split them, as this introduces additional freedom without any clear disadvantages.

–   Finally, if none of the tests can give a definite answer, we decide to keep the original array together, as the split offers no clear advantages and would only increase the placement complexity.

After this decision, we proceed with the placement step as described above.

In practice, the number of opportunities for splitting arrays is relatively small (i.e. usually only a few arrays are candidates for splitting), and if an array is a candidate, the number of possible sub-arrays is also usually small (e.g. 2 to 4). Therefore the effect of the possible additional splitting on the overall run-time is relatively small in practice (see subsection 12.3.8), whereas it may provide some relevant additional optimization opportunities. Extra splitting opportunities may also be derived based on basic set analysis (see section 6.2).

12.3.8   Experimental results

In Table 12.3 we present some of the results that were obtained by a prototype tool for the (multimedia) applications and application kernels that we also used in subsection 12.2.8. We also made the same assumptions (all arrays in one memory and a sequential execution order) and the experiments were run under similar conditions.

The table contains the results for the dynamic windowed strategies (i.e. with the additional placement step described in this subsection), with and without the array splitting option enabled respectively.

When we compare the results in this table with the multiple- and single-assign-ment figures in Table 12.1 on page 283, we can see that for most examples the potential reduction in storage size requirements is considerable.

Table 12.3. Experimental results: required storage sizes for our test vehicles after a dynamic placement phase.

Application	Dynamic Windowed [words]	Dynamic Windowed + Array Split [words]
Updating SVD	312	312
2-D wavelet	2846	2846
	(832)	(832)
Edge detection	189	189
3-D volume rendering	25603	25603
Voice coder	1130	1120
GSM autocorrelation	248	223

Moreover, a comparison with the results in Table 12.2 on page 284 reveals that the dynamic aspect of our approach (i.e. the sharing of memory locations between different arrays) can result in a large gain compared to a static approach with no memory reuse between arrays, an approach that is also taken by other researchers [430]. In most cases, we even come close to the scalar minimum (see Table 12.1 on page 283), without the run-time and implementation overhead associated with a scalar expansion. This clearly demonstrates that our pragmatic approach is very valuable as it tractable and leads to near optimal results (in an acceptable time, see further).

Also, the table indicates that the additional gain obtained by allowing array splits may be worthwhile (e.g. a 10% reduction for the GSM application).

As already mentioned in subsection 12.2.8, for the 2D wavelet kernel our simple projection test could not detect a relevant projection opportunity for 2 of the arrays. The results between brackets indicate the results that would be obtained if we used an extended projection test [430].

In Fig. 12.29 we present the optimization run-times that were required by our prototype tool. As in subsection 12.2.8 we have again investigated four possible configurations (pruning and speed-up heuristics enabled or disabled). The figure contains the results with and without the optional splitting step enabled. The bars in the figure contain 2 components: the lower parts correspond to the time needed for the window calculation and the upper parts correspond to the time needed for the compatibility and mergability calculations and the actual placement. For the window optimization the B&B search strategy was used.

The run-times are certainly acceptable in an embedded system design exploration context, especially when both pruning and speed-up heuristics have been applied. As in subsection 12.2.8, the figures indicate that the speed-up heuristics are more effective than the pruning, but also that both of them are necessary to reduce the possibility of failure to an acceptable level.

For most of our test vehicles, the window calculation consumed the largest amount of time. Only for the voice coder application the second substep was dominant. This can be explained by the fact that this application has a relatively large number of arrays (> 200). The complexity of the window calculation substep increases linearly with the number of arrays, but the the complexity of the compatibility and mergability calculations in the second substep increases quadratically with the number of arrays. Therefore, when the number of arrays increases, which can be expected in complex multimedia applications like MPEG-type video compression, the second substep is likely to become dominant.

The figures also illustrate that the effect of the optional additional step on the overall run-time is limited (a maximal increase of 50% when pruning and heuristics have been applied). Considering the fact that the splitting can result in a relevant storage size reduction, this optional splitting step is certainly worthwhile.

In Fig. 12.30 we also present the memory layout obtained for the GSM autocorrelation application (extracted from a public domain GSM specification). All of the arrays in this example have 16-bit

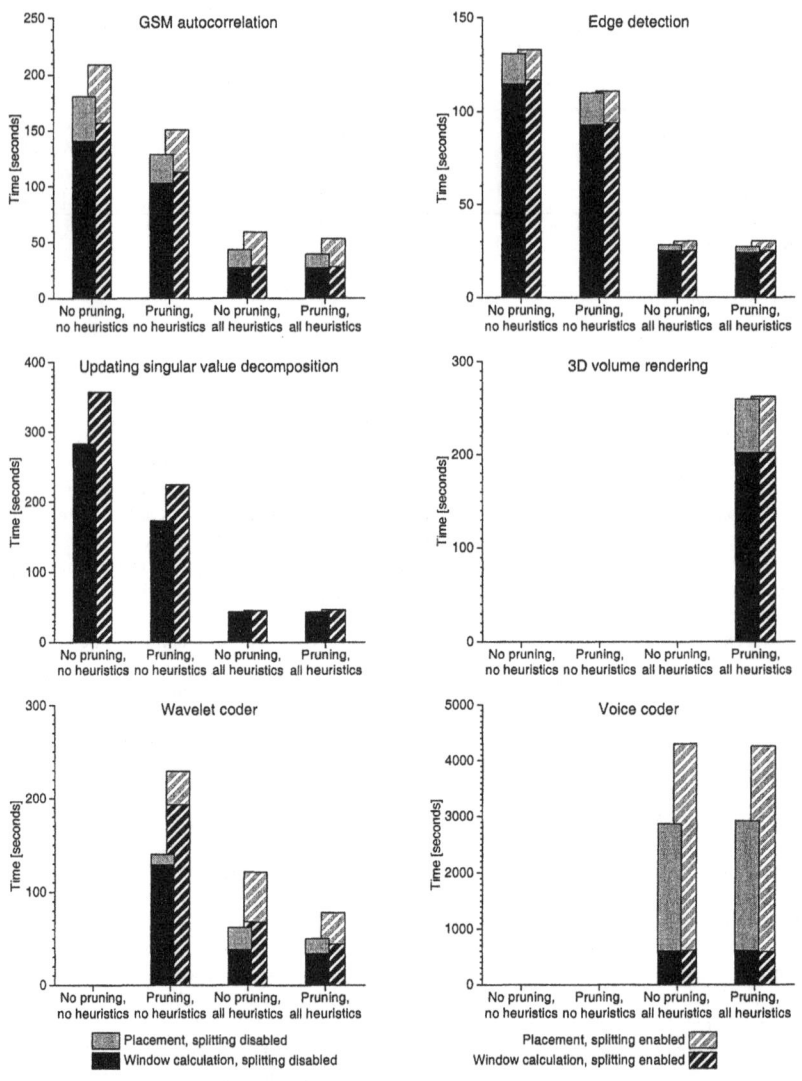

Figure 12.29. Optimization run-times for the overall optimization strategy, with and without array splitting enabled. Missing bars correspond to experiments that failed (except for the wavelet coder: there the missing bars are relatively large (up to > 4000) and they are left out because they would distort the rest of the graph too much).

word size, except for array y32, which has 32-bit word size. We assumed a 16-bit memory, and so we modeled the 32-bit array as a 16-bit array with an extra dimension (for storing the lower and higher 16-bit words). Note that in case we allow array splitting, our prototype tools detects that this extra dimension can be split and decides to do this, because it enables merging with array dn. The tool also detects that array rrv actually consist of 4 independent sub-arrays, which are interleaved in one dimension with a period of 5. The array is split and each of the sub-arrays is then contracted. It also detects that the resulting sub-arrays are alternately alive, i.e. that they are compatible, and that they can have overlapping address ranges. The reader may verify that this split results in an additional gain on top of the one obtained by splitting y32.

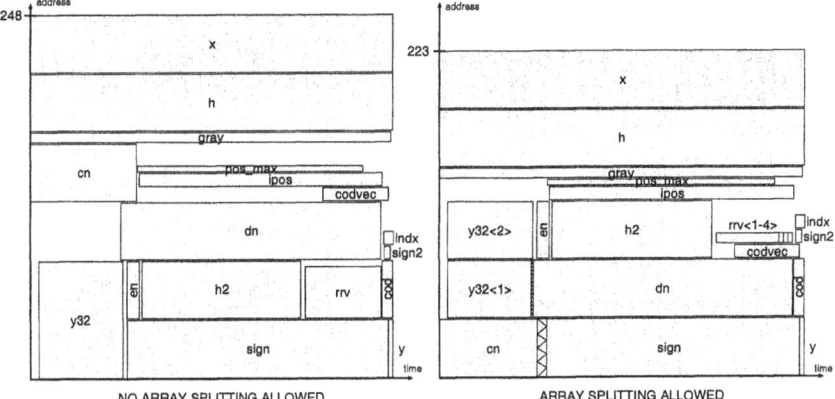

NO ARRAY SPLITTING ALLOWED   ARRAY SPLITTING ALLOWED

Figure 12.30.   Memory layout for the GSM autocorrelation example with the dynamic windowed place-ment strategy with and without array splitting. Note that the (virtual) overlap between merged arrays is indicated by zigzag lines.

For arrays ipos, indx, and dn our tool also identified sub-arrays (not shown on the figure), but decided not to split them because there was no clear indication that this could result in a better overall solution.

Also note that in the second case arrays cn and sign are merged, whereas this did not happen in the first case (although it is also possible). The reader may verify that in the first case merging these two arrays would increase the total memory size.

12.4 CONCLUSIONS

In this chapter, we have first defined the problem of reducing storage size requirements through storage order optimization and situated it in the ATOMIUM context. Despite several simplifying assumptions, the problem is very difficult to solve due to its highly non-linear and non-continuous nature. Therefore we had to come up with a pragmatic solution strategy that can obtain a reasonable solution, but which is still based on the exact modeling techniques presented in chapter 5. We have presented several alternative strategies and selected the most promising one, which consist of two major substeps. In the first substep we try to optimize the intra-array storage order (through the equivalent of data-reverse and data-interchange transformations) in order to obtain an as small as possible address reference window for each array. In the second substep, we approximate the shapes of the occupied address/time domains of the different arrays by rectangles. This allows us to reduce the second substep to a relatively simple placement.

Next, we have presented an exact technique for calculation the address reference window sizes of the arrays in a program for given execution and intra-array storage orders. In most practical cases these window sizes can be determined by solving a limited set of reasonably simple ILP problems. We have also indicated how we can calculate upper and lower bounds on the window size in case the storage order is only partially fixed. This allows us to traverse the search space of intra-array storage orders in a very efficient way. Moreover, we have shown how we can reduce the size of the search space in many cases in a simple way by means of projections of invariant dimensions. A simple test is able to detect the vast majority of these projection opportunities. The experimental results for several applications have demonstrated that these techniques are feasible in practice and that even a static windowed approach can already result in a meaningful storage size reduction compared to what would be obtained by standard allocation techniques. In the next section we show that adding a dynamic aspect to the allocation yields even better results.

Finally, we have discussed the second substep of our storage size reduction strategy in more detail. First we have derived a set of criteria that allow us to decide whether a certain placement of arrays is valid or not. These criteria are then used by our placement algorithm to obtain an optimal offset in memory for each array. An additional array splitting step allows us to reduce the storage size requirements even further in certain cases. The effectiveness and feasibility of this strategy have been demonstrated on several relevant multimedia algorithms and the reduction in size is clearly important and is very valuable in an embedded multimedia application context.

13 APPLICATION DEMONSTRATORS

In this chapter, our DTSE approach will be demonstrated on several realistic applications, both in the domain of video and image processing, as well as in the ATM domain. We will show the effect on the system power budget, memory size and system simulation time. The absolute power and area figures used below are estimated for a particular library and technology. As such they should only be used as indications. What is most important are the relative differences in the data for the different design alternatives. For instance, in a custom optimized architecture for a complete H.263 video conferencing decoder, one of our in-depth design studies has shown that data-path and controller related power is negligable compared to the contribution of the frame memory accesses. Moreover, we have been able to reduce the maximal power consumption related to these frame memory accesses by a factor 9 for the worst-case mode (PB) of the video decoder. This has been achieved solely by applying our system-level DTSE exploration methodology and it is complementary to other power savings feasible at lower abstraction levels by voltage and technology scaling. These substantial effects are confirmed in the other demonstrators.

13.1 TYPICAL IMAGING DEMONSTRATOR: MEDICAL BACK-PROJECTOR

A first realistic demonstrator (designed in 1993) is the kernel of a medical back-projection algorithm based on a Siemens application [213]. The main features which are relevant for the data-dominated application domain (in terms of memory storage size and amount of transfers) have been captured in this test-vehicle. The principle of the back-projection is illustrated in figure 13.1.
Pseudo code for this 2D reconstruction kernel is described below:

```
FOR θ := 1 TO N_pr − 1 DO
    BEGIN
    FOR i := 0 TO N − 1 DO
        FOR j := 0 TO N − 1 DO
            BEGIN
            pds_data[i, j, θ] = f(proj_data[input, θ]);
            /* Accumulation for all θ in figure 13.1 */
            I[i, j, θ] = I[i, j, θ − 1] + interpol(pds_data[i, j, θ]);
            END;
    /* Add to background image to produce image for display */
    FOR i := 0 TO N − 1 DO
```

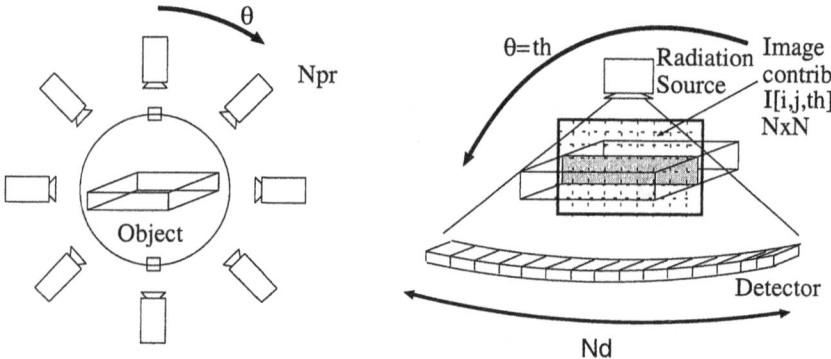

Figure 13.1. Geometry of a fan-beam scanner for medical imaging. The original 2D object slice has to be reconstructed from the 1D detector responses under different angles θ (referred to as th in text).

FOR j := 0 TO N − 1 DO
$\quad I_{out}[i,j] = I_{in}[i,j] + I[i,j,N_{pr} - 1];$
END;

The parameters are $N = 512$ (image size), $N_{pr} = 12$ (number of projection angles θ), and $N_d = 1024$ (size of 1D projection vector $proj\_data[]$ for 1 angle). This specification shows a 3-level nested loop followed by the final accumulation for the output image. Note that the $proj\_data$ signal is addressed in a data-dependent way (for the i and j indices).

We have used our DTSE methodology to optimize this application towards power and area [54]. The first design alternative will be directly derived from the procedurally interpreted code. This means that the outer angle loop θ is evaluated the "slowest". For the given parameters of image size and number of projection angles, and for a cycle budget of the inner loop which has to be 1 for the intended design, the memory organisation then requires 2 external memories. A dual-port memory is needed to store and update the intermediate image values I, which are in the end overwritten with the displayed image I_{out}. The intended clock rate is at least 60 MHz which means that a fast SRAM can be a good choice, even though some special video DRAMs are available too. Some advanced SRAMs are targeting low-power [365], but still require about 0.26 W at these clock rates. More conventional SRAMs can easily consume 3 to 6 times more per MHz. Moreover, this is only a single-port RAM and the communication between the ASIC and the external memory (including the data and address pads) is not incorporated yet.

The total storage required for the 18-bit image information is about 4.5 Mbit, which means five 1 Mbit SRAMs. Assuming that a dual-port version requires about 1.5 times more power per access (due to the larger internal capacity which can be partly compensated by reorganizing the memory partitioning), the power budget for this is around $2 \times 1.5 \times 5 \times 0.26 = 3.9$ W. So the total DTSE related power is quite large in this system.

We will now consider also a second, more optimal memory organisation. It is indeed better to split up the image memory into 2 parts: one with even and one with odd pixel addresses. This allows to use 2 single-port memories which are alternatingly addressed and which perform an alternating sequence of reads and writes. They each require half the size, namely 2.25 Mbit, which results in an increase of the number of 1 Mbit SRAMs to six. In this way, the power consumption can be reduced to about $0.26 \times 6 = 1.56$ W but at the cost of a more complex memory access scheme and potentially more pins on the chip if external data bus logic should be avoided. This design alternative is illustrated on the left side in figure 13.2.

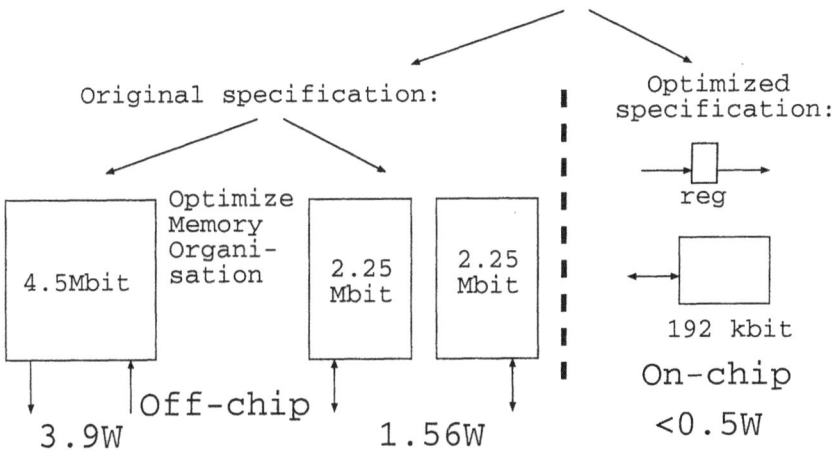

Figure 13.2. Estimated power cost for back-projector with $N_{pr} = 12$. Notice that both system-level loop transformations and memory organisation decisions are needed to support these steps. The loop transformations have been automatically obtained with our MASAI tool.

In addition, the projection detector data $proj\_data[]$ has to be buffered, which requires an additional memory of about 16 Kbit per angle. As the angle loop is the outer one, these can be transferred to the ASIC in a stream from the scanner. This means that a small buffer memory can be located on-chip which is overwritten with new projection data for every th iteration. The power consumption related to this buffer access is relatively small compared to the image information access though still significant compared to the data-path related power.

We can still do much better however. Using our systematic global loop transformation strategy, partly supported with our early prototype tool called MASAI [144], we have derived a more optimal specification as input for the subsequent memory organisation decisions. Here, the image loops have become the outer ones and the internal th loop has been sequenced to indicate the correct ordering for maximizing the in-place storage of signals and for reducing the number of accesses. In-place storage means that the I data can be overwritten one at a time in a simple register and that the $Iout$ for a particular pixel location can be directly transferred to the image display when all angle contributions have been accumulated. In this way, everything is stored in-place and no background memory is needed. As a result, we do not require an external memory for intermediate storage. On the other hand, if we impose the same parameters and cycle budget, we now require a storage of all projection data for the 12 angles in the single-ported projection buffer, i.e. 192 Kbit. This is a relatively large memory but still feasible to implement on-chip in an advanced submicron technology. Even if it has to be stored externally, the power consumption due to the accesses will be at most a fraction of the 1.56 W in the initial specification. This is illustrated on the right side in figure 13.2.

We will now show that these considerations do depend on context and parameters. The effect of the optimizing transformations is illustrated on the same algorithm with a larger number of projection data ($N_d = 4096$) and angles ($N_{pr} = 1200$). This is also more realistic for future high-resolution scanners. In this case, the second specification above — with the inner θ loop — would require a single-port projection data buffer of no less than 77 Mbit. This would require an enormous power budget of about 20 W, which is an order of magnitude larger. A better solution in that case, which is also automatically found by MASAI when starting from the reused "wrong" specification, is to choose the θ loop as outer again. In that case, the main memory requirement is then again due to the

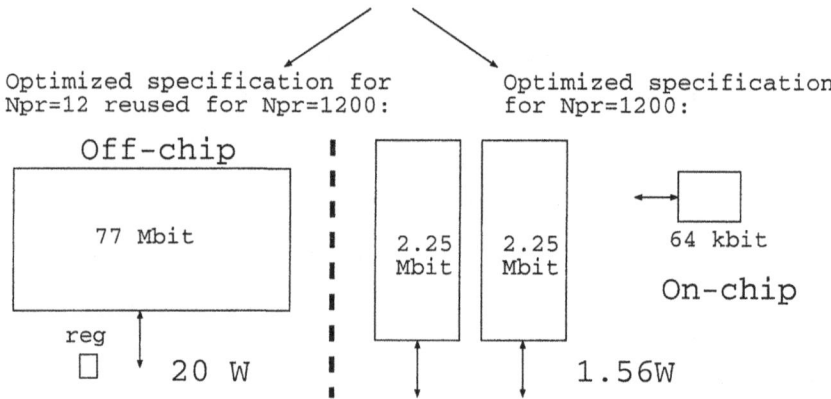

Figure 13.3.   Estimated power cost for high-performance back-projector with $N_{pr} = 1200$. Notice that a different system transformation and memory organisation are needed to arrive at a good result. This clearly illustrates that a parameterized specification with fixed realisation directives stored in a system module library would lead to unacceptable results in terms of memory power and size.

storage of the intermediate image information I which requires the same number of SRAMs.[1] So the original power estimate of about 1.56 W is still valid. The on-chip projection detector data buffer is now increased to about 64 Kbit, which is still very acceptable and which contributes only in a minor way to the system power budget. These alternatives are illustrated in figure 13.3.

In this analysis, we have not yet incorporated the effect of the number of bits which are really active (the worst-case power budget has been assumed for the RAM's which is the only figure given on the data sheets) or the interconnect capacity between the memory chips and the ASIC for the data-path operations (which is very difficult to characterize). Taking these considerations into account would change the absolute value of the power contributions but the conclusions will remain the same, given the large relative differences for the alternative memory organisations.

In addition, we have coupled the DTSE transformation tools to our simulation engine, with excellent results for the virtual memory usage and run-times on mid-size workstations. An example of the effect for the medical imaging application is shown in figure 13.4. For relatively small parameter values $N_{pr} = 6$, the simulation of the original description (BEFORE) becomes infeasible due to memory shortage even with a workstation swap space of 80 Megabyte. Also for lower N_{pr} values, the CPU times are becoming rapidly dominated by the memory swapping. After automatic optimization of the description (AFTER), the simulation run-time (on a HP715/50) and the memory requirements are drastically reduced as shown in figure 13.4. The simulation becomes then feasible even for large N_{pr} values in the range of 300 to 1200 as required for realistic modern scanners. Unrolling the loops, as done in conventional simulation tools, results in an infeasible number of scalar signals (several billion words).

13.2   TYPICAL VIDEO DEMONSTRATORS: 2D MOTION ESTIMATION KERNEL AND FULL H.263 VIDEO CONFERENCING DECODER

We will now also show the effect of our approach on memory size/area in a video coding context. This will be illustrated first on a motion estimation kernel, one of the memory-dominated submodules

---

[1]These signals now need to be slightly larger because of the increased dynamic range during accumulation but simulations have indicated that a word-length for I of 20 is probably sufficient. As a result, the total memory size can still fit within 6 external single-port 1 Mbit SRAMs, again divided in odd and even pixels.

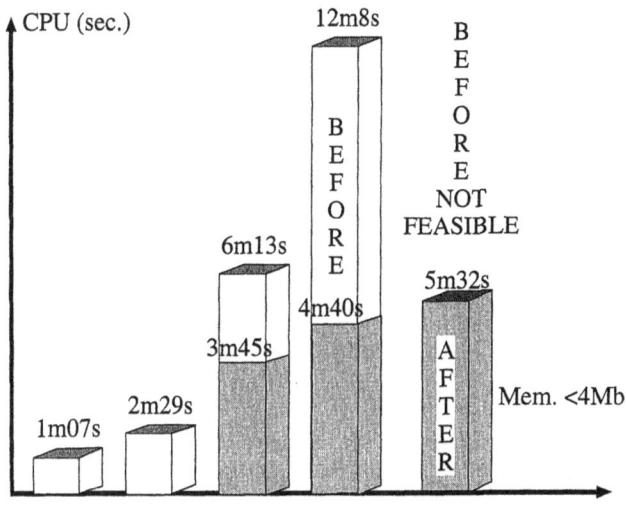

Npr = 2   Npr = 3   Npr=4   Npr=5   Npr=6   System Param

Figure 13.4.   Execution times for system-level simulation (run of 1 image frame of 256 × 256 pixels) of medical tomography (CT) back-projection subsystem both before and after automatic optimisation. For N_{pr} larger than 5, the 80 Mb swap space of the mid-size HP station was not sufficient any longer to start the actual simulation originally.

of the MPEG2 video codec standard.   The principle has been discussed in section 8.6. The version we consider here is commonly referred to as the "full-search full-pixel" implementation [216].

The algorithm is typically executed in 6 nested loops. A C-like description of the algorithm template is given next.

```
FOR g:= 1 TO H DO              [ vertical CB counter ]
 FOR h:= 1 TO W DO             [ horizontal CB counter ]
  FOR i:= -m TO m-1 DO         [ horizontal searching of RW ]
   FOR j:=-m TO m-1 DO         [ vertical searching of RW ]
    FOR k:= 1 TO n DO {        [ horizontal traversal of CB ]
     FOR l:= 1 TO n DO         [ vertical traversal of CB ]
     [ Basic Operation 1 ] (BO1)
     [ Basic Operation 2 ] (BO2)
    }
```

For the motion estimation kernel considered here, BO1 consists of an accumulation of pixel differences, while BO2 consists of the calculation of the new minimum and its location. However, in general, BO1 and BO2 can contain other operations, such that a large class of similar algorithms is covered by the same template. The parameters we used for our implementation are the following: W=720 pixels, H=576 pixels, m=8 pixels, n=8 pixels. The pixels are 8-bit gray scale values.

The result of the memory allocation (automatically obtained with our early prototype tool HI-MALAIA in 1994) for several cycle budgets in the inner loop body is provided in figure 13.5 [28]. Note that the frame memory which is needed to store the delayed frames is not included. This large frame memory needs to be stored off-chip. What is included here are the smaller background (and partly also foreground) memories for storing the intermediate data on-chip, relatively close to the data-paths. The area figures have been obtained for an accurate RAM layout model, instantiated for a fast 1- or 2-port embedded SRAM generator in a 1.2μ CMOS technology. So these absolute results

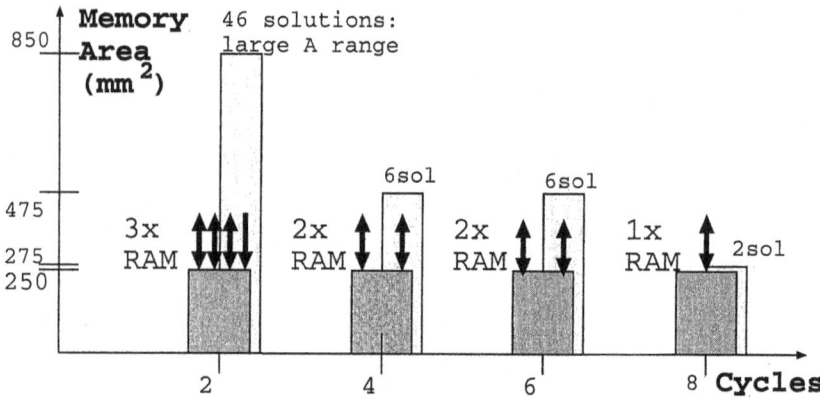

Figure 13.5.   Results for automated memory allocation for several cycle budgets in inner loop of a video motion estimation kernel. The cross-hatched boxes show the area required for the optimal memory organisation found, including the number of RAMs needed and their Read or Read/Write port requirements (arrows on top of boxes). The dashed boxes represent the area range for the feasible solution space.

are quite large (and for a by now outdated technology), but what is most important in the sequel are the relative figures.

It can be seen that the memory organisation is heavily influenced by the allowed cycle count. For instance for 2 cycles, 3 RAMs are needed, two of which have 1 R/W port and the third one has a R port and a R/W port. In contrast, if we allow 6 clock cycles (with a shorter period) only 2 RAMs are needed, both with 1 R/W port. However, the area which is needed in the end is not really smaller. This is true because in the 2-cycle solution, the data can be efficiently distributed over several memories by our tools. These type of area-cycle trade-offs are very useful for the system architect to decide on the type of RAMs which are chosen: either slower but cheaper ones, or faster but more expensive ones. For this application, it is clearly more effective to choose 3 slower inexpensive on-chip memory modules, with other parameters for the layout and area estimation model. The latter require less area and exhibit a higher yield than the 2 memories in the 6-cycle solution using the relatively fast default SRAM generator.

The dashed boxes in the background of figure 13.5 represent the number of feasible solutions which have been found by the tool and their range of area costs. For instance, for the 2 cycle case, 46 feasible solutions have been identified in the large search space and these require an area cost ranging from $850mm^2$ to about $250mm^2$ (optimal case found). Suppose the designer has manually selected a solution (e.g. with $500mm^2$) during memory allocation. This can easily lead to the wrong conclusion that the 6-cycle solution would be better, whereas this requires much faster and more expensive RAMs to achieve the same real-time throughput constraint. This clearly shows the usefulness of accurate estimators and near-optimal exploration tools to support these decisions.

Also for the combined power/area exploration, a number of interesting alternatives have been derived with our DTSE methodology (see e.g. motion estimation test-vehicle in [437]).

For a full, very demanding video test-vehicle, namely the entire decoder algorithm in a H.263 video conferencing application [177], we have obtained even larger reductions of the maximal power consumption associated with the background memory accesses [291]. The starting point for our exploration has been a C specification of the video decoder, available in the public domain from

Telenor Research. We have transformed the data transfer scheme in the initial system specification and have optimized the distributed memory organization. This results in a memory architecture with significantly reduced power consumption. For the worst-case mode using Predicted (P) frames, memory power consumption is reduced by a factor of 7 when compared to a good direct implementation of the reference specification. For the worst-case mode using Predicted and Bi-directional (PB) frames, the maximum memory power consumption is even reduced by a factor of 9. To achieve these results, we have used our formalized high-level memory management methodology, partly supported in the ATOMIUM environment and script.

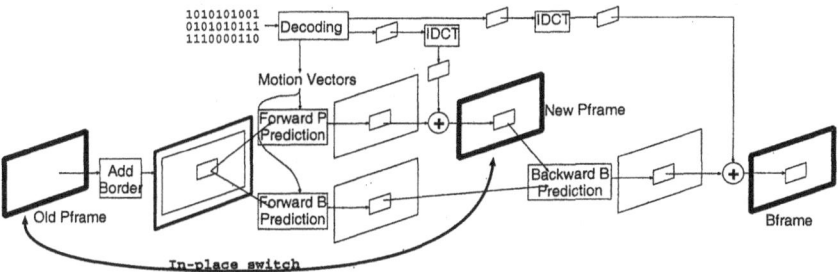

Figure 13.6.   Initial data-flow for H.263 reference design

The initial data-flow of the decoder is illustrated in figure 13.6. Both a "normally" predicted frame (P) and a bidirectionally predicted frame (B) are derived. The algorithms in the functions *forward_P*, *forward_B* and *backward_B* are very complicated and involve many complex array accesses.

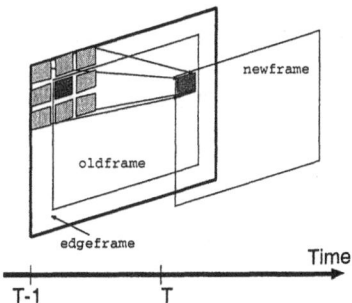

Figure 13.7.   Border dependencies in H.263 reference design

Data-flow transformations emphasizing especially the global propagation of conditions through the code (figure 13.7) allow to remove all dependencies which are required to deal with the border macro-blocks which are normally needed in the H.263 algorithm. The result is a new data-flow as indicated in figure 13.8, corresponding to a decrease of 16% in power related to frame accesses and 9% in total area of the frame memories.

Next, global loop and function hierarchy transformations have allowed to reorganize the code in such a way that more locality of data access was achieved, resulting in a major reductions of the memory related power. The result is that the typical system-level buffers present between large functions in the overall data-flow (see figure 13.9) are largely gone. Combining the backward/forward predictions allowed to reduce the power to 51% of the reference, whereas the additional effect of also involving the IDCT computations reduced this further to 38%. The result is a new data-flow as indicated in figure 13.10.

As the third global step in our methodology we introduce extra memory hierarchy for subarray data which are read multiple times. The motivation for this is that if most accesses take place on

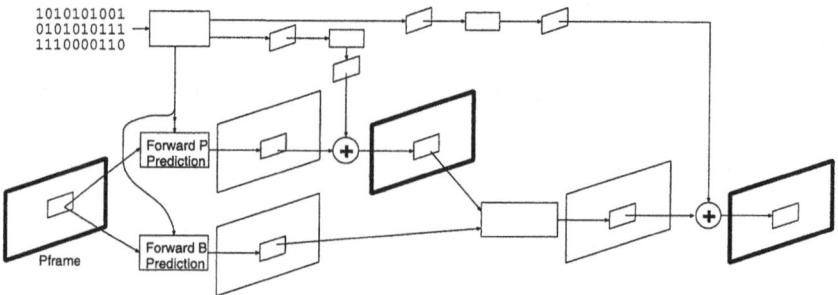

Figure 13.8.   Data-flow after data-flow transformations for optimized H.263 design

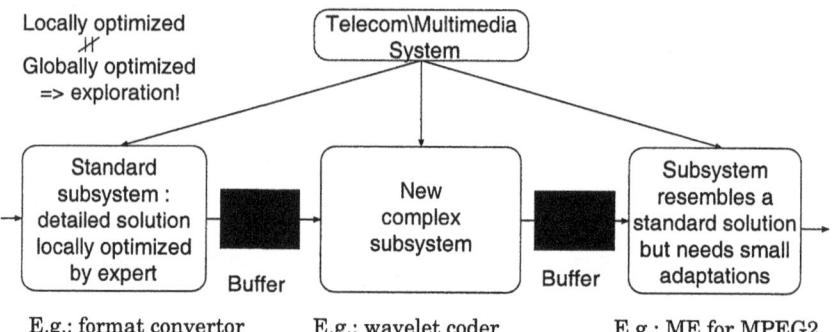

Figure 13.9.   Removal of accesses and storage of system-level buffers by global loop reorganisation

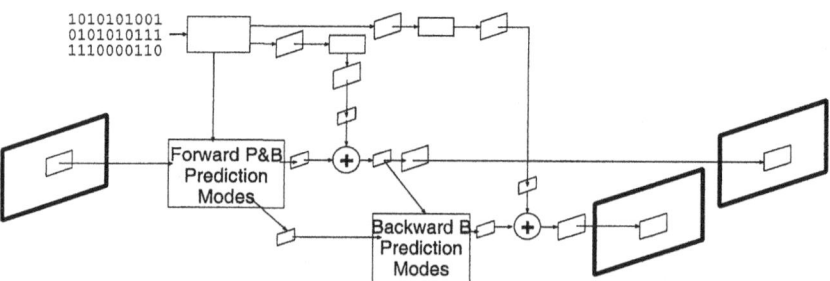

Figure 13.10.   Data-flow after global loop and function hierarchy transformations for optimized H.263 design

the smallest memories, the saving in power is sufficient to motivate the extra transfers between the levels in the memory hierarchy. This can happen even over several layers(see figure 3.2). Note that the end result is typically a much more customized memory organisation than the traditional caching schemes in microprocessors. Performing this on the complex data accesses related to the $OBMC$ (Overlapped-Block Motion Compensation) mode lead to an overall power budget of 24% of the reference. By adding an extra hierarchy level for the interpolation related accesses at the lowest level of the motion compensation, a saving up to 11% was achieved. The result is a new data-flow as indicated in figure 13.11.

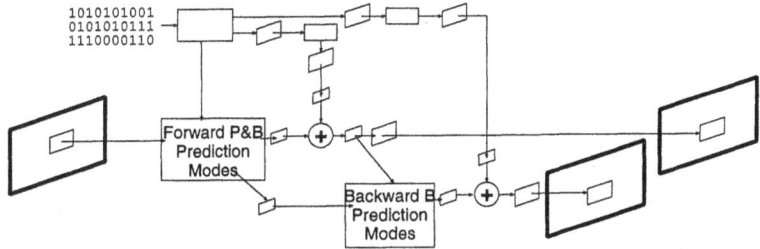

Figure 13.11.   Data-flow after memory hierarchy introduction for optimized H.263 design

oldframe   newframe   old/newframe

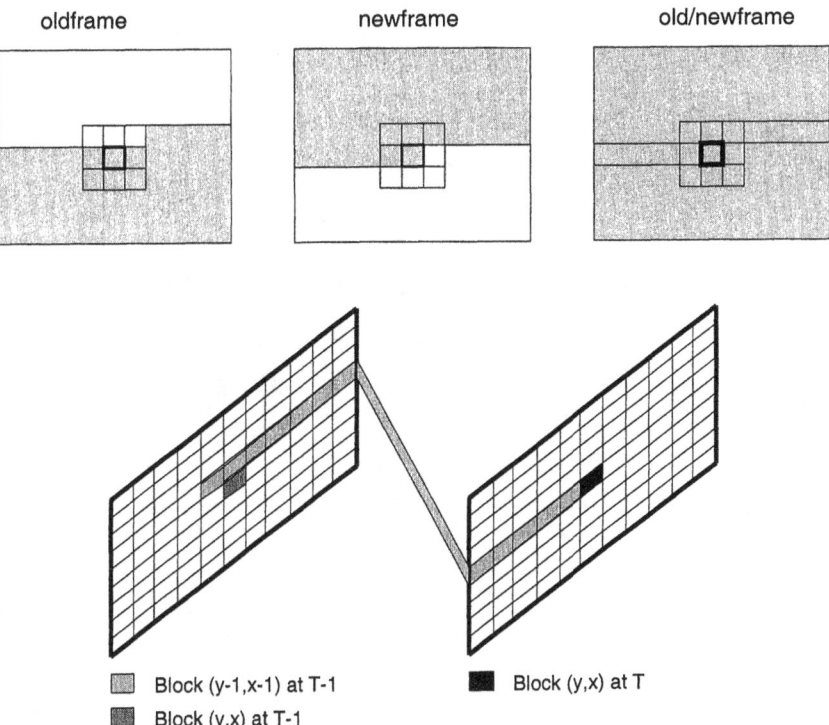

▦ Block (y-1,x-1) at T-1   ■ Block (y,x) at T

▦ Block (y,x) at T-1

Figure 13.12.   In-place optimisation: principle on combining the new and old image frames for the P sequence

Finally, we have performed a global in-place optimisation step to reduce the area cost of the memories while maintaining the power consumption. The principle is shown in figure 13.12. The total area decreased to 60% of the reference design. The end result is a new data-flow as indicated in figure 13.13.

Figure 13.14 gives an overview of the relative power consumption for each major optimization stage for the PB mode. This is the power consumed by the frame memories when decoding bidirectional B frames with unrestricted motion vectors and overlapped-block motion compensation. The power consumption is normalized with respect to the power consumption for the reference design. For the same design, we have also determined the global power associated with the most demanding

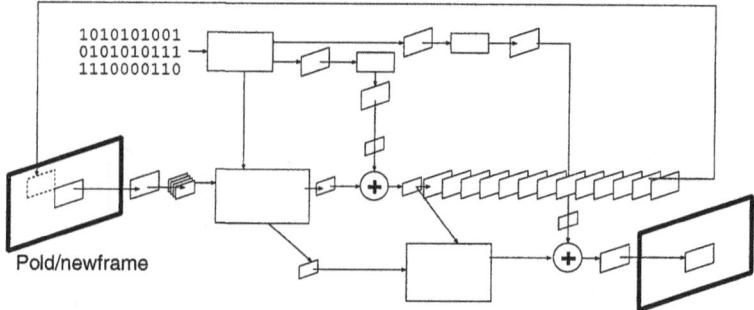

Figure 13.13.   Data-flow after in-place optimisation between new and old P frames for optimized H.263 design

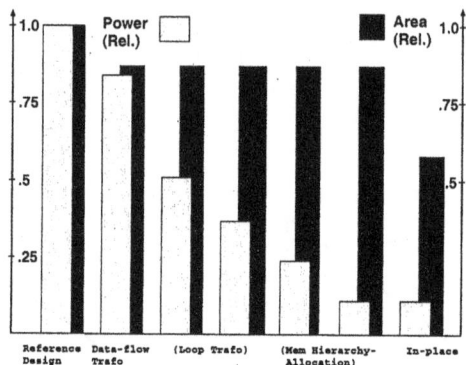

Figure 13.14.   Relative power in continuous PB mode for each optimization stage of H.263 frame access

arithmetic module namely the IDCT component. Here, the combined power of the unoptimized data-path, including its local controller and the local memories was more than a factor 150 lower than the unoptimized background memory contribution. This clearly motivates our strategy to initially ignore the data-path effects in our system-level exploration.

We believe these results to be very convincing for the effect on the global system power of the proposed approach.

13.3 TYPICAL NETWORK COMPONENT DEMONSTRATOR: SEGMENT PROTOCOL PROCESSOR IN ATM

In the power related studies for memory organisations we have obtained very good results for table-driven network components too [435, 436, 376]. An example for a segment protocol processor within the common adaptation layer (3-4) of ATM [389] is provided in figure 13.15. The SRAM memory power model has been adopted from U.C.Berkeley for a $1.2\mu m$ CMOS technology. The large search space for possible data structure realisations (horizontal axis) can be scanned with a formalized methodology [436, 439] which allows to efficiently select the optimum for a given set based specification without exhaustively searching all candidates. Examples are pointer arrays (PA) and linked lists (LL). Notice also the very large range of power budgets for different alternatives, for which the right-hand side solutions obviously lead to infeasible system realisations.

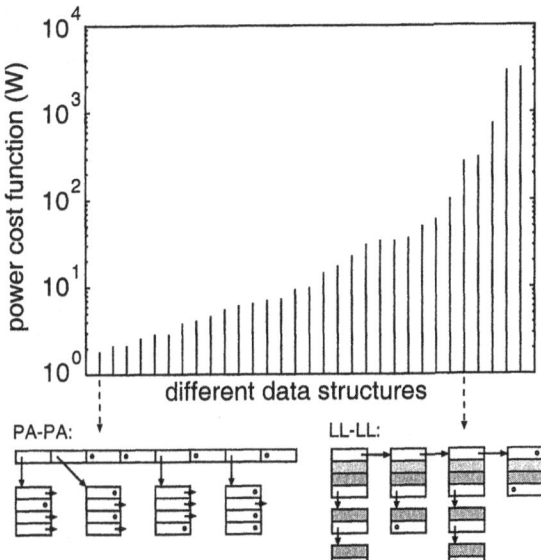

Figure 13.15. Estimated power cost for a large range of candidate pointer data structure realisations for a given set based specification of an ATM network component. With our design exploration methodology, the optimal solution can be found without exhaustively scanning the search space.

In addition, we have successfully studied ways to reduce the number of address transitions in ATM communication network components (with up to 60%) [435].

13.4   OTHER CUSTOM APPLICATION DRIVERS AND DEMONSTRATORS

Other major driver vehicles during the period 1993-96 have included a voice coder in a mobile terminal application (auto-correlation, Schur decomposition and Viterbi coder), a regularity detector kernel for vision [144], a quadrature-structured differential PCM kernel [200] for image processing, a non-linear diffusion algorithm for image enhancement [162, 349], updating singular value decomposition for beam-forming [28, 279, 278], memory-dominated submodules of the MPEG2 video decoder/encoder (including motion estimation [28, 102, 104, 290, 348] and block/field/frame formatting), and a 2D wavelet coder (forward and backward) for image coding [226, 225, 227, 228].

13.5   POWER CONSUMPTION REDUCTION FOR MOTION ESTIMATION KERNEL ON
PROGRAMMABLE PARALLEL PROCESSOR

In this section, we will continue our study of mapping the motion estimation kernel on programmable processors, which was started in section 8.8. As stated there, our goal is to reduce both the area *and* the power consumption. We will now summarize the impact of the other DTSE steps. By further reducing the memory size and the accesses to large memories, we can save both area and power. The power model we will use here is a simplification of the one introduced in section 4.4:

$$P_{memory} = f(\#words, \#bits) \times F_{mathitreal} \qquad (13.1)$$

where we assume a linear dependence of $f()$ on the number of words for illustration purposes. In practice, a more accurate model should be used.

If real-time operation is required, a lower number of accesses also results in reduced memory bandwidth requirements. Higher bandwidths are normally obtained through an increase of the number of memory ports or even a duplication of memories, which are very costly solutions in terms of area. Therefore, bandwidth reduction may also have a positive effect on the area cost.

13.5.1 Data reuse decision for optimized custom caching

Given this power model, it must be clear that the only way to obtain a drastic power reduction is to decrease the number of accesses to the large memories as they have the largest cost. In this case, we mainly focus on the input frame buffer, as it is the largest (even after our transformations) and most frequently accessed buffer.

The obvious technique to reduce the number of accesses to a large buffer is caching. Although caching techniques are very widespread, they are mostly used to enhance the processing speed. It is less well known that they can also have a considerable effect on the power consumption. Even by introducing small caches, the number of expensive accesses to the large frame memories can be drastically reduced.

We have opted for a software-controlled caching strategy, in which we insert explicit transfers to and from the cache memory, for several reasons. Firstly, hardware controlled caching methods, which can obtain good results in a general-purpose processing environment, are generally less suited for the regular class of algorithms that we are considering. For instance, on a cache miss, unnecessary data may be transferred to the cache because data are typically cached per line, or data that are still needed may be selected for replacement. Both phenomena have a negative effect on the power consumption. For this class of algorithms a custom caching strategy derived from compile-time analysis can achieve much better results. Secondly, and even more important, hardware caching is in general unpredictable, unless one has complete knowledge of the caching algorithms implemented in the hardware. This would make it very difficult to satisfy the stringent synchronization constraints. Therefore, we have chosen to explicitly introduce cache transfers in the program to be executed. This gives us full control of what data are in the cache. Moreover, many DSPs and even modern video processors such as the TMS320C80 have no hardware-controlled cache. In those cases, it is crucial to select an efficient software-controlled caching scheme.

Naturally, in our optimization strategy, this kind of caching decisions can only be taken after the loop hierarchy has been (more or less) fixed, because our buffer size reducing transformations may drastically change the order of execution, and hence seriously affect the caching decisions.

For our demonstrator algorithm, there are two good candidates for caching: the blocks of the current frame, and the reference windows of the previous frame, as both of them are extensively used in the inner loops of the program.

13.5.1.1 Current blocks. If we look at the inner loops of Fig. 8.20, and more in particular at the transfers from the current frame (image(xCB,yCB,T)), we can see that the coordinates for the transfers are independent of i, j, i1, and i2. This means for each execution of the T, X, and Y loops, a block of N^2 pixels is transferred $4M^2$ times from the frame buffer. So this is certainly a good candidate for caching. If we transfer these blocks once from the frame memory to a N^2 word cache, and then read them from the cache $4M^2$ times, we can reduce the number of frame memory accesses for the current frame by a factor of $4M^2$ (typically 256). Provided that the required cache is small enough, which is usually the case (N typically equals 8), this results in a substantial reduction of the power consumption.

However, we should be very careful when implementing this caching strategy. A naive implementation of our caching decision would result in the code presented in Fig. 13.16. Unfortunately, if we have a closer look, we can see that we have introduced a relatively large amount of code in between the X and i loops, which eventually results in a large amount of dummy code overhead in the inner loops (if we want to obtain perfect synchronization).

Therefore, we should try to distribute the frame memory-to-cache transfers equally over the different sample intervals. Here, this is relatively easy as the number of transfers equals the number of input samples. So for each input sample, we should also perform one transfer. We cannot do this in a straightforward way. Simply interleaving the k1 and l1 loops of Fig. 13.16 with the i and j loops

```
for ( T = -infinity; T < +infinity; ++T )
  for ( Y = 0; Y < H/N; ++Y )
    for ( X = 0; X < W/N; ++X )
    {
      Init1();
      for ( k1 = 0; k1 < N; ++k1 )
        for ( 11 = 0; 11 < N; ++11 )
        {
          xCB = X*N + k1;
          yCB = Y*N + 11;
          READ(image[xCB,yCB,T]);
          WRITE(CBcache[k1,11]);
        }
      for ( i = 0; i < N; ++i )
        for ( j = 0; j < N; ++j )
        {
          Xin = (X*N*N + i*N+j) mod W;
          Yin = ((Y+1) mod (H/N))*N + (X*N*N + i*N+j) div W;
          Tin = (Y+1) div (H/N) + T;
          INPUT(image[Xin,Yin,Tin]);
          for ( i1 = -M/N; i1 < M/N; ++i1 )
            for ( j1 = -M/N; j1 < M/N; ++j1 )
            {
              Init2();
              for ( k = 0; k < N; ++k )
                for ( l = 0; l < N; ++l )
                {
                  READ(CBcache[k,l]);
                  xRW = X*N+k+i1*N;
                  yRW = Y*N+l+j+j1*N;
                  if (WithinBoundaries(xRW,yRW))
                    READ(image[xRW,yRW,T-1]);
                  Process();
                }
              PostProcess();
            }
        }
      WRITE(result[X,Y,T]);
      Xout = (X-1) mod (W/N);
      Yout = (Y + (X-1) div (W/N)) mod (H/N);
      Tout = T + (Y + (X-1) div (W/N)) div (H/N);
      OUTPUT(result[Xout,Yout,Tout]);
    }
```

Figure 13.16.   A naive caching implementation.

would result in the cache being filled too late. So, we have to put the filling of the cache forward in time. This requires a doubling of the CB cache size, because while we are processing the pixels of a CB, we need space to store the pre-fetched pixels of the next CB. Due to the relatively small size of the CB buffer, this doubling won't affect the power budget too much and the overall effect is very positive. The resulting code is shown in Fig. 13.17.

Note that the code of Fig. 13.17 can be obtained from the code of Fig. 13.16 through a number of loop transformations and a data transformation to double the cache. Also note that we have used the CB cache in a circular way (by means of the mod 2 operation) and that we have carefully avoided any irregularities, even at the frame borders. While we are processing the last block of a frame, the first block of the next frame is already being pre-fetched. Failing to do so would result in some extra code at the beginning of each frame, i.e. extra code in between loop boundaries.

13.5.1.2 Reference windows. Just like the CBs, the RWs are intensively accessed during the execution of the inner loops, so applying caching to these data is also worthwhile. However, the caching decision is somewhat more complex in this case. Unlike the CBs, which are all disjoint, there is a lot of overlap between neighbouring RWs. This is indicated in Fig. 13.18a.

Because of this overlap, we can reuse a lot of RW pixels when we go from one RW to one of its neighbours, such that we don't have to re-fetch the common pixels from the frame buffers. As shown in Fig. 13.18b, there are two ways in which we can reuse RW pixels: horizontally and vertically. If we opt to reuse only pixels in the horizontal direction, we need a relatively small cache: $(2M + N)^2$ pixels for the RW currently being used and an extra $(2M + N)N$ pixels as a pre-fetch buffer (for

```
for ( T = -infinity; T < +infinity; ++T )
  for ( Y = 0; Y < H/N; ++Y )
    for ( X = 0; X < W/N; ++X )
    {
      Init1();
      for ( i = 0; i < N; ++i )
        for ( j = 0; j < N; ++j )
        {
          Xin = (X*N*N + i*N+j) mod W;
          Yin = ((Y+1) mod (H/N))*N + (X*N*N + i*N+j) div W;
          Tin = (Y+1) div (H/N) + T;
          INPUT(image[Xin,Yin,Tin]);
          XCB = ((X+1) mod (W/N))*N + i;
          YCB = ((Y + (X+1) div (W/N)) mod (H/N))*N + j;
          TCB = T + (Y + (X+1) div (W/N)) div (H/N);
          READ(image[XCB,YCB,TCB]);
          WRITE(CBcache[i,j,(((T*H/N+Y)*W/N)+X+1) mod 2]);
          for ( i1 = -M/N; i1 < M/N; ++i1 )
            for ( j1 = -M/N; j1 < M/N; ++j1 )
            {
              Init2();
              for ( k = 0; k < N; ++k )
                for ( l = 0; l < N; ++l )
                {
                  READ(CBcache[k,l,(((T*H/N+Y)*W/N)+X) mod 2]);
                  xRW = X*N+k+i+i1*N;
                  yRW = Y*N+l+j+j1*N;
                  if (WithinBoundaries(xRW,yRW))
                    READ(image[xRW,yRW,T-1]);
                  Process();
                }
              PostProcess();
            }
        }
      WRITE(result[X,Y,T]);
      Xout = (X-1) mod (W/N);
      Yout = (Y + (X-1) div (W/N)) mod (H/N);
      Tout = T + (Y + (X-1) div (W/N)) div (H/N);
      OUTPUT(result[Xout,Yout,Tout]);
    }
```

Figure 13.17.   A better caching implementation.

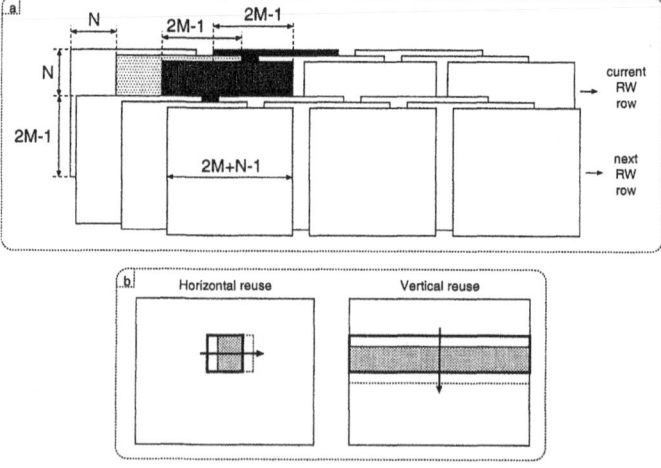

Figure 13.18.   Reference window overlap and reuse.

similar reasons as for the CB cache[2]). In case we also want to exploit vertical reuse between different rows of RWs, we need a cache of $W(2M + N) + N^2$ pixels, N^2 of which are due to a pre-fetch buffer. Note that the pre-fetch buffer is even smaller than in the previous case, because we have to pre-fetch the pixels less often.

Deciding which one of these 2 caching alternatives is the best one in terms of power consumption, is not straightforward. In the case of horizontal reuse only, we have to pre-fetch the data a few times from the large frame memories, but we have the advantage that we can keep the cache very small (typically a few hundred words). For the other alternative, every pixel fetched only once from the large frame memories, but on the other hand, the cache becomes much larger (typically a few tens of Kwords) and therefore also consumes a lot more power when accessed. In Table 13.1, the sizes and transfer counts for the frame buffer[3] and RW cache are shown for both cases.

Table 13.1.   Buffer sizes, read access counts, and power estimations due to different RW caching strategies.

	Horizontal reuse only	Horizontal & vertical reuse
Frame buffer size	$W(H + N)$	$W(H + N)$
Frame buffer reads	$WH(2M/N + 1)$	WH
RW cache size	$(2M + 2N)(2M + N)$	$W(2M + N) + N^2$
RW cache reads	$4WHM^2$	$4WHM^2$
RW cache writes	$WH(2M/N + 1)$	WH
Frame buffer power	$\sim 3K_1 W^2 H^2$	$\sim K_1 W^2 H^2$
RW cache power	$\sim 48K_2 WHN^4$	$\sim 12K_2 W^2 HN^3$

If we have to decide between both alternatives, we need to know what the effect on the power consumption is. The power consumption for these buffers is highly dependent on the chosen technology, but as mentioned at the start of this section, for illustration purposes we use a simple model in which the power consumption is linearly dependent on the memory size and the number of accesses.[4] The resulting power consumption formulas, under the assumption that $W \gg N$ and $M \approx N$, are shown at the bottom of Table 13.1. From these, we can see that for the horizontal reuse only case the power consumption in the frame memories is approximately 3 times larger, but the power consumption in the cache is approximately $W/4N$ times smaller due to its much smaller size. Since $W/4N$ is typically much larger than 3 (e.g. 22.5 for our parameters W and N), it is likely that the vertical reuse is more costly in terms of power (and also area).

In case the algorithm has to be implemented on a fixed processor, with a fixed available cache size, the tradeoff becomes quite different. In that case, we cannot influence the power consumption through the cache size, as it is fixed, and we can only try to reduce the power consumption by keeping the number of frame memory transfers as low as possible, i.e. try to implement the vertical reuse, if the cache size allows it. If not, more complex transformations are needed. The steering of these complex transformations is a topic of current research [223, 222].

In this chapter we assume that we implement the horizontal reuse only. The resulting code, which can be obtained by means of a similar set of transformations as for the CB cache, is shown in Fig. 13.19.

Note that we have introduced an additional i1 loop which seemingly disturbs the regularity of the code. However, no I/O operations are located in the inner loops of the program, and hence the regularity of the I/O operations is not affected.

---

[2]Note that due to the RW overlap, we do not need to double the cache in size in order to be able to pre-fetch pixels. Only the difference between two neighbouring RWs needs to be pre-fetched.
[3]We do not take into account the frame buffer transfers for the CBs.
[4]This is a good approximation for small memories, but not for larger ones.

```
for ( T = -infinity; T < +infinity; ++T )
  for ( Y = 0; Y < H/N; ++Y )
    for ( X = 0; X < W/N; ++X )
    {
      Init1();
      for ( i = 0; i < N; ++i )
        for ( j = 0; j < N; ++j )
        {
          Xin = (X*N*N + i*N+j) mod W;
          Yin = ((Y+1) mod (H/N))*N + (X*N*N + i*N+j) div W;
          Tin = (Y+1) div (H/N) + T;
          INPUT(image[Xin,Yin,Tin]);
          XCB = ((X+1) mod (W/N))*N + i;
          YCB = ((Y + (X+1) div (W/N)) mod (H/N))*N + j;
          TCB = T + (Y + (X+1) div (W/N)) div (H/N);
          READ(image[XCB,YCB,TCB])
          WRITE(CBcache[i,j,(((T*H/N+Y)*W/N)+X+1) mod 2]);
          for ( i1 = -M/N; i1 <= M/N; ++i1 )
          {
            XRW = ((X+M/N+1) mod (W/N))*N + i;
            YRW = ((Y+i1 + (X+M/N+1) div (W/N)) mod (H/N))*N + j;
            TRW = T-1 + ((Y+i1) + (X+M/N+1) div (W/N)) div (H/N);
            READ(image[XRW,YRW,TRW]);
            xRWc = (((T*H/N+Y)*W/N+X)*N+M+N+i1)mod(2M+2N);
            yRWc = M+i1*N+j
            WRITE(RWcache[xRWc,yRWc]);
          }
          for ( i1 = -M/N; i1 < M/N; ++i1 )
            for ( j1 = -M/N; j1 < M/N; ++j1 )
            {
              Init2();
              for ( k = 0; k < N; ++k )
                for ( l = 0; l < N; ++l )
                {
                  READ(CBcache[k,l,(((T*H/N+Y)*W/N)+X) mod 2]);
                  xRW = X*N+k+i+i1*N;
                  yRW = Y*N+l+j+j1*N;
                  if (WithinBoundaries(xRW,yRW))
                  {
                    xRWc = (((T*H/N+Y)*W/N+X)*N+i1*N+i+k)mod(2M+2N);
                    yRWc = M+j1*N+l+j;
                    READ(RWcache[xRWc,yRWc]);
                  }
                  Process();
                }
              PostProcess();
            }
        }
      WRITE(result[X,Y,T]);
      Xout = (X-1) mod (W/N);
      Yout = (Y + (X-1) div (W/N)) mod (H/N);
      Tout = T + (Y + (X-1) div (W/N)) div (H/N);
      OUTPUT(result[Xout,Yout,Tout]);
    }
```

Figure 13.19.   RW caching.

13.5.2 Further DTSE steps

In order to illustrate the global effect on the programmable processor system, we will now also apply the most important remaining steps of our DTSE methodology for our particular target domain.

In [102], we have outlined a strategy on how to allocate and organize the PEs and memories in order to obtain real time operation for our demonstrator class of algorithms. This has resulted in an area and power efficient architectural template whose memory organization has been obtained via the strategy outlined in this chapter. An overview of this template is presented in Fig. 13.20. The storage management related issues are discussed briefly now.

Given the transformed algorithm that we have obtained in the previous subsection and the architectural template shown Fig. 13.20, we still have to perform several tasks to complete the mapping process. These tasks are also part of the DTSE methodology outlined in chapter 3. They are mostly dealing with the management of the data storage and are more briefly described next.

Storage cycle budget distribution.   Once we know what data have to be stored in what memories, we have to make sure that the data get in and out of the memories in time. This requires a careful scheduling of the transfers to keep the required number of buses and memory ports minimal. Several

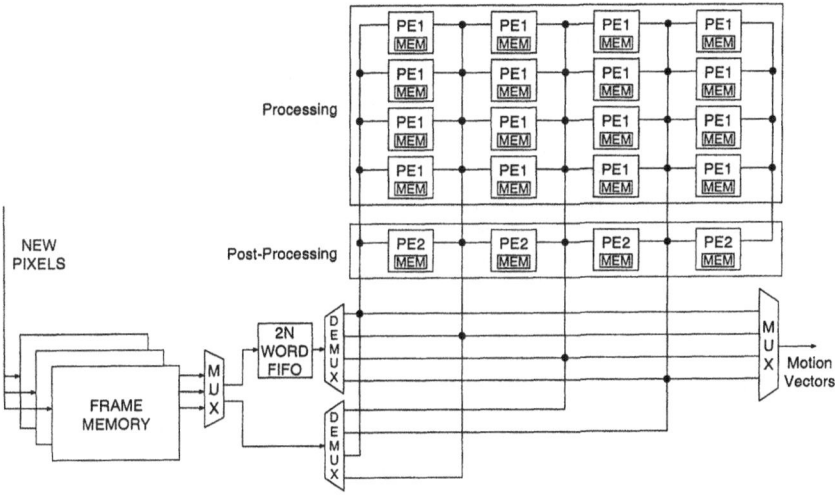

Figure 13.20.   A parallel architectural template for regular block-wise algorithms.

optimizations are possible [437]. For instance, it may be possible to share transfers between processors that require the same data [118].

Memory allocation and signal-to-memory assignment.   In section 8.8 we have presented our techniques for the reduction of the buffer sizes. However, these techniques only influence the life-time of the data. At that point no decision is taken on how to distribute the data over the different memories. Especially the distribution of the CB and RW pixels over the local (cache) memories of the different processors requires special care. It is important that each processor has (local) access to the data that it needs.

Also the organization of the frame memory can be optimized. In Fig. 13.20 we have opted for a split of the frame memory in 3 memory banks[5], which allows to make use of standard 1-port memories instead of expensive multi-port ones.

In-place optimisation.   Given that we know what data are stored in what memories and when these data enter and leave the memories, we still have to decide on *how* we store these data in the memories. As already indicated in chapter 1, the storage order can have a large effect on the storage size cost.

In the case of the motion-estimation kernel, the decisions on the storage order of the data are still relatively simple, as there are only 3 kinds of data: the frame memory, the current blocks and the reference windows. However, as stated before, for most real-life applications the amount of multi-dimensional data to be processed and stored is very large, which makes it far less trivial to choose an overall optimal storage order for all the data. In those cases, methodology and even automated tool support are almost indispensable to arrive at a cost-efficient solution. Note that during the entire DTSE trajectory, we have postponed any decisions about the allocation of memories and processing elements (PEs). We have been able to do this because we have shifted most of the processing to the inner loops of the algorithm. These loops still contain a lot of potential parallelism, so we can exploit this and allocate a number of PEs and associated memories such that the processing times are small enough and fit inside the sample intervals. More information on mapping algorithms onto (weakly) parallel processors can be found in [89], but we also briefly discuss some of the issues here.

---

[5]The total size is still about the size of 1 frame.

Based on this entire strategy, we have defined a memory and power efficient programmable architectural template, which has been presented in [102]. Especially in a multimedia context, where area and power are considered to be the main cost factors, our strategy can be very valuable. The possible gain in area and power consumption due to memories can easily outweigh the potential increase in processor data-path area.

13.5.3 Performance Measurements

The transformations applied during the storage size and power reduction steps have a serious impact on the application code complexity. Consequently one might fear that much more arithmetic and control processing power is required to achieve real-time operation, which might destroy the positive effect of the optimizations. This would mean that the ATOMIUM approach, where the allocation of data-paths is postponed to till the end, would not be valid.

In order to measure the effect of our optimization strategy on the performance, we have performed some experiments in which we started from a fixed architecture consisting of only one processor with a limited amount of on-chip data memory and one external buffer. We used three different fixed-point DSPs: the Texas Instruments TMS320C50, the Analog Devices ADSP-2101 and the Motorola DSP56116. Naturally none of these processors is fast enough to achieve real-time operation on its own, but the experiments allow us to estimate the difference in the number of required processors to achieve real-time operation. The performance results for the three different processors are similar and a summary is presented in Fig. 13.21.

In our first experiment, we have implemented a classical frame-synchronous mode of operation (including some caching) as a reference. In the other three experiments, we implemented the perfect synchronization. In case we would not have altered the processing order as indicated in subsection 8.8.1 and would have sticked to the original column-wise processing order, we would not have been able to reduce the frame buffer size. This is also shown in the figure (we did not measure the cycle count). The rightmost 2 measurements result from experiments in which we have chosen a row-wise processing order, which enabled use to reduce the (large) frame buffer by a factor of three in size. A naive caching implementation however, as described in subsection 13.5.1.1 resulted in a large increase in the cycle budget (about a factor of five) due to dummy code overhead. By implementing the caching in a slightly different way, i.e. by pre-fetching new cache data while older data are being processed, we were able to reduce the dummy code overhead to a reasonable level (less than 40%), at the cost of a small increase of the cache size.

The difference in cycle count between the first and the last alternative is partly caused by the more complex address equations resulting from our transformation-based strategy. However, more modern video processors such as the Texas Instruments TMS320C80 tend to have more powerful address generation hardware, such that the address generation overhead can be reduced considerably. Also, the availability of multiple level hardware DO-loops on these modern processors can reduce the dummy code overhead even further, probably to less than 10%. Moreover, the motion-estimation algorithm that we implemented requires only very simple data-processing in the inner loop. Therefore the relative importance of the overhead in the address generation and loop structure is quite large. For algorithms that require more complex data-processing, the importance of this overhead is likely to be smaller. This is confirmed by current research results in our group [223, 280]: due to the increased temporal data access locality the number of cache hits increases, which may even reduce the overall cycle count.

So in case real-time operation is required, this overhead can easily be compensated by (slightly) increasing the number of PEs, if necessary. The potential gain in buffer sizes (a reduction of a factor of about three for our demonstrator algorithm) almost certainly outweighs the area increase due to the extra processing elements. This illustrates that the ATOMIUM approach of postponing data-path related issues is certainly valid in this case.

13.6 OTHER RESULTS FOR PREDEFINED (MULTI-)PROCESSOR REALISATIONS

Since end 1994 also predefined processors (e.g. DSP cores) have been addressed by us in terms of DTSE, with very interesting results [55, 102, 104, 103, 110, 171, 280, 292]. Experiments have

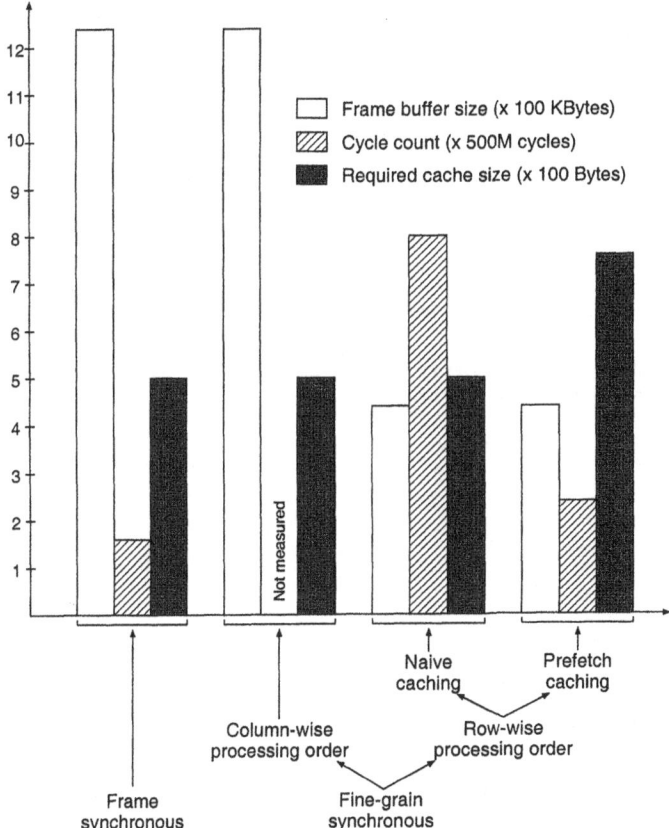

Figure 13.21.   Experimental results for different mapping alternatives of the motion estimation algorithm on a single DSP.

shown for instance that our methodology allows to heavily reduce the power of the dominant storage components in a GSM voice coder application (a factor 2 or more) and this without a penalty in processor instruction count. For this purpose, we have performed aggressive transformations on the data-flow and the loop structure, we have introduced a custom memory hierarchy (see figure 13.22) and we have performed careful in-place optimization of array signals on top of one another in the memory space.

Other experiments on processor based realisations for data-dominated submodules, oriented to instruction count reduction and power or external memory size, have been performed too. These have shown that also the cycle budget on (programmable) application-specific instruction-set processors (ASIPs) is positively affected by our methodology, especially due to the reduced data transfer time to/from the storage units and the more efficient address generation. When mapping video coding test-vehicles on an ARM processor, at the same time also significant power reductions can be obtained still [280, 292]. Preliminary results for a study of the main motion estimation routines in the MPEG4 standard [374, 373] being defined lead to similar observations.

Finally, since 1996 also work focussed on data transfer and storage exploration in a multi-processor context has become a full-fledged activity. This area will also be the focus of much of our future

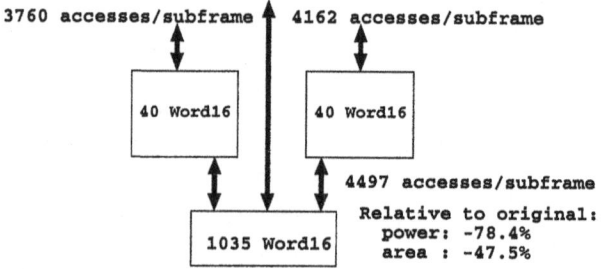

Figure 13.22.   Power optimized GSM memory architecture.

research, as addressed in the ACROPOLIS project (see summary in [1]). Our novel techniques (almost) do not interfere with traditional parallelization techniques and are in that sense complementary to them.

14 CONCLUSIONS AND FUTURE WORK

Power efficient design has become a very important issue for a broad class of applications. Many of the applications for low-cost, low-power design turn out to be data-dominated. Experiments have shown that for such data-dominated applications, a very large part of the power consumption is due the data storage and data transfer. Also the area cost is for a large part dominated by the memories. However, until recently, outside our group little research has been done on systematically optimizing the storage and transfer of data. In contrast, since the late 80's our research group is convinced that the memory architecture should be optimized as a first step in the design methodology for this type of applications. This has been formalized in the DTSE methodology which is presented for the first time with its overall trajectory in this book.

14.1 CONTRIBUTIONS

The current starting point of our DTSE methodology [53, 58, 435], partly supported with prototype tools developed in our *Atomium* research project, is a system specification with accesses on multi-dimensional (M-D) signals which can be statically ordered. The output is a netlist of memories and address generators combined with a transformed specification which is the input for the architecture (high-level) synthesis when custom realizations are envisioned, or for the software compilation stage in the case of predefined processors. The address generators are produced by a separate methodology [270, 272].

The major tasks in the methodology and the main research contributions on supporting techniques and tools are briefly discussed now (see also figure 3.1). More details about our past and current research (including Postscript files of papers) are available at our web site:

http://www.imec.be/vsdm/projects/atomium/

- The foundation of much of our work lies in the use of appropriate data representation models[1], in particular an extended polyhedral model [419]. This is extracted during a *memory oriented data flow analysis* preprocessing task at the system specification level. Originally it was developed

---

[1] All of our prototype tools operate on models which allow run-time complexities to be dependent only in a limited way on system parameters like the size of the loop iterators, as opposed to the scalar-based methods published in conventional high-level synthesis literature.

to support irregular nested loops with manifest, affine iterator bounds and index expressions. Extensions are however possible towards WHILE loops and to data-dependent and regular piecewise linear (modulo) indices [143].

■ We have classified and developed a systematic, formalized methodology for the set of system-level data-flow transformations that have the most crucial effect on the system exploration decisions [59]. During the global data-flow transformations (*data-flow trafo*) task, we apply advanced signal substitution (which especially includes moving conditional scopes), modification of computation order in associative chains, shifting of "delay lines" through the algorithm, and recomputation issues in order to remove redundant background transfers. In addition, we remove data-flow bottle-necks.

■ Next, the global loop and control-flow transformations (*loop trafo*) task aims at improving the data access locality for M-D signals and at removing the intermediate buffers introduced due to mismatches in production and consumption ordering. The main emphasis lies on loop manipulations including both affine (loop interchange, reversal and skewing) and non-affine (e.g. loop splitting and merging) cases. An interactive loop transformation engine has been developed that allows both interactive and automated (script based) steering of language-coupled source code transformations [354]. In addition, research has been performed on loop transformation steering methodologies. For power, we have developed a systematic design methodology oriented to removing the global buffers which are typically present between subsystems and on creating more data locality [103].

■ In the *data reuse decision* task, we decide on the exploitation of the memory hierarchy [118, 438]. Important considerations here are the distribution of the data (copies) over the hierarchy levels as these determine the access frequency and the size of each resulting memories. Obviously, the most frequently accessed memories should be the smallest ones. We have proposed a formalized methodology to steer this partitioning of the background transfers, which is driven by high-level estimates on bandwidth and in-place storage cost [118]. At this stage of the methodology, the transformed behavioral description can already be used as input for compilation or for more efficient simulation, or it can be further optimized in the next tasks.

■ Several additional tasks are involved to define the physical memory organisation. Given the cycle budget, the goal is to allocate memory units and ports (including their types) from a memory library and to assign the data to the best suited memory units. In a first task, we determine the bandwidth requirements and the balancing of the available cycle budget over the different memory accesses in the algorithmic specification (*storage cycle budget distribution* or SCBD) [438]. Next, the necessary number of memory units and their type are determined during *memory allocation* [32], matching the access conflicts in this balanced flow-graph. Then, the multi-dimensional signals are assigned to these background memory units and ports (*memory assignment*) [32, 376]. Again, this results in an updated flow-graph specification. A prototype tool supporting part of this functionality has been developed to illustrate the feasibility of our approach [32]. It has been extended to incorporate conflict information from the cycle budget distribution task and to deal with more powerful memory models [376].

■ Finally, the *in-place optimisation* task exploits the fact that for data with (partially) exclusive lifetimes, the memory space can be (partially) shared [402]. Based on a solid theoretical foundation a prototype tool has been developed incorporating a heuristic 2-stage approach [107]. intra-signal windowing [108], followed by inter-signal placement [106].

In addition, several other memory organization and storage management related contributions have been made by us which are not detailed in this book. They are briefly described below (with appropriate pointers):

■ *Set data type optimization:*

In the context of ADT refinement for power, we have developed a model for set data types. Using this model and a power cost function we have developed a methodology for determining an optimized set data structure without exhaustively scanning the search space [436, 439].

- *Virtual management exploration:*

  For dynamically allocated data, we have also developed a search space representation to efficiently explore different virtual memory management schemes. In addition, effective power and main memory size oriented exploration strategies are being developed [96].

- *Reduction of address bus activity:*

  A considerable amount of system power is dissipated due to data transfers over large buses both on-chip and off-chip. This power dissipation can be reduced by reducing the amount of transitions on the buses. This work has contributed to reducing the address bus activity, mainly in the context of network component applications. Even though in this type of applications most of the data is dynamically allocated and accessed via data-dependent indexing, we have shown how a substantial reduction of the address bus activity can be achieved by separating the manifest part from the data-dependent part of the addressing. The transition activity of the manifest part can then be optimized by means of an optimized mapping of data on the address space, by applying specialized address filters, and by optimally scheduling the data accesses [435].

- *Application of DTSE to many real-life driver and demonstrators:*

  Very promising results on real-life multi-media and telecom network applications have been obtained for all of the DTSE tasks. The power optimizing DTSE stage is illustrated in detail on a motion estimation application in [437, 58, 59]. This allows to follow all the steps and their effect. Many other applications have been studied, the results for most of which have been partly published, as indicated in chapter 13.

14.2 RELATION WITH MATISSE RESEARCH PROJECT

In the context of the MATISSE compiler [111, 95, 97, 416], which aims at applications with dynamically allocated data, as found e.g. in network components like ATM, the work described in this book work has contributed in defining the different stages allowing to do storage management for these highly dynamic and data-dependent applications.

This is illustrated in figure 14.1. It shows the storage management part of our design flow for RMSP applications, targeted by ATOMIUM (left), and for network components, targeted by MATISSE (right). Two distinct stages are shown: the *data type refinement* stage and the *DTSE* stage. Each stage is further subdivided into a number of steps. The areas marked in grey show which steps are covered by the ATOMIUM and MATISSE exploration environments. Some steps are only important for one application domain. Consequently, they are only supported by one of our exploration environments as indicated in the figure. The storage-bandwidth optimization for flat graphs (part of SCBD) and memory allocation plus assignment steps discussed in this book are the key steps here. They have been successfully incorporated in the back-end of the MATISSE compiler [111, 376, 97].

14.3 FUTURE WORK

It is clear from the above that much progress has been made in the past decade but many relevant research issues still need to be solved in order to obtain a full design support for the complex DTSE task. Some of the most relevant objectives are:

- further extension of the data-flow analysis and model extraction tasks.

- a further refined methodology and design tool support for system-level data-flow transformations, the global loop and reindexing transformations, and for the data-reuse decision task.

- better high-level estimators of the power and area costs in early tasks of the methodology.

- extensions towards other application domains (see e.g. [376] for network component applications).

- support for system-level design validation using formal methods to void as much as possible the very costly resimulation after the complex code transformations (see e.g. [355, 86]).

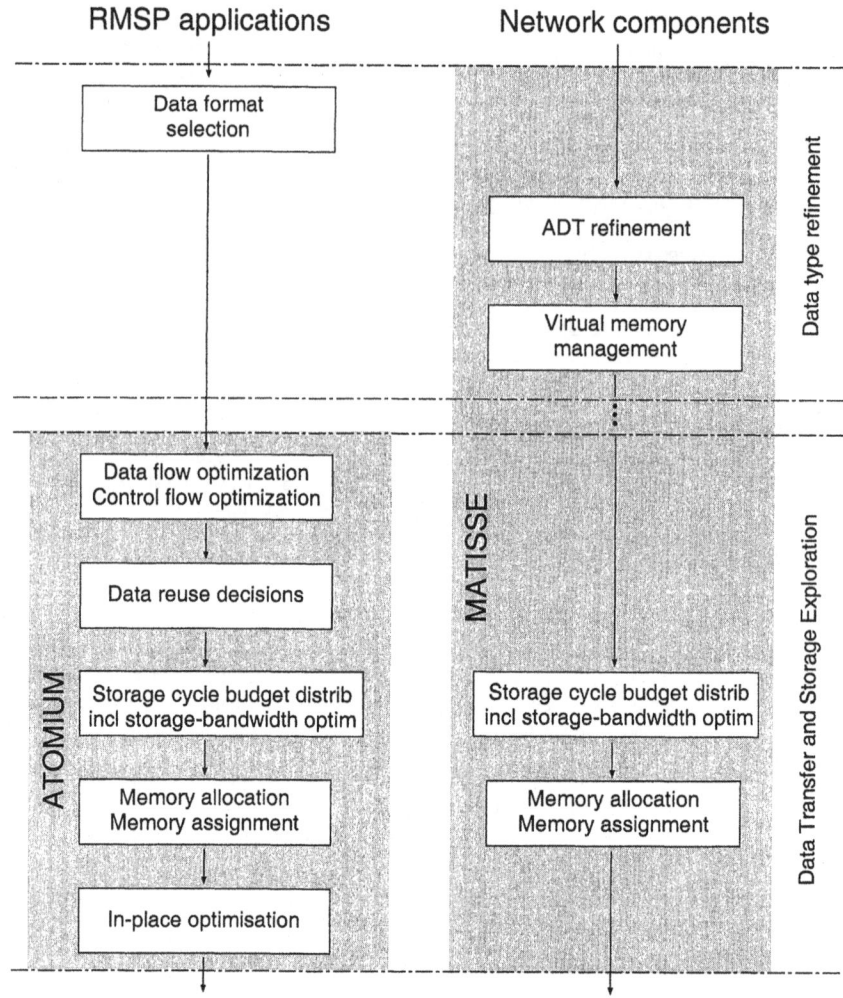

Figure 14.1.   Storage management for RMSP and network component applications.

In addition, further extensions of the concepts and methodologies are required to go beyond custom single-threaded custom hardware target architectures:

■ extension of DTSE methodology and tasks to fully accommodate predefined processor architectures, including software or hardware controlled caching (see e.g. [223]).

■ study of programmable processor architecture modifications to alleviate the power and speed inefficiencies, ranging from low design impact with reasonable saving (short-term) to heavy design impact with potentially larger savings (long-term).

- extensions to deal with even higher abstraction levels where concurrent tasks are communicating and exchanging or sharing data (see e.g. [89]).

References

[1] Multi-media compilation project (Acropolis) at IMEC
http://www.imec.be/vsdm/domains/designtechno/Welcome.html

[2] M.Adé, R.Lauwereins, J.Peperstraete, "Buffer memory requirements in DSP applications", *IEEE Proc. 5th IEEE Intnl. Wsh. on Rapid System Prototyping*, Los Alamitos, CA, pp.108-123, June 1994.

[3] M.Adé, R.Lauwereins, J.Peperstraete, "Implementing DSP applications on heterogeneous targets using minimal size data buffers", *IEEE Proc. 7th Intnl. Wsh. on Rapid System Prototyping* Porto Carras, Greece, pp.166-172, June 1996.

[4] S.Adve, D.Burger, R.Eigenmann, A.Rawsthorne, M.Smith, C.Gebotys, M.Kandemir, D.Lilja, A.Choudhary, J.Fang, P-C.Yew, "The interaction of architecture and compilation technology for high-performance processor design", *IEEE Computer Magazine*, Vol.30, No.12, pp.51-58, Dec. 1997.

[5] A.Agarwal, D.Krantz, V.Natarajan, "Automatic partitioning of parallel loops and data arrays for distributed shared-memory multiprocessors", *IEEE Trans. on Parallel and Distributed Systems*, Vol.6, No.9, pp.943-962, Sep. 1995.

[6] R.Agrawal, H.Jagadish, "Partitioning techniques for large-grained parallelism", *IEEE Trans. on Computers*, Vol.37, No.12, pp.1627-1634, Dec. 1988.

[7] I.Ahmad, C.Y.R.Chen, "Post-processor for data path synthesis using multiport memories", *Proc. IEEE Int. Conf. Comp. Aided Design*, Santa Clara CA, pp.276-279, Nov. 1991.

[8] A.V.Aho, R.Sethi, J.D.Ulmann, "Compilers: principles, techniques and tools", Addison-Wesley, 1986.

[9] R.Allen, K.Kennedy, "Vector register allocation," *IEEE Trans. on Computers*, Vol.41, No.10, pp.1290-1316, Oct. 1992.

[10] M.Al-Mouhamed, "Analysis of macro-dataflow dynamic scheduling on nonuniform memory access architectures", *IEEE Trans. on Parallel and Distributed Systems*, Vol.4, No.8, pp.875-888, Aug. 1993.

[11] M.Al-Mouhamed, S.Seiden, "A Heuristic Storage for Minimizing Access Time of Arbitrary Data Paterns", *IEEE Trans. on Parallel and Distributed Systems*, Vol. 8, No. 4, pp.441-447, Apr. 1997.

[12] E.Altman, G.Gao, "Optimal software pipelining through enumeration of schedules", *Proc. EuroPar Conference*, Lyon, France, August 1996. "Lecture notes in computer science" series, Springer Verlag, pp.833-840, 1996.

[13] S.Amarasinghe, J.Anderson, M.Lam, and C.Tseng, "The SUIF compiler for scalable parallel machines", in *Proc. of the 7th SIAM Conf. on Parallel Proc. for Scientific Computing*, 1995.

[14] B.Amrutur, M.Horowitz, "Techniques to reduce power in fast wide memories", *Symp. on Low-power Electronics*, 1994.

[15] C.Amza, A.Cox, S.Dwarkadas, P.Keleher, H.Lu, R.Rajmony, W.Yu, W.Zwaenepoel, "Treadmarks: shared memory computing on network of workstations", *IEEE Computer Magazine*, Vol.29, No.2, pp.18-28, Feb. 1996.

[16] C.Ancourt and F.Irigoin, "Scanning polyhedra with DO loops", in *3th ACM SIGPLAN Symposium on Principles and Practice of Parallel Programming*, pp.39-50, April 1991.

[17] C.Ancourt, F.Irigoin and Y.Yang, "Minimal data dependence abstractions for loop transformations", *Int. Journal of Parallel Programming*, Vol.23, No.4, pp.359-388, 1995.

[18] C.Ancourt, D.Barthou, C.Guettier, F.Irigoin, B>Jeannet, J.Jourdan, J.Mattioli, "Automatic data mapping of signal processing applications", *Proc. Intnl. Conf. on Applic.-Spec. Array Processors*, Zurich, Switzerland, pp.350-362, July 1997.

[19] J.Anderson, S.Amarasinghe, M.Lam, "Data and computation transformations for multiprocessors", in *5th ACM SIGPLAN Symposium on Principles and Practice of Parallel Programming*, pp.39-50, August 1995.

[20] A.Antola, A.Avai, L.Breveglieri, "Modular design methodologies for image processing architectures", *IEEE Trans. on VLSI Systems*, Vol.1, No.4, pp.408-414, Dec. 1993.

[21] G.Araujo, S.Malik, M.T.Lee, "Using register-transfer paths in code generation for heterogeneous memory-register architectures", *Proc. 33rd ACM/IEEE Design Automation Conf.*, Las Vegas NV, pp.597-600, June 1996.

[22] O.Arregi, C.Rodriguez, A.Ibarra, "Evaluation of the optimal strategy for managing the register file", *Microprocessing and microprogramming*, No.30, pp.143-150, 1990.

[23] D.Bailey, "RISC Microprocessors and scientific computing", Technical Report, RNR-93-004, NASA Ames Research Center, March 1993.

327

[24] D.Bairagi, S.Kumar, D.Agrawal, "Parallelizing OO programs: precise call graph in the presence of virtual functions", *IEEE TC on Computer Architecture Newsletter*, special issue on "Interaction between Compilers and Computer Architectures", pp.43-45, June 1997.

[25] R.Bajwa, M,Hiraki, H.Kojima, D.Gorny, K.Nitta, A.Shridhar, K.Seki, K.Sasaki, "Instruction buffering to reduce power in processors for signal processing", *IEEE Trans. on VLSI Systems*, Vol.5, No.4, pp.417-424, Dec. 1997.

[26] M.Balakrishnan, A.K.Majumdar, D.K.Banerji, J.G.Linders, J.C.Majithia, "Allocation of multiport memories in data path synthesis" *IEEE Trans. on Comp.-aided Design*, Vol.CAD-7, No.4, pp.536-540, April 1988.

[27] F.Balasa, F.Catthoor, H.De Man, "Exact Evaluation of Memory Area for Multi-dimensional Processing Systems", *Proc. IEEE Int. Conf. Comp. Aided Design*, Santa Clara CA, pp.669–672, Nov. 1993.

[28] F.Balasa, F.Catthoor, H.De Man, "Dataflow-driven Memory Allocation for Multi-dimensional Signal Processing Systems", *Proc. IEEE Int. Conf. Comp. Aided Design*, San Jose CA, pp.31-34 Nov. 1994.

[29] F.Balasa, F.Franssen, F.Catthoor, H.De Man, "Transformation of Nested Loops with Modulo Indexing to Affine Recurrences", in special issue of *Parallel Processing Letters* on "Parallelization techniques for uniform algorithms", C.Lengauer, P.Quinton, Y.Robert, L.Thiele (eds.), World Scientific Pub., pp.271-280, 1994.

[30] K.Balmer, N.Ing-Simmons, P.Moyse, I.Robertson, J.Keay, M.Hammes, E.Oakland, R.Simpson, G.Barr, D.Roskell, "A single chip multimedia video processor" *Proc. IEEE Custom Integrated Circuits Conf.*, San Diego CA, pp.91-94, May 1994.

[31] F.Balasa, "Background memory allocation for multi-dimensional signal processing", *Doctoral dissertation*, ESAT/EE Dept., K.U.Leuven, Belgium, Nov. 1995.

[32] F.Balasa, F.Catthoor, H.De Man, "Background Memory Area Estimation for Multi-dimensional Signal Processing Systems", *IEEE Trans. on VLSI Systems*, Vol.3, No.2, pp.157-172, June 1995.

[33] F.Balasa, F.Catthoor, H.De Man, "Practical Solutions for Counting Scalars and Dependences in ATOMIUM – a Memory Management System for Multi-dimensional Signal Processing", *IEEE Trans. on Comp.-aided Design*, Vol.CAD-16, No.2, pp.133-145, Feb. 1997.

[34] U.Banerjee, "Dependence Analysis for Supercomputing", Kluwer Academic Publishers, Boston, 1988.

[35] U.Banerjee, "Loop Transformations for Restructuring Compilers: the Foundations", Kluwer, Boston, 1993.

[36] U.Banerjee, R.Eigenmann, A.Nicolau, D.Padua, "Automatic program parallelisation", *Proc. of the IEEE*, invited paper, Vol.81, No.2, Feb. 1993.

[37] P.Banerjee, J.Chandy, M.Gupta, E.Hodges, J.Holm, A.Lain, D.Palermo, S.Ramaswamy, E.Su, *IEEE Computer Magazine*, Vol.28, No.10, pp.37-47, Oct. 1995.

[38] M.Barreteau, P.Feautrier, "Efficient mapping of interdependent scans", *Proc. EuroPar Conference*, Lyon, France, August 1996. "Lecture notes in computer science" series, Springer Verlag, pp.463-466, 1996.

[39] Bellcore, "Generic System Requirements in Support of Switched Multi-megabit Data Service", *TR-TSV-0007772*, No.1, May 1992.

[40] D.Bernstein, M.Rodeh, I.Gertner, "On the complexity of scheduling problems for parallel/pipelined machines", *IEEE Trans. on Computers*, Vol.C-38, No.9, pp.1308-1313, Sep. 1989.

[41] M.Berkelaar. "Lp_solve Mixed Integer Linear Programming solver 2.0, 1995", Available at ftp://ftp.es.ele.tue.nl/pub/lp_solve/.

[42] V.Bhaskaran, R.B.Lee, and J.P.Beck, "Algorithmic and architectural enhancements for real-time MPEG-1 decoding on a general purpose RISC workstation", *IEEE Trans. on Circuits and Systems for Video Technology*, Vol.5, No.5, pp.380–386, Oct. 1995.

[43] V.Bhaskaran and K. Konstantinides, "Image and Video Compression Standards : Algorithms and Architectures", Kluwer Academic Publishers, Boston, 1995.

[44] W.Blume, R.Eigenmann, J.Hoeflinger, D.Padua, P.Petersen, L.Rauchwerger, P.Tu, "Automatic detection of parallelism", *IEEE Parallel and Distributed Technology*, Vol.2, No.3, pp.37-47, Fall 1994.

[45] F.Bodin, W.Jalby, D.Windheiser, C.Eisenbeis, "A quantitative algorithm for data locality optimization", Technical Report IRISA/INRIA, Rennes, France, 1992.

[46] L.Bolcioni, M.Borgatti, M.Felici, R.Rambaldi, R.Guerrieri, "A 1 V 350 μW voide-controlled H.263 video decoder for portable applications", *Proc. IEEE Int. Solid-State Circ. Conf.*, San Francisco CA, pp.112-113, Feb. 1998.

[47] D.Bradlee, S.Eggers, R.Henry, "Integrated register allocation and instruction scheduling for RISCs", *Proc. 4th Int. Conf. on Architectural Support for Prog. Lang. and Operating Systems*, pp.122-131, April 1991.

[48] G.Braceras, D.Evans, J.Sousa, J.Connor, "A 350 MHz 3.3V 4Mb SRAM fabricated in a 0.3 μm CMOS process", *Proc. IEEE Int. Solid-State Circ. Conf.*, San Francisco CA, pp.404-405, Feb. 1997.

[49] E.Brockmeyer, F.Catthoor, J.Bormans, H.De Man, "Code transformations for reduced data transfer and storage in low power realisation of MPEG-4 full-pel motion estimation", accepted for *Proc. IEEE Int. Conf. on Image Proc.*, Chicago, IL., Oct. 1998.

[50] J.Bu, E.Deprettere, P.Dewilde, "A design methodology for fixed-size systolic arrays", in *Application-specifc array processors* (S.Y.Kung, E.Swartzlander, J.Fortes eds.), IEEE computer society Press, pp.591-602, 1990.

[51] T.Burd and R.Brodersen, "Processor Design for Portable Systems," *Journal of VLSI Signal Processing*, No.13, Kluwer, Boston, pp.203-211, 1996.

[52] G.Cabillic, I.Puaut, "Dealing with heterogeneity in Stardust: an environment for parallel programming on networks of heterogeneous workstations", *Proc. EuroPar Conference*, Lyon, France, August 1996. "Lecture notes in computer science" series, Springer Verlag, pp.114-119, 1996.

[53] F.Catthoor, F.Franssen, S.Wuytack, L.Nachtergaele, H.De Man, "Global communication and memory optimizing transformations for low power signal processing systems", *IEEE workshop on VLSI signal processing*, La Jolla CA, Oct. 1994. Also in *VLSI Signal Processing VII*, J.Rabaey, P.Chau, J.Eldon (eds.), IEEE Press, New York, pp.178-187, 1994.

[54] F.Catthoor, W.Geurts, H.De Man, "Loop transformation methodology for fixed-rate video, image and telecom processing applications", *Proc. Intnl. Conf. on Applic.-Spec. Array Processors*, San Francisco, CA, pp.427-438, Aug. 1994.

[55] F.Catthoor, M.Moonen (eds.), "Parallel Programmable Architectures and Compilation for Multi-dimensional Processing", special issue in *Microprocessors and Microprogramming*, Elsevier, Amsterdam, pp.333-439, Oct. 1995.

[56] F.Catthoor, M.Janssen, L.Nachtergaele, H.De Man, "System-level data-flow transformations for power reduction in image and video processing", *Proc. Intnl. Conf. on Electronic Circuits and Systems*, Greece, pp.1025-1028, Oct. 1996.

[57] F.Catthoor, K.Danckaert, "Summary of system-level data transfer and storage methodology and critical tasks in multi-media multi-processor context", IMEC, Internal Report, Dec. 1996.

[58] F.Catthoor, S.Wuytack, E.De Greef, F.Franssen, L.Nachtergaele. H.De Man, "System-level transformations for low power data transfer and storage", in paper collection on "Low power CMOS design" (eds. A.Chandrakasan, R.Brodersen), IEEE Press, pp.609-618, 1998.

[59] F.Catthoor, S.Wuytack, E.De Greef, F.Balasa, P.Slock, "System exploration for custom low power data storage and transfer", chapter in "Digital Signal Processing for Multimedia Systems" (eds. K.Parhi, T.Nishitani), Marcel Dekker, Inc., New York, 1998.

[60] F.Catthoor, "Energy-delay efficient data storage and transfer architectures: circuit technology versus design methodology solutions", invited paper, *Proc. 1st ACM/IEEE Design and Test in Europe Conf.*, Paris, France, pp.709-714, Feb. 1998.

[61] F.Catthoor, M.Janssen, L.Nachtergaele, H.De Man, "System-level data-flow transformation exploration and power-area trade-offs demonstrated on video codecs", special issue on "Systematic trade-off analysis in signal processing systems design" (eds. M.Ibrahim, W.Wolf) in *Journal of VLSI Signal Processing*, Vol.18, No.1, Kluwer, Boston, pp.39-50, 1998.

[62] G.Chaitin, M.Auslander, A.Chandra, J.Cocke, M.Hopkins, P.Markstein, "Register allocation and spilling via graph coloring", *Computer languages*, No.6, pp.47-57, 1981.

[63] A.Chandrakasan, S.Scheng, R.W.Brodersen, "Low power CMOS digital design", *IEEE J. of Solid-state Circ.*, Vol.SC-27, No.4, pp.473-483, April 1992.

[64] A.Chandrakasan, R.W.Brodersen, "Low power CMOS digital design", Kluwer Academic Publishers, 1995.

[65] J-M.Chang, M.Pedram, "Energy minimization using multiple supply voltages", *IEEE Trans. on VLSI Systems*, Vol.5, No.4, pp.436-443, Dec. 1997.

[66] A.Chandrakasan, M.Potkonjak, R.Mehra, J.Rabaey, R.W.Brodersen, "Optimizing power using transformations". *IEEE Trans. on Comp.-aided Design*, Vol.CAD-14, No.1, pp.12-30, Jan. 1995.

[67] B.Chapman, H.Zima, P.Mehrotra, "Extending HPF for advanced data-parallel applications", *IEEE Parallel and Distributed Technology*, Vol.2, No.3, pp.59-71, Fall 1994.

[68] Y-Y.Chen, Y-C.Hsu, C-T.King, "MULTIPAR: behavioral partition for synthesizing multiprocessor architectures", *IEEE Trans. on VLSI Systems*, Vol.2, No.1, pp.21–32, March 1994.

[69] T-S.Chen, J-P.Sheu, "Communication-free data allocation techniques for parallelizing compilers on multicomputers", *IEEE Trans. on Parallel and Distributed Systems*, Vol.5, No.9, pp.924-938, Sep. 1994.

[70] W.Chin, J.Darlington, Y.Guo, "Parallelizing conditional recurrences", *Proc. EuroPar Conference*, Lyon, France, August 1996. "Lecture notes in computer science" series, Springer Verlag, pp.579-586, 1996.

[71] A.Choi, M.Ruschitzka, "Managing locality sets: the model and fixed-size buffers", *IEEE Trans. on Computers*, Vol.42, No.2, pp.190-204, Feb. 1993.

[72] L.Choi, P-C.Yew, "A compiler-directed cache coherence scheme with improved intertask locality", *Proc. Supercomputing*, Washington DC, Nov. 1994.

[73] L.O.Chua, A.-C.Deng, "Canonical piecewise-linear representation," *IEEE Trans. on Circuits and Systems*, Vol.35, No.1, pp.101-111, Jan. 1988.

[74] M.Cierniak, W.Li, "Unifying Data and Control Transformations for Distributed Shared-Memory Machines", *Proc. of the SIGPLAN'95 Conf. on Programming Language Design and Implementation*, La Jolla, pp.205-217, Feb. 1995.

[75] L.Claesen, F.Catthoor, H.De Man, J.Vandewalle, S.Note, K.Mertens, "A CAD Environment for the thorough Analysis, Simulation and Characterisation of VLSI implementable DSP Systems", *Proc. IEEE Int. Conf. on Computer Design*, Port Chester NY, pp.72-75, Oct. 1986.

[76] J.F.Collard, "Space-time transformations of while-loops using speculative execution", *Proc. IEEE Scalable High-Performance Computing Conf.*, SHPCC'94, Knoxville TN, pp.429-436, May 1994.

[77] J.F.Collard, D.Barthou, P.Feautrier, "Fuzzy data flow analysis", *Proc. ACM Principles and Practice of Parallel Programming, PPoPP'95*, Santa Barbara CA, July 1995.

[78] J.F.Collard, M.Griebl, "Array dataflow analysis for explicitly parallel programs", *Proc. EuroPar Conference*, Lyon, France, August 1996. "Lecture notes in computer science" series, Springer Verlag, pp.406-413, 1996.

[79] R.Comerford, G.Watson (eds.), "Memory catches up", *IEEE Spectrum*, pp.34-57, Oct. 1992.

[80] T.Conte and C.Gimarc (eds.), "Fast simulation of computer architectures", Kluwer Academic Publishers, Boston, 1995.

[81] W.Cook, M.Hartmann, R.Kannan, C.McDiarmid, "On Integer Points in Polyhedra", *Combinatorica*, 1989.

[82] C.Cook, C.Pancake, R.Walpole, 'Are expectations for parallelism too high? A survey of potential parallel users", *Proc. Supercomputing*, Washington DC, Nov. 1994.

[83] B.Creusillet, F.Irigoin, "Interprocedural array region analysis", Technical report CRI, Ecole des Mines de Paris, 1995.

[84] B.Creusillet, "Array region analysis and applications", *Doctoral dissertation*, Ecole des Mines de Paris, 1997.

[85] R.E.Crochiere and L.R.Rabiner, "Interpolation and decimation of Digital signals - A tutorial overview", *Proc. of the IEEE*, Vol.69, No.3, pp.300-331, March 1981.

[86] M.Cupak, F.Catthoor, "Efficient functional validation of system-level loop transformations for multi-media applications", *Proc. Electronic Circuits and Systems Conference*, Bratislava, Slovakia, pp.39-43, Sep. 1997.

[87] R.Cypher, J.Sanz, "SIMD architectures and algorithms for image processing and computer vision", *IEEE Trans. on Acoustics, Speech and Signal Processing*, Vol.ASSP-37, No.12, pp.2158-2174, Dec. 1989.

[88] G.B.Dantzig, B.C.Eaves, "Fourier-Motzkin Elimination and Its Dual", *J. of Combinatorial Theory (A)*, Vol.14, pp.288-297, 1973.

[89] K.Danckaert, F.Catthoor, H.De Man, "System-level memory management for weakly parallel image processing", *Proc. EuroPar Conference*, Lyon, France, August 1996. "Lecture notes in computer science" series, Vol.1124, Springer Verlag, pp.217-225, 1996.

[90] K.Danckaert, F.Catthoor, H.De Man, "System level memory optimization for hardware-software co-design", *Proc. IEEE Intnl. Workshop on Hardware/Software Co-design*, Braunschweig, Germany, pp.55-59, March 1997.

[91] K.Danckaert, K.Masselos, F.Catthoor, H.De Man, "Strategy for power efficient combined task and data parallelism exploration illustrated on a QSDPCM video codec", accepted for special issue on "Parallel Image Processing" in *J. of Systems Architecture*, Elsevier, Amsterdam, 1998.

[92] K.Danckaert, K.Masselos, F.Catthoor, H.De Man, C.Goutis, "Strategy for power efficient design of parallel systems", (almost) accepted for *IEEE Trans. on VLSI Systems*, 1999.

[93] A.Darte, Y.Robert, "Mapping uniform loop nests onto distributed memory architectures", Internal report 93-03, ENSL/IMAG, Jan. 1993.

[94] A.Darte, T.Risset, Y.Robert, "Loop nest scheduling and transformations", in *Environments and Tools for Parallel Scientific Computing*, J.J.Dongarra et al. (eds.), Advances in Parallel Computing 6, North Holland, Amsterdam, pp.309-332, 1993.

[95] J.L.da Silva Jr, C.Ykman-Couvreur, G.de Jong, B.Lin, H.De Man, "A system design methodology for telecom network applications", Urbana-Champaig, March 1997.

[96] J.L.da Silva Jr, F.Catthoor, D.Verkest, H.De Man, "Power Exploration for Dynamic Data Types through Virtual Memory Management Refinement", accepted for *Proc. IEEE Intnl. Symp. on Low Power Design*, Monterey CA, Aug. 1998.

[97] J.L.da Silva Jr, C.Ykman-Couvreur, M.Miranda, K.Croes, S.Wuytack, G.de Jong, F.Catthoor, D.Verkest, P.Six, H.De Man, "Efficient System Exploration and Synthesis of Applications with Dynamic Data Storage and Intensive Data Transfer", accepted for *Proc. 35th ACM/IEEE Design Automation Conf.*, San Francisco CA, June 1998.

[98] R.Davoli, L-A.Giachini, O.Baboglu, A.Amoroso, L.Alvisi, "Parallel computing in networks of workstations with Paralex", *IEEE Trans. on Parallel and Distributed Systems*, Vol.7, No.4, pp.371-384, April 1996.

[99] E.de Angel, E.Swartzlander, "Survey of low power techniques for ROMs", *Proc. IEEE Intnl. Symp. on Low Power Design*, Monterey CA, pp.7-11, Aug. 1997.

[100] L.De Coster, M.Engels, R.Lauwereins, J.Peperstraete, "Global approach for compiled bit-true simulation of DSP systems", *Proc. EuroPar Conference*, Lyon, France, August 1996. "Lecture notes in computer science" series, Springer Verlag, pp.236-239, 1996.

[101] E.De Greef, "Pruning in a High-level Memory Management Context" IMEC, Internal Report, Dec. 1993.

[102] E.De Greef, F.Catthoor, H.De Man, "A memory-efficient, programmable multi-processor architecture for real-time motion estimation type algorithms", *Intnl. Workshop on Algorithms and Parallel VLSI Architectures*, Leuven, Belgium, August 1994. Also in "Algorithms and Parallel VLSI Architectures III" (eds. M.Moonen, F.Catthoor), Elsevier, pp.191-202, 1995.

[103] E.De Greef, F.Catthoor, H.De Man, "Memory organization for video algorithms on programmable signal processors", *Proc. IEEE Int. Conf. on Computer Design*, Austin TX, pp.552-557, Oct. 1995.

[104] E.De Greef, F.Catthoor, H.De Man, "Mapping real-time motion estimation type algorithms to memory-efficient, programmable multiprocessor architectures", *Microprocessors and Microprogramming*, special issue on "Parallel Programmable Architectures and Compilation for Multi-dimensional Processing" (eds. F.Catthoor, M.Moonen), Elsevier, pp.409-423, Oct. 1995.

[105] E.De Greef, F.Catthoor, H.De Man, "Reducing storage size for static control programs mapped onto parallel architectures", presented at *Dagstuhl Seminar on Loop Parallelisation*, Schloss Dagstuhl, Germany, April 1996.

[106] E.De Greef, F.Catthoor, H.De Man, "Array Placement for Storage Size Reduction in Embedded Multimedia Systems", *Proc. Intnl. Conf. on Applic.-Spec. Array Processors*, Zurich, Switzerland, pp.66-75, July 1997.

[107] E.De Greef, F.Catthoor, H.De Man, "Memory Size Reduction through Storage Order Optimization for Embedded Parallel Multimedia Applications", special issue on "Parallel Processing and Multi-media" (ed. A.Krikelis), in *Parallel Computing* Elsevier, Vol.23, No.12, Dec. 1997.

[108] E.De Greef, F.Catthoor, H.De Man, "Memory Size Reduction through Storage Order Optimization for Embedded Parallel Multimedia Applications", *Intnl. Parallel Proc. Symp.(IPPS)* in Proc. Workshop on "Parallel Processing and Multimedia", Geneva, Switzerland, pp.84-98, April 1997.

[109] E.De Greef, "Storage size reduction for multimedia applications", *Doctoral dissertation*, ESAT/EE Dept., K.U.Leuven, Belgium, Jan. 1998.

[110] E.De Greef, F.Catthoor, H.De Man, "Program transformation strategies for reduced power and memory size in pseudo-regular multimedia applications", accepted for publication in *IEEE Trans. on Circuits and Systems for Video Technology*, 1998.

[111] G.de Jong, B.Lin, C.Verdonck, S.Wuytack, F.Catthoor, "Background memory management for dynamic data structure intensive processing systems", *Proc. IEEE Int. Conf. Comp. Aided Design*, San Jose CA, pp.515-520, Nov. 1995.

[112] H.De Man, I.Bolsens, B.Lin, K.Van Rompaey, S.Vercauteren, D.Verkest, "Co-design of DSP systems", NATO Advanced Study Institute on "Hardware/Software Co-design", Tremezzo, Italy, June 1995.

[113] T.Denk, K.Parhi, "Lower bounds on memory requirements for statically scheduled DSP programs", *Journal of VLSI Signal Processing*, No.12, Kluwer, Boston, pp.247-263, 1996.

[114] M.De Prycker, "Asynchronous Transfer Mode, solution for Broadband ISDN", Ellis Horwood, 1991.

[115] C.Dezan, H.Le Verge, P.Quinton, and Y.Saouter, "The Alpha du CENTAUR experiment", in *Algorithms and parallel VLSI architectures II*, P.Quinton and Y.Robert (eds.), Elsevier, Amsterdam, pp.325-334, 1992.

[116] -, "DFL user manual", Frontier Design, Abdijstraat 34, 3000 Leuven.

[117] C.Diderich, M.Gengler, "Solving the constant-degree parallelism alignment problem", *Proc. EuroPar Conference*, Lyon, France, August 1996. "Lecture notes in computer science" series, Springer Verlag, pp.451-454, 1996.

[118] J.P.Diguet, S.Wuytack, F.Catthoor, H.De Man, "Formalized methodology for data reuse exploration in hierarchical memory mappings", *Proc. IEEE Intnl. Symp. on Low Power Design*, Monterey CA, pp.30-35, Aug. 1997.

[119] M.Dion, Y.Robert, "Mapping affine loop nests: new results", *Intnl. Conf. on High-performance computing and networking*, Milan, Italy, pp.184-189, May 1995.

[120] M.Dion, T.Risset, Y.Robert, "Resource-constrained scheduling of partitioned algorithms on processor arrays", *Integration, the VLSI journal*, Elsevier, Amsterdam, No.20, pp.139-159, 1996.

[121] M.Dubois, J-C.Wang, "Analytical modeling of data sharing in cache based multiprocessors", Technical report CENG 89-18, U.S.C, June 1989.

[122] R.J. Duffin, "On Fourier's analysis of linear inequality systems," *Mathematical Programming Study*, Vol.1, pp.71-95, 1974.

[123] M.Dyer, "On Counting Lattice Points in Polyhedra", *SIAM J. on Computing*, Vol.20, No.4, pp.695-707, Aug. 1991.

[124] U.Eckhardt, R.Merker, "Scheduling in co-partitioned array architectures", *Proc. Intnl. Conf. on Applic.-Spec. Array Processors*, Zurich, Switzerland, pp.219-228, July 1997.

[125] C.Eisenbeis, W.Jalby, D.Windheiser, F.Bodin, "A Strategy for Array Management in Local Memory", *Proc. of the 4th Workshop on Languages and Compilers for Parallel Computing*, Aug. 1991.

[126] C.Eisenbeis, O.Temam, H.Wijshoff, "Fast enumeration of solutions for data dependence analysis and data locality optimization," *Int. Conf. on Parallel Processing (ICPP'93)*, Aug. 1993.

[127] H.El-Rewini, H.Ali, T.Lewis, "Task scheduling in multiprocessing systems", *IEEE Computer Magazine*, Vol.28, No.12, pp.27-37, Dec. 1995.

[128] R.Evans, "Energy consumption modeling and optimization for SRAMs", *Doctoral Dissertation*, North Carolina State Univ., June 1993.

[129] R.Evans, P.Franzon, "Energy consumption modeling and optimization for SRAMs", *IEEE J. of Solid-state Circ.*, Vol.SC-30, No.5, pp.571-579, May 1995.

[130] J.Z.Fang, M.Lu, "An iteration partition approach for cache or local memory thrashing on parallel processing", *IEEE Trans. on Computers*, Vol.C-42, No.5, pp.529-546, May 1993.

[131] A.Faruque, D.Fong, "Performance analysis through memory of a proposed parallel architecture for the efficient use of memory in image processing applications", *Proc. SPIE'91, Visual communications and image processing*, Boston MA, pp.865-877, Oct. 1991.

[132] P.Feautrier. "Parametric integer programming", *Operations research*, Vol.22, No.3, pp.243–268, 1988.

[133] P.Feautrier. "Array expansion", *ACM Intnl. Conf. on Supercomputing*, pp.429–441, St.Malo, France, 1988.

[134] P.Feautrier, "Dataflow analysis of array and scalar references", *Int. J. of Parallel Programming*, Vol.20, No.1, pp.23-53, 1991.

[135] P.Feautrier, "Some efficient solutions to the affine scheduling problems", *Int. J. of Parallel Programming*, Vol.21, No.5, pp.389-420, 1992.

[136] P.Feautrier, "Toward automatic partitioning of arrays for distributed memory computers", *Proc. 7th ACM Int. Conf. on Supercomputers*, pp.175-184, Tokyo, Japan, 1993.

[137] P.Feautrier, "Compiling for massively parallel architectures: a perspective", *Intnl. Workshop on Algorithms and Parallel VLSI Architectures*, Leuven, Belgium, August 1994. Also in "Algorithms and Parallel VLSI Architectures III" (eds. M.Moonen, F.Catthoor), Elsevier, pp.259-270, 1995.

[138] A.Fernandez, J.Llaberia, J.Navarro, M.Valero-Garcia, "Transformation of systolic algorithms for interleaving partitions", *Proc. Intnl. Conf. on Applic.-Spec. Array Processors*, Barcelona, Spain, pp.56-70, Sep. 1991.

[139] G.Fettweiss, L.Thiele, "Algebraic recurrence transformations for massive parallelism", Also in *VLSI Signal Processing V*, K.Yao, R.Jain, W.Przytula (eds.), IEEE Press, New York, pp.332-341, 1992.

[140] S.Fitzgerald, R.Oldehoeft, "Update-in-place analysis for true multidimensional arrays", *Scientific Programming*, J.Wiley & Sons Inc, Vol.5, pp.147-160, 1996.

[141] I.Foster, "High-performance distributed computing: the I-WAY experiment and beyond", *Proc. EuroPar Conference*, Lyon, France, August 1996. "Lecture notes in computer science" series, Springer Verlag, pp.3-10, 1996.

[142] F.Franssen, M.van Swaaij, F.Catthoor, H.De Man, "Modelling piece-wise linear and data dependent signal indexing for multi-dimensional signal processing", *Proc. 6th Int. Workshop on High-Level Synthesis*, Laguna Beach CA, Nov. 1992.

[143] F.Franssen, F.Balasa, M.van Swaaij, F.Catthoor, H.De Man, "Modeling Multi-Dimensional Data and Control flow", *IEEE Trans. on VLSI systems*, Vol.1, No.3, pp.319-327, Sep. 1993.

[144] F.Franssen, L.Nachtergaele, H.Samsom, F.Catthoor, H.De Man, "Control flow optimization for fast system simulation and storage minimization", *Proc. 5th ACM/IEEE Europ. Design and Test Conf.*, Paris, France, pp.20-24, Feb. 1994.

[145] M.A.Frumkin, "Polynomial time algorithms in the theory of linear Diophantine equations," *Fundamentals of Computation Theory*, Lecture Notes in Comp. Sc., Vol.56, Springer-Verlag, pp.386-392, 1977.

[146] T.Fujii, N.Ohta, "A load balancing technique for video signal processing on a multicomputer type DSP", *Proc. IEEE Int. Conf. on Acoustics, Speech and Signal Processing*, New York, pp.1981-1984, April 1988.

[147] B.Furht, "Parallel computing: glory and collapse", *IEEE Computer Magazine*, Vol.27, No.11, pp.74-75, Nov. 1994.

[148] D.Gajski, N.Dutt, A.Wu, "High-level synthesis: introduction to chip and system design", Kluwer Academic Publishers, Boston, 1992.

[149] D.Gannon, W.Jalby, K.Gallivan, "Strategies for cache and local memory management by global program transformations", *Journal of Parallel and Distributed Computing*, Vol.5, pp.568-586, 1988.

[150] G.Gao, C.Eisenbeis, J.Wang, "Introduction to Instruction Level Parallelism workshop", *Proc. EuroPar Conference*, Lyon, France, August 1996. "Lecture notes in computer science" series, Springer Verlag, pp.745-746, 1996.

[151] M.R.Garey, D.S.Johnson, "Computers and Intractability: a Guide to the Theory of NP-completeness", W.H.Freeman and Co., New York, 1979.

[152] C.H.Gebotys, "Low energy memory component design for cost-sensitive high-performance embedded systems", *Proc. IEEE Custom Integrated Circuits Conf.*, San Diego CA, pp.397-400, May 1996.

[153] J.Gee, A.Smith, "'Analysis of Multiprocessor Memory Reference Behavior", *Proc. IEEE Int. Conf. on Computer Design*, Port Chester NY, pp.53-59, Oct. 1994.

[154] R.Gerber, S.Hong, M.Saksena, "Guaranteeing real-time requirements with resource-based calibration of periodic processes", *IEEE Trans. on Software Engineering*, Vol.SE-21, No.7, pp.579-592, July 1995.

[155] W.Geurts, F.Catthoor, H.De Man, "Heuristic techniques for the synthesis of complex functional units", *Proc. 4th ACM/IEEE Europ. Design Automation Conf.*, Paris, France, pp.552-556, Feb. 1993.

[156] W.Geurts, "Synthesis of accelerator data-paths for high-throughput signal processing applications", *Doctoral dissertation*, ESAT/EE Dept., K.U.Leuven, Belgium, March 1995.

[157] W.Geurts, F.Catthoor, H.De Man, "Quadratic Zero-one Programming-based Synthesis of Application-Specific Data Paths", *IEEE Trans. on Comp.-aided Design*, Vol.CAD-14, No.1, pp.1-11, Jan. 1995.

[158] W.Geurts, F.Catthoor, S.Vernalde, H.De Man, "Accelerator data-paths synthesis for high-throughput signal processing applications", Kluwer Academic Publishers, Boston, 1996.

[159] K.Gharachorloo, A.Gupta, J.Hennessy, "Performance evaluation of memory consistency models for shared-memory multiprocessors", *Fourth Intnl. Conf. on Arch. Support for Progr. Lang. and Oper. Systems*, pp.245-257, April 1991.

[160] S.Ghosh, M.Martonosi, S.Malik, "Cache miss equations: an analytical representation of cache misses", *IEEE TC on Computer Architecture Newsletter*, special issue on "Interaction between Compilers and Computer Architectures", pp.52-54, June 1997.

[161] T.Gijbels, F.Catthoor, L.Van Eycken, A.Oosterlinck, H.De Man, "An application-specific architecture for the RBN-coder with efficient memory organization", *Journal of VLSI signal processing*, special issue on "Video/image signal processing", T.Nishitani, P.Ang, F.Catthoor (eds.), Vol.5, No.2-3, Kluwer, Boston, pp.221-236, April 1993.

[162] T.Gijbels, P.Six, L.Van Gool, F.Catthoor, H.De Man, A.Oosterlinck, "A VLSI architecture for parallel non-linear diffusion with applications in vision", *IEEE workshop on VLSI signal processing*, La Jolla CA, Oct. 1994. Also in *VLSI Signal Processing VII*, J.Rabaey, P.Chau, J.Eldon (eds.), IEEE Press, New York, pp.398-407, 1994.

[163] G.Golub, C.Van Loan, "Matrix Computations, Second Edition", The John Hopkins Univ. Press, Baltimore, 1989.

[164] R.Gonzales, M.Horowitz, "Energy dissipation in general-purpose microprocessors", *IEEE J. of Solid-state Circ.*, Vol.SC-31, No.9, pp.1277-1283, Sep. 1996.

[165] G.Goossens, J.Rabaey, J.Vandewalle, H.De Man, "An efficient microcode-compiler for custom DSP-processors", *Proc. IEEE Int. Conf. Comp. Aided Design*, Santa Clara CA, pp.24-27, Nov. 1987.

[166] J.Goodman, W.Hsu, "Code scheduling and register allocation in large basic blocks", *Proc. 2th ACM Int. Conf. on Supercomputers*, pp.442-452, 1988.

[167] G.Goossens, "Optimisation Techniques for Automated Synthesis of Application-Specific Signal-Processing Architectures," *Doctoral dissertation*, ESAT/EE Dept., K.U.Leuven, Belgium, June 1989.

[168] G.Goossens, J.Rabaey, J.Vandewalle, H.De Man, "An efficient microcode compiler for application-specific DSP processors", *IEEE Trans. on Comp.-aided Design*, Vol.9, No.9, pp.925-937, Sep. 1990.

[169] G.Goossens, D.Lanneer, M.Pauwels, F.Depuydt, F.Catthoor, H.De Man, "Synthesis of flexible IC architectures for medium throughput real-time signal processing", *Journal of VLSI signal processing*, special issue on "Synthesis for real-time digital signal processing", Vol.5, No.4, Kluwer, Boston, 1993.

[170] G.Goossens, I.Bolsens, B.Lin, F.Catthoor, "Design of heterogeneous ICs for mobile and personal communication systems", *Proc. IEEE Int. Conf. Comp. Aided Design*, San Jose CA, pp.524-531, Nov. 1994.

[171] G.Goossens, D.Lanneer, M.Pauwels, F.Depuydt, K.Schoofs, A.Kifli, M.Cornero, P.Petroni, F.Catthoor, H.De Man, "Integration of medium-throughput signal processing algorithms on flexible instruction-set architectures", *Journal of VLSI signal processing*, special issue on "Design environments for DSP" (eds.I.Verbauwhede, J.Rabaey), No.9, Kluwer, Boston, pp.49-65, Jan. 1995.

[172] M.Griebl, C.Lengauer, "On the space-time mapping of WHILE-loops", in special issue of *Parallel Processing Letters* on "Parallelization techniques for uniform algorithms", C.Lengauer, P.Quinton, Y.Robert, L.Thiele (eds.), World Scientific Pub., pp.221-232, 1994.

[173] C.Guerra, R.Melhem, "Synthesizing non-uniform systolic designs", *Proc. of Intnl. Conf. on Parallel Processing*, pp.765-771, 1986.

[174] N.Guil, E.Zapata, "A parallel pipelined Hough transform", *Proc. EuroPar Conference*, Lyon, France, August 1996. "Lecture notes in computer science" series, Springer Verlag, pp.131-138, 1996.

[175] P.Gupta, A.Parker, "SMASH: a program for scheduling memory-intensive application-specific hardware", *Proc. 7th ACM/IEEE Intnl. Symp. on High-Level Synthesis*, Niagara-on-the-Lake, Canada, May 1994.

[176] V.Gutnik, A.Chandrakasan, "Embedded power supply for low-power DSP", *IEEE Trans. on VLSI Systems*, Vol.5, No.4, pp.425-435, Dec. 1997.

[177] ITU-H.263, "Video coding for narrow telecommunications channels at less than 64 kbits/s", *http://www.nta.no/brukere/DVC/h263_wht/* .

[178] Digital Video Coding at Telenor R & D, "Telenor's H.263 Software", Version 3.1, *http://www.nta.no/brukere/DVC/h263_software/* .

[179] J.L.Hafner, K.S.McCurley, "Asymptotically fast triangularization of matrices over rings," *SIAM J. on Comput.*, Vol.20, pp.1068-1083, Dec.1991.

[180] T.Halfhill, J.Montgomery, "Chip fashion: multi-media chips", *Byte Magazine*, pp.171-178, Nov.1995.

[181] M.Hall, J.Anderson, S.Amarasinghe, B.Murphy, S.Liao, E.Bugnion, M.Lam, "Maximizing multiprocessor performance with the SUIF compiler", *IEEE Computer Magazine*, Vol.30, No.12, pp.84-89, Dec. 1996.

[182] W.Hardt, W.Rosenstiel, "Prototyping of tightly coupled hardware/software systems", *Design Automation for Embedded Systems*, Vol.2, Kluwer Academic Publishers, Boston, pp.283-317, 1997.

[183] A.Hashimoto, J.Stevens, "Wire routing by optimizing channel assignment within large apertures," *Proc. 8th Design Automation Wsh.*, pp.155-169, 1971.

[184] L.Haynes (ed.), "Highly parallel computing", *IEEE Computer Magazine*, Vol.15, No.1, pp.7-104, Jan. 1982.

[185] N.Hedenstierna, K.Jeppson, "CMOS Circuit Speed and Buffer Optimization", *IEEE Trans. on Computer-Aided Design*, Vol.CAD-6, No.3., pp.270-281, March 1987.

[186] S.Heemstra de Groot, O.Herrman, "Evaluation of some multiprocessor scheduling techniques of atomic operations for recursive DSP filters", *Proc. Europ. Conf. on Circ. Theory and Design*, ECCTD, Brighton, U.K., pp.400-404, Sep. 1989.

[187] P.N.Hilfinger, J.Rabaey, D.Genin, C.Scheers, H.De Man, "DSP specification using the Silage language", *Proc. Int. Conf. on Acoustics, Speech and Signal Processing*, Albuquerque, NM, pp.1057-1060, April 1990.

[188] P.Hoang, J.Rabaey, "Program partitioning for a reconfigurable multiprocessor system", presented at *IEEE workshop on VLSI signal processing*, San Diego CA, Nov. 1990.

[189] P.Hoang, J.Rabaey, "Scheduling of DSP programs onto multiprocessors for maximum throughput", *IEEE Trans. on Signal Processing*, Vol.SP-41, No.6, pp.2225-2235, June 1993.

[190] C-H.Huang, P.Sadayappan, "Communication-free hyperplane partitioning of nested loops", in "Languages and compilers for parallel computing" (U.Banerjee et al., eds.), 1991.

[191] C.T.Hwang, J-H.Lee, Y-C.Hsu, "A formal approach to the scheduling problem in high-level synthesis", *IEEE Trans. on Comp. aided Design*, Vol.10, No.4, pp.464-475, April 1991.

[192] Y-T.Hwang, Y-H.Hu, "A unified partitioning and scheduling scheme for mapping multi-stagwe regular iterative algorithms onto processor arrays", *Journal of VLSI Signal Processing*, No.11, Kluwer, Boston, pp.133-150, 1995.

[193] K.Hwang, Z.Xu, "Scalable parallel computers for real-time signal processing", *IEEE Signal Processing Magazine*, No.4, pp.50-66, July 1996.

[194] H.Igura, S.Narita, Y.Naito, K.Kazama, I.Kuroda, M.Motomura, M.Yamashina, "An 800 MOPS 110 mW 1.5V parallel DSP for mobile multi-media processing", *Proc. IEEE Int. Solid-State Circ. Conf.*, San Francisco CA, pp.292-293, Feb. 1998.

[195] K.Itoh, K.Sasaki, Y.Nakagome, "Trends in low-power RAM circuit technologies", special issue on "Low power electronics" of the *Proceedings of the IEEE*, Vol.83, No.4, pp.524-543, April 1995.

[196] K.Itoh, "Low Voltage Memory Design", in tutorial on "Low voltage technologies and circuits", *IEEE Intnl. Symp. on Low Power Design*, Monterey CA, Aug. 1997.

[197] E.Iwata et al, "A 2.2 GOPS video DSP with 2-RISC MIMD, 6-PE SIMD architecture for real-time MPEG2 video coding/decoding", *Proc. IEEE Int. Solid-State Circ. Conf.*, San Francisco CA, pp.258-259, Feb. 1997.

[198] B.Jacob, P.chen, S.Silverman, T.Mudge, "An analytical model for designing memory hierarchies", *IEEE Trans. on Computers*, Vol.C-45, No.10, pp.1180-1193, Oct. 1996.

[199] J.JaJa, K.W.Ryu, "The block distributed memory model", *IEEE Trans. on Parallel and Distributed Systems*, Vol.7, No.8, pp.830-840, Aug. 1996.

[200] J.M.Janssen, F.Catthoor, H.De Man, "Memory management aspects in the architecture design of a QSDPCM coder for video-phone", *Proc. Picture Coding Symposium*, Lausanne, Switzerland, March 1993.

[201] J.M.Janssen, F.Catthoor, H.De Man, "A Specification Invariant Technique for Operation Cost Minimisation in Flow-graphs". *Proc. 7th ACM/IEEE Intnl. Symp. on High-Level Synthesis*, Niagara-on-the-Lake, Canada, pp.146-151, May 1994.

[202] J.M.Janssen, F.Catthoor, H.De Man, "A Specification Invariant Technique for Regularity Improvement between Flow-Graph Clusters", *Proc. European Design Automation Conf.*, Paris, France, pp.138-143, Feb. 1996.

[203] A.Jantsch, P.Ellervee, J.Oberg, A.Hemani, H.Tenhunen, "Hardware/software partitioning and minimizing memory interface traffic", *Proc. ACM EuroDAC'94*, pp.226-231, Sep. 1994.

[204] M.Jimenez, J.Llaberia, A.Fernandez, E.Morancho, "A unified transformation technique for multi-level blocking" *Proc. EuroPar Conference*, Lyon, France, August 1996. "Lecture notes in computer science" series, Springer Verlag, pp.402-405, 1996.

[205] L.John, R.Radhakrishnan, "c.ICE: A compiler-based instruction cache exclusion scheme", *IEEE TC on Computer Architecture Newsletter*, special issue on "Interaction between Compilers and Computer Architectures", pp.61-63, June 1997.

[206] T.Juan, T.Lang, J.Navarro, "Reducing TLB power requirements", *Proc. IEEE Intnl. Symp. on Low Power Design*, Monterey CA, pp.196-201, Aug. 1997.

[207] M.Kamble, K.Ghose, "Analytical Energy Dissipation Models for Low Power Caches", *Proc. IEEE Intnl. Symp. on Low Power Design*, Monterey CA, pp.143-148, Aug. 1997.

[208] R.Kannan, A.Bachem, "Polynomial algorithms for computing the Smith and Hermite Normal Forms of an integer matrix," *SIAM J. on Comput.*, Vol.8, No.4, pp.499-507, Nov. 1979.

[209] M.Kandemir, A.Choudhary, J.Ramanujam, R.Bordawekar, "Optimizing out-of-core computations in uniprocessors", *IEEE TC on Computer Architecture Newsletter*, special issue on "Interaction between Compilers and Computer Architectures", pp.25-27, June 1997.

[210] W.Kelly, W.Pugh, "Generating schedules and code within a unified reordering transformation framework", Technical Report UMIACS-TR-92-126, CS-TR-2995, Institute for Advanced Computer Studies Dept. of Computer Science, Univ. of Maryland, College Park, MD 20742, 1992.

[211] B.Kim, T.Barnwell III, "Resource allocation and code generation for pointer based pipelined DSP multiprocessors", *Proc. IEEE Int. Symp. on Circuits and Systems*, New Orleans, pp., May 1990.

[212] U.Ko, P.Balsara, A.Nanda, "Energy optimization of multi-level processor cache architectures", *Proc. IEEE Intnl. Workshop on Low Power Design*, Laguna Beach CA, pp.45-50, April 1995.

[213] S.Koeppe, M.Boehner, "Architectural study for backprojection processor for medical applications in computed tomography", SPRITE deliverable report C3.f/Siemens/Y5m6/1, June 1993.

[214] D.Kolson, A.Nicolau, N.Dutt, "Minimization of memory traffic in high-level synthesis", *Proc. 31st ACM/IEEE Design Automation Conf.*, San Diego, CA, pp.149-154, June 1994.

[215] D.Kolson, A.Nicolau, N.Dutt, "Elimination of redundant memory traffic in high-level synthesis", *IEEE Trans. on Comp.-aided Design*, Vol.15, No.11, pp.1354-1363, Nov. 1996.

[216] T.Komarek, P.Pirsch, "Array Architectures for Block Matching Algorithms", *IEEE Trans. on Circuits and Systems*, Vol.36, No.10, Oct. 1989.

[217] K.Konstantinides, R.Kaneshiro, J.Tani, "Task allocation and scheduling models for multi-processor digital signal processing", *IEEE Trans. on Acoustics, Speech and Signal Processing*, Vol.ASSP-38, No.12, pp.2151-2161, Dec. 1990.

[218] A.Krikelis, "Workshop on parallel processing and multimedia: a new research forum", *IEEE Concurrency Magazine*, pp.5-7, April-June 1997.

[219] H.W.Kuhn, "Solvability and consistency for linear equations and inequalities," *Amer. Math. Monthly*, Vol.43, 1956.

[220] D.Kulkarni, M.Stumm, "Linear loop transformations in optimizing compilers for parallel machines", Technical report, Comp. Systems Res. Inst. Univ. of Toronto, Canada, Oct. 1994.

[221] D.Kulkarni, M.Stumm, R.Unrau, "Implementing flexible computation rules with subexpression-level loop transformations", Technical report, Comp. Systems Res. Inst. Univ. of Toronto, Canada, 1995.

[222] C.Kulkarni, F.Catthoor, H.De Man, "Hardware cache optimization for parallel multimedia applications", accepted for *Proc. EuroPar Conference*, Southampton, UK, Sep. 1998.

[223] C.Kulkarni, F.Catthoor, H.De Man, "Cache Optimization for Multimedia Compilation on Embedded Processors for Low Power", *Proc. Intnl. Parallel Proc. Symp.(IPPS)*, Orlando FL, pp.292-297, April 1998.

[224] F.J.Kurdahi, A.C.Parker, "REAL: a program for register allocation", *Proc. 24th ACM/IEEE Design Automation Conf.*, Miami FL, pp.210-215, June 1987.

[225] G.Lafruit, "Architectures for subband coding of images" *Doctoral dissertation*, EE Dept., V.U.Brussel, Belgium, Sep. 1995.

[226] G.Lafruit, F.Catthoor, J.Cornelis, H.De Man, "A VLSI architecture for the 2-D wavelet transform with novel image scan", *Proc. Intnl. Conf. on Digital Signal Processing*, Limassol, Cyprus, June 1995.

[227] G.Lafruit, P.Schelkens, F.Catthoor, J.Cornelis, "Reduction of the Memory Requirements for the VLSI Implementation of the 2D-Inverse Fast Wavelet Transform, Using A Space-Filling Curve", *Proc. Intnl. Conf. on Electronic Circuits and Systems*, Greece, pp.836-839, Oct. 1996.

[228] G.Lafruit, F.Catthoor, J.Cornelis, H.De Man, "An efficient VLSI architecture for the 2-D wavelet transform with novel image scan", accepted for *IEEE Trans. on VLSI Systems*, 1998.

[229] L.Lamport, "The parallel execution of DO loops", *Communications of the ACM*, Vol.17, No.2, pp.83-93, Feb. 1974.

[230] D.Lanneer, M.Cornero, G.Goossens, H.De Man, "Data routing: a paradigm for efficient data-path synthesis and code generation", *Proc. 7th ACM/IEEE Intnl. Symp. on High-Level Synthesis*, Niagara-on-the-Lake, Canada, May 1994.

[231] E.Lee, D.Messerschmitt, "Pipeline interleaved programmable DSP's: synchronous data-flow programming", *IEEE Trans. on Acoustics, Speech and Signal Processing*, Vol.35, No.9, pp.1334-1346, Sep. 1987.

[232] E.Lee, D.Messerschmitt, "Synchronous data flow", *Proc. of the IEEE*, Vol.75, No.9, pp.1235-1245, Sep. 1987.

[233] M.T-C.Lee, V.Tiwari, S.Malik, M.Fujita, "Power analysis and minimization techniques for embedded DSP software", *IEEE Trans. on VLSI Systems*, Vol.5, No.1, pp.123-135, March 1997.

[234]   V.Lefebvre, P.Feautrier, "Optimizing storage size for static control programs in automatic parallelizers", *Proc. EuroPar Conference*, Passau, Germany, August 1997. "Lecture notes in computer science" series, Springer Verlag, Vol.1300, 1997.

[235]   D.Le Gall. "MPEG: A video compression standard for multimedia applications", *Commun. of the ACM*, Vol.34, No.4, pp.46–58, April 1991.

[236]   C.Lengauer. "Loop parallelization in the polytope model", *Proc. of the Fourth Intnl. Conf. on Concurrency Theory (CONCUR93)*, Hildesheim, Germany, Aug. 1993.

[237]   S-T.Leung, J.Zahorjan, "Restructuring arrays for efficient parallel loop execution", Technical Report, Dep. of CSE, Univ. of Washington, Feb. 1994.

[238]   W.Li, K.Pingali. "Access normalization: loop restructuring for NUMA compilers", *Proc. 5th Int. Conf. on Architectural Support for Prog. Lang. and Operating Systems*, April 1992.

[239]   W.Li, K.Pingali. "A singular loop transformation framework based on non-singular matrices", *Proc. 5th Annual Workshop on Languages and Compilers for Parallelism*, New Haven CN, August 1992.

[240]   D.Lidsky, J.Rabaey, "Low power design of memory intensive applications - case study: vector quantisation", *IEEE Intnl. Symp. on Low Power Design*, San Diego CA, Aug. 1994.

[241]   D.Lidsky, J.Rabaey, "Early power exploration - a world wide web application", *33rd ACM/IEEE Design Automation Conf.*, Las Vegas NV, pp.27-32, June 1996.

[242]   C.Liem, T.May, P.Paulin, "Register assignment through resource classification for ASIP microcode generation", *Proc. IEEE Int. Conf. Comp. Aided Design*, San Jose CA, pp.397-402, Nov. 1994.

[243]   D.Lilja, "The impact of parallel loop scheduling strategies on prefetching in a shared memory multi-processor", *IEEE Trans. on Parallel and Distributed Systems*, Vol.5, No.6, pp.573-584, June 1994.

[244]   P.Lippens, J.van Meerbergen, A.van der Werf, W.Verhaegh, B.McSweeney, "Memory synthesis for high speed DSP applications", *Proc. IEEE Custom Integrated Circuits Conf.*, 1991.

[245]   P.Lippens, J.van Meerbergen, W.Verhaegh, A.van der Werf, "Allocation of multiport memories for hierarchical data streams", *Proc. IEEE Int. Conf. Comp. Aided Design*, Santa Clara CA, Nov. 1993.

[246]   D.Liu, C.Svensson, "Power Consumption Estimation in CMOS VLSI Chips", *IEEE J. of Solid-state Circ.*, Vol.SC-29, No.6, pp.663-670, June 1994.

[247]   M.Liu, "MPEG decoder architecture for embedded applications", *IEEE Trans. on Consumer Electronics*, Vol.42, No.4, pp.1021–1028, Nov. 1996.

[248]   R.Lo, S.Chan, J.Dehnert, R.Towle, "Aggregate operation movement: a min-cut approach to global code motion", *Proc. EuroPar Conference*, Lyon, France, August 1996. "Lecture notes in computer science" series, Springer Verlag, pp.801-814, 1996.

[249]   W.Löwe, J.Eisenbiegler, W.Zimmermann, "Optimization of parallel programs on machines with expensive communication", *Proc. EuroPar Conference*, Lyon, France, August 1996. "Lecture notes in computer science" series, Springer Verlag, pp.602-610, 1996.

[250]   D.G.Luenberger, "Introduction to Linear and Nonlinear Programming", Addison-Wesley, 1973.

[251]   J.Ma, E.Deprettere, K.Parhi, "Pipelined CORDIC based QRD-RLS adaptive filtering using matrix lookahead", *Proc. IEEE Wsh. on Signal Processing Systems (SIPS)*, Leicester, UK, pp.131-140, Nov. 1997.

[252]   M.Mace, "Memory storage patterns in parallel processing", Kluwer Academic Publishers, Boston, 1987.

[253]   N.Manjiakian, T.Abdelrahman, "Reduction of cache conflicts in loop nests", Technical report CSRI-318, Comp. Systems Res. Inst. Univ. of Toronto, Canada, March 1995.

[254]   N.Manjiakian, T.Abdelrahman, "Fusion of loops for parallelism and locality", Technical report CSRI-315, Comp. Systems Res. Inst. Univ. of Toronto, Canada, Feb. 1995.

[255]   M.Martonosi, K.Shaw, "Interactions between application write performance and compilation techniques: a preliminary view", *IEEE TC on Computer Architecture Newsletter*, special issue on "Interaction between Compilers and Computer Architectures", pp.16-18, June 1997.

[256]   V.Maslov, "Lazy array data-flow dependence analysis", *21st ACM Symp. on Principles of Programming Languages*, pp.311-325, Jan. 1994.

[257]   C.Mauras, P.Quinton, S.Rajopadhye, Y.Saouter, "Scheduling Affine Parameterized Recurrences by means of Variable Dependent Timing Functions", *Proc. Intnl. Conf. on Applic.-Spec. Array Processors*, Princeton NJ, Sep. 1990.

[258]   D.E.Maydan, "Accurate analysis of array references", *Doctoral Dissertation*, Stanford, Sep. 1992.

[259]   D.E.Maydan, S.Amarasinghe, M.Lam, "Array data-flow analysis and its use in array privatization", *20th ACM Symp. on Principles of Programming Languages*, pp.2-15, Jan. 1993.

[260]   D.McCrackin, "Eliminating interlocks in deeply pipelined processors by delay enforced multistreaming", *IEEE Trans. on Computers*, Vol.C-40, No.10, pp.1125-1132, Oct. 1991.

[261]   K.McKinley, M.Hall, T.Harvey, K.Kennedy, N.McIntosh, M.Paleczny, and G.Roth, "Experiences using the ParaScope editor: an interactive parallel programming tool", in *4th ACM SIGPLAN Symposium on Principles and Practice of Parallel Programming*, San Diego, USA, May 1993.

[262]   K.McKinley, S.Carr, C-W.Tseng, "Improving data locality with loop transformations", *ACM Trans. on Programming Languages and Systems*, Vol.18, No.4, pp.424-453, July 1996.
Proc. 7th Int. Conf. on Architectural Support for Prog. Lang. and Operating Systems, Boston MA, Oct. 1996.

[263] G.Mei, W.Liu, "Parallel algorithms for computer vision primitives", *Proc. IEEE Int. Conf. on Computer Design*, Port Chester NY, pp.506-509, Oct. 1986.

[264] T.H.Meng, B.Gordon, E.Tsern, A.Hung, "Portable video-on-demand in wireless communication", special issue on "Low power electronics" of the *Proceedings of the IEEE*, Vol.83, No.4, pp.659-680, April 1995.

[265] S.Meyers, "More effective C++", Addison Wesley, 1996.

[266] P.Middelhoek, G.Mekenkamp, B.Molenkamp, T.Krol, "A transformational approach to VHDL and CDFG based high-level synthesis: a case study", *Proc. IEEE Custom Integrated Circuits Conf.*, Santa Clara CA, pp.37-40, May 1995.

[267] M.Miki, G.Fujita, T.Onoye, I.Shirakawa, "Low power H.263 video codec dedicated to mobile computing", *Proc. IEEE Intnl. Symp. on Low Power Design*, Monterey CA, pp.80-83, Aug. 1997.

[268] M.Minoux, "Mathematical Programming – Theory and Algorithms", John Wiley & Sons, 1986.

[269] M.Miranda, F.Catthoor, H.De Man, "Address equation optimization and hardware sharing for real-time signal processing applications", *IEEE workshop on VLSI signal processing*, La Jolla CA, Oct. 1994. Also in *VLSI Signal Processing VII*, J.Rabaey, P.Chau, J.Eldon (eds.), IEEE Press, New York, pp.208-217, 1994.

[270] M.Miranda, F.Catthoor, M.Janssen, H.De Man, "ADOPT: Efficient Hardware Address Generation in Distributed Memory Architectures", *Proc. 9th ACM/IEEE Intnl. Symp. on System-Level Synthesis*, La Jolla CA, pp.20-25, Nov. 1996.

[271] M.Miranda, M.Kaspar, F.Catthoor, H.De Man, "Architectural Exploration And Optimization for Counter Based Hardware Address Generation", *Proc. European Design Automation Conf.*, Paris, France, pp.293-298, March 1997.

[272] M.Miranda, F.Catthoor, M.Janssen, H.De Man, "High-level Address Optimisation and Synthesis Techniques for Data-Transfer Intensive Applications", accepted for *IEEE Trans. on VLSI Systems*, Vol., No., pp., 1998.

[273] N.Mitchell, L.Carter, J.Ferrante, "A compiler perspective on architectural evolutions", *IEEE TC on Computer Architecture Newsletter*, special issue on "Interaction between Compilers and Computer Architectures", pp.7-9, June 1997.

[274] E.Miyagoshi et al., "A 100 mm^2 0.95 W single-chip MPEG2 MP@ML video encoder with a 128 GOPS motion estimator and a multi-tasking RISC type controller", *Proc. IEEE Int. Solid-State Circ. Conf.*, San Francisco CA, pp.30-31, Feb. 1998.

[275] M.Mizuno et al, "A 1.5 W single-chip MPEG2 MP@ML encoder with low-power motion-estimation and clocking", *Proc. IEEE Int. Solid-State Circ. Conf.*, San Francisco CA, pp.256-257, Feb. 1997.

[276] Y.S.Mo, W.S.Lu and A.Antoniou, "Embedded coding for 1-D signals using zerotrees of wavelet coefficients", *IEEE Pacific Rim Conference on Communications, Computers and Signal Processing*, Victoria, pp.306–309, Aug. 1997.

[277] S-M.Moon, K.Ebcioglu, "A study on the number of memory ports in multiple instruction issue machines", *Micro'26*, pp.49-58, Nov. 1993.

[278] M.Moonen, F.Catthoor, (eds.), "Algorithms and parallel VLSI architectures", special issue in *Integration, the VLSI journal*, Elsevier, Amsterdam, pp.1-120, Dec. 1995.

[279] M.Moonen, F.Catthoor (eds.), "Algorithms and Parallel VLSI Architectures III", Elsevier Academic Publishers, Amsterdam, 1995.

[280] D.Moolenaar, L.Nachtergaele, F.Catthoor, H.De Man, "System-level power exploration for MPEG-2 decoder on embedded cores : a systematic approach", *Proc. IEEE Wsh. on Signal Processing Systems (SIPS)*, Leicester, UK, Nov. 1997. Also in VLSI Signal Processing X, M.Ibrahim et al. (eds.), IEEE Press, New York, pp.395-404, 1997.

[281] L.J.Mordell, "Lattice Points in a Tetrahedron and Generalized Dedekind Sums", *J. Indian Math.*, Vol.15, pp.41-46, 1951.

[282] L.J.Mordell, "Diophantine Equations", Academic Press, 1969.

[283] H.Morimura, N.Shibata, "A 1V 1MB SRAM for portable equipment", *Proc. IEEE Intnl. Symp. on Low Power Design*, Monterey CA, pp.61-66, Aug. 1996.

[284] T.S.Motzkin, "Beidrage zur Theorie der Linearen Ungleichungen", *Doctoral Dissertation*, Univ. of Basel, 1936.

[285] T.Motzkin, H.Raiffa, G.Thompson, R.Thrall, "The double description method," in *Contributions to the Theory of Games – Annals of Mathematics Studies*, (H.Kuhn, A.Tucker eds.), Nr.28, pp.51-73, Princeton University Press, Princeton, 1953.

[286] J.M.Mulder, N.T.Quach, M.J.Flynn, "An Area Model for On-Chip Memories and its Application", *IEEE J. of Solid-state Circ.*, Vol.SC-26, No.1, pp.98-105, Feb. 1991.

[287] K.Murakami, S.Shirakawa, H.Miyajima, "Parallel processing RAM chip with 256 Mb DRAM and quad processors", *Proc. IEEE Int. Solid-State Circ. Conf.*, San Francisco CA, pp.228-229, Feb. 1997.

[288] L.Nachtergaele, "An attempt to tackle the inplace reduction problem", Technical report, IMEC, Leuven, Belgium, December 1990.

[289] L.Nachtergaele, I.Bolsens, H.De Man, "Specification and simulation front-end for hardware synthesis of digital signal processing applications", *Intnl. J. in Computer Simulation*, Vol.2, No.2, pp.213–230, 1992.

[290] L.Nachtergaele, F.Catthoor, F.Balasa, F.Franssen, E.De Greef, H.Samsom, H.De Man, "Optimisation of memory organisation and hierarchy for decreased size and power in video and image processing systems", *Proc. Intnl. Workshop on Memory Technology, Design and Testing*, San Jose CA, pp.82-87, Aug. 1995.

[291] L.Nachtergaele, F.Catthoor, B.Kapoor, D.Moolenaar, S.Janssens, "Low power storage exploration for H.263 video decoder", *IEEE workshop on VLSI signal processing*, Monterey CA, Oct. 1996. Also in *VLSI Signal Processing IX*, W.Burleson, K.Konstantinides, T.Meng, (eds.), IEEE Press, New York, pp.116-125, 1996.

[292] L.Nachtergaele, D.Moolenaar, B.Vanhoof, F.Catthoor, H.De Man, "System-level power optimization of video codecs on embedded cores : a systematic approach", accepted for special issue on *Future directions in the design and implementation of DSP systems* (eds. Wayne Burleson, Konstantinos Konstantinides) of *Journal of VLSI Signal Processing*, Vol.18, No.2, Kluwer, Boston, pp., Feb. 1998.

[293] L.Nachtergaele, F.Catthoor, B.Kapoor, D.Moolenaar, S.Janssens, "Low power data transfer and storage exploration for H.263 video decoder system", accepted for special issue on *Very low-bit rate video coding* (eds. Argy Krikelis et al.) of *IEEE Journal on Selected Areas in Communications*, Vol.15/16, No.12/1, Dec. 1997 - Jan. 1998.

[294] K.Nakamura, et al., "A 500 MHz 4Mb CMOS pipeline-burst cache SRAM with point-to-point noise reduction coding I/O", *Proc. IEEE Int. Solid-State Circ. Conf.*, San Francisco CA, pp.406-407, Feb. 1997.

[295] M.Nakamura, "Challenges in semiconductor technology for multi-megabit network services", Plenary paper in *Proc. IEEE Int. Solid-State Circ. Conf.*, San Francisco CA, pp.16-20, Feb. 1998.

[296] G.L.Nemhauser, L.A.Wolsey, "Integer and Combinatorial Optimization", J.Wiley&Sons, New York, N.Y., 1988.

[297] C.Neugebauer, R.Carlson, "Comparison of Wafer Scale Integration with VLSI Packaging Approaches", *IEEE Transactions on Components, Hybrids, and Manufacturing Technology*, Vol.CHMT-10, No.2, pp.184-189, June 1987.

[298] S.Note, W.Geurts, F.Catthoor, H.De Man, "Cathedral III: Architecture driven high-level synthesis for high throughput DSP applications", *Proc. 28th ACM/IEEE Design Automation Conf.*, San Francisco CA, pp.597-602, June 1991.

[299] S.Note, "Mapping high throughput signal processing algorithms into dedicated data-path architectures" *Doctoral dissertation*, ESAT/EE Dept., K.U.Leuven, Belgium, March 1991.

[300] R.J.Offen (ed.), "VLSI Image Processing", McGraw Hill, 1985.

[301] K.Okamoto et al, "A DSP for DCT-based and wavelet-based video codecs for consumer applications", *Proc. IEEE Custom Integrated Circuits Conf.*, San Diego CA, pp.359-362, May 1996.

[302] Y.Okada et al, "An 80 mm^2 MPEG2 audio/video decode LSI", *Proc. IEEE Int. Solid-State Circ. Conf.*, San Francisco CA, pp.264-265, Feb. 1997.

[303] A.V.Oppenheim (ed.), "Applications of Digital Signal Processing", Prentice Hall, Englewood Cliffs, NJ, 1978.

[304] D.A.Padua, M.J.Wolfe. "Advanced compiler optimizations for supercomputers", *Communications of the ACM*, Vol.29, No.12, pp.1184-1201, 1986.

[305] M.Palis, J.Liou, D.Wei, "Task clustering and scheduling for distributed memory parallel architectures", *IEEE Trans. on Parallel and Distributed Systems*, Vol.7, No.1, pp.46-55, Jan. 1996.

[306] P.R.Panda, N.D.Dutt, A.Nicolau, " Memory data organization for improved cache performance in embedded processor applications", *Proc. 9th ACM/IEEE Intnl. Symp. on System-Level Synthesis*, La Jolla CA, pp.90-95, Nov. 1996.

[307] P.Panda, N.Dutt, "Low power mapping of behavioral arrays to multiple memories", *Proc. IEEE Intnl. Symp. on Low Power Design*, Monterey CA, pp.289-292, Aug. 1996.

[308] P.R.Panda, H.Nakamura, N.D.Dutt and A.Nicolau, "A data alignment technique for improving cache performance", *Proc. IEEE Int. Conf. on Computer Design*, Santa Clara CA, pp.587-592, Oct. 1997.

[309] P.R.Panda, N.D.Dutt, A.Nicolau, "Efficient utilization of scratch-pad memory in embedded processor applications", *Proc. 5th ACM/IEEE Europ. Design and Test Conf.*, Paris, France, pp., March 1997.

[310] P.R.Panda, "Memory optimizations and exploration for embedded systems", *Doctoral Dissertation*, U.C.Irvine, April 1998.

[311] P.R.Panda, N.D.Dutt, A.Nicolau, "Data cache sizing for embedded processor applications", *Proc. 1st ACM/IEEE Design and Test in Europe Conf.*, Paris, France, pp.925-926, Feb. 1998.

[312] K.Parhi, "Rate-optimal fully-static multiprocessor scheduling of data-flow signal processing programs", *Proc. IEEE Int. Symp. on Circuits and Systems*, Portland OR, pp.1923-1928, May 1989.

[313] K.Parhi, "Calculation of minimum number of registers in arbitrary life time chart," *IEEE Trans. on Circuits and Systems - II: Analog and Digital Signal Processing*, Vol.41, No.6, pp.434-436, June 1994.

[314] K.Parhi, "High-level algorithm and architecture transformations for DSP synthesis", *Journal of VLSI signal processing*, special issue on "Design environments for DSP" (eds.I.Verbauwhede, J.Rabaey), Vol.9, No.1, Kluwer, Boston, pp.121-143, Jan. 1995.

[315] N.Passos, E.Sha, "Full parallelism of uniform nested loops by multi-dimensional retiming", *Proc. Int. Conf. on Parallel Processing*, Vol.2, pp.130-133, Aug. 1994.

[316] N.Passos, E.Sha, "Push-up scheduling: optimal polynomial-time resource constrained scheduling for multi-dimensional applications", *Proc. IEEE Int. Conf. Comp. Aided Design*, San Jose CA, pp.588-591, Nov. 1995.

[317] N.Passos, E.Sha, L-F.Chao, "Multi-dimensional interleaving for time-and-memory design optimization", *Proc. IEEE Int. Conf. on Computer Design*, Austin TX, pp.440-445, Oct. 1995.

[318] N.Passos, E.Sha, "Synchronous circuit optimization via multi-dimensional retiming", *IEEE Trans. on Circuits and Systems II: Analog and Digital Signal Processing*, Vol.CAS-43, No.7, pp.507-519, July 1996.

[319] Y.Patt, S.Patel, M.Evers, D.Friendly, J.Stark, "One billion transistors, one uniprocessor, one chip", *IEEE Computer Magazine*, Vol.30, No.9, pp.51-58, Sep. 1997.

[320] P.G.Paulin, J.P.Knight, "Algorithms for high-level synthesis", *IEEE Design and Test of Computers*, Vol.6, No.6, pp.18-31, Dec. 1989.

[321] W.B.Pennebaker and J.L.Mitchell, "JPEG Still Image Data compression Standard", Van Nostrand Reinhold, 1993.

[322] P.Petersen, D.Padua, "Static and dynamic evaluation of data dependence analysis techniques", *IEEE Parallel and Distributed Technology*, Vol.7, No.11, pp.1121-1132, Nov. 1996.

[323] S.Pinter, "Register Allocation with Instruction Scheduling: a New Approach", *ACM SIGPLAN Notices*, Vol.28, pp.248-257, June 1993.

[324] P.Pirsch, N.Demassieux, W.Gehrke, "VLSI architectures for video compression - a survey", *Proc. of the IEEE*, invited paper, Vol.83, No.2, pp.220-246, Feb. 1995.

[325] C.Polychronopoulos, "Compiler optimizations for enhancing parallelism and their impact on the architecture design", *IEEE Trans. on Computers*, Vol.37, No.8, pp.991-1004, Aug. 1988.

[326] R.Potter, G.Steven, "Investigating the limits of fine-grained parallelism in a statically scheduled superscalar architecture", *Proc. EuroPar Conference*, Lyon, France, August 1996. "Lecture notes in computer science" series, Springer Verlag, pp.779-788, 1996.

[327] B.Prince, "Memory in the fast lane", *IEEE Spectrum*, pp.38-41, Feb. 1994.

[328] S.Przybylski, "New DRAM architectures", tutorial at *IEEE Int. Solid-State Circ. Conf.*, San Francisco CA, Feb. 1997.

[329] W.Pugh, "The Omega Test: a fast and practical integer programming algorithm for dependence analysis", *Proc. Supercomputing'91*, Nov. 1991.

[330] W.Pugh, "The Omega Test: a fast and practical integer programming algorithm for dependence analysis", *Communications of the. ACM*, Vol.35, No.8, Aug. 1992.

[331] W.Pugh, D.Wonnacott. "Nonlinear array dependence analysis", Technical Report UMIACS-TR-94-123, CS-TR-3372, Inst. for Advanced Computer Studies, CS Dept., Univ. of Maryland MD, 1994.

[332] W.Pugh, "Counting solutions to Presburger formulas: how and why", *Proc. ACM-SIGPLAN*, Orlando FL, pp.121-134, May 1994.

[333] P.Quinton, V.van Dongen. "The mapping of linear recurrence equations on regular arrays", *Journal of VLSI Signal Processing*, No.1, Kluwer, Boston, pp.95-113, 1989.

[334] F.Quillere, S.Rajopadhye, "Optimizing memory usage in the polyhedral model", *Proc. Massively Parallel Computer Systems Conf.*, April 1998.

[335] J.Rabaey, "Digital integrated circuits: a design perspective", Prentice Hall, Englewood Cliffs NJ, 1996.

[336] J.Rabaey, "System-level power estimation and optimization - challenges and perspectives", *Proc. IEEE Intnl. Symp. on Low Power Design*, Monterey CA, pp.158-160, Aug. 1997.

[337] L.Rabiner, R.Schafer, "Digital processing of speech signals", Prentice Hall, Englewood Cliffs NJ, 1978.

[338] C.Ramachandran, F.Kurdahi, D.D.Gajski, V.Chaiyakul, A.Wu, "Accurate layout area and delay modeling for system level design," *Proc. IEEE Int. Conf. Comp.-Aided Design*, pp.355-361, Santa Clara CA, Nov. 1992.

[339] L.Ramachandran, D.Gajski, V.Chaiyakul, "An algorithm for array variable clustering", *Proc. 5th ACM/IEEE Europ. Design and Test Conf.*, Paris, France, pp.262-266, Feb. 1994.

[340] F.Catthoor, "Evolution of massive storage organisation in SRAM/DRAM", IMEC, Internal Report, Oct. 1996.

[341] L.Rapanotti, G.Megson, "Uniformisation techniques for reducible integral recurrence equations", *Intnl. Workshop on Algorithms and Parallel VLSI Architectures*, Leuven, Belgium, August 1994. Also in "Algorithms and Parallel VLSI Architectures III" (eds. M.Moonen, F.Catthoor), Elsevier, pp.283-294, 1995.

[342] C.Reffay, G-R.Perrin, "From dependence analysis to communication code generation: the "look-forwards" model", *Intnl. Workshop on Algorithms and Parallel VLSI Architectures*, Leuven, Belgium, August 1994. Also in "Algorithms and Parallel VLSI Architectures III" (eds. M.Moonen, F.Catthoor), Elsevier, pp.341-352, 1995.

[343] K.Rijkse, "Video coding for narrow telecommunication channels at < 64 kbit/s", *Technical Report*, Telenor R&D, 1995.

[344] SIA, "The National Technology Roadmap for Semiconductors", *SIA Semiconductor Industry Associator*, 1994.

[345] J.Robinson, "Efficient General-Purpose Image Compression with Binary Tree Predictive Coding", *IEEE Trans. on Image Processing*, Vol.6, No.4, pp.601-608, Apr. 1997.

[346] K.Roenner, J.Kneip, "Architecture and applications of the HiPar video signal processor", to appear in *IEEE Trans. on Circuits and Systems for Video Technology*, special issue on "VLSI for video signal processors" (eds. B.Ackland, T.Nishitani, P.Pirsch), 1998.

[347] J.Rosseel, F.Catthoor, H.De Man, "The exploitation of global operations in affine space-time mapping", *Proc. IEEE workshop on VLSI signal processing*, Napa Valley CA, Oct. 1992. Also in *VLSI Signal Processing V*, K.Yao, R.Jain, W.Przytula (eds.), IEEE Press, New York, pp.309-319, 1992.

[348] J.Rosseel, F.Catthoor, H.De Man, "An optimisation methodology for array mapping of affine recurrence equations in video and image processing applications", *Proc. Intnl. Conf. on Applic.-Spec. Array Processors*, San Francisco, CA, pp.415-425, Aug. 1994.

[349] J.Rosseel, F.Catthoor, T.Gijbels, P.Six, L.Van Gool, H.De Man, "An optimisation methodology for mapping a diffusion algorithm for vision into a modular and flexible array architecture", *Proc. Intnl. Workshop on Algorithms and Parallel VLSI Architectures*, Leuven, Belgium, August 1994. Also in "Algorithms and Parallel VLSI Architectures III" (eds. M.Moonen, F.Catthoor), Elsevier, pp.131-142, 1995.

[350] B.Rouzeyre, G.Sagnes, "A new method for the minimization of memory area in high level synthesis", *Proc. Euro-ASIC Conf.*, Paris, France, pp.184-189, May 1991.

[351] I.Saeed, "Integrated memory ups speed, saves power", *Electronic Engineering Times*, pp.94, April 28, 1997.

[352] S.Sait, H.Youssef, "VLSI Physical Design Automation - Theory and Practice", McGrawhill Europe, Berkshire, England, 1995

[353] J.Saltz, H.Berrymann, J.Wu, "Multiprocessors and runtime compilation", *Proc. Int. Wsh. on Compilers for Parallel Computers*, Paris, France, 1990.

[354] H.Samsom, L. Claesen, H. De Man, "SynGuide: an environment for doing interactive correctness preserving transformations", *IEEE workshop on VLSI signal processing*, Veldhoven, The Netherlands, Oct. 1993. Also in *VLSI Signal Processing VI*, L.Eggermont, P.Dewilde, E.Deprettere, J.van Meerbergen (eds.), IEEE Press, New York, pp.269-277, 1993.

[355] H.Samsom, F.Franssen, F.Catthoor, H.De Man, "Verification of loop transformations for real time signal processing applications", *IEEE workshop on VLSI signal processing*, La Jolla CA, Oct. 1994. Also in *VLSI Signal Processing VII*, J.Rabaey, P.Chau, J.Eldon (eds.), IEEE Press, New York, pp.208-217, 1994.

[356] H.Samsom, "Formal verification and transformation of video and image processing applications", *Doctoral dissertation*, ESAT/EE Dept., K.U.Leuven, Belgium, Oct. 1995.

[357] H.Samsom, F.Franssen, F.Catthoor, H.De Man, "System-level Verification of Video and Image Processing Specifications", *Proc. 8th ACM/IEEE Intnl. Symp. on System-Level Synthesis*, Cannes, France, pp.144-149, Sep. 1995.

[358] H.Samsom, F.Catthoor, "SynGuide, Reference Manual, Version 2.0", IMEC, Internal report, August 1996.

[359] H.Sasaki (NEC), "Multimedia complex on a chip", Plenary paper in *Proc. IEEE Int. Solid-State Circ. Conf.*, San Francisco CA, pp.16-19, Feb. 1996.

[360] H.Schmit and D.Thomas. "Address generation for memories containing multiple arrays", *Proc. IEEE Int. Conf. Comp. Aided Design*, San Jose CA, pp.510–514, Nov. 1995.

[361] H.Schmit and D.Thomas. "Synthesis of application-specific memory designs", *IEEE Trans. on VLSI Systems*, Vol.5, No.1, pp.101-111, March 1997.

[362] M.Schönfeld, M.Schwiegershausen, P.Pirsch, "Synthesis of intermediate memories for the data supply to processor arrays," in *Algorithms and Parallel Architectures II*, P.Quinton, Y.Robert (eds.), Elsevier, Amsterdam, pp.365-370, 1992.

[363] A.Schrijver, "Theory of linear and integer linear programming", J.Wiley&Sons, 1986.

[364] D.A.Schwartz, T.P.Barnwell, "Cyclo-static multiprocessor scheduling for shift-invariant flow graphs", *Proc. IEEE Int. Conf. on Acoustics, Speech and Signal Processing*, Tampa, Florida, pp.1384–1387, March 1985.

[365] T.Seki, E.Itoh, C.Furukawa, I.Maeno, T.Ozawa, H.Sano, N.Suzuki, "A 6-ns 1-Mb CMOS SRAM with Latched Sense Amplifier", *IEEE J. of Solid-state Circuits*, Vol.SC-28, No.4, pp.478-483, Apr. 1993.

[366] O.Sentieys, D.Chillet, J.P.Diguet, J.Philippe, "Memory module selection for high-level synthesis", *Proc. IEEE workshop on VLSI signal processing*, Monterey CA, Oct. 1996.

[367] R.Sethi, "Complete register allocation problems," *SIAM J. on Comput.*, Vol.4, pp.224-248, Sep.1975.

[368] W.Shang, M.O'Keefe, J.Fortes, "Generalized cycle shrinking", presented at workshop on "Algorithms and Parallel VLSI Architectures II", Bonas, France, June 1991. Also in *Algorithms and parallel VLSI architectures II*, P.Quinton and Y.Robert (eds.), Elsevier, Amsterdam, pp.131-144, 1992.

[369] W.Shang, J.Fortes, "Independent partitioning of algorithms with uniform dependencies", *IEEE Trans. on Computers*, Vol.41, No.2, pp.190-206, Feb. 1992.

[370] W.Shang, Z.Shu, "Data alignment of loop nests without nonlocal communications", *Proc. Intnl. Conf. on Applic.-Spec. Array Processors*, San Francisco, CA, pp.439-451, Aug. 1994.

[371] W.Shang, E.Hodzic, Z.Chen, "On uniformization of affine dependence algorithms", *IEEE Trans. on Computers*, Vol.45, No.7, pp.827-839, July 1996.

[372] J.M.Shapiro, "Embedded image coding using zerotrees of wavelet coefficients", *IEEE Trans. on Signal Processing*, Vol.SP-41, No.12, pp.3445-3462, Dec. 1993.

"Globally asynchronous, locally synchronous systems",

[373] T.Sikora, "MPEG digital video coding standards", *IEEE Signal Processing Magazine*, No.5, pp.82–100, Sep. 1997.

[374] T.Sikora, "The MPEG-4 video standard verification model", *IEEE Trans. on Circuits and Systems for Video Technology*, Vol.7, No.1, pp.19–31, Feb. 1997.

[375] D.Singh, J.Rabaey, M.Pedram, F.Catthoor, S.Rajgopal, N.Sehgal, T.Mozdzen, "Power conscious CAD tools and methodologies: a perspective", special issue on "Low power electronics" of the *Proceedings of the IEEE*, Vol.83, No.4, pp.570-594, April 1995.

[376] P.Slock, S.Wuytack, F.Catthoor, G.de Jong, "Fast and extensive system-level memory exploration for ATM applications", *Proc. 10th ACM/IEEE Intnl. Symp. on System-Level Synthesis*, Antwerp, Belgium, pp.74-81, Sep. 1997.

[377] M.Stan, W.Burleson, "Low-power encodings for global communication in CMOS VLSI", *IEEE Trans. on VLSI Systems*, Vol.5, No.4, pp.444-455, Dec. 1997.

[378] P.Stenström, "A survey of cache coherence schemes for multiprocessors", *Computer*, Vol.23, No.6, pp.12-24, June 1990.

[379] P.Stenström, E.hagersten, D.Lilja, M.Martonosi, M.Venugopal, "Trends in shared memory multiprocessing", *IEEE Computer Magazine*, Vol.30, No.12, pp.44-50, Dec. 1997.

[380] L.Stok, R.van den Born, "EASY: multiprocessor architecture optimisation", *Proc. Int. Workshop on Logic and Architecture Synthesis for Silicon Compilers*, Grenoble, France, May 1988.

[381] L.Stok, J.Jess, "Foreground memory management in data path synthesis" *Int. Journal on Circuit Theory and Appl.*, Vol.20, pp.235-255, 1992.

[382] L.Stok, "Data path synthesis", *Integration, the VLSI journal*, Vol.18, pp.1-71, June 1994.

[383] J.Subhlok, D.O'Hallaron, T.Grosss, P.Dinda, J.Webb, "Communication and memory requirements as the basis for mapping task and data parallel programs", *Proc. Supercomputing*, Washington DC, Nov. 1994.

[384] A.Sudarsanam, S.Malik, "Memory bank and register allocation in software synthesis for ASIPs',, *Proc. IEEE Int. Conf. Comp. Aided Design*, San Jose CA, pp.388–392, Nov. 1995.

[385] T.Sugibayashi et al., "A 30 ns 256 Mb DRAM with a multi-divided array structure", *IEEE J. of Solid-state Circ.*, Vol.SC-28, No.11, pp.1092-1096, Nov. 1993.

[386] M.Takahashi et al., "A 60 mW MPEG4 video codec using clustered voltage scaling with variable supply-voltage scheme", *Proc. IEEE Int. Solid-State Circ. Conf.*, San Francisco CA, pp.36-37, Feb. 1998.

[387] N.Tawbi, "Parallelisation automatique: estimation des durées d'execution et allocation statique de processeurs", *Doctoral dissertation*, Inst. Blaise Pascal, Univ. Paris, Sep. 1991.

[388] J.Teich, L.Thiele, "Partitioning of processor arrays: a piecewise regular approach", *Integration: The VLSI Journal*, Elsevier, Amsterdam, Vol.14, No.3, pp.297-332, 1993.

[389] Y.Therasse, G.H.Petit, M.Delvaux, "VLSI architecture of a SMDS/ATM router", *Annales des Télécommunications*, Vol.48, No.3-4, pp.166-180, 1993.

[390] L.Thiele, "On the design of piecewise regular processor arrays", *Proc. IEEE Int. Symp. on Circuits and Systems*, Portland OR, pp.2239-2242, May 1989.

[391] L.Thiele. "Compiler techniques for massive parallel architectures", in *State of the art in computer science* (ed. P.Dewilde), chapter 1, pp.19-50, Kluwer Academic Publishers, May 1992.

[392] D.E.Thomas, E.Dirkes, R.Walker, J.Rajan, J.Nestor, R.Blackburn, "The system architect's workbench", *Proc. 25th ACM/IEEE Design Automation Conf.*, San Francisco CA, pp.337-343, June 1988.

[393] V.Tiwari, S.Malik, A.Wolfe, "Power analysis of embedded software: a first step towards software power minimization", *Proc. IEEE Int. Conf. Comp. Aided Design*, Santa Clara CA, pp.384-390, Nov. 1994.

[394] V.Tiwari, S.Malik, A.Wolfe, M.Lee, "Instruction-level power analysis and optimization of software", *Journal of VLSI Signal Processing*, No.13, Kluwer, Boston, pp.223-238, 1996.

[395] E.Torrie, M.Martonosi, C-W.Tseng, M.Hall, "Characterizing the memory behavior of compiler-parallelized applications", *IEEE Trans. on Parallel and Distributed Systems*, Vol.7, No.12, pp.1224-1236, Dec. 1996.

[396] R.Touzeau, "A Fortran compiler for the FPS-164 scientific computer", in *ACM SIGPLAN Symposium on Compiler Construction*, pp.48-57, June 1984.

[397] D.N.Truong, F.Bodin, A.Seznec, "Accurate data distribution into blocks may boost cache performance", *IEEE TC on Computer Architecture Newsletter*, special issue on "Interaction between Compilers and Computer Architectures", pp.55-57, June 1997.

[398] C-J.Tseng, D.Siewiorek, "Automated synthesis of data paths in digital systems", *IEEE Trans. on Comp.-aided Design*, Vol.CAD-5, No.3, pp.379-395, July 1986.

[399] T.Tzen, L.Ni, "Trapezoid self-scheduling: a practical scheduling scheme for parallel compilers", *IEEE Trans. on Parallel and Distributed Systems*, Vol.4, No.1, pp.87-98, Jan. 1993.

[400] T.Tzen, L.Ni, "Dependence uniformization: a loop parallelization technique", *IEEE Trans. on Parallel and Distributed Systems*, Vol.4, No.5, pp.547-557, May 1993.

[401] L.Valiant, "A bridging model for parallel computation", *J. CACM*, Vol.33, No.8, pp.103-111, Aug. 1990.

[402] I.Verbauwhede, F.Catthoor, J.Vandewalle, H.De Man, "Background memory management for the synthesis of algebraic algorithms on multi-processor DSP chips", *Proc. VLSI'89, Int. Conf. on VLSI*, Munich, Germany, pp.209-218, Aug. 1989.

[403] I.Verbauwhede, F.Catthoor, J.Vandewalle, H.De Man, "High-level memory management for real-time signal processing of algebraic algorithms on application-specific micro-coded processors", *Proc. Intnl. Workshop on Algorithms and Parallel VLSI Architectures*, Pont-a-Mousson, France, June 1990.

[404] I.Verbauwhede, F.Catthoor, J.Vandewalle, H.De Man, "In-place memory management of algebraic algorithms on application-specific IC's", in *Algorithms and Parallel VLSI Architectures*, Vol.B, E.Deprettere, A.Van der Veen (eds.), Elsevier, Amsterdam, pp.353-362, 1991.

[405] I.Verbauwhede, C.Scheers, J.Rabaey, "Memory estimation for high-level synthesis", *Proc. 31st ACM/IEEE Design Automation Conf.*, San Diego, CA, pp.143-148, June 1994.

[406] J.Vandewalle, L.Vandenberghe, "Piecewise-linear circuits and piecewise linear analysis," in *Circuits and Filters Handbook*, CRC Press, 1995.

[407] A.van der Werf, F.Bruls, R.Kleihorst, E.Waterlander, M.Verstraelen, T.Friedrich, "I.McIC: a single-chip MPEG2 video encoder for storage", *Proc. IEEE Int. Solid-State Circ. Conf.*, San Francisco CA, pp.254-255, Feb. 1997.

[408] W.Verhaegh, P.Lippens, E.Aarts, J.Korst, J.van Meerbergen, A.van der Werf, "Modelling periodicity by PHIDEO streams", *Proc. 6th Int. Workshop on High-Level Synthesis*, Laguna Beach CA, Nov. 1992.

[409] W.Verhaegh, P.Lippens, E.Aarts, J.Korst, J.van Meerbergen, A.van der Werf, "Improved Force-Directed Scheduling in High-Throughput Digital Signal Processing", *IEEE Trans. on CAD and Systems*, Vol.14, No.8, Aug. 1995.

[410] W.Verhaegh, "Multi-dimensional periodic scheduling", *Doctoral dissertation*, T.U.Eindhoven, Dec. 1995.

[411] W.Verhaegh, P.Lippens, E.Aarts, J.van Meerbergen, A.van der Werf, "Multi-dimensional periodic scheduling: model and complexity", *Proc. EuroPar Conference*, Lyon, France, August 1996. "Lecture notes in computer science" series, Springer Verlag, pp.226-235, 1996.

[412] W.Verhaegh, P.Lippens, E.Aarts, J.van Meerbergen, A.van der Werf, "Multidimensional Periodic Scheduling: A solution approach", *Proc. European Design Automation Conf.*, Paris, France, pp.468-474, March 1997.

[413] J.Vanhoof, I.Bolsens, H.De Man, "Compiling multi-dimensional data streams into distributed DSP ASIC memory", *Proc. IEEE Int. Conf. Comp. Aided Design*, Santa Clara CA, pp.272-275, Nov. 1991.

[414] J.Vanhoof, K.Van Rompaey, I.Bolsens, G.Goossens, H.De Man, "High-level synthesis for real-time digital signal processing" Kluwer Academic Publishers, Boston, 1993.

[415] B.Vanhoof, M.Kaspar, P.Schaumont, "Address generation within Cathedral-2/3", SPRITE deliverable report C3.g/IMEC/Y5m12/1, Dec. 1993.

[416] D.Verkest, J.da Silva, C.Ykman, K.Croes, M.Miranda, S.Wuytack, G.de Jong, F.Catthoor, H.De Man, "Matisse: A system-on-chip design methodology emphasizing dynamic memory management", Proc. IEEE CS Workshop on VLSI, Orlando FL., April 1998.

[417] J.Van Meerbergen, P.Lippens, W.Verhaegh, A.van der Werf, "PHIDEO: high-level synthesis for high throughput applications", Journal of VLSI signal processing, special issue on "Design environments for DSP" (eds.I.Verbauwhede, J.Rabaey), Vol.9, No.1/2, Kluwer, Boston, pp.89-104, Jan. 1995.

[418] F.Vermeulen, F.Catthoor, D.Verkest, H.De Man, "A System-Level Reuse Methodology for Embedded Data-Dominated Applications", accepted for Proc. IEEE Wsh. on Signal Processing Systems (SIPS), Boston MA, pp., Oct. 1998.

[419] M.van Swaaij, F.Franssen, F.Catthoor, H.De Man, "Modelling data and control flow for high-level memory management", Proc. 3rd ACM/IEEE Europ. Design Automation Conf., Brussels, Belgium, pp.8-13, March 1992.

[420] M.van Swaaij, F.Franssen, F.Catthoor, H.De Man, "Automating high-level control flow transformations for DSP memory management", Proc. IEEE workshop on VLSI signal processing, Napa Valley CA, Oct. 1992. Also in VLSI Signal Processing V, K.Yao, R.Jain, W.Przytula (eds.), IEEE Press, New York, pp.397-406, 1992.

[421] M.van Swaaij, "Data-flow geometry: exploiting regularity in system-level synthesis", Doctoral dissertation, ESAT/EE Dept., K.U.Leuven, Belgium, Dec. 1992.

[422] M.van Swaaij, F.Franssen, F.Catthoor, H.De Man, "High-level modelling of data and control flow for signal processing systems", in Design Methodologies for VLSI DSP Architectures and Applications, M.Bayoumi (ed.), Kluwer, Boston, pp.219-259, 1994.

[423] D.Wang, Y.Hu, "Multiprocessor implementation of real-time DSP algorithms", IEEE Trans. on VLSI Systems, Vol.3, No.3, pp.393-403, Sep. 1995.

[424] C.Y.Wang, K.Parhi, "High-level DSP synthesis using concurrent transformations, scheduling and allocation", IEEE Trans. on Comp.-aided Design, Vol.14, No.3, pp.274-295, March 1995.

[425] N.Weste, K.Eshraghian, "Principles of CMOS VLSI Design: A System Perspective", Addison Wesley, 1988.

[426] N.Weste, K.Eshraghian, "Principles of CMOS VLSI design", Addison-Wesley, pp.231-237, 1991.

[427] D.Wilde, "A Library for Doing Polyhedral Operations", M.Sc. thesis, Oregon State Univ., Dec. 1993.

[428] D.Wilde, S.Rajopadhye, "Allocating memory arrays for polyhedra", Technical Report, IRISA/INRIA 749, Rennes, France, July 1993.

[429] P.R.Wilson, M.Johnstone, M.Neely, D.Boles, "Dynamic Storage Allocation: A Survey and Critical Review", Proc. Intnl. Wsh. on Memory Management, Kinross, Scotland, UK, Sep. 1995.

[430] D.Wilde, S.Rajopadhye, "Memory reuse analysis in the polyhedral model" Proc. EuroPar Conference, Lyon, France, August 1996. "Lecture notes in computer science" series, Springer Verlag, pp.463-466, 1996.

[431] M.Wolf, M.Lam, "A loop transformation theory and an algorithm to maximize parallelism", IEEE Trans. on Parallel and Distributed Systems, Vol.2, No.4, pp.452-471, Oct. 1991.

[432] M.Wolf, M.Lam, "A data locality optimizing algorithm", Proc. of the SIGPLAN'91 Conf. on Programming Language Design and Implementation, Toronto ON, Canada, pp.30-43, June 1991.

[433] M.Wolfe, "Data dependence and program restructuring", J. of Supercomputing, No.4, Kluwer, pp.321-344, 1990.

[434] M.Wolfe, "The Tiny loop restructuring tool", Proc. of Intnl. Conf. on Parallel Processing, pp.II.46-II.53, 1991.

[435] S.Wuytack, F.Catthoor, F.Franssen, L.Nachtergaele, H.De Man, "Global communication and memory optimizing transformations for low power systems", IEEE Intnl. Workshop on Low Power Design, Napa CA, pp.203-208, April 1994.

[436] S.Wuytack, F.Catthoor, H.De Man, "Transforming Set Data Types to Power Optimal Data Structures", Proc. IEEE Intnl. Workshop on Low Power Design, Laguna Beach CA, pp.51-56, April 1995.

[437] S.Wuytack, F.Catthoor, L.Nachtergaele, H.De Man, "Power Exploration for Data Dominated Video Applications", Proc. IEEE Intnl. Symp. on Low Power Design, Monterey CA, pp.359-364, Aug. 1996.

[438] S.Wuytack, F.Catthoor, G.De Jong, B.Lin, H.De Man, "Flow Graph Balancing for Minimizing the Required Memory Bandwidth", Proc. 9th ACM/IEEE Intnl. Symp. on System-Level Synthesis, La Jolla CA, pp.127-132, Nov. 1996.

[439] S.Wuytack, F.Catthoor, H.De Man, "Transforming Set Data Types to Power Optimal Data Structures", IEEE Trans. on Comp.-aided Design, Vol.CAD-15, No.6, pp.619-629, June 1996.

[440] S.Wuytack, J.P.Diguet, F.Catthoor, H.De Man, "Formalized methodology for data reuse exploration for low-power hierarchical memory mappings", accepted for IEEE Trans. on VLSI Systems, Vol.6, No., pp., 1998.

[441] S.Wuytack, F.Catthoor, G.De Jong, H.De Man, "Minimizing the Required Memory Bandwidth in VLSI System Realizations", accepted for IEEE Trans. on VLSI Systems, Vol.6, No., pp., 1998.

[442] Y.Yaacoby, P.Cappello, "Scheduling a system of nonsingular affine recurrence equations onto a processor array", Journal of VLSI Signal Processing, No.1, Kluwer, Boston, pp.115-125, 1989.

[443] Y.Yaacoby, P.Cappello, "Converting affine recurrence equations to quasi-uniform recurrence equations", Journal of VLSI Signal Processing, No.11, Kluwer, Boston, pp.113-131, 1995.

[444] T.Yamada (Sony), "Digital storage media in the digital highway era", Plenary paper in Proc. IEEE Int. Solid-State Circ. Conf., San Francisco CA, pp.16-20, Feb. 1995.

[445] J.Yano et al, "23 GOPS programmable systolic array DSP for video signal processing", *Proc. IEEE Int. Solid-State Circ. Conf.*, San Francisco CA, pp.268-269, Feb. 1997.

[446] T.Yoshida et al, "A 2V 250 MHz multimedia processor", *Proc. IEEE Int. Solid-State Circ. Conf.*, San Francisco CA, pp.266-267, Feb. 1997.

[447] J.Zeman, G.Moschytz, "Systematic design and programming of signal processors, using project management techniques", *IEEE Trans. on Acoustics, Speech and Signal Processing*, Vol.31, No.12, pp., Dec. 1983.

Index